T0269268

Advances in Intelligent Systems and Computing

Volume 424

Series editor

Janusz Kacprzyk, Polish Academy of Sciences, Warsaw, Poland
e-mail: kacprzyk@ibspan.waw.pl

About this Series

The series "Advances in Intelligent Systems and Computing" contains publications on theory, applications, and design methods of Intelligent Systems and Intelligent Computing. Virtually all disciplines such as engineering, natural sciences, computer and information science, ICT, economics, business, e-commerce, environment, healthcare, life science are covered. The list of topics spans all the areas of modern intelligent systems and computing.

The publications within "Advances in Intelligent Systems and Computing" are primarily textbooks and proceedings of important conferences, symposia and congresses. They cover significant recent developments in the field, both of a foundational and applicable character. An important characteristic feature of the series is the short publication time and world-wide distribution. This permits a rapid and broad dissemination of research results.

Advisory Board

More information about this series at http://www.springer.com/series/11156

Václav Snášel · Ajith Abraham
Pavel Krömer · Millie Pant
Azah Kamilah Muda
Editors

Innovations in Bio-Inspired Computing and Applications

Proceedings of the 6th International
Conference on Innovations in Bio-Inspired
Computing and Applications (IBICA 2015)
held in Kochi, India during
December 16–18, 2015

 Springer

Editors
Václav Snášel
Department of Computer Science
VŠB-Technical University of Ostrava
Ostrava
Czech Republic

Ajith Abraham
Machine Intelligence Research Labs
Scientific Network Innovation and Research
  Excellence
Auburn, Washington
USA

Pavel Krömer
Faculty of Electrical Engineering and
  Computer Science
VŠB-Technical University of Ostrava
Ostrava-Poruba
Czech Republic

Millie Pant
Department of Paper Technology
Indian Institute of Technology Roorkee
Roorkee
India

Azah Kamilah Muda
Faculty of Information and Communication
  Technology, Computational Intelligences
  and Technologies Lab
Universiti Teknikal Malaysia Melaka
Durian Tunggal
Malaysia

ISSN 2194-5357   ISSN 2194-5365 (electronic)
Advances in Intelligent Systems and Computing
ISBN 978-3-319-28030-1   ISBN 978-3-319-28031-8 (eBook)
DOI 10.1007/978-3-319-28031-8

Library of Congress Control Number: 2015957800

Printed on acid-free paper

This Springer imprint is published by SpringerNature
The registered company is Springer International Publishing AG Switzerland

Preface

IBICA'15 is the 6th International Conference on Innovations in Bio-Inspired Computing and Applications. The aim of IBICA is to provide a platform for world research leaders and practitioners, to discuss the Òfull spectrumÓ of current theoretical developments, emerging technologies, and innovative applications of bio-inspired computing, emerging technologies. Bio-inspired computing is currently one of the most exciting research areas, and it is continuously demonstrating exceptional strength in solving complex real-life problems. The main driving force of the conference is to further explore the intriguing potential of bio-inspired computing.

The conference includes the keynote address and contributed papers. The papers were solicited in the following areas:

Topics include, but are not limited to:

Bio-inspired Computing

Ant Colony System
Artificial Immune Systems
Artificial Intelligence
Artificial Neural Networks
Cellular Automaton
Cognitive Modeling
DNA Computing
Differential Evolution
Emergent Systems
Evolutionary Computations
Evolutionary Strategies/Programming
Fuzzy Logic
Genetic Algorithms/Programming
Granular Computing

Neutrosophic Systems
Organic Computing
P Systems
Particle Swarm Optimization

Emerging Technologies and Applications

Biomedicine
Bioinformatics
Business Intelligence
Cloud Computing
Data Mining and Knowledge Discovery
E-learning
Financial Computing
Healthcare
Life Sciences Multimedia Applications
Network Management
Robotics
Social Networks Analysis and Computing
System Control and Optimization
Ubiquitous/Pervasive Computing
Web Intelligence
Wireless/Sensor Networks
Other Applications

Intelligent Distributed and High-Performance Architecture

Hybrid systems involving software agents and human actors
Intelligent cloud infrastructures
Agent-based wireless sensor networks
Distributed frameworks and middleware for the Internet of Things
GPU, multicore, and many-core intelligent computing
Intelligent grid
Intelligent high-performance architectures
Context-aware intelligent computing
Virtualization infrastructures for intelligent computing

The aim is to bring together worldwide leading researchers, developers, practitioners, and educators interested in advancing the state of the art in biologically inspired computing for exchanging knowledge that encompasses a broad range of disciplines among various distinct communities. The conference will provide an exceptional platform to researchers to meet and discuss the utmost solutions, scientific results, and methods in solving intriguing problems with people who are actively involved in these evergreen fields.

IBICA 2015 received contributions from more than 15 countries. Each paper was sent to at least 5 reviewers from our International Program Committee in a standard peer-review evaluation, and based on the recommendations 51 papers were included in the proceedings. IBICA 2015 is technically supported by the IEEE SMC Technical Committee on Soft Computing. Many people have collaborated and worked hard to make IBICA 2015 a success. First and foremost, we would like to thank all the authors for submitting their papers to the conference, and for their presentations and discussions during the conference. Our thanks to the Program Committee members and reviewers, who carried out the most difficult work by carefully evaluating the submitted papers. We also thank the Springer Series on Advances in Intelligent Systems and Computing Editorial Team: Prof. Janusz Kacprzyk, Dr. Thomas Ditzinger, and Mr. Holger Schaepe for the wonderful support to publish this volume so quickly.

We wish all IBICA 2015 delegates an exciting meeting and a pleasant stay in Kochi, India. Enjoy the conference!

General Chairs

Vaclav Snasel, VSB-Technical University of Ostrava, Czech Republic
Vincent H Wilson, Toc H Institute of Science and Technology, India
Ajith Abraham, Machine Intelligence Research Labs (MIR Labs), USA

Program Chairs

Millie Pant, Indian Institute of Technology, Roorkee, India
Pavel Kršmer, VSB-Technical University of Ostrava, Czech Republic

Contents

Multiset Genetic Algorithm Approach to Grid Resource Allocation ...   1
Absalom E. Ezugwu, Daniel I. Yakmut, Paschal A. Ochang,
Seyed M. Buhari, Marc E. Frincu and Sahalu B. Junaidu

Secure Cloud Multi-tenant Applications with Cache in PaaS   15
K.R. Remesh Babu, S. Saranya and Philip Samuel

**HAWK EYE: Intelligent Analysis of Socio Inspired Cohorts
for Plagiarism**   29
Preeti Mulay and Karuna Puri

Mechanism of Fuzzy *ARMS* on Chemical Reaction   43
P. Helen Chandra, S.M. Saroja Theerdus Kalavathy,
A. Mary Imelda Jayaseeli and J. Philomenal Karoline

The Power of Hybridity and Context Free in *HP* System.   55
S.M. Saroja T. Kalavathy, P. Helen Chandra and M. Nithya Kalyani

**Enhanced Bee Colony Algorithm for Efficient Load Balancing
and Scheduling in Cloud**   67
K.R. Remesh Babu and Philip Samuel

**Robust Optimized Artificial Neural Network Based PEM Fuelcell
Voltage Tracking**.   79
R. Vinu and Paul Varghese

**Gravitational Search Algorithm to Solve Open Vehicle
Routing Problem**   93
Ali Asghar Rahmani Hosseinabadi, Maryam Kardgar,
Mohammad Shojafar, Shahab Shamshirband and Ajith Abraham

**PIRIDS: A Model on Intrusion Response System
Based on Biologically Inspired Response Mechanism in Plants**.   105
Rupam Kumar Sharma, Hemanta Kr Kalita and Biju Issac

**Particle Swarm Optimization Method Based Consistency
Checking in UML Class and Activity Diagrams** 117
Renu George and Philip Samuel

Data Centric Text Processing Using MapReduce................. 129
N. Sandhya and Philip Samuel

Request Reply Detection Mechanism for Malicious MANETs 139
S. Sreelakshmi and K.G. Preetha

**Ensemble of Flexible Neural Trees for Predicting Risk
in Grid Computing Environment** 151
Sara Abdelwahab, Varun Kumar Ojha and Ajith Abraham

A Novel Approach for Malicious Node Detection in MANET........ 163
K.P. Anjana and K.G. Preetha

Improving the Productivity in Global Software Development........ 175
D. Manoj Ray and Philip Samuel

**Prediction of Heart Disease Using Random Forest and Feature
Subset Selection**.. 187
M.A. Jabbar, B.L. Deekshatulu and Priti Chandra

**Software Requirement Elicitation Using Natural
Language Processing**....................................... 197
Murali Mohanan and Philip Samuel

**Application of Hexagonal Coordinate Systems for Searching
the K-NN in 2D Space**...................................... 209
Vojtěch Uher, Petr Gajdoš, Tomáš Ježowicz and Václav Snášel

Locality Aware MapReduce............................... 221
Reema Rhine and Nikhila T. Bhuvan

**Hybrid Feature Selection Using Correlation Coefficient
and Particle Swarm Optimization on Microarray Gene
Expression Data** 229
Arunkumar Chinnaswamy and Ramakrishnan Srinivasan

**Resource Aware Adaptive Scheduler for Heterogeneous
Workload with Task Based Job Sampling**...................... 241
Athira V. Panicker and G. Jisha

**Generating Picture Arrays Based on Grammar Systems
with Flat Splicing Operation** 251
K.G. Subramanian, G. Samdanielthompson, N. Gnanamalar David
and Atulya K. Nagar

Identification of Multimodal Human-Robot Interaction Using Combined Kernels 263
Saith Rodriguez, Katherín Pérez, Carlos Quintero, Jorge López,
Eyberth Rojas and Juan Calderón

Debris Detection and Tracking System in Water Bodies Using Motion Estimation Technique 275
T. Senthil Kumar, K.S. Gautam and H. Haritha

Understanding the Consequences of Social Isolation Using Fireworks Algorithm 285
Lourdes Margain, Alberto Ochoa, Teresa Padilla, Saúl González,
Jorge Rodas, Odalid Tokudded and Julio Arreola

Ant Pheromone Evaluation Models Based Gateway Selection in MANET 297
Naveen Kumar Gupta, Rakesh Kumar, Amit Kumar Gupta
and Prakash Srivastava

A Review on How Human Aging Influences Facial Expression Recognition (FER) 313
Robson Mary and T.V. Jayakumar

An Intelligent Approach for Diabetes Classification, Prediction and Description 323
Tarik A. Rashid, Saman M. Abdullah and Rezhna Mirza Abdullah

A Novel Algorithm for Utility-Frequent Itemset Mining in Market Basket Analysis 337
M.A. Jabbar, B.L. Deekshatulu and Priti Chandra

Exact Computation of 3D Geometric Moment Invariants for ATS Drugs Identification 347
Satrya Fajri Pratama, Azah Kamilah Muda, Yun-Huoy Choo
and Ajith Abraham

D-MBPSO: An Unsupervised Feature Selection Algorithm Based on PSO 359
K. Umamaheswari and M. Dhivya

Delay Scheduling with Reduced Workload on JobTracker in Hadoop 371
Krishan Kumar Sethi and Dharavath Ramesh

Reducing Travel Time in VANETs with Parallel Implementation of MACO (Modified ACO) 383
Vinita Jindal and Punam Bedi

FS-EHS: Harmony Search Based Feature Selection Algorithm for Steganalysis Using ELM 393
Veenu Bhasin, Punam Bedi, Neha Singh and Charu Aggarwal

A Hybrid Dimension Reduction Technique for Document Clustering 403
Cynthia Marea Nebu and Sumy Joseph

2D Image Reconstruction After Removal of Detected Salient Regions Using Exemplar-Based Image Inpainting 417
Hima Anns Roy and V. Jayakrishna

Solution to Constrained Test Problems Using Cohort Intelligence Algorithm 427
Apoorva S. Shastri, Priya S. Jadhav, Anand J. Kulkarni and Ajith Abraham

Placement Strategies for Faulty Cells in Module Relocation Based BISR Approach 437
Madhuri Elsa Eapen, C. Pradeep, Anila Ann Varghese and Jisha M. Nair

Fetal Heart Rate Variability: Multiple Regression Models Using Autoregressive Analysis and Fast Fourier Transform 447
Manoj S. Sankhe and Kamalakar D. Desai

Restricted Boltzmann Machine Based Energy Efficient Cognitive Network 463
P. MohanaPriya, S. Mercy Shalinie and Tulika Pandey

An Overview of Computational Intelligence Technique in Drug Molecular Structure Identification 473
Yee Ching Saw and Azah Kamilah Muda

An Effective Bio-Inspired Methodology for Optimal Estimation and Forecasting of CO_2 Emission in India 481
A. Sangeetha and T. Amudha

Conceptual Voice Based Querying Support Model for Relevant Document Retrieval 491
Olufade F.W. Onifade and Ayodeji O.J. Ibitoye

Online Pairwise Ranking Based on Graph Edge–Connectivity 499
Carlos Quintero, Reinaldo Uribe, Juan Calderón and Fernando Lozano

Fuzzy Variable Stiffness in Landing Phase for Jumping Robot 511
Juan M. Calderón, Wilfrido Moreno and Alfredo Weitzenfeld

Contents

An Extended Study on the Association Between Elicitation Issues and Software Project Performance: A Theoretical Model 523
S. Neetu Kumari and S. Pillai Anitha

Compact Design of Rectangular Patch Antenna with Symmetrical U Slots on Partial Ground for UWB Applications 535
Sandeep Toshniwal, Somesh Sharma, Sanyog Rawat, Pushpendra Singh and Kanad Ray

Big Data Challenges and Solutions in Healthcare: A Survey 543
Prabha Susy Mathew and Anitha S. Pillai

Experimental Study on Bound Handling Techniques for Multi-objective Particle Swarm Optimization 555
Devang Agarwal and Deepak Sharma

Role of Bio-Inspired Optimization in Disaster Operations Management Research 565
R. Dhveya and T. Amudha

Optimal Reservoir Release for Hydropower Generation Maximization Using Particle Swarm Optimization 577
D. Kiruthiga and T. Amudha

Author Index 587

Multiset Genetic Algorithm Approach to Grid Resource Allocation

Absalom E. Ezugwu, Daniel I. Yakmut, Paschal A. Ochang,
Seyed M. Buhari, Marc E. Frincu and Sahalu B. Junaidu

Abstract This paper investigates an alternative way of efficiently matching and allocating grid resources to user jobs, in such a way that the resource demand of each grid user job is met. A proposal of resource selection method that is based on the concept of Genetic Algorithm, using populations based on Multisets is presented. For the proposed resource allocation method, an additional mechanism (populations based on multiset) is introduced into the genetic algorithm components, to enhance its search capability in a large problem space. A computational experiment is presented in order to show the importance of operator improvement on traditional genetic algorithms. The preliminary performance results show that the introduction of an additional operator fine-tuning is efficient in both speed and precession, and can keep up with the high job arrival rates.

A.E. Ezugwu (✉) · P.A. Ochang
Department of Computer Science, Federal University Lafia, Lafia
Nasarawa State, Nigeria
e-mail: ezugwu.absalom@fulafia.edu.ng

P.A. Ochang
e-mail: paschal.Ochang@fulafia.edu.ng

D.I. Yakmut
Directorate of ICT, Federal University Lafia, Lafia, Nasarawa State, Nigeria
e-mail: daniel.yakmut@fulafia.edu.ng

S.M. Buhari
Department of Information Technology, King Abdulaziz University,
Jeddah, Saudi Arabia
e-mail: mesbukary@kau.edu.sa

M.E. Frincu
Department of Electrical Engineering, University of Southern California,
Los Angeles, USA
e-mail: frincu@usc.edu

S.B. Junaidu
Institute of Computing & ICT, Ahmadu Bello University, Zaria, Nigeria
e-mail: sahalu@abu.edu.ng

© Springer International Publishing Switzerland 2016   1
V. Snášel et al. (eds.), *Innovations in Bio-Inspired Computing and Applications*,
Advances in Intelligent Systems and Computing 424,
DOI 10.1007/978-3-319-28031-8_1

Keywords Genetic algorithm · Multiset · Multipopulation · Resource allocation

1 Introduction

Grid computing is said to have a layered architecture that is divided into two-level hierarchy of scheduling [1]. In the first level, the global scheduler assigns submitted user applications to remote resources, after which the local schedulers would generate schedules for the local resources that they manage. The global scheduler has limited knowledge of the potential capacity of the available local resources and likewise has no control over them. On the other hand, the local scheduler seems to be unaware of other resources available to the user applications and has also no knowledge of other submitted applications that subsequently require the services of these available re-sources.

The essence of the above paragraph is to emphasize on the difference that exist between resource allocation and scheduling. Resource allocation here simply means the assignment of jobs and specific tasks to the corresponding suitable resources. Resource allocation may involve data staging and binary code transferring before the job starts execution. Furthermore, scheduling is the allocation of jobs to resources over a period of time [1]. It is also interesting to note that grid scheduling procedure roughly comprises of four phases: information collection, resource selection, job mapping and resource allocation [2]. These concepts will be explained in detail in the subsequent section.

In this paper, a proposal is made to apply the concept of Multiset and Genetic Algorithm (GA), to address the problem of resource allocation in the Grid computing environments. The main essence of this hybridization is to enhance genetic algorithm performance to solve combinatorial optimization problem, specifically grid resource allocation. Another advantage of the proposed hybridization is the use of population based multiset to control the growing population pool of individuals that is maintained by GA. Relevant discussions on genetic algorithm and their applications are presented in [3–5].

In this work, a computational grid is considered to be made up of clusters of computing nodes. A cluster is said to consist of multiplicities of execution points or machines that are either homogeneous or heterogeneous in terms of system and performance configurations. Based on this assumption of the grid properties, the proposed resource allocation model is addressed in this paper.

The remainder of the paper is organized as follows. In Sect. 2, we present the related work. Section 3 covers basic multisets theory in relation to the proposed work. Section 4 describes the proposed system. Section 5, presents the proposed approach to address the resource allocation problem using modified genetic algorithm parameters. Section 6 reports the experimental configuration and computational results. Finally, in Sect. 7 we briefly discuss our conclusions and relevant future work.

2   Related Work

Genetic Algorithms (GA) for scheduling are addressed in several works (Abdulal et al. [6, 7], Carretero et al. [8, 9], Zhang et al. [10], Martino and Mililotti [11], and Priya et al. [12], Gao et al. [13]). Some hybrid heuristic approaches have also been reported for the problem. Abraham et al. [14] addressed the hybridization of GA, SA and TS heuristics for dynamic job scheduling on large-scale distributed systems. In these hybridizations, a population-based heuristic such as GAs, is combined with two other Local Search heuristics, such as TS and SA, which deal with only one solution at a time.

The motivation behind this paper came from the concept of using population based on Multisets to address the problem of traditional representation of population that is often used in evolutionary algorithms. The two main problems identified, are the loss of genetic diversity during evolutionary process and evaluation of redundant individuals [1]. In this paper, the authors proposed a computational representation of populations based on multisets and the adaptation of a genetic algorithm, referred to as Multiset Genetic Algorithm (MGA) to deal with this type of representations. Similar work that minimizes the problem of traditional representation of populations used in evolutionary algorithms can be seen in [15, 16]. Each of this paper presented new formal models for multiset representation of individuals and their populations, which address the aforementioned problems.

3   Multisets Fundamentals

Multisets are generalizations of sets that capture the multiplicity of elements [17]. In a much more simplified tune, a multiset (mset for short) can be expressed as a collection of objects in which those objects have multiple occurrences. A finite mset over a set X is an mset M formed with finitely many elements from X such that each element has a finite multiplicity of occurrence in M. A valuable reference for multisets and their multi-faceted applications can be seen in [2].

Multiset in terms of data structure can be seen as an unordered collection of mathematical objects with repetitions allowed. We consider the following definition of multiset.

Definition A multiset on a data object X can be defined as function $M: X \rightarrow \mathbb{N}$. The number of times an object x occurs in the data structure M is called the multiplicity of x in M and denoted by $M(x)$. If $M(x) > 0$ we say that x is an element of M. In general, a multiset M can be represented by the set of pair as follows:

$$M = \{\langle m_M(x_1), x_1 \rangle, \ldots, \langle m_M(x_i), x_i \rangle, \ldots\}, \forall x_i | x_i \in M \tag{1}$$

3.1   Relations and Operations with Multisets

Lorem If we denote the multiplicity of an object x in the data structure M by $m_M(x)$. Then, two multisets M_1 and M_2 are said to be either equal or the same if for any object x

$$m_{M_1}(x) = m_{M_2}(x) \qquad (2)$$

The multisets M_1 and M_2 are said to be unequal if $M_1 \neq M_2$, similarly the following follows if for any object x,

$$\left. \begin{array}{ll} m_{M_1}(x) > m_{M_2}(x), & \textit{if } M_1 > M_2 \text{ and } M_1 \neq M_2 \\ m_{M_1}(x) < m_{M_2}(x), & \textit{if } M_1 \leq M_2 \text{ and } M_1 \neq M_2 \\ m_{M_1}(x) \leq m_{M_2}(x), & \textit{if } M_1 \leq M_2 \end{array} \right\} \qquad (3)$$

Cardinality of a multiset is the sum of the multiplicities of its elements [17, 18]. Let X be a finite set with cardinality n. Let $X^n(x)$ denote the set of all multisets each having x objects occuring with multiplicities of at most n times. Similarly, a multiset can be defined as a set of an ordered pairs $\langle n, e \rangle$, where n is the cardinality of the element e. In this definition, the set $\{a, a, a, a, b, b, b, c, c, d\}$, would be represented as $\{\langle 4, a \rangle, \langle 3, b \rangle, \langle 2, c \rangle, \langle 1, d \rangle\}$.

3.2   Multiset and Genetic Algorithm

One of the earlier applications of Genetic Algorithms to scheduling problem was first suggested in [1]. GA is considered to be a stochastic search algorithm that borrows some concepts from nature [2, 3]. In other words, GA is seen as a meta-heuristic searching technique which mimics the principles of evolution and natural genetics. This principle is a guided random search which scans through the entire sample space and therefore provides reasonable solutions in all situations. The attractiveness of using a stochastic search method is due to the size of the search space. GA maintains a population pool of individuals in the form of collections. In [19] an individual was defined as a pair $\langle c, m \rangle$ where c is a chromosome (a point of the search space) and m is its associated memory.

In the traditional representation it is common to have repeated individuals within the population [1]. Since multiset allows repetition of objects, as its member elements, it implies that a multiset can therefore be used to efficiently represent a population pool in GA. Multipopulation (MP) as regards to multiset are populations where the individuals are represented by ordered pairs $\langle n, g \rangle$, where n is the number of copies of the genom g. The ordered pair $\langle n, g \rangle$ is referred to as multiindividual (MI), while a set of $\langle n, g \rangle$ with number of copies greater than zero is called multipopulation. In [1] population was defined as a dynamic collection where individuals can be searched by following the concepts of similar operations in a traditional data structure. Area of interest here is the searching of resources, which can be compared to the searching of elements in multisets.

3.3 Search Mechanism

Compare set of resources grouped into clusters, which can be indexed. Then, the indexing of each resource can be expressed as a range from the sum of the multiplicity of all previous clusters to the value added to its own multiplicity. Therefore, searching for a resource would be to search for the index of that specific resource. One main advantage of the resource indexing process is for efficiency.

However, it is also very important to note that the traditional GA when using MP and the traditional GA operators, benefits automatically from the reduced number of evaluations and increased genetic diversity. This is possible simply because of the observed decrease in the number of evaluation, as a result of many individual being able to be evaluated at once, while a remarkable increase in the genetic diversity of the population is recorded because all the MI in the support set have different genotypes.

4 System Description

The proposed efficient resource selection model is presented and described in this section. The model comprises of three major phases of scheduling tasks. The first phase is the job analysis and requirement gathering phase, the second phase is the resource selection and allocation phase, while the third phase is the job execution phase. Figure 1 gives the detailed description of the schematic flow of the resource allocation model. Next is the explanations of each of the three main phases aforementioned.

Job Analysis Phase: In this paper, two classes of jobs are considered, CPU bound jobs and Memory bound jobs. The system first task is to analyze and generate requirement report for the user submitted job (job preparation), based on these two categories of jobs. The process starts when a user sends job request to the job analyzer system. The job analyzer, analyses the job and generates requirement report according to the job's resource needs. In order to harmonize the implementation with the bio-inspired genetic algorithm (genotype coding), which technique is explained later, the requirement values are set to either 1 or 0. For example, if the values of processing speed "PS", memory size "MS" and the associated I/O bandwidth "BW" generated from the requirement report is 1, then the job needs high speed, high memory resources, or high network bandwidth. The job analysis report generated in this phase is as shown in Table 1.

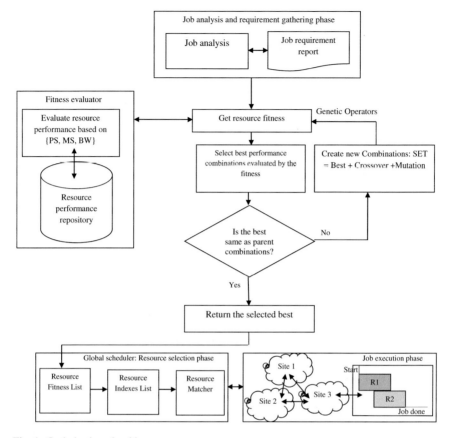

Fig. 1 Optimization algorithm operators

Job ID	PS	MS	BW
J1	1	1	1
J2	1	1	0
J3	0	1	1
J4	1	1	0
J5	1	0	1

Table 1 Job requirement description

Job resource requirement report is generated by using the descriptions for each value codes that are generated for jobs and resources. Suppose we consider a set of cluster machines characteristics such as processing speed, memory size, and network bandwidth. Let us choose two levels of ordering pattern say high (*H*) and low (*L*) to make the characterization decisions. Here we have selected three (3) machine configuration specifications and two (2) ordering decision levels, high and low, to characterize the classification of resources for a particular cluster. If the value of

Table 2 Resource capability
description

Resource ID	PS	MS	BW
R1	1	1	1
R2	1	1	0
R3	0	1	1
R4	1	1	0
R5	1	0	1

configuration capability of a resource (processor, memory and bandwidth) is greater than a certain capacity *Threshold*, then it is considered as High (*H*) otherwise Low (*H*). The binary value 1 is assigned to *H* and 0 is assigned to *L*.

Resource Fitness Evaluation Phase: The resource fitness evaluation phase is responsible for generating the resource capability table with respect to the three resource configuration parameters, that is the resource processing speed, memory size and network bandwidth. The value of each column will be either 1 or 0. If the resource is good in memory and processing speed, but have low bandwidth, then the values will be given as 1 for memory, 1 for processing speed and 0 for bandwidth. The resource capability report (Table 2) is very essential for the general operation of the Multiset Genetic Algorithm (MGA).

Resource Selection Phase: This phase consists of three main components, which includes: Resource Fitness component, Resource Indexing component and Resource Matcher component. The resource fitness component is responsible for generating lists of highly ranked resources sorted according to their computing fitness. The resource fitness component receives its report from the output generated by the computation from the GA operators tuning. The resource indexing component generates the list of indices for all the sorted resources, thereby preparing the resources for easier access, searching, insertion, deletion and selection processes. The resource matcher component receives the resource capability report, job requirement report, and indexing profile and uses this information to map the respective resources with the jobs and provide the optimal list of resources.

Job Execution Phase: The job execution phase provides an avenue where after selecting the respective resources, jobs are scheduled in that selected resource by the local scheduler. During job execution, if any of the jobs fail to proceed, then the job will be reallocated to the next available resource which has the capacity to execute it from the resource fitness list table. However, during the process of job break off period or interrupt, the local scheduler takes care of the selection of next capable resource and submits the job from its last saved state and continues to monitor the job till it reaches its completion stage. After which, the local scheduler returns the feedback to the user through the global scheduler. Since the Fitness Evaluation Phase is linked with the GA mechanism, it is better left out for later discussion under the GA operator functionalities presented in subsequent section.

5   Genetic Algorithm Parameters

Algorithm: The genetic algorithm starts with a population that is made up of individuals, representing basic solution to the problem usually in the form of a bit vector. The outlined solutions are only possible where the minimum attribute specification requirement of each resource has been chosen to be allocated. The solutions are generated using the best-fit heuristics from the bin-packing problem. The heuristic is strictly deterministic: it starts by considering the first resource to see whether it can be allocated to the user job, and if not, it moves on to the second resource. This process is repeated until the most suitable resource is discovered. Furthermore, in order to generate more different solutions, the resource selection process is changed by doing a random permutation. To avoid population flooding, a rescaling operator is introduced as part of the genetic operator to control the number of copies of the individuals.

Chromosomes formulation: Three variables were considered for the chromosome formulation, (*ps, ms, and bw*). The first part of a chromosome contains the binary representation of the first variable (*ps*: x_1), the second part contains the second variable (*ms*:x_2), while the third part represents the third variable (*bw*: x_3). Let a_i and b_i be the lower and upper bound of the *i*th variable. The range of the *i*th variable (r_i) becomes ($b_i - a_i$). If *d* is the number of decimal places on each variable, then the number of genes required for the binary representation of the *i*th variable (g_i) is:

$$g_i = lin(r_i * 10^d) / lin(2) \qquad (4)$$

The chromosome consists of a total of ($g_1 + g_2 + g_3$) genes, since there are only three variables ($x_1, x_2,$ and x_3). The first generation of chromosome is randomly created by filling all the gene slots with 0 or 1. For each gene slot, a random number (between zero and one) is generated. If this number is less than 0.5, the value 0 is assigned into the gene slot, otherwise the value 1 is assigned.

Population Selection mechanisms: Tournament and rank based selection mechanisms have been used to select individuals for the genetic operations [20]. A tournament size of k = 5 was employed, which is selected arbitrarily; this is a rank based selection in which chromosomes in the population are sorted according to their fitness values, and then the first k chromosomes are selected for the best chromosome crossover method. The probability that a resource r_i is selected from the tuple ($r_1, r_2, ..., r_m$) of a node to be a member of the next generation of highly ranked node at each experiment is given by

$$P\{r_i \text{ is selected}\} = \frac{f(r_i)}{\sum_{j=1}^{m} f(r_j)} > 0 \qquad (5)$$

Crossover: The crossover operator is one of the most important genetic functions in evolutionary algorithms. This is a process where chromosomes exchange segments via recombination. An investigation of the standard one point crossover method and a "eugenic" best chromosome crossover method was carried out, in which a certain percentage of offspring was forced to be generated from the chromosomes with the highest fitness values. The new individual that emerges will be defined as the crossover of them. Instead of adopting a random crossover a uniform crossover is rather chosen, where the probability that the ith variable of the new individual is equal to the ith variable of the first or second parent is proportional to the fitness of the first or second parent. See [21] for further details.

Mutation: mutation is a genetic operator used to maintain genetic diversity from one generation of a population of chromosomes to the next. Mutation is an important part of the genetic search, which help to prevent the population from stagnating at any local optima. Two mutation operators (swap and random) are used to introduce some extra variability into the population. A swap mutation swaps two selected genes from randomly selected genes in the chromosome; this is the primary method used to introduce variability. Random mutation changes a randomly selected gene value to another possible value; this method generally has an adverse effect on fitness due to poor 'distribution', but it can introduce 'lost' gene values back into the chromosome, thus helping to prevent premature convergence. A combination of these two methods shows better results than using them individually. See also [22] for further details.

Rescaling: in addition to the aforementioned genetic algorithm parameters, a rescaling genetic operator is introduced in this paper to control the number of copies of the individuals, based on this, the schema defined in [16, 22] is adapted. Also in the work of Manso and Correia [1], it was stated that the introduction of repeated elements in the MP causes an increase in the number of copies of the corresponding MI and the rescaling operator avoids the best individuals getting too many copies. Therefore a control measure is introduced to control the number of repeated elements. The rescaling function r implements a linear fitness scaling algorithm as described by Goldberg [19]. The rescaling operator therefore ensures that each MI has at least one copy and that the total number of individuals in MP is not greater than a constant defined as the double of the number of MI. Experimental results show that this value is a good compromise between selection pressure and genetic diversity.

Fitness function: The user is allowed to specify the relative importance of fitness constraints: such as CPU speed, memory size, associated network bandwidth and so on, for a specific resource object. The fitness function of a schedule is computed by determining the individual resource capacity and isolating the resource with the highest capacity for allocation. In this work, the proposed hybrid GA algorithm uses the current state of the resource configuration to compute the node fitness and chooses the resource with the highest performance capability on the basis of its quantitative evaluation. The fitness function estimates the current processing capability for each potential node. With this metric, the algorithm selects and distributes jobs to the most qualified and freely available grid nodes.

In this study, three resource attributes are considered and used for estimating the fitness of nodes. These sets of attributes include; processing speed (ps), memory size (ms), and network bandwidth (bw). Each computed node j, is sorted in ascending order according to their respective computation capability, first, the aggregate fitness of all machines $r_1 \in j$ per node are estimated. Therefore, we express the fitness of each resource *fitness* of a node j as follows:

$$fitness(j) = \sum_{i=1}^{n} fitness(r_i) \tag{6}$$

$$fitness(r_i) = k_1 \times ps_i + k_2 \times ms_i + k_3 \times bw_i \tag{7}$$

where

$$k_1 = \frac{\beta_1}{\sum_{\forall r_i \in R_j} ps_i}, k_2 = \frac{\beta_2}{\sum_{\forall r_i \in R_j} ms_i}, k_3 = \frac{\beta_3}{\sum_{\forall r_i \in R_j} bw_i},$$

where:

- R_j is the set of all candidate machines r_i.
- ps_i is the processor or CPU clock speed of machine r_i.
- ms_i is the memory size of machine r_i and it is a constant attribute.
- bw_i is the communication speed (Mbps) of machine r_i.
- β_1, β_2, and β_3 are the weights of the first to third terms respectively, the three weight parameters must sum up to one. e.g. $\beta_1 = 0.5$, $\beta_2 = 0.4$, and $\beta_3 = 0.1$.

Finally, the selection is sorted in descending order based on aggregated resource fitness and the resource with the largest fitness function as shown in Eq. 8 will be recognized as the best.

$$best(j) = \max_{j \in \{1, 2, ..., n\}} fitness(j) \tag{8}$$

6   Experimental Configuration

For A small scale simulation and evaluation environment is constructed to evaluate the MGA resource allocation technique. The simulation process was performed on MATLAB platform with several resource state repositories. One type of resource was considered for this experiment, which is the compute server (with attributes: processing speed, memory size and associated bandwidth). Six of the compute server resource instance was defined. The resource instances belonging to the same resource type differ in their total and available capacities that can be consumed by a user submitted job.

The input consists of jobs whose sizes are uniformly distributed between 10 and 160 MFLOPS. The number of jobs submitted was varied at a time in a batch. The

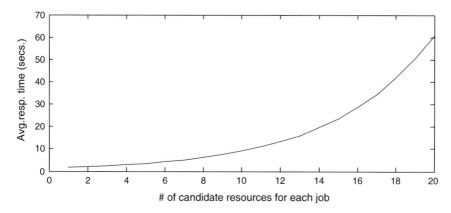

Fig. 2 Average response time for locating best (fitness) resources with MGA

time between the submission of a batch of jobs and the return of matched results was measured. The simulation was performed for 10 batches of jobs evaluated for 20 iterations. The fitness function returns the best available resource that matches the user job. An individual resource in the population pool of resources is randomly selected and any two jobs in that chromosome are randomly selected and swapped. This approach ensures that all the solutions in the search space are more thoroughly examined. The evaluation is determined by taking the average over the 20 iterations as shown in Fig. 2.

In related simulation experiment conducted, the new scheduling heuristic is compared with the performance of a Classical GA (CGA) hybrid heuristic algorithm. The simulation has been executed for three randomly generated Grid jobs which appear in batches of 20, 60 and 100 respectively. The gird resource environment consists of 30 resources (processors) from cluster of computational node. The two search heuristics were subjected to the same number of resources, iteration and execution time epoch for each of the problem sets and the average execution time taken to converge is computed as shown in Fig. 3.

Fig. 3 Average convergence time for CGA and MGA search heuristics

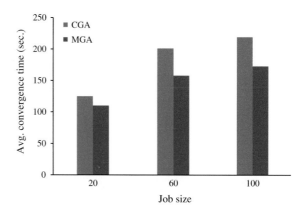

The simulation results for the four search heuristics execution for each of the two problem sets are computed, compared and are as shown in Fig. 3. Analysis of the two results show that the classical GA took more time to reach its best solution with fitness of 125, while the MGA had its best solution at 110 for the first problem. For the second problem with 60 job pool, the CGA is still outperformed by the MGA hybrids GA, with MGA having the least average execution time of 158. For the third problem with job pool of 100, CGA shows its convergence in best solution fitness of 219, whereas MGA achieved 173 s in the same time and continues its reduction.

7 Conclusions and Future Work

In this work, a formal model for the selection and allocation of highly ranked grid resources in a distributed computing environment is proposed. Similarly, a search technique to find an approximate solution is also developed. In the simulation experiment carried out, the MGA search heuristics performed best compared to the classical GA solution method, on the problems of optimal resource allocation with complex resource mix. The additional operators introduced into the system did not alter the genetic algorithm performance. In fact, it made the problem easier for the genetic algorithm to expand its search horizon within a limited time frame. This suggests that the MGA algorithm is well-suited for more-complicated optimization problems, as the case may be with heterogeneous grid resource allocation. Future work will consider finding structure characterization of real distributed system, to evaluate and fine-tune the additional genetic operators introduced in this paper and then compare it against the proposed heuristic.

Acknowledgments We would like to thank the corps member, Nneoma Okoroafor and Bridget Pwajok, who worked with us during the initial preparation of this paper. We also thank all the anonymous reviewers for their comments and recommendations, which have been crucial to improving the quality of this work.

References

1. Manso, A., Correia, L.: Genetic algorithms using populations based on multisets. In: New Trends in Artificial Intelligence, EPIA, pp. 53–64 (2009)
2. Singh, D., Ibrahim, A., Yohanna, T., Singh, J.: An overview of the applications of multisets. Novi Sad J. Math. **37**(3), 73–92 (2007)
3. Davis, L.: Job shop scheduling with genetic algorithms. In: Proceedings of an International Conference on Genetic Algorithms and their Applications. Lawrence Erlbaum Associates, Pittsburgh (1985)
4. Davis, E.W., Heidorn, G.E.: An algorithm for optimal project scheduling under multiple resource constraints. Manage. Sci. 17.12 (1971): B-803–b817

5. Davis, E.W., Patterson, J.H.: A comparison of heuristic and optimum solutions in resource-constrained project scheduling. Manage. Sci. **21**(8), 944–955 (1975)
6. Abdulal, W., Ramachandram, S.: Reliability-aware genetic scheduling algorithm in grid environment. In: IEEE International Conference on Communication Systems and Network Technologies, pp. 673–677. IEEE, Katra, Jammu, India, June (2011). ISBN: 978-0-7695-4437-3/11
7. Abdulal, W., Jadaan, O. A., Jabas, A., Ramachandram, S.: Mutation based simulated annealing algorithm for minimizing makespan in grid computing systems. In: IEEE International Conference on Network and Computer Science (ICNCS 2011), vol. 6, pp. (90–94). IEEE, Kanyakumari, India, April (2011). ISBN: 978-1-4244-8679-3
8. Carretero, J. Xhafa, F.: Using genetic algorithms for scheduling jobs in large scale grid applications. J. Technol. Econ. Dev. **12**, 11–17 (2006). http://citeseer.ist.psu.edu, ISSN: 1392-8619 print/ISSN: 1822-3613 Online
9. Carretero, J., Xhafa, F., Abraham, A.: Genetic algorithm based schedulers for grid computing systems. Int. J. Innovative Comput. Inf. Control **3**(6), 1–19 (2007)
10. Zhang, L., Chen, Y., Sun, R., Jing, S., Yang, B.: A task scheduling algorithm based on PSO for grid computing. Int. J. Comput. Intell. Res. **4**(1), 37–43 (2008)
11. Di Martino, V., Mililotti, M.: Sub optimal scheduling in a grid using genetic algorithms. Parallel Comput. **30**, 553–565 (2004)
12. Priya, S.B., Prakash, M., Dhawan, K.K.: Fault tolerance-genetic algorithm for grid task scheduling using check point. In: Sixth International Conference on Grid and Cooperative Computing, GCC 2007, pp. 676–680. IEEE (2007)
13. Gao, Y., Rong, H., Huang, J.Z.: Adaptive grid job scheduling with genetic algorithms. Future Gener. Comput. Syst. **21**(1), 151–161 (2005)
14. Abraham, A., Buyya, R., Nath, B.: Nature's heuristics for scheduling jobs on computational grids. In: The 8th IEEE International Conference on Advanced Computing and Communications (ADCOM 2000), India (2000)
15. Wieder, T.: Generation of all possible multiselections from a multiset. Prog. Appl. Math. **2**(1), 61–66 (2011)
16. Aparício, J.N., Correia, L., Moura-Pires, F.: Expressing population based optimization heuristics using PLATO. EPIA **1999**, 367–383 (1999)
17. Ibrahim, A.M., Ezugwu, A.E.S., Abdulsalami, A.: Computational model for cardinality bounded multiset space. Int. J. Appl. Math. Res. **1**(3), 330–341 (2012)
18. Blizard, W.: Multiset theory. Notre Dame J. Formal Logic **30**(1), 36–66 (1989)
19. Golberg, D.E.: Genetic algorithms in search, optimization, and machine learning. Addion wesley, Boston (1989)
20. Michalewicz, Z.: Genetic algorithms + data structures = evolution programs. Springer Science & Business Media, New York (2013)
21. Campegiani, P.: A genetic algorithm to solve the virtual machines resources allocation problem in multi-tier distributed systems. In: Second International Workshop on Virtualization Performance: Analysis, Characterization, and Tools (VPACT 2009), Boston, Massachusett (2009)
22. Hugh, M.C.: Getting the timing right—the use of genetic algorithms in scheduling. In: Proceeding of Adaptive computing and information processing conference (UNISYS 1994), pp 393–411. Brunel, London (1994)

Secure Cloud Multi-tenant Applications with Cache in PaaS

K.R. Remesh Babu, S. Saranya and Philip Samuel

Abstract Multi-tenant applications come into existence in clouds, which aims "better resource utilization" for application provider. Today most of the present application optimizations are based on Service Level Agreements which focuses on virtual machine (VM) based computing service, while other services such as storage and cache are often neglected. This paper mainly focuses on cache based approach for multi-tenant application on PaaS. Currently in multi-tenant cloud applications data are often evicted mistakenly by cache service, which is managed by existing algorithms such as LRU. It keeps the query information to reload the evicted data from storage which might be sensitive. Hence there is a possibility of data breach when these data are accessed improperly by other tenants. For faster access caching of the data is common in cloud based applications while the security is an important area that should not be neglected when these systems uses other third party systems/networks as caching servers. Also security of the tenant's data/information is also a crucial component of the SLA between cloud service provider and tenant. So this paper proposes a DES based information security framework within Platform as a Service (PaaS) for better security and Quality of Service (QoS).

Keywords Cloud computing · Multi-tenant · Cache protection · Qos

K.R. Remesh Babu (✉) · S. Saranya
Government Engineering College Idukki, Idukki, Kerala, India
e-mail: remeshbabu@yahoo.com

S. Saranya
e-mail: saranyasasangan3@gmail.com

P. Samuel
Cochin University of Science and Technology, Kochi, India
e-mail: philipcusat@gmail.com

© Springer International Publishing Switzerland 2016
V. Snášel et al. (eds.), *Innovations in Bio-Inspired Computing and Applications*,
Advances in Intelligent Systems and Computing 424,
DOI 10.1007/978-3-319-28031-8_2

15

1 Introduction

A multi-tenant application supports tenants to share one application and database instance while allowing them to configure the application to fit their needs as if it runs on a dedicated environment [1]. To achieve QoS for each tenant, more and more people advocate these applications should abide by SLAs to perform computation. Several works has offered solutions to help these applications to improve resource utilization during runtime. Currently most of these works focus on virtual machine (VM) resources.

Multi-tenant database schema is used for storing meta-data in cloud storage. It is designed and implemented with five techniques [2]. Meta-data driven data-sharing storage model is proposed to implement multitenant applications on top of a standard relational database. This approach works by splitting up the "common tables" shared by each tenant, and mapping the data to "meta data tables" and "data tables".

The rapidly increasing gap between the relative speeds of processor and main memory has made the need for advanced caching mechanisms more intense than ever. Cache service in cloud computing area provides faster access to users. It store small chunks of arbitrary data (strings, objects) from results of database calls, API calls, or page rendering [3].

Cloud distributed cache [4, 5] service plays a vital role in improving cloud application performance. It reduces latency and improves user experience greatly. In Google App Engine (GAE) Memcache service is used for multitenant application. Memcached service is open source and widely used by many high-traffic sites, such as Facebook, Live Journal, Wikipedia and Fotolog. In cloud, the time for reading/writing data from/into cache T_tcache and reading/writing data from/into data store T_tround includes transfer time between different nodes. Where T_tround time is much greater than T_tcache [5]. Hence effective utilization of cache service leads to high hit rate and low response time. Therefore, how to improve cache hit rate by reasonably appointing data to cache becomes a key to multi-tenant application success.

Security is considered as one of the top ranked open issues in adopting the cloud computing model. Internal security is the main issue in multi-tenant application with cache. This is due to improper access of memory by other tenants. Different security models are used in multi-tenant application on SaaS has some limitations. Currently no security mechanism is used in multi-tenant application on PaaS.

This paper introduces a security oriented cache approach that tries to help the tenants by caching data securely. It provides better cloud cache service utilization. Since public key cryptography requires significant computing power than the symmetric key method DES is used for encrypt the data securely on PaaS. It provides integrity of original cache service and improves application portability for different users.

Fig. 1 Multi-tenant
application

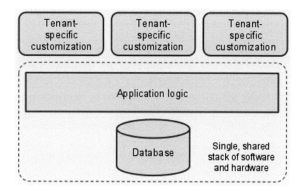

2 Multi-tenant Application

Multi-tenant applications development and adoption are greatly promoted by cloud computing [1], which aims for "better resource utilization" for application provider. Figure 1 show the overview of the multi-tenant application, where tenants share one application and database instance.

A multi-instance approach [6], in which each tenants gets its own instance of the application (and possibly also of the database). Multi-tenant application can be built on top of Software as a Service (SaaS) and PaaS. Service Level Agreement (SLA) can be incorporated to the tenants [7, 8] to ensure required QoS.

3 Data Store

3.1 i-Meta Data

There are five techniques on design and implementation multi-tenant database schema [2], which can extend default data model to address tenant's unique needs. However, there are some drawbacks or limitation in these techniques. Table 1 shows the five techniques and its shortcomings.

To overcome these limitations a novel meta-data driven data-sharing storage model can be used to implement multitenant applications. It is built on the top of a standard relational database. In this approach splitting up of the "common tables" shared by each tenant is carried out. Then mapping of the data to "meta data tables" and to "data tables" are done.

Table 1 Different techniques and drawbacks

Techniques	Limitation
Extension table [2]	Number of tables will be increased by increasing the number of tenants
Universal table [2]	Rows need to be very wide, even for narrow source tables indexes are not supported
Pivot table [2]	It has more columns of meta-data. Higher runtime overhead for interpreting the metadata
Chunk folding [2]	Lack of an effective vertical partitioning algorithm to get the most appropriate results
XML table [2]	Limited to extend fields in a table

3.2 Cache and Replication

In the proposed method Memcached is taken for cloud cache service. Figure 2 illustrates on how Memcached works in web application. Request sent by clients for a dynamic web page usually contains data query to database server. Web server will find and get the data in the form of object in Memcached through application server. If the object is found, it will return to web server and respond to the client. However if object is not found in Memcached or 'cache miss', the data will be fetched from

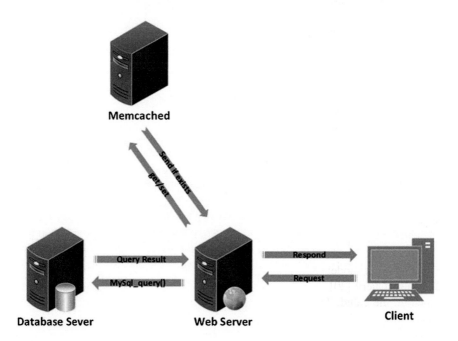

Fig. 2 Memcache operation in web application

database server and it will be set to Memcached as a new item before it is returned to the client. The same process will be repeating all over again.

How can manage the cache in shared environment? Cache replacement strategy is at the heart of cache management. A machine learning approach [9] is used to reconfigure the cache strategy online, which is off-line training coupled with online system monitoring. In [4, 5] introduced an approach to select optimal cache strategy dynamically using trace-driven simulations, so as to differentiate caching and replication policies for each document based on its most recent trace. Authors in [10] evaluated the most likely strategies rather than the entire set of candidate strategies, capturing the history of transitions between different cache strategies. Cache replacement algorithm [11] in adaptive processor is presented for different workloads. These workloads switch between any two replacement algorithms such as LIRS, LRU, LFU and Random. The feedback control theory to allocate cache space was proposed in paper [12]. This self-adaptive multi-tenant memory management achieves each tenant's SLA requirement. It minimizing the memory consumption and dynamically generates a series of cache replacement units according to the current access model was done in [13]. In [14] a Proportional Hit Rate method is introduced to meet clients' SLAs.

Since security is major concern in cloud computing platform the researchers are proposed several methods to address it. An Authorization model [15] is introduced for controlling access to resources in a cloud system. This describes a multi-tenancy authorization system suitable to middleware service in the PaaS layer. The paper [16] introduced a new cloud security management framework based on aligning the NIST-FISMA standard. It based on collaboration among cloud stakeholders to develop a cloud security specification and enforcement covering all of their needs. This approach trying to overcome lack of security constraints in the SLA between the cloud providers and consumers results in a loss of trust as well.

Cryptographic Trust agreement [17] is proposed in order to increase trust in the answers given by services during the negotiation process. It enables SLA approach to ensure that the data of cloud service users are not processed or stored at undesired locations. If the chosen service violates its assured service quality, the service user is enabled to use alternative services. For this, a trusted agreement is provided in order to prevent malicious services; the analysis request is kept secret by encrypting it with the public key of the certified program analysis.

RandTest method [18] is a lightweight and robust dataflow integrity checking method. This approach use a trusted third party to pre-compute results of some randomly generated testing data. Next introduced a method called TOSSMA [19], a Tenant-oriented security management architecture for multi-tenant SaaS applications. TOSSMA is based on "Tenant-Oriented Security", which overcomes the existing classic model "Service-Oriented Security".

Signed Query [20] technique used to improve the confidentiality of users' data stored on the cloud. The usage of a signature to sign the tenant's request, so the server can recognize the requesting tenant and ensure that the data to be accessed is belonging to this tenant. It uses a custom HTTP scheme based on a keyed Hash

Message Authentication Code (keyed-HMAC) [21] for authentication. This approach uses the tenant's secret key to create the HMAC of that string.

A scalable, flexible, resilient, and cost-effective Hybrid Security Architecture (HAS) solution for data center security service is proposed in [22]. It decouples security service provisioning from network routing, thus facilitating operation and management. From these reviews, it is clear that there is no information security mechanism proposed for tenant caching method for multi-tenant applications.

4   Security Oriented Cache Approach

The main issue in multi-tenant cloud application with cache is internal security due to improper memory access by other tenants. In this proposed method information security is provided to multi-tenant application with cache by avoiding improper data access by other tenant's. Public-key cryptography requires significantly more computing power than symmetric cryptography, i.e. strong key pair can take hundreds or even thousands of times as long to encrypt and decrypt data as a symmetric key of similar quality. So in the proposed method more-efficient symmetric DES is used to encrypt the tenant's critical data inorder to prevent improper access by the other active tenants. DES requires lesser computational power compared to other public key encryption methods.

Figure 3 illustrates security oriented cache for multi-tenant application on PaaS. It is built on PaaS, so available to all devices that are connected to the Internet. Tenants uses web browser to access applications. Tenants can send request to the controller for getting data or inserting data into database. Controller will map the user request to manager. Data encryption and decryption is done by manager.

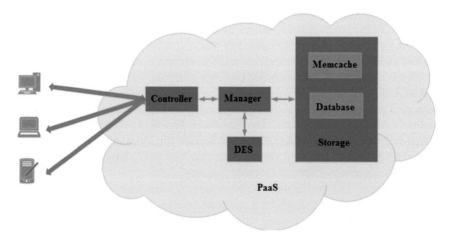

Fig. 3 Proposed security oriented cache approach

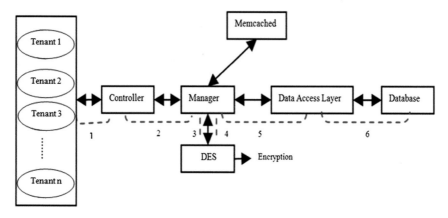

Fig. 4 Insertion of user data into database

For inserting data into database, manager first encrypts the user data and passes it to storage unit. Only encrypted data is stored in database. For getting data of a particular user, the Manager will find it and get the data in the form of object in Memcached. After getting data from Memcached, manager will decrypt the data and returned it to the user.

If the object is not found, data is fetched from the database server and it will be set to Memcached as a new item. Then decrypt the data from database and returned to the user. The same process will be repeating all over again.

Figure 4 illustrates the insertion of data item into database. User first sends a request to the controller with data, key and its id (1). Controller will map the request

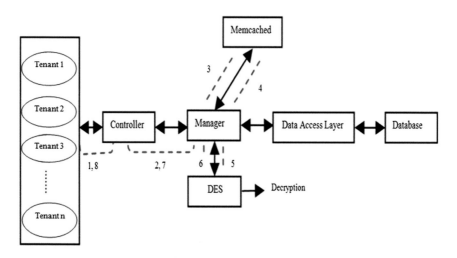

Fig. 5 Data retrieval from Memcached

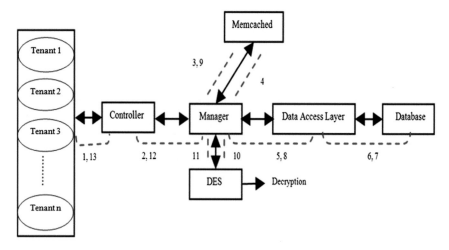

Fig. 6 Data retrieval from database if not found in Memcached

to manager (2). Using DES manager will encrypt the data and it is passed to data access layer (3, 4, 5). Data access layer store these details in database securely (6).

Figure 5 illustrates Data Retrieval from Memcached, when user sends a request to controller for retrieving data (1). Controller will map the request to manager (2). The Manager will check and get the data from Memcached (3, 4). After getting data from Memcached, the manager will decrypt it and returned to the user (5, 6).

If data is not found in cache, then it is fetched from database and it will be set to Memcached before decryption. Then manager will decrypt the data and returned to the user. This procedure is shown in Fig. 6.

5   Experimental Setup and Results

This section presents the simulation experiments conducted to evaluate performance of the proposed approach. The experiment adopted Memcached to simulate cache service in GAE [3]. Memcached is a powerful distributed cache management system which already integrates powerful caching policies in its architecture and widely applied in industry such as Facebook.

5.1 Experimental Design

The experiment environment is constructed by machines with cloud service. Memcached version v 1.4.15 is installed on these machines. The experiment test cases is conducted by using Tomcat version 7.0.39, and available to all the devices

Table 2 User performance with and without cache

Experiment cycle	Without cache (s)	With cache (s)
Tenant 1	14.61	4.38
Tenant 2	15.433	4.4
Tenant 3	13.65	3.48
Tenant 4	14.2	3.9

Table 3 Number of miss count for tenants

Experiment cycle	Tenant 1	Tenant 2	Tenant 3
1	1	2	0
2	1	3	1
3	2	3	1
4	3	4	2
5	3	5	2
6	3	6	2

that are connected to this network. The test cases are implemented as a web service. This helps the tenants to access the data through web browser. The business logic using cache service is as follows. The users send requests for data by keys, and the server first looks up to the cache and then the data store otherwise. Decrypted data is stored in cache and database. To investigate the relation between with and without cache, we consider up to four tenants, each of which owns identical amount of users and workloads. The Table 2 shows the response time required for tenants using with and without cache for data retrieval.

The experiment considered up to 100 users for every tenant. The workload reaching 4000 request per second is considered, which is far heavier than that in the real world condition. Every tenant has its own key. This key is same for all users under one tenant. The proposed method controls the data characteristics by using this pre-generated key.

To provide further evidence, we consider the miscount rate for all tenants in each experiment cycle. Miscount gives number of time the data not present in cache, i.e., miscount increment when data is not present in cache. Hit count gives the number of times data present in cache, i.e. Hit count increments when data present in cache. Table 3 shows the miss count for three tenants in each experiment cycle.

5.2   Performance Calculation

The performance analysis indicates whether the multi-tenant application with cache captures trends in average response time. Figure 7 shows the response time with and without cache for one tenant with one user. The slope of the graph for with cache is lesser compared to without cache. Response time of tenant 1 is decreasing in every cycle. That means tenants can retrieve data quickly after every cycle.

Fig. 7 Response time with
and without cache for one
tenant

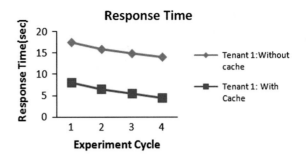

Fig. 7 Response time with and without cache for one tenant

Response time for tenant 1 is higher in without cache and less in with cache experiments. It shows that there is significant improvement in the response time of tenant with cache than tenant without cache, which increases user satisfaction.

To provide further evidence on the performance of the proposed method, consider the experiment with multiple tenants and their response time for with and without cache. Cloud cache service provides a greater difference in response time as compared with database storage. In the case of without cache the response time is high due to direct fetch from the database. The graphical representation of the comparison is shown in Fig. 8.

Figure 9 shows Response time versus number of users. Number of users within the tenants is incremented to study the relation with response time. Tenant 2 is

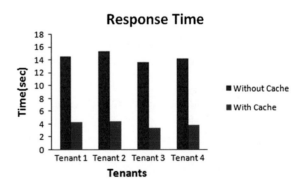

Fig. 8 Response time with and without cache for multiple tenants

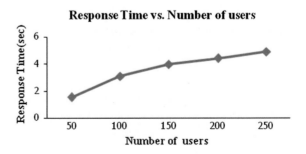

Fig. 9 Response time versus number of users

Fig. 10 Miss count versus
number of tenants

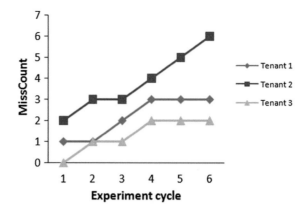

taken for this. Number of users in tenant 2 is incremented periodically and checks
their corresponding response time. Here a high number of users will leads to high
response time.

Figure 10 shows the miscount rate for all tenants in each experiment life cycle.
Miscount incremented after every experiment cycle, that means, corresponding
tenants data is frequently requested. This leads to the caching of tenant data for
faster access. Also no changes in miscount rate after a experiment cycle means that,
the request for that tenant data is not reached.

6 Conclusion

In cloud multi-tenant applications most of the research works focus on VM
migrations and load balancing issues. Efficient and secure data storage is an
important issue to improve the resource utilization and for better response time. In
this work multi-tenant application with cache mechanism is proposed for effective
and faster service. Security is another major concern in these cloud application
platforms. In order to address the security issues, the proposed cache approach for
multi-tenant application employs DES algorithm to avoid improper access of
memory by other tenants. This provides an internal security mechanism to tenant
data.

In future, this work can be extended to increase the cache performance and
cost-effectiveness through cache optimization. Response time can also be reduced
through an internal cache mechanism that may provide better performance to
tenants.

References

1. Peter, M., Grance, T.: The NIST definition of cloud computing. NIST special publication 800.145 **7**, 1–3 (2011)
2. Xuxu, Z., Qingzhong, L., Lanju, K.: A data storage architecture supporting multi-level customization for SaaS. In: IEEE 7th International Conference on Web Information Systems and Applications (WISA), pp. 106–109 (2010)
3. Dormando.: Memcached. In: Internet. www.memcached.org. Accessed on 6 March 2015
4. Pierre, Guillaume, Tanenbaum, A.S.: Differentiated strategies for replicating web documents. Elsevier. Comput. Commun. **24**(2), 232–240 (2001)
5. Guillaume, P., Van Steen, M., Tanenbaum, A.S.: Dynamically selecting optimal distribution strategies for web documents. IEEE Trans. Comput. **51**(6), 637–651 (2002)
6. Bezemer, Z.: Multi-tenant SaaS applications: maintenance dream or nightmare?. In: ACM 4th Joint ERCIM Workshop on Software Evolution (EVOL) and International Workshop on Principles of Software Evolution (IWPSE), pp. 88–92 (2010)
7. Wu, L., Garg, S.K. Buyya, R.: SLA-based resource allocation for software as a service provider in cloud computing environments. In: 11th IEEE/ACM International Symposium on Cluster, Cloud and Grid Computing (CCGrid), pp. 195–204 (2011)
8. Nandi, B.B.: Dynamic SLA based elastic cloud service management: a SaaS perspective. In: IFIP/IEEE International Symposium on Integrated Network Management (IM2013), pp. 60–67 (2013)
9. Xiulei, Q.: On-line cache strategy reconfiguration for elastic caching platform: a machine learning approach. In: IEEE 35th Annual Conference on Computer Software and Applications Conference (COMPSAC), pp. 523–534 (2011)
10. Swaminathan, S., Pierre, G., van Steen, M.: A case for dynamic selection of replication and caching strategies, Web content caching and distribution, pp. 275–282. Springer, Netherlands (2004)
11. Subramanian, R., Yannis, S., Loh, G.H.: Adaptive caches: effective shaping of cache behavior to workloads. In: 39th IEEE/ACM International Symposium on Microarchitecture, pp. 385–396 (2006)
12. Prabhakar, R.: Adaptive QoS decomposition and control for storage cache management in multi-server environments. In: 11th IEEE/ACM International Symposium on Cluster, Cloud and Grid Computing (CCGrid), pp. 402–413 (2011)
13. Yao, J.C., Shi-Dong, Z., Yu-Liang, S., Qing-Zhong, L.: Multi-tenant database memory management mechanism based on chunk folding. Chin. J. Comput. **34**(12), 2320–2331 (2011)
14. Ang, G., Dejun M., Yansu H.: A QoS control approach in differentiated web caching service. J. Netw. **6.1**, 62–70 (2011)
15. Jose, M., Calero, A., Edwards, N., Kirschnick, J., Wilcock, L., Wray, M.: Towards a multi-tenancy authorization system for cloud services. IEEE Secur. Priv. **8**(6), 48–55 (2010)
16. Almorsy, M., Grundy, J., Ibrahim, A.S.: Collaboration-Based cloud computing security management framework. In: IEEE International Conference on Cloud Computing (CLOUD), pp. 364–371 (2011)
17. Mandy, W., Zimmermann, W.: Controlling data-flow in the cloud. In: 3rd International Conference on Cloud Computing, GRIDs, and Virtualization, pp. 24–29 (2012)
18. Liang, Y., Hao, Z., Yu, N., Liu, B.: RandTest: towards more secure and reliable dataflow processing in cloud computing. In: International Conference on Cloud and Service Computing, pp. 180–184 (2011)
19. Almorsy, M., Grundy, J., Ibrahim, A.S.: TOSSMA: a tenant-oriented SaaS security management architecture. In: IEEE 5th International Conference on Cloud Computing (CLOUD), pp. 981–989 (2012)

20. Saleh, E., Takouna, I., Meinel, C.: SignedQuery: protecting users data in multi tenant SaaS environments. In: IEEE International Conference on Advances in Computing, Communications and Informatics (ICACCI), pp. 213–218 (2013)
21. HMAC RFC 2104.: http://tools.ietf.org/html/rfc2104. Accessed on April 2015
22. Lam, H.Y., Zhao, S., Xi, K., Chao, H.J.: Hybrid security architecture for data center networks. In: IEEE International Conference on Communications (ICC), pp. 2939–2944 (2012)

HAWK EYE: Intelligent Analysis of Socio Inspired Cohorts for Plagiarism

Preeti Mulay and Karuna Puri

Abstract Plagiarism has become a major cause of concern that has even spread its root across academic area. Universities are becoming more concerned about it because of growing development of internet especially socio media and thereby increasing opportunity among students to copy and paste the electronic content. Students in today's digital era follow the trend of exchange and copying of information in order to maintain their socio integrity among their circle without considering its long term negative social impact especially from their career perspective. 'They feel The more you exchange, the more social you are'. To avoid this kind of plagiarism especially in Universities labs, *Hawk Eye* an innovative mobile plagiarism detection system was an initiative in this regard. *Hawk Eye* combination with *Cohort Intelligence (CI)* represents higher state of vision to see things even with more clarity in ordinary experiences, by using *Hawk's* keen and observant eyesight, and *CI* self-supervising nature. This would also help to take appropriate preventive measures to avoid plagiarism from its root among students. 'Hawk Eye for Cohort (HEC)' based on comparative analysis of various algorithms like *CI* and Genetic Search *GA* can play an important role in formulation of behavioral distribution patterns of students. *CI* algorithm deploys its self-supervising mechanism to improvise an individual behavior in a cohort and by observing these behavioral patterns, decisions can be taken by teachers in regard of re-design of appropriate evaluation systems to check and stop plagiarism among students. The final outcome of HEC would be an incrementally learning evaluation systems which would iteratively grow with evolving cohort behavioral patterns with every upcoming batch of students. This evolving behavioral patterns search process can be opti-

P. Mulay (✉) · K. Puri
Computer Science and Engineering Department, Symbiosis Institute of Technology,
Lavale, Mulshi, Pune 412115, Maharashtra, India
e-mail: Preeti.Mulay@sitpune.edu.in

K. Puri
e-mail: Karuna.Puri@sitpune.edu.in

© Springer International Publishing Switzerland 2016
V. Snášel et al. (eds.), *Innovations in Bio-Inspired Computing and Applications*,
Advances in Intelligent Systems and Computing 424,
DOI 10.1007/978-3-319-28031-8_3

29

mized using *GA*. HEC really would be a concrete evaluation system for analyzing percentage of plagiarism among students, understanding real time reasons behind the growing percentage and coming up with suitable prevention measures in order to cure plagiarism. The concept of study of cohort behavioral distribution pattern using algorithms like *GA* and *CI* for plagiarism detection based on student's socio thinking using different Cohort Analysis Tools is indeed an entirely new idea which is being discussed in this paper in detail.

Keywords HEC · CI · OCR · IWR · Plagiarism · Optimization · Genetic search

1 Introduction

Hawk Eye [1] is an innovative mobile plagiarism detection system. This system uses Mobile Scanner OCR (*Optical Character Recognition*) Engine to convert the clicked snapshots into the text format. The OCR Engine [2] would preprocess the clicked image in order to remove noise and disturbance from it and extract relevant keywords from image. Then the system uses Plagiarism detection Algorithms to remove unnecessary details like comments, or changing variables names. In end the mobile version of plagiarism detection applications like Viper or Plagiarism Checker can be used to detect plagiarism. Figure 1 represents the flowchart for Hawk Eye System.

After successfully deploying Hawk Eye System, the concept of *GA* and *CI* [3] comes into play. Based on proposed Genetic search and *CI* procedure can be deployed in order to formulate variable and optimized behavioral patterns of different students. Then suitable remedial measures can be taken and appropriate evaluation system can be designed in order to prevent plagiarism among students. The flow for analysis of cohorts (students) and Cohort Intelligence System is represented in Fig. 2.

Hence the complete System is a hybrid combination of following procedures as given below:

$$\text{Hawk Eye} + (\text{Genetic Search} \times \text{Cohort Intelligence}) = (\text{Efficient} + \text{Flexible} + \text{Preventive})$$
$$\times \text{PLAGIARISM DETECTION SYSTEM}$$

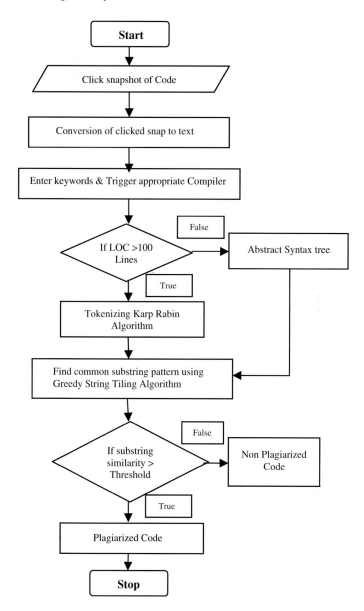

Fig. 1 Part I: Flowchart for *Hawk Eye* system

Fig. 2 Part-II: Flowchart for
Cohort intelligence system

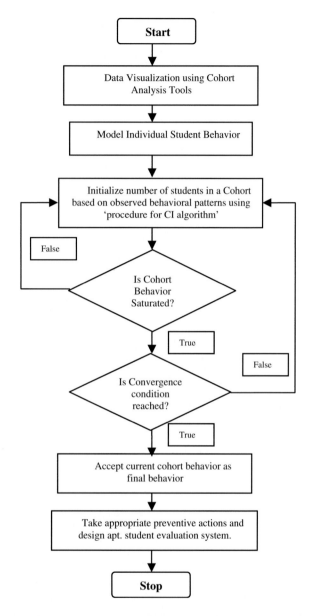

2   Literature Review

A number of studies have been reported on the feasibility of Detection of Plagiarism using different tools in different context.

Overview and Comparison of Plagiarism Detection tools [4]: Asim et al. have examined various plagiarism detection tools with respect to tools features and performance. The comparison of tools proves that there is no tool that can detect or prove plagiarism of a document 100 % because each tool has its own advantages and limitations. Due to the limitations of these tools some set of parameters and rules can be suggested that need to be considered in order to overcome plagiarism in academic areas.

Computer-Based Plagiarism Detection Methods and Tools [5]: Romans et al. introduces different ways to reduce plagiarism in terms of both prevention and detection. Analysis of few of the known plagiarism detection tools like Turnitin, Wordcheck, Moss, JPlag and many more shows although these tools provides excellent services in detecting plagiarism but these advance tools can't detect plagiarism in the manner that manually a human can do. Thus this paper concludes that Human brain is a universal plagiarism detection tool, which analyzes document using various statistical and semantical methods. Thus it is able to operate with textual and non-textual information. At the present such abilities are not available with advance plagiarism detection software tools.

Software Plagiarism Detection using Model Driven Software Development in Eclipse Platform [6]: Pierre has described a concept, the design and the development of a software plagiarism detection application based on the Eclipse Platform. It is a generic front-end approach which converts the source program from different programming languages into generic models in order to detect source code plagiarism. The results of this approach highlight the fact that application doesn't provide absolute results like any other plagiarism detection software but signals the source code submitted by student requires further investigation in regard of plagiarism.

Plagiarism Detection in Java Code [7]: Ahmad Gull and Aijaz focuses on Java programs that could assist teachers in detecting Plagiarism in Java programming. They also highlight properties of different approaches in detecting plagiarism in different Java code files and recommend accordingly best approaches for future work and study.

3 Our Approach and Contributions

In above studies it is clearly visible that there is no plagiarism tool that can detect plagiarism in a document or source code 100 % accurately. Every tool suffers from some or other limitations which makes its approach of detection not very concrete and absolute in terms of confidence in a particular plagiarism detection tool.

Hawk Eye is an initiative in this regard which tries to overcome some of the limitations of already available tools using its keen and observant eyesight. It takes into account various parameters into consideration while detecting plagiarism. To name a few of the parameters are Database Checking, Structure Checking, Supported Languages, using concept of Tokens rather than just strings, Abstract Syntax

Tree and various Hybrid Approaches. In order to strike a balance between already available some of the best features in various detection tools like Moss, JPlag, Turnitin etc. we have studied and taken into consideration various limitations in existing tools and how can these limitations, can be overcome.

Hawk Eye extension with Genetic Search and Cohort Intelligence concept will continue to make it a more concrete system especially in terms of confidence as a software for plagiarism detection. With continuously evolving evaluation system which is the final objective of *HEC* as new behavioral patterns of a cohort will emerge, *HEC* will strengthen itself in terms of an absolute evaluation and prevention system for plagiarism detection.

4 Methodology for Hawk Eye

A. Mobile Scanner OCR Engine Working

Smartphone's capability of scanning using various scanning apps like OCR Instantly, Cam-Scanner [8] can be used in proposed Plagiarism Detection System for extracting relevant text from clicked snapshot. The complete process can be summarized as:-

1. Enhance Image (i.e. Image Preprocessing) for better image quality and reduction of noise as far as possible.
2. Use OCR to extract relevant text in editable form from captured image.
3. Save As (default options available—PDF/JPEG Format) in appropriate format.
4. Share (as available with scanners like OCR Instantly) using various means instantly just with one click.

This concept of OCR [2] can be further extended to IWR [9] (Intelligent Word Recognition) that can be used for detection of handwritten plagiarized codes by students.

B. Plagiarism Detection Methods
B.1 Abstract Syntax Tree

An Abstract Syntax Tree or AST [10, 11] is a hierarchical representation of a program. Each node represents a programming language construct and its children are the parameters of this construct. The nodes of an AST [11] can be mathematical operators, function calls or other programming structures, the leaves are variables or constants. Compilers perform optimizations on AST before generating lower-level code because of this property; AST can be used in plagiarism detection.

B.2 Tokenizing String Based System

Consider the program as a normal text. The pre-processing phase removes all comments, white spaces and punctuation, and renames variables into a common token. Then a string sequences comparison is performed. It performs a string-based comparison using the Karp-Rabin algorithm [12–14]. This algorithm uses concept of hash function which can compute hash value. The main advantage of tokens is that they discard all unnecessary information therefore, token-based systems are insensible to "search and replace" changes.

5 Methodology for Genetic Search and Cohort Intelligence

CI [15] is a novel methodology inspired from the candidates' self-supervised learning behavior in a cohort. Cohort analysis allows identifying relationships between the different characteristics and behavior of a population.

There can be different possible sources which can contribute to source code plagiarism ranging from low, medium to high level. A possible means is required to validate and optimize these range of cases of plagiarism. *GA* can be used to prove the validity and appropriateness of selected cases. Thereafter the proposed Cohort intelligence procedure can be applied to same cases and incrementally evolving dataset which will grow iteratively with every upcoming new batch of students. This will strengthen and prove the concreteness about assumed cases.

In this way we can optimize our search for possible sources of plagiarism and thereafter by applying cohort Intelligence procedure can formulate different cohort buckets categorizing different behavioral characteristics of students.

5.1 Possible Cases of Source Code Plagiarism

Low Level Code Plagiarism Sources

Case 1. *Attribute based Code Plagiarism*—Number of variables, functions, classes, size of code and others could be one of the possible sources of plagiarism. Code Metrics calculation could be one of the means to prevent this type of plagiarism.

Case 2. *Token based Code Plagiarism*—Renaming or changing methods, fields, class, identifiers or replacing the expressions by equivalent, changing comments or indentation could be another possible cause. To avoid this type of plagiarism the code can be tokenized and hashed thus creating the fingerprints of code and reducing the growing incidence of plagiarism.

5.2 Procedure for Genetic Search Algorithm [16]

1. **Start/Initialization**—Generate random initial population of desired size.
2. **Fitness Criteria Evaluation**—Each individual of population is evaluated for fitness of the individual with respect to desired requirement which may range from simple to complex requirements.
3. **Generation of New Population**
   (a) **Selection**—Improve population fitness by selecting only best individuals and discarding bad designs among population i.e. Darwin's Theory of Natural Selection—"Survival of the fittest individual among others".
   (b) **Crossover**—Create new individuals by crossover i.e. combining the aspects of two or more individuals. By combining traits from two or more individuals would generate even fitter off springs.
   (c) **Mutation**—Making small changes at random to individuals in order to avoid combinations of solutions created to be from initial population only.
4. **Loop**—Use new generation of population for further run of algorithm until termination condition is reached i.e. Repeat again from Step 2.

5.2.1 Implementation of GA on Different Cases of Source Code Plagiarism

Details of GA implementation on different cases of source code plagiarism is discussed in Tables 1, 2, 3, 4 and 5 using WEKA run time environment. GA gives the most promising and likely predictive attribute (i.e. optimized solution) from a given set of attributes in each of the cases of source code plagiarism as a possible cause of overall increase in impact of plagiarism.

Table 1 WEKA GA run time information common to CASE 1 and 2 of source code plagiarism

=== WEKA run information for different cases of source code plagiarism ===

Evaluator: weka.attributeSelection.CfsSubsetEval

Search: weka.attributeSelection.GeneticSearch -Z 20 -G 20 -C 0.6 -M 0.033 -R 20 -S 1

Evaluation mode: evaluate on all training data

=== Attribute Selection on all input data ===

Search Method:

Genetic search

Population size	Number of generations	Probability of crossover	Probability of mutation	Report frequency	Random number seed
20	20	0.6	0.033	20	1

Table 2 CASE 1: attribute based code plagiarism

S. no.	Attributes	Type	Possible values		
1.	Maintainability index	Numeric	87	81	100
2.	Cyclomatic complexity	Numeric	2	1	1
3.	Depth of inheritance	Numeric	1	0	0
4.	Class coupling	Numeric	1	1	0
5.	Lines of code	Numeric	4	3	1

Table 3 WEKA Run information for CASE 1

Run information for attribute based code plagiarism				
Relation: codeMetric				
Instances: 4				
Attributes: 5				
Includes locally predictive attributes				
Attribute 1	Attribute 2	Attribute 3	Attribute 4	Attribute 5
Selected attributes: Class coupling	Selected attributes: Depth of inheritance, Lines of code	Selected attributes: Cyclomatic complexity, Lines of code	Selected attributes: Maintainability Index, Cyclomatic complexity, Depth of inheritance, Lines of code	Selected attributes: Depth of inheritance Class coupling

Table 4 CASE 2: token based code plagiarism

S. no.	Attributes	Type	Possible values		
1.	Delimiters	Nominal	5	8	4
2.	Operators	Nominal	2	5	2
3.	Identifiers	Nominal	3	4	6
4.	Keywords	Nominal	2	1	2
5.	Constants	Nominal	0	10	20
6.	Literals	Numeric	1	2	2

Table 5 WEKA Run information for CASE 2

Run information for token based code plagiarism					
Relation: CodeTokens-weka.filters.unsupervised.attribute.Discretize-B10-M-1.0-Rfirst-last					
Instances: 5					
Attributes: 6					
Includes locally predictive attributes					
Attribute 1	Attribute 2	Attribute 3	Attribute 4	Attribute 5	Attribute 6
Selected attributes: Identifiers, Keywords, Constants, Literals	Selected attributes: Keywords	Selected attributes: Delimiters, Keywords, Constants, Literals	Selected attributes: Operators, Identifiers, Constants, Literals	Selected attributes: Delimiters, Identifiers, Keywords, Literals	Selected attributes: Keywords

5.3   Procedure for Cohort Intelligence

The details of procedure for Cohort Intelligence [15] for computation of a student behavioral pattern distribution are explained using algorithmic steps below:

1. Initialize the Cohort C (students) whose behavior has to be analyzed.
2. Initialize all other parameters like convergence parameter ϵ, number of iterations n, sampling interval Si, sampling interval reduction factor r, and number of variation t.
3. Calculate the probability of every cohort c being selected that's associated with the behavior being followed by every student in cohort.
4. Apply roulette wheel approach to decide the behavior to follow and qualities of the student to follow from C available choices.
5. Every student shrinks/grow its sampling interval of quality based on:-

   (a) Calculate behavioral fitness value for each particle and initialize best behavioral fitness value as C_b among the cohort.
   (b) Compare the current behavioral fitness value with best value C_b. If current value is better than best value C_b, then update the best value with current value. Otherwise continue with the best value C_b.
   (c) Find the cohort in neighborhood with best fitness value so far and consider this value as a global best value G_b.

6. Every cohort samples behavior from updated interval and associated behavior can be found.
7. If there is no change in behavior of each cohort, the cohort can be considered as saturated.
8. If cohort converges to same saturated behavior even after maximum number of attempts then current cohort behavior can be accepted as final behavior.
9. Stop if number of iterations equals to cohort or cohort is saturated.

5.3.1   Results Based on Cohort Analysis Tools

Based on analysis of different Cohort Analysis Tools like RJMetrics, Excel and many more as visible from Tables 6, 7, 8 and 9 and using a combination of appropriate *CI* Algorithms suitable remedial measures can be taken and evaluation system can be devised to prevent plagiarism among students of different streams of study.

Table 6 Cohort bucket for engineering stream of study

Cohort bucket 1—engineering students				
Students who: copy		Plagiarism content	Plagiarism %	Nature of cloning
Category	Student ID			
CS/IT	14021	Lab session-code	65	Variables rename and code reordering
CAD/CAM	14041	Drawing sheets	19	Materials, processes, formulas, and products, units, mathematics, and designs
E&TC	14061	Circuit logic/code/simulation	2	Formulas, calculations, circuits parameters

Table 7 Cohort analysis based on Cohort bucket 1

Cohort analysis table based on bucket1					
Student ID	Expected %	Actual %	Student behavior	Faculty action— (suggested remedial measures/design of assignments)	Analysis tool
14021	70	65	**Quality of work**— Work does not meet expectations, has more than the expected number of errors	Assign more no. of assignments on particular topic till clarity of topic is attained	Excel
14041	20	19	**Ability to learn/Social** (Whether exchanges content without knowing the long term effects)—Sometimes slow to become proficient at new tasks or work processes	Indulge more in real time tasks or group presentations or seminars etc.	RJMetrics
14061	0	2	**Problem solving**— The student's demonstrated ability to analyze problems or procedures, evaluate alternatives, and select the best course of action	Increase the complexity of task and test using it. Assign more real time based assignments to enhance problem solving skills	Datapine

Table 8 Cohort bucket for commerce stream of study

Cohort bucket 2—commerce students

Students who: copy		Plagiarism content	Plagiarism %	Nature of cloning
Category	Student ID			
Law—B.A/B. Com/B.Sc. + LLB/LLM	20810	Case studies, legal aspects handling tricks—strategies	85	Copyright infringements, exchange of information, communication, unethical means
Marketing/Finance —B.com + MBA	20210	Management techniques, plan of action, implemented and tested strategies	53	Social communication, Continuous interaction, unfair means, Sniffing (monitoring data flowing over network links) organization websites
Journalism— B.Com + Mass Communication	20412	News, new story, headline	5	Use of malicious tricks— techniques to get the access

Table 9 Cohort analysis based on Cohort bucket 2

Cohort analysis table based on bucket 2

Students ID	Expected %	Actual %	Student behavior	Faculty action— (Suggested remedial measures/design of assignments)	Analysis tool
20810	90	85	**Dependability/Social** (*The manner in which the students conducts his or herself in the working environment*) —Displays an inconsistent work ethic and does not always report to work on time or has attendance issues	Assign assignments deadline based and mark every assignment individually based on given timeline	RJMetrics
20210	40	53	**Interest in word**— Shows little enthusiasm for assigned work, infrequently requests additional tasks	Motivate student using different styles of teaching and assign more research oriented tasks	Datapine
20412	3	5	**Resourcefulness**— The student's demonstrated ability to develop innovative solutions and display flexibility in unique or demanding circumstances	Involve as representative in group assignments	RJMetrics

6   Why Genetic Search and Cohort Intelligence [15, 17]

Cohort Intelligence is a branch of behavioral analytics which plays an important role in big data analysis like students records of different streams of studies in Universities. It is also in use in various data mining applications. *CI* uses its self-supervising mechanism in order to improvise a student behavior in a cohort. Genetic Search [16] algorithm is a process of natural selection that is used to generate useful and optimized solutions to complex search problems.

Algorithms like Particle Swarm Optimization (PSO), Ant Colony Optimization (ACO), and Honey Bee Mating Algorithm are inspired from natural behavior of the living organisms whereas *GA* and *CI* considers natural selection human tendency to solve complex optimization problems. Because of *CI* self-supervised nature and *GA* optimized search procedure they proves to be better evaluation strategy compared to others. This approach is also reasonable with respect to computation cost and it gives more edge compared to other contemporary approaches already in use.

7   Conclusion

'HEC' intelligently uses different Cohort analysis tools like RJMetrics, datapine etc. to evaluate code cloning. By using procedure for *GA* and *CI* Algorithm the system evaluates different types of code plagiarism done by students. Cohort analysis along with Genetic Search procedure acts as a trigger/activator to Hawk Eye system to generate student's different behavioral distribution patterns. Based on modeling individual student behavior, teachers can design individual assignments for students.

The proposed evaluation system design that would be the outcome of HEC, specific to a particular student plagiarism behavior this evaluation system design can be exchanged among other teachers. This reflects socially inspired behavior of teacher's community. Students in order to maintain their socio integrity among their groups would continue with their behavior of exchanging and cloning of information thereby reflecting their socially inspired behavior. As students are more receptive to use of e-media for learning than traditional reference books.

HEC as a system would discourage the overall concept of plagiarism among students of social digital era. The evolving evaluation systems can act as a prevention measure to stop cloning and will continue to re-evolve as new cohort behavioral attributes will emerge. *HEC* as a concrete initiative can contribute significantly to improve the socio economic development of the country as well as help universities, teachers to understand the growing socio impact among today's digital student generation.

References

1. Mulay, P., Puri, K.: Hawk Eye: a plagiarism detection system. In: Proceedings of the Second International Conference on Computer and Communication Technologies (IC3T), vol. 379, CMR Technical Campus, Hyderabad, Advances in Intelligent System Computing: AISC Series of Springer, Ch. 20, 24–26 July 2015
2. Comparison of optical character recognition Software. http://en.wikipedia.org/wiki/Comparison_of_optical_character_recognition_software. Accessed 28 Jan 2015
3. Cohort Analysis. http://cohortanalysis.com/. Accessed 26 July 2015
4. El Tahir, A.M., et al: Overview and comparison of plagiarism detection tools, pp. 161–172. Department of Computer Science, VˇSB-Technical University of Ostrava, 17, listopadu 15, Ostrava, Poruba, Czech Republic (2011). ISBN: 978-80-248-2391-1
5. Lukashenko, R., et al.: Computer-based plagiarism detection methods and tools: an overview. In: International Conference on Computer Systems and Technologies—CompSysTech (2007)
6. Pierre, Cornic: Software Plagiarism Detection using Model-Driven Software Development in Eclipse Platform. University of Manchester, School of Computer Science (2008)
7. Liaqat, A.G., Ahmad, A.: Plagiarism detection in java code. Linnaeus University, School of Computer Science, Physics and Mathematics (2011)
8. Cam Scanner—Phone PDF creator. https://play.google.com/store/apps/details?id=com.intsig.camscanner&hl=en. Accessed 25 Jan 2015
9. Intelligent Character Recognition Software. http://www.cvisiontech.com/ocr/text-ocr/intelligent-character-recognition-software.html?lang=eng. Accessed 2 Feb 2015
10. Baxter, I.D., Yahin, A., Moura, L., Sant'Anna, M., Bier, L.: Clone detection using abstract syntax trees. In: Proceedings of the International Conference on Software Maintenance, vol. 98, pp. 368–377 (1998)
11. Poongodi, D., TholkkappiaArasu, G.: An automatic method or statement level plagiarism detection in source code using abstract syntax tree. Research Scholar, Manonmaniam Sundaranar University, Tirunelveli
12. Karp, R.M., Rabin, M.O.: Efficient randomized pattern-matching algorithms. IBM J. Res. Dev. **31**(2), 249–260 (1987)
13. Rabin—Karp Algorithm. http://en.wikipedia.org/wiki/Rabin%E2%80%93Karp_algorithm. Accessed 20 Jan 2015
14. Karp-Rabin Algorithm. http://www-igm.univ-mlv.fr/~lecroq/string/node5.html. Accessed 20 Jan 2015
15. Kulkarni, A.J.: Cohort intelligence: a self supervised learning behavior. In: IEEE International Conference on Systems, Man, and Cybernetics, pp. 1396–1400 (2013)
16. Genetic Algorithm.: https://en.wikipedia.org/wiki/Genetic_algorithm. Accessed 26 July 2015
17. Bhosale, M.S., Mane, R.V.: study and analysis of cluster optimization algorithms: particle swarm optimization and Cohort intelligence. Int. J. Mod. Trends Eng. Res. **2**(3), 567–571 (2015)

Mechanism of Fuzzy *ARMS* on Chemical Reaction

P. Helen Chandra, S.M. Saroja Theerdus Kalavathy, A. Mary Imelda Jayaseeli and J. Philomenal Karoline

Abstract A new computing model Fuzzy abstract rewriting system on multisets is designed that is closer to reality by introducing fuzziness on computation. The mechanism of Artificial Cell System with hierarchically structurable membrane on chemical reaction is developed with fuzzy multiset evolution rules and fuzzy data. Significance of a parameter on the Fuzzy Artificial Cell System which describes the behaviour of the membrane structure is studied.

Keywords Membrane structure · Multiset · Abstract rewriting system · Artificial Cell System · Chemical reaction · Fuzzy data · Fuzzy abstract rewriting rules

1 Introduction

Uncertainty is an inherent property of all living systems. Fuzzy set was introduced by Zadeh and it has application in many fields [12]. Formal languages are precise while natural languages are quite imprecise. To reduce a gap between these two constructs [3], it becomes advantageous to introduce fuzziness into the structures of formal languages [8]. Since rigid mathematical models employed in life sciences are not

---

P.H. Chandra (✉) · S.M.S.T. Kalavathy · A.M.I. Jayaseeli · J.P. Karoline
Jayaraj Annapackiam College for Women (Autonomous),
Periyakulam, Theni, Tamilnadu, India
e-mail: chandrajac@yahoo.com
URL: http://www.annejac.com

S.M.S.T. Kalavathy
e-mail: kalaoliver@gmail.com

A.M.I. Jayaseeli
e-mail: imeldaxavier@gmail.com

J.P. Karoline
e-mail: philoharsh@gmail.com

© Springer International Publishing Switzerland 2016
V. Snášel et al. (eds.), *Innovations in Bio-Inspired Computing and Applications*,
Advances in Intelligent Systems and Computing 424,
DOI 10.1007/978-3-319-28031-8_4

completely adequate for the interpretation of biological information, there have been various proposals to use fuzzy sets in the modeling of biological systems. Thus, studies have been made on the use of the theory of fuzzy sets in P system, which is a computing model proposed in the area of membrane computing [7].

The last years have witnessed an increasing interest in the development of uncertain mathematical approaches to membrane computing. The reasons have been, among others, to keep close to the development of new formal computational paradigms dealing with fuzzy information, and the possibility of applying P systems to model real biological processes where handling with uncertainty, are necessary. A first contribution to this line of research was given by Obtulowicz and Paun [4], by extending the classical model to several probabilistic ones. Obtulowicz [5] also discussed several possible rough set based mathematical models of uncertainty that could be used in membrane computing. In a similar vein, several fuzzy approaches have been introduced. In one of them, fuzzy mathematics are used to handle the uncertainty in the number of copies of the reactives in the membranes [6]. An orthogonal approach to the fuzzification of both multisets and hybrid sets is presented by Apostolos Syropoulos [11]. Y. Suzuki and H. Tanaka have introduced the multiset rewriting system, "Abstract rewriting System on multisets" (*ARMS*). Based on this system, they have developed a molecular computing model called Artificial Cell System which consists of a multiset of symbols, a set of rewriting rules and membranes [9, 10]. These correspond to a class of P systems which is a parallel molecular computing model proposed by Gh. Paun and is based on the processing of multisets of objects in cell-like membrane structures [7].

In [2], the authors consider the phenomenon of Iron(III)salen complexes catalyzing the H_2O_2 oxidation of aryl methyl sulfides and sulfoxides and propose the possible mechanisms based on the kinetic and spectral studies. Recently, in [1], based on membrane computing, a computational study of the work in [2] is done and a model, called *Kinetic ARMS* in Artificial Cell System with hierarchically structurable membrane (*KACSH*), is developed.

In this present study we introduce a new mechanism of computing system called as *FACSH* with fuzzy multiset evolution rules and fuzzy data. i.e., *Fuzzy ARMS* in Artificial Cell System with hierarchically structurable membrane to study significantly the biochemical reactions to understand the nature of interaction between the synthesised complexes and biomolecules. We also have analysed a parameter on the system which describe the behaviour of the membrane structure.

2 Preliminaries

We first recall the basic structural ingredients of the computing device.

2.1 P System with Fuzzy Data [11]

A *P* system with fuzzy data is a construct $\Pi_{FD} = (O, \mu, w^{(1)}, \dots, w^{(m)}, R_1, \dots, R_m, i_0, \lambda)$ where O is an alphabet (i.e., a set of distinct entities) whose elements are called objects; μ is the membrane structure of degree $m \geq 1$; membranes are injectivelly labeled with succeeding natural numbers starting with one; $w^{(i)} : O \rightarrow N_0 \times I, 1 \leq i \leq m$, are functions that represent multi-fuzzy sets over O associated with each region i; N_0 is the set of all natural numbers including 0, $I = [0, 1]$; $R_i, 1 \leq i \leq m$, are finite sets of multiset rewriting rules (called evolution rules) over O. An evolution rule is of the form $u \rightarrow v, u \in O^*$ and $v \in O^*_{TAR}$, where $O_{TAR} = O \times TAR, TAR = \{here, out\} \cup \{in_j | 1 \leq j \leq m\}$. The effect of each rule is the removal of the elements of the left-hand side of each rule from the current compartment and the introduction of the elements of right-hand side to the designated compartments; $i_0 \in \{1, 2, \dots, m\}$ is the label of an elementary membrane (i.e., a membrane that does not contain any other membrane), called the output membrane; and $\lambda \in [0, 1]$ is a threshold parameter, which is used in the final estimation of the computational result.

2.2 P System with Fuzzy Multiset Rewriting Rules [11]

A P system with fuzzy multiset rewriting rules and crisp data is just an ordinary P system that has, in addition, a corresponding fuzzy set for each set R_i of multiset rewriting. A P system with multiset fuzzy rewriting rules will compute a number to some degree. Clearly, such systems must also obey the so called maximal parallelism principle, that is the rules should be selected in such a way that only optimal output will be yielded. Thus, P systems with fuzzy data differ fundamental from P systems with probabilistic rewriting rules in that there is no bias in the selection of the rules. When a P system with fuzzy multiset rewriting rules halts, the result of the computation up to some degree is equal to the cardinality of the multiset contained in the output compartment. Clearly, it is also necessary to know how to compute the truth degree that is associated with the computational result.

2.3 ARMS *(Abstract Rewriting System on Multisets)* [9]

ARMS is like a chemical solution in which molecules floating on it can interact with each other according to reaction rules. Technically, a chemical solution is a finite multiset of elements denoted by $A^k = \{a, b, .., \}$; these elements correspond to molecules. Reaction rules that act on the molecules are specified in *ARMS* by rewriting rules. In fact, this system can be thought of as an underling algorithmic chemistry [1].

Let A be an alphabet whose elements are called objects; the alphabet itself is called a set of objects. A multiset over a set of objects A is a mapping $M : A \to N_0$. The number $M(a)$, for $a \in A$, is the multiplicity of object a in the multiset M. We do not accept here an infinite multiplicity. We denote by $A^\#$ the set of all multisets over A including the empty multiset ϕ defined by $\phi(a) = 0$ for all $a \in A$. A multiset rewriting rule (evolution rule) over a set A of objects is a pair (M_1, M_2) of elements in $A^\#$ (which can be represented as a rewriting rule $w_1 \to w_2$, for two strings $w_1, w_2 \in A^\#$ such that $M_{w_1} = M_1$ and $M_{w_2} = M_2$). We use to represent such a rule in the form $M_1 \to M_2$.

An abstract rewriting system on multisets (ARMS) is a pair $\Gamma = (A, (R, \rho))$ where A is a set of objects; R is a finite set of multiset evolution rules over A; ρ is a partial order relation over R, specifying a priority relation among rules of R.

With respect to an ARMS Γ, we can define over $A^\#$ a relation ($\Rightarrow$): for $M, M' \in A^\#$ we write $M \Rightarrow M'$ iff $M' = (M - (M_1 \cup \cdots \cup M_k)) \cup (M'_1 \cup \cdots \cup M'_k)$, for some $M_i \to M'_i \in R, 1 \le i \le k, k \ge 1$, and there is no rule $M_s \to M'_s \in R$ such that $M_s \subseteq (M - (M_1 \cup \cdots \cup M_k))$; atmost one of the multisets $M_i, 1 \le i \le k$, may be empty. A multiset $M \in A^\#$ is dead if there is no $M' \in A^\#$ such that $M \Rightarrow M'$. This is equivalent to the fact that there is no rule $M_1 \to M_2 \in R$ such that $M_1 \subseteq M$. A multiset $M \in A^\#$ is initial if there is no $M' \in A^\#$ such that $M' \Rightarrow M$.

In ARMS, all the reaction rules are applied in parallel. In every step, all the rules are applied to all objects in every membrane that can be applied. If there are more than one applicable rule that can be applied to an object then one rule is selected randomly.

3   Fuzzy ARMS

Now we propose a new computing device that based on Abstract Rewriting systems on multisets which is closely related to P system with fuzzy multiset rewriting rules and fuzzy data which is called as FARMS.

3.1   Definition

A Fuzzy ARMS (FARMS) is a quintuple $\Gamma = \{A, (R, \rho), J, \mu\}$ where A is a set of objects, R is a finite set of multiset rewriting rules over A, ρ is a partial order relation over R, specifying a priority relation among the rules of R, $J = \{r_j / j = 1 \text{ to } n, n = \text{cardinality of } R\}$ i.e. the number of multiset rewriting rules over A, $\mu : J \to [0, 1]$ is the membership function in R such that $\mu(r_j) = i, i \in [0, 1]$.

In FARMS, reaction rules are applied in parallel. When there are more than one applicable-rules then one rule is selected randomly.

A *Fuzzy ARMS* generates a *Fuzzy ARMS language* $L(FARMS)$ as follows. An object $x \in A^*$ is said to be in $L(FARMS)$ iff it is derivable from any object $S \in A$ and the grade of membership $\mu_{(L(FARMS))}(x)$ is greater than 0, where

$$\mu_{L(FARMS)}(x) = \left(\begin{array}{c} max \\ 1 \le k \le n \end{array} \right) \left[\left(\begin{array}{c} min \\ 1 \le i \le l_k \end{array} \right) \mu(r_i^k) \right], x \in A^*$$

where n is the number of different derivatives that x has in *FARMS*; l_k is the length of the kth derivative chain; r_i^k denotes the label of the ith multiset rewriting rule used in the kth derivative chain, $i = 1, 2, \dots, l_k$.

Clearly, $\mu_{L(FARMS)}(x) =$ Strength of the strongest derivative chain for S to x for all $x \in A^*$.

3.2 Example

Consider the *Fuzzy ARMS*

$$\Gamma = \{A, (R, \rho), J, \mu\}$$

where $A = \{a, b, c, d, f\}$

$$R = \left\{ \begin{array}{l} r_1 : a^m, f \to b^m, c^m \text{ with } \mu(r_1) = 0.8 \\ r_2 : c^m, d \to a^m, c^m \text{ with } \mu(r_2) = 0.5 \end{array} \right\},$$

$J = \{r_j \in R/j = 1 \text{ to } 2\}$, $\mu : J \to I$ is the membership function s.t $\mu(r_j) = i, i \in [0, 1]$ and $\rho = \phi$.

The set of the rewriting rules, R is $\{r_1, r_2\}$. We do not assume priority among these rules. In *FARMS*, reaction rules are applied in parallel. When there are more than one applicable rule, then one rule is selected randomly. Let us take $\{a^m, f, d : m \ge 1\}$ as an initial state.

If $m = 1$, the rule r_1 is applied in parallel and $\{a, f, d\}$ is transformed into $\{b, c, d\}$ with $\mu(r_1) = 0.8$. Since r_1 cannot be applied on this multiset, r_2 is applied, resulting into the multiset $\{a, b, c\}$ with $\mu(r_2) = 0.5$. As there are no rules that can transform the multiset further, the system is in a dead state. Thus the grade of membership value is

$$\mu_{L(FARMS)}(a, b, c) = \left(\begin{array}{c} max \\ 1 \le k \le n \end{array} \right) \left[\left(\begin{array}{c} min \\ 1 \le i \le l_k \end{array} \right) \{0.8, 0.5\} \right] = 0.5$$

If $m = 2$, the rule r_1 is applied in parallel and $\{a, a, f, d\}$ is transformed into $\{b, b, c, c, d\}$ with $\mu(r_1) = 0.8$. Since r_1 cannot be applied on this multiset, r_2 is applied, resulting into the multiset $\{a, a, b, b, c, c\}$ with $\mu(r_2) = 0.5$. As before, there are no rules that can transform the multiset further. So, the system is in a dead state.

Thus the grade of membership value is

$$\mu_{L(FARMS)}(a,a,b,b,c,c) = \left(\begin{array}{c} max \\ 1 \leq k \leq n \end{array}\right)\left[\left(\begin{array}{c} min \\ 1 \leq i \leq l_k \end{array}\right)\{0.8, 0.5\}\right] = 0.5$$

proceeding like this, we obtain the language as

$$L(FARMS) = \{a^n, b^n, c^n / n \geqq 1\} \quad with \quad \mu_{L(FARMS)}(a^n, b^n, c^n) = 0.5$$

4 *FARMS* in *ACS* with Hierarchically Structurable Membrane

We now introduce *FACSH*, a mechanism of Artificial Cell System with hierarchically structurable membrane with fuzzy multiset evolution rules and fuzzy data.

4.1 *Definition*

A *Fuzzy ACSH (FACSH)* is a construct

$$\Gamma = \{A, \mu, M_1, M_2, M_3, ., M_m, (R_p, \rho), i_0, J, \omega\}$$

where A is the set of objects; μ is the membrane structure; M_i are the multisets associated with the regions $1, 2, \ldots, m$ of μ, where $i = 1$ to m; R_p is a set of Fuzzy multiset evolution rules over $A, p = 1$ to m of μ; ρ is the partial order relation over R_p, $i_0 \in \{1, 2, \ldots, m\}$ is the elementary membrane (output); $J = \{R_{pq} \in R_p / p = 1, \ldots, m, q \geq 1\}$, $q = $ cardinality of R_p; $\omega : J \rightarrow [0, 1]$ is the membership function s.t. $\omega(R_{pq}) = i, i \in [0, 1]$.

Reaction rules are applied in the following manner:
The same rules are applied to every membrane. There are no rules specific to a membrane. All the rules are applied in parallel. In every step all the rules are applied to all objects in every membrane that can be applied. If there are more than one applicable rules that can be applied to an object then one rule is selected randomly. If a membrane dissolves then all the objects in its region are left free in the region immediately above it. All objects and membranes not specified in a rule and which do not evolve are passed unchanged to the next step.

A *Fuzzy ACS* generates a language *L(FACSH)* as follows: An object $x \in A^*$ is said to be in *L(FACSH)* iff it is derivable from any object $S \in A$ and the grade of membership $\omega_{L(FARMS)}(x)$ is greater than 0, where

$$\omega_{L(FACSH)}(x) = \left(\begin{array}{c} max \\ 1 \leq k \leq n \end{array}\right)\left[\left(\begin{array}{c} min \\ 1 \leq i \leq l_k \end{array}\right)\omega(R_i^k)\right], x\ in A^*$$

where n is the number of different derivatives that x has in *FACSH*, l_k is the length of the kth derivative chain, R_i^k denotes the label of the ith multiset evolution rule used in the kth derivative chain, $i = 1, 2, \ldots, l_k$.

Clearly, $\omega_{L(FACSH)}(x) =$ Strength of the strongest derivative chain for S to x for all $x \in A^*$.

4.2 Mechanism for Sulfides Oxidation in FACSH

4.2.1 Process A

First we describe the complex formation between the oxidant and the substrate.

$(a) Z + X(F3)X \rightarrow X(F4O)X;$
  $X(F4O)X + RSR' \rightarrow X(F3)X + RSOR'$

A simple abstract reaction scheme is followed. Following convention is used to do the computation. When $X = H = L, X = Cl = M, X = Br = N, X = CH_3 = P$ and $X = OCH_3 = Q, (a)$ will have the following reaction rules

1. $Z + L(F3)L \rightarrow L(F4O)L; \quad L(F4O)L + RSR' \rightarrow L(F3)L + RSOR'$

2. $M(F3)M + Z \rightarrow M(F4O)M; \quad M(F4O)M + RSR' \rightarrow M(F3)M + RSOR'$

3. $N(F3)N + Z \rightarrow N(F4O)N; \quad N(F4O)N + RSR' \rightarrow N(F3)N + RSOR'$

4. $P(F3)P + Z \rightarrow P(F4O)P; \quad P(F4O)P + RSR' \rightarrow P(F3)P + RSOR'$

5. $Q(F3)Q + Z \rightarrow Q(F4O)Q; \quad Q(F4O)Q + RSR' \rightarrow Q(F3)Q + RSOR'$

$(b) Z + XY(F3)XY \rightarrow XY(F4O)XY;$
  $XY(F4O)XY + RSR' \rightarrow XY(F3)XY + RSOR'$

Following convention is used to do the computation. When $X = Y = Cl = M$ and $X = Y = t - Butyl = T, (b)$ will have the following reaction rules

6. $Z + MM(F3)MM \rightarrow MM(F4O)MM;$
   $MM(F4O)MM + RSR' \rightarrow MM(F3)MM + RSOR'$

7. $Z + TT(F3)TT \rightarrow TT(F4O)TT;$
   $TT(F4O)TT + RSR' \rightarrow TT(F3)TT + RSOR'$

4.2.2 Behaviour of FACSH − A

Consider the *FACSH*

$$\Gamma = (A, \mu, M_1, M_2, M_3, (R_p, \rho), i_0, J, \omega)$$

where $A = \{Z, S, A_i, B_i, P_i, i = 1, \ldots, 7\}$, $\mu = [_1 [_2 [_3\ ]_3]_2]_1$,
M_1, M_2, M_3 are the multisets associated with the regions $1, 2, 3$ of μ,
$M_1 = \{Z, A_i, i = 1, \ldots, 7\}, M_2 = \{S\}, M_3 = \{\phi\}$,
$i_0 = 3$ is the output membrane and $\rho = \phi$,
R_p is a set of Fuzzy multiset evolution rules over $A; p = 1$ to 3
$J = \{R_{pq} \in R_p / p = 1 \text{ to } 3, q = 1 \text{ to } 7\}, q = $ cardinality of R_p,
$\omega : J \to [0, 1]$ is the membership function s.t. $\omega(R_{pq}) = i, i \in [0, 1]$, where

$$\omega_{L(FACSH)}(x) = \begin{pmatrix} max \\ 1 \leq k \leq n \end{pmatrix} \left[\begin{pmatrix} min \\ 1 \leq i \leq l_k \end{pmatrix} \omega(R_i^k) \right], x \in A^*$$

$R_p = \{R_1, R_2, R_3\}$ consists the following evolution rules.

$$R_1 = \begin{cases} R_{11} : Z + A_1 \to B_{1in} & with & \omega(R_{11}) = 0.005 \\ R_{12} : Z + A_2 \to B_{2in} & with & \omega(R_{12}) = 0.006 \\ R_{13} : Z + A_3 \to B_{3in} & with & \omega(R_{13}) = 0.001 \\ R_{14} : Z + A_4 \to B_{4in} & with & \omega(R_{14}) = 0.0009 \\ R_{15} : Z + A_5 \to B_{5in} & with & \omega(R_{15}) = 0.001 \\ R_{16} : Z + A_6 \to B_{6in} & with & \omega(R_{16}) = 0.006 \\ R_{17} : Z + A_7 \to B_{7in} & with & \omega(R_{17}) = 0.001 \end{cases}$$

$$R_2 = \begin{cases} R_{21} : B_1 + S \to A_{1out}, +P_{1in} & with & \omega(R_{21}) = 0.0009 \\ R_{22} : B_2 + S \to A_{2out} + P_{2in} & with & \omega(R_{22}) = 0.003 \\ R_{23} : B_3 + S \to A_{3out} + P_{3in} & with & \omega(R_{23}) = 0.02 \\ R_{24} : B_4 + S \to A_{4out} + P_{4in} & with & \omega(R_{24}) = 0.03 \\ R_{25} : B_5 + S \to A_{5out} + P_{5in} & with & \omega(R_{25}) = 0.02 \\ R_{26} : B_6 + S \to A_{6out} + P_{6in} & with & \omega(R_{26}) = 0.003 \\ R_{27} : B_7 + S \to A_{7out} + P_{7in} & with & \omega(R_{27}) = 0.0009 \end{cases}$$

$R_3 = \phi$

Initially the value of k is 1. In *FACSH*, all the rules are applied to all objects in every membrane that can be applied. If there are more than one applicable rule that can be applied to an object then one rule is selected randomly. In the initial state, we have 7 objects ($A_i, i = 1$ to 7) and an object Z in membrane 1. Any one of the 7 objects and Z are processed by the rule R_1. Let the object A_1 and the object Z are processed by the rule $R_{11} : Z + A_1 \to B_{1in}$ with $\omega(R_{11}) = 0.005$. It evolves the object B_1 to membrane 2. Objects in membrane 2 cannot be processed by the rules in membrane 2. There is no rule and object in membrane 3. In the next state, the object B_1 and the object S are processed only by the rule $R_{21} : B_1 + S \to A_{1out}, +P_{1in}$ with $\omega(R_{21}) = 0.0009$ in membrane 2. It evolves the object P_1 to membrane 3 and the object A_1 to membrane 1. Objects in membrane 1 cannot be processed by the rules in membrane 1. There is no object in membrane 2. Since there is no rule that can transform the object in membrane 3 further, the process halts. The resulting object in the output membrane 3 is P_1 (Fig. 1).

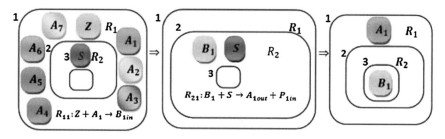

Fig. 1 *FACSH -A*

$$\begin{matrix} max \\ 1 \le k \le n \end{matrix} \left[\begin{matrix} min \\ 1 \le i \le l_1 \end{matrix} (0.005, 0.0009) \right] = 0.0009; \quad \mu_{L(FACSH)}(P_1) = 0.0009$$

Similarly, we have the possible seven transformations and the rules that can be applied in parallel to transform the objects to the desired one. As a result, we have

$$\mu_{L(FACSH)}P_i = \begin{cases} 0.0009 & if \ i = 1, 4, 7 \\ 0.001 & if \ i = 3, 5 \\ 0.003 & if \ i = 2, 6 \end{cases}; \quad L(FACSH) = \{P_i/i = 1, 2, \dots, 7\}$$

4.3 Mechanism for Sulfoxides Oxidation in FACSH

4.3.1 Process B

First we describe the complex formation between the oxidant and the substrate.

$(a) \ Z + X(F3)X \rightarrow X(F4O)X;$
$\quad X(F4O)X + RSOR' \rightarrow X(F3)X + RSO_2R'$

$(b) \ Z + XY(F3)XY \rightarrow XY(F4O)XY;$
$\quad XY(F4O)XY + RSOR' \rightarrow XY(F3)XY + RSO_2R'$

A simple convention is used to do the computation on reactions in Process B as in process A.

4.4 FARMS in ACS with Hierarchically Structurable Membrane (FACSH − B)

Next we present the Fuzzy Abstract Rewriting System on multisets based on Artificial Cell System with Hierarchically structurable membrane to describe the complex formation between the oxidant and the substrate. Shortly we call the system as *FACSH − B*.

4.5   *Behaviour of FACSH − B*

Consider the *FACSH*

$$\Gamma = (A, \mu, M_1, M_2, M_3, (R_p, \rho), i_0, J, \omega)$$

where $A = \{Z, SO, A_i, B_i, P_i, i = 1, \ldots, 7\}$, $\mu = [_1[_2[_3]_3]_2]_1$,
M_1, M_2, M_3 are the multisets associated with the regions 1, 2, 3 of μ,
$M_1 = \{Z, A_i, i = 1, \ldots, 7\}, M_2 = \{SO\}, M_3 = \{\phi\}$,
$i_0 = 3$ is the output membrane and $\rho = \phi$,
$R_p = \{R_1, R_2, R_3\}$ consists the set of Fuzzy multiset evolution rules over A,

$$R_1 = \left\{ \begin{array}{llll} R_{11} : Z + A_1 \rightarrow B_{1in} & with & \omega(R_{11}) = 0.2 \\ R_{12} : Z + A_2 \rightarrow B_{2in} & with & \omega(R_{12}) = 0.01 \\ R_{13} : Z + A_3 \rightarrow B_{3in} & with & \omega(R_{13}) = 0.006 \\ R_{14} : Z + A_4 \rightarrow B_{4in} & with & \omega(R_{14}) = 0.005 \\ R_{15} : Z + A_5 \rightarrow B_{5in} & with & \omega(R_{15}) = 0.006 \\ R_{16} : Z + A_6 \rightarrow B_{6in} & with & \omega(R_{16}) = 0.01 \\ R_{17} : Z + A_7 \rightarrow B_{7in} & with & \omega(R_{17}) = 0.006 \end{array} \right\}$$

$$R_2 = \left\{ \begin{array}{llll} R_{21} : B_1 + SO \rightarrow A_{1out}, +P_{1in} & with & \omega(R_{21}) = 0.0002 \\ R_{22} : B_2 + SO \rightarrow A_{2out} + P_{2in} & with & \omega(R_{22}) = 0.0007 \\ R_{23} : B_3 + SO \rightarrow A_{3out} + P_{3in} & with & \omega(R_{23}) = 0.0008 \\ R_{24} : B_4 + SO \rightarrow A_{4out} + P_{4in} & with & \omega(R_{24}) = 0.0009 \\ R_{25} : B_5 + SO \rightarrow A_{5out} + P_{5in} & with & \omega(R_{25}) = 0.0008 \\ R_{26} : B_6 + SO \rightarrow A_{6out} + P_{6in} & with & \omega(R_{26}) = 0.0007 \\ R_{27} : B_7 + SO \rightarrow A_{7out} + P_{7in} & with & \omega(R_{27}) = 0.0002 \end{array} \right\}$$

$$R_3 = \phi$$

$J = \{R_{pq} \in R_p / p = 1 \text{ to } 3, q = 1 \text{ to } 7\}, q = \text{cardinality of } R_p$,
$\omega : J \rightarrow [0, 1]$ is the membership function s.t. $\omega(R_{pq}) = i, i \in [0, 1]$, where

$$\omega_{L(FACSH)}(x) = \left(\begin{array}{c} max \\ 1 \leq k \leq n \end{array} \right) \left[\left(\begin{array}{c} min \\ 1 \leq i \leq l_k \end{array} \right) \omega(R_i^k) \right], x \in A^*$$

As we have done in *FACSH − A* the computation is done. We have the possible 7 transformations and the rules that can be applied in parallel to transform the objects to the desired one. As a result,

$$\mu_{L(FACSH)}(P_i) = \left\{ \begin{array}{l} 0.0002 \ if \ i = 1, 7 \\ 0.0007 \ if \ i = 2, 6 \\ 0.0008 \ if \ i = 3, 5 \\ 0.0009 \ if \ \ i = 4 \end{array} \right\}; \quad L(FACSH) = \{P_i / i = 1, 2, \ldots, 7\}$$

5   Conclusion

A new membrane computing model on *Fuzzy ARMS* is introduced and the mechanism on Artificial Cell system is proposed. We have studied the significance of a parameter on the Fuzzy Artificial Cell system with hierarchically structurable membrane. It is decided to extend the work to investigate the correlation between *ACS* and *P* System. It is a preliminary research work on the study of binding of metal complexes with protein which may pave the way for drug designing further to design a computing system. As an application, it is worth to find out the power of this computing system and the characteristics of the concentration of chemical compounds. Further application and properties of the proposed system could be studied. We believe that the new system would be of use in the modeling of living organisms.

References

1. Helen Chandra, P., Saroja Theerdus Kalavathy, S.M., Jayaseeli, A.M.I.: Mechanism of sulfoxidation in artificial cell system. In: Proceedings of Asian Conference on Membrane Computing (ACMC), IEEE (2014)
2. Mary Imelda Jayaseeli, A., Rajagopal, S.: [Iron(III)-salen] ion catalyzed H_2O_2 oxidation of organic sulfides and sulfoxides. J. Mol. Catal, A: Chem. **309**, 103–110 (2009)
3. Mordeson, J.N., Malik, D.S., Kuroki, N.: Regular fuzzy expressions. In: Chapter 10: Fuzzy Semigroups, Studies in Fuzziness and Soft Computing, Springer (2003)
4. Obtulowicz, A., Paun, Gh: (In search of) probabilistic P systems. Bio Syst. **70**(2), 107–121 (2003)
5. Obtulowicz, A.: Mathematical models of uncertainty with a regard to membrane systems. Nat. Comput. **2**, 251–263 (2003)
6. Obtułowicz, A.: General multi-fuzzy sets and fuzzy membrane systems. In: Giancarlo, M., Gheorghe, P., Pérez-Jímenez, M.J., Rozenberg, G., Salomaa, A. (eds.) WMC 2004. LNCS, vol. 3365, pp. 359–372. Springer, Heidelberg (2005)
7. Paun, Gh: Membrane Computing: An Introduction. Springer, Berlin (2002)
8. Rozenberg, G., Salomaa, A. (eds.): Handbook of Formal Languages, vol. 1–3. Springer, Berlin (1997)
9. A New molecular Computing Model Artificial Cell System. http://www.cs.bham.ac.uk/wb/biblio/gecco/AA218.pdf. Accessed 29 Nov 2015
10. Yasuhiro, S., Yoshi, F., Junji, T., Hiroshi, T.: Artificial life applications of a class of p systems: abstract rewriting systems on multisets. In: Calude, Cristian S., Pun, G., Rozenberg, G., Salomaa, A. (eds.) Multiset Processing. LNCS, vol. 2235, pp. 299–346. Springer, Heidelberg (2001)
11. Syropoulos, A.: Fuzzyfying P systems. Comput. J. **49**(5), 619–628 (2006)
12. Zadeh, L.A.: Fuzzy sets. Inf. Control **8**, 338–353 (1965)

The Power of Hybridity and Context Free in *HP* System

S.M. Saroja T. Kalavathy, P. Helen Chandra and M. Nithya Kalyani

Abstract In this paper a new computing model called Hybrid P System is introduced and examined. The aim is to elaborate the power of hybrid P system with prescribed teams and context-free puzzle grammar rules in generating characters. Various non-context free sets of arrays that can be generated in a simple way by hybrid context-free puzzle grammar system with prescribed teams working in different modes are presented in [12]. We show the power of the mechanism of hybridity for picture description on P system.

Keywords Context-free puzzle grammar · Hybrid prescribed team *CD* grammar · Hybrid prescribed team *CFPG* · Array P System · Hybrid P system

1 Introduction

In Image Analysis, synthetic techniques have been employed in various studies. Motivated by problems of tiling in the two-dimensional plane, one such syntactic method was proposed by Nivat et al. [8] for generating pictures of connected finite arrays of symbols in the two-dimensional plane. This model called puzzle grammar has been investigated in [8] for its properties by comparing with array grammars. A subclass of puzzle grammars called context-free puzzle grammars with rules of a

S.M.S.T. Kalavathy · P.H. Chandra (✉) · M.N. Kalyani
Jayaraj Annapackiam College for Women (Autonomous), Periyakulam,
Theni, Tamilnadu, India
e-mail: chandrajac@yahoo.com
URL: http://www.annejac.com

S.M.S.T. Kalavathy
e-mail: kalaoliver@gmail.com

M.N. Kalyani
e-mail: rnithraj@gmail.com

V. Snášel et al. (eds.), *Innovations in Bio-Inspired Computing and Applications*,
Advances in Intelligent Systems and Computing 424,
DOI 10.1007/978-3-319-28031-8_5

55

specific nature was introduced by Subramanian et al. [15] and has been studied in [7, 13, 16]. In the area of grammar systems, Dassow et al. [3] have introduced cooperating array grammar system extending the notion of cooperating distributed (string) grammar system to arrays. Subramanian et al. [17] have dealt with cooperating array grammar systems with the components having basic puzzle grammar rules and have shown that the generative capacity of cooperating array grammar systems with basic puzzle rules is strictly greater than that of context-free array grammars.

The notion of a team *CD* grammar system was introduced and investigated by removing the restriction in the *CD* grammar system that at each moment only one component is enabled [2, 6, 9]. Henning and Maurice H. ter Beek [4, 11] studied hybrid (prescribed) team *CD* grammar system allowing work to be done in teams while at the same time assuming these teams have different capabilities. They have dealt with hybrid prescribed team cooperating array grammar systems with the components having context-free or regular array grammar rules.

On the other hand, the area of membrane computing was initiated by Paun [10] introducing a new computability model called as *P* system, which is a distributed, highly parallel theoretical computing model based on the membrane structure and the behavior of the living cells. Among a variety of applications of this model, the problem of handling array languages using *P* systems has been considered by Ceterchi et al. by introducing array rewriting *P* System [1] and thus linking the two areas of membrane computing and picture grammars. A kind of array *P* system with objects in the regions as arrays and the productions as hybrid prescribed team of *CD* grammar rules was introduced in [5] which allow work to be done in team with the possibility of different teams having different modes of derivation. Rewriting is done in parallel in a team.

In this paper a new computing model called Hybrid *P* System is introduced by considering context-free puzzle grammar rules instead of context-free or regular array rewriting rules. The generative power of the mechanism of hybrid prescribed teams for picture description on P system is examined by comparing with the other models.

2   Preliminaries

In this section, some prerequisites necessary for understanding the sequel are defined.

2.1   *Context-Free Puzzle Grammar (CFPG) [15]*

A basic puzzle grammar (*BPG*) is a structure $G = (N, T, P, S)$ where N and T are finite sets of symbols; $N \cap T = \phi$. Elements of N are called non-terminals and elements of T, terminals. The start symbol or the axiom is $S \in N$. The set P consists of rules of the following forms:

$$A \to \text{(a)}\, B \quad , \quad A \to a\, \text{(B)} \quad , \quad A \to B\, \text{(a)}$$

$$A \to \text{(B)}\, a \quad , \quad A \to \genfrac{}{}{0pt}{}{a}{\text{(B)}} \quad , \quad A \to \genfrac{}{}{0pt}{}{\text{(a)}}{B}$$

$$A \to \genfrac{}{}{0pt}{}{B}{\text{(a)}} \quad , \quad A \to \genfrac{}{}{0pt}{}{\text{(B)}}{a} \quad , \quad A \to \text{(a)}$$

where $A, B \in N$ and $a \in T$.

Derivations begin with S written in a unit cell in the two-dimensional plane, with all the other cells containing the blank symbol #, not in $N \cup T$. In a derivation step, denoted $\to$, a non-terminal A in a cell is replaced by the right-hand member of a rule whose left-hand side is A. In this replacement, the circled symbol of the right-hand side of the rule used, occupies the cell of the replaced symbol and the non-circled symbol of the right side occupies the cell to the right or the left or above or below the cell of the replaced symbol depending on the type of rule used. The replacement is possible only if the cell to be filled in by the non-circled symbol contains a blank symbol.

A context-free puzzle grammar (*CFPG*) is a structure $G = (N, T, P, S)$ where N, T, S are as above and P the set of rules of the form $A \to \alpha$ where α is a finite, connected array of one or more cells, each cell containing a nonterminal or a terminal symbol, with a symbol in one of the cells of α being circled. Derivations are defined in a similar manner.

2.2 Hybrid Prescribed Team CD Grammar System [11]

A hybrid prescribed team *CD* grammar system is a construct

$$\Gamma = (N, T, S, P_1, \dots, P_n, (Q_1, f_1), (Q_2, f_2), \dots, (Q_m, f_m)),$$

where $N, T, , P_1, \dots, P_n$ are defined as in the cooperating array grammar system [3]. $Q_1, Q_2, \dots Q_m$ are teams over $N \cup T$, multiset of sets of productions $P_1, \dots, P_n$ and $f_1, f_2, \dots, f_n$ are modes of derivation.

For a team $Q_i, 1 \le i \le m, Q_i = \{P_{ij} | 1 \le j \le m_i\}$, and two arrays D_1 and $D_2 \in (N \cup T)^+$ a direct derivation step is defined by $D_1 \vdash_{Q_i} D_2$ if and only if there are array productions $p_j \in P_{ij}, 1 \le j \le m_i$, such that in D_1 we can find m_k non-overlapping areas such that the sub-patterns of D_1 located at these areas coincide with the left-hand sides of the array productions p_j and yield D_2 by replacing them by the right-hand sides of the array productions p_j.

The language generated by Γ is

$$L(\Gamma) = \{X \in T^{**}/S \Rightarrow^{\,f_1}_{\,Q_1} X_1 \Rightarrow^{\,f_2}_{\,Q_2} X_2 \Rightarrow \cdots \Rightarrow^{\,f_m}_{\,Q_m} X_m = X\}$$

2.3 Array P Systems [1]

The array P system (of degree $m \geq 1$) is a construct

$$\Pi = (V, T, \#, \mu, F_1, ., ., F_m, R_1, \ldots, R_m, i_o),$$

where V is the total alphabet, $T \subseteq V$ is the terminal alphabet, # is the blank symbol, μ is a membrane structure with m membranes labeled in a one-to-one way with $1, 2, ., m$, $F_1, ., F_m$ are finite sets of arrays over V associated with the m regions of $\mu, R_1, \ldots, R_m$ are finite sets of array rewriting rules over V associated with the m regions of μ ; the rules have attached targets *here, out , in* (in general, *here* is omitted), hence they are of the form $A \rightarrow B(tar)$; finally, i_o is the label of an elementary membrane of μ (the output membrane).

The set of all arrays generated by a system Π is denoted by $AL(\Pi)$. The family of all array languages $AL(\Pi)$ generated by systems Π as above, with at most m membranes, with rules of type $\alpha \in \{REG, CF, \#CF\}$ is denoted by $EAP_m(\alpha)$. If non-extended systems are considered, then we write $AP_m(\alpha)$.

2.4 Hybrid Prescribed Teams Context-Free Puzzle Grammar System [12]

A Hybrid context-free puzzle grammar system with prescribed teams (*PTHCFPGS*) is a construct

$$\Gamma = (N, T, P_1, \ldots, P_n, S, (Q_1, f_1), (Q_2, f_2), \ldots, (Q_m, f_m))$$

where N, T, S and (Q_i, f_i) , $i = 1, 2, \ldots, n$ are defined as in the Hybrid prescribed team *CD* grammar system and $P_i, i = 1, 2, \ldots, n$ are non-empty finite sets of context-free puzzle grammar rules over $N \cup T$.

For a Hybrid Context-free puzzle grammar system with prescribed teams Γ, the array language generated by Γ is

$$L(\Gamma) = \{X \in T^{**}/S \Rightarrow^{\,f_1}_{\,Q_1} X_1 \Rightarrow^{\,f_2}_{\,Q_2} X_2 \Rightarrow \cdots \Rightarrow^{\,f_m}_{\,Q_m} X_m = X , m \geq 1\}$$

The family of array languages generated by a *PTHCFPGS* with atmost n components is denoted by $PTH_n(CFPGL), n \geq 1$.

3 Hybrid P System with *CFPG*

Now we introduce Hybrid P system with context-free puzzle grammar which allows work to be done in team with the possibility of different teams having different modes of derivation.

3.1 *Definition*

A Hybrid P system with *CFPG* of degree $m(m > 1)$ is a construct

$$\Pi = (V, T, \#, \mu, F_1, \ldots, F_m, R_1, \ldots, R_m, i_o)$$

where V is the total alphabet, $T \subseteq V$ is the terminal alphabet, # is the blank symbol, μ is a membrane structure with m membranes labeled in a one-to-one way with $1, 2, \ldots, m$; $F_1, F_2, \ldots, F_m$ are finite sets of arrays over V initially associated with the m regions of μ; $R_1, R_2, \ldots, R_m$ are finite sets of prescribed teams of context - free puzzle grammar rules with the derivation modes associated with the m regions of μ; the rules have attached targets, *here, out, in*; finally, i_o is the label of an elementary membrane of μ (the output membrane).

A computation in the Hybrid *P* systems with context-free puzzle grammar is defined in the same way as in an array rewriting P system with the successful computations being the halting ones; each array, from each region of the system, which can be rewritten by a team of rules associated with that region (membrane), in a specific derivation mode. The array obtained by rewriting is placed in the region indicated by the target associated with the rule used. The term *here* means that the array remains in the same region, *out* means that the array exits the current membrane, thus, if the rewriting was done in the skin membrane, then it can exit the system; arrays leaving the system are *"lost"* in the environment, and *in* means that the array is immediately sent to one of the directly lower membranes, non-deterministically chosen of several exist (if no internal membrane exists, then a rule with the target indication *in* cannot be used). A computation is successful only if it stops and a configuration is reached where no rule can be applied to the existing arrays. The result of a halting computation consists of the arrays composed only of symbols from T placed in the membrane with label i_o in the halting configuration.

The set of all such arrays computed (or generated) by a system Π is denoted by $HPCFPL(\Pi)$. The family of all array languages $HPCFPL(\Pi)$ generated by system Π as above, with at most m membranes with prescribed teams of context - free puzzle grammar rules is denoted by $HP_m(CFPL)$.

3.2 Example

Consider the Hybrid P system with *CFPG*

$$\Pi_2 = (\{S, A, B, C, D, E, a\}, \{a\}, \#, [_1[_2[_3[_4]_4]_3]_2]_1, S, \emptyset, \emptyset, \emptyset, R_1, R_2, R_3, R_4, 4)$$

where $R_1 = (Q_1, t)_{in}$, $R_2 = \{(Q_2, *)_{here}, (Q_2, *)_{in}\}$,

$R_3 = \{(Q_3, t)_{here}, (Q_3, t)_{in}\}$, $R_4 = \emptyset$,

$Q_1 = \{P_1\}$, $Q_2 = \{P_2\}$, $Q_3 = \{P_3, P_4, P_5, P_6\}$

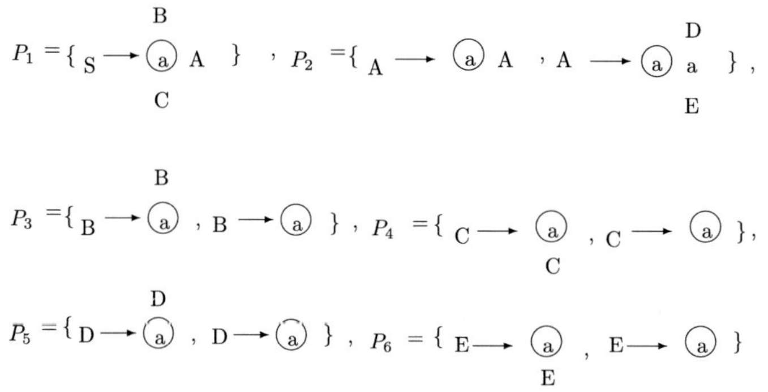

$$P_1 = \left\{ S \longrightarrow \begin{matrix} B \\ \textcircled{a} \ A \\ C \end{matrix} \right\} , \quad P_2 = \left\{ A \longrightarrow \textcircled{a} \ A \ , \ A \longrightarrow \begin{matrix} D \\ \textcircled{a} \ a \\ E \end{matrix} \right\} ,$$

$$P_3 = \left\{ B \longrightarrow \begin{matrix} B \\ \textcircled{a} \end{matrix} \ , \ B \longrightarrow \textcircled{a} \right\} , \quad P_4 = \left\{ C \longrightarrow \textcircled{a} \ , \ C \longrightarrow \begin{matrix} \textcircled{a} \\ C \end{matrix} \right\} ,$$

$$P_5 = \left\{ D \longrightarrow \begin{matrix} D \\ \textcircled{a} \end{matrix} \ , \ D \longrightarrow \widetilde{\textcircled{a}} \right\} , \quad P_6 = \left\{ E \longrightarrow \begin{matrix} \textcircled{a} \\ E \end{matrix} \ , \ E \longrightarrow \textcircled{a} \right\}$$

The axiom array is initially in the region 1 and the other regions do not have objects. The array generated on application of the rules in the team Q_1, with the derivation mode t is sent to region 2. In region 2, the rules in the team Q_2 with the derivation mode $*$ is applied and the process is repeated as the target attached to the rule is *here*. The generated array is sent to inner region 3. In region 3, if the first rule in the team Q_3 is applied in the t-mode, one pixel is grown upwards and downwards both in the right side and left side. The process is repeated as the target attached to the rule is *here*. If the second rules in the team Q_3, i.e., P_3, P_4, P_5 and P_6 are applied in parallel, then the array of H shape is obtained and sent to the inner region 4. If the final array productions are not applied synchronously taken from the team Q_3 then the computation in region 3 is blocked without any possibility to yield a terminal array any more.

The picture language generated by Π_2 consists of H shapes with the horizontal line at the middle of the vertical ones as in Fig. 1.

Fig. 1 Array describing
pattern H

$$\begin{matrix} a & & a \\ a & & a \\ a & a & a & a \\ a & & a \\ a & & a \end{matrix}$$

4 Generative Power

4.1 Theorem

Token T with equal arms can be generated by $HP_4(CFPL)$.

Proof

Consider the Hybrid P system with *CFPG*

$$\Pi_3 = (\{S, A, D, a\}, \{a\}, \#, [_1[_2[_3[_4]_4]_3]_2]_1, S, \emptyset, \emptyset, \emptyset, R_1, R_2, R_3, R_4, 4)$$

where $R_1 = (Q_1, t)_{in}$, $R_2 = \{(Q_2, *)_{here}, (Q_2, *)_{in}\}$,
$R_3 = \{(Q_3, t)_{here}, (Q_3, t)_{in}\}$, $R_4 = \emptyset$,
where $Q_1 = \{P_1\}$, $Q_2 = \{P_2, P_3, P_4\}$, $Q_3 = \{P_5, P_6\}$

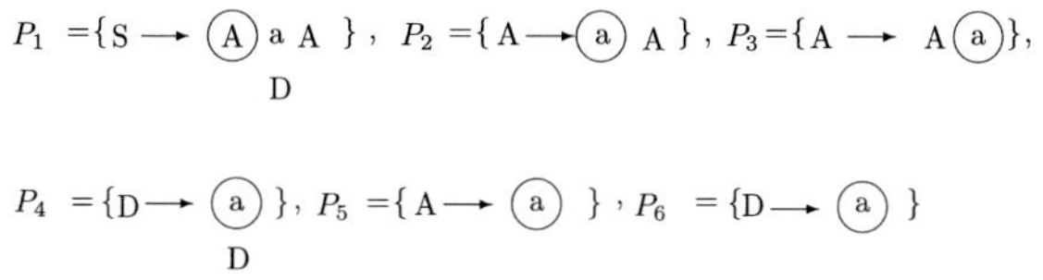

$$P_1 = \{S \longrightarrow \text{(A) a A} \atop \text{D}\}, \quad P_2 = \{A \longrightarrow \text{(a) A}\}, \quad P_3 = \{A \longrightarrow \text{A (a)}\},$$

$$P_4 = \{D \longrightarrow \text{(a)} \atop \text{D}\}, \quad P_5 = \{A \longrightarrow \text{(a)}\}, \quad P_6 = \{D \longrightarrow \text{(a)}\}$$

It generates picture token T with all three 'arms' of equal length as in Fig. 2.

4.2 Theorem

The set of all solid rectangles of size $n \times m$ with $n, m \geq 2$ are generated by the Hybrid P systems of degree 3 .

Proof Let $\Pi_4 = (\{S, A, B, C, a\}, \{a\}, \#, [_1[_2[_3]_3]_2]_1, S, \emptyset, \emptyset, R_1, R_2, R_3, 3)$
where $R_1 = \{(Q_1, *)_{here}, (Q_1, *)_{in}\}$, $R_2 = \{(Q_2, \geq k)_{in}$, $R_3 = \{(Q_3, t)_{here}$,
$\quad Q_1 = \{P_1\}$, $Q_2 = \{P_2, P_3\}$, $Q_3 = \{P_4\}$

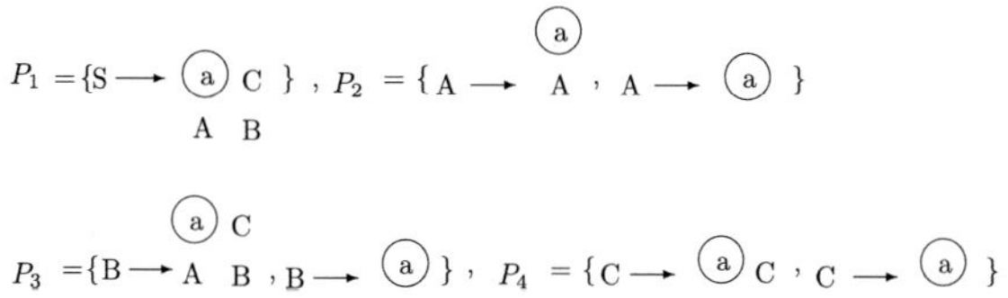

$$P_1 = \{S \longrightarrow \text{(a) C} \atop \text{A B}\}, \quad P_2 = \{A \longrightarrow \text{(a)} \atop \text{A}, \quad A \longrightarrow \text{(a)}\}$$

$$P_3 = \{B \longrightarrow \text{(a) C} \atop \text{A B}, \quad B \longrightarrow \text{(a)}\}, \quad P_4 = \{C \longrightarrow \text{(a) C}, \quad C \longrightarrow \text{(a)}\}$$

Fig. 2 Token T with arm length 3

$$\begin{matrix} a\ a\ a\ a\ a\ a\ a \\ a \\ a \\ a \end{matrix}$$

Fig. 3 Rectangles of size
3×4

$$
\begin{array}{cccc}
a & a & a & a \\
a & a & a & a \\
a & a & a & a
\end{array}
$$

$L(\Pi_3)$ is the set of all solid rectangles of size $n \times m$ with $n, m \geq 2$ as depicted in Fig. 3.

4.3 Theorem

The classes $HP_3(CFPL)$ and PAP_3 [18] have non-empty intersection.

Proof

Parallel array P system (PAP) has been introduced in [18] and in this system the regions have rectangular array objects and tables of context-free rules. We now compare our hybrid model with PAP.

Consider the parallel array P system

$$\Pi_1 = (V, V, [_1[_2[_3]_3]_2]_1, M_0, \phi, \phi, \mathfrak{I}_1, \mathfrak{I}_2, \phi, 3), \text{ where } V = \{X, .\},$$

$$
M_0 = \begin{array}{l} X \ . \\ X \ . \\ X \ X \end{array} \qquad \mathfrak{I}_1 = \{(R_1, in)\}, \qquad \mathfrak{I}_2 = \{(U, out), (R_2, in)\}
$$

$R_1 = \{X \rightarrow XX, \quad . \rightarrow ..\}, \quad R_2 = \{X \rightarrow X, \quad . \rightarrow .\}$ are right tables.

$U = \{X \rightarrow \begin{array}{l} X \\ X \end{array}, \quad . \rightarrow \begin{array}{l} . \\ . \end{array}\}$ is an up table.

The axiom rectangular array M_0 is initially in the region 1. When the rules of the table R_1 are applied to this array it grows one column in the right and the generated array M_1 is sent to region 2. If R_2 is applied, then this array is sent to region 3 where it remains forever and the language collects this array. If U is applied to M_1 in region 2 the array grows upwards and is sent back to region 1. The derivation then continues. The array language generated consists of arrays of the form in Fig. 4. where the array represents token L (. is represented as blank) with equal "arms".

This language also can be generated by $HP_3(CFPL)$.

Consider the Hybrid P system with context-free puzzle grammar

$$\Pi_1 = (\{S, U, R, a\}, \{a\}, \#, [_1[_2[_3]_3]_2]_1, S, \emptyset, \emptyset, R_1, R_2, R_3, 3)$$

where $R_1 = (Q_1, t)_{in}, \quad R_2 = \{(Q_2, *)_{here}, (Q_3, t)_{in}\}, \quad R_3 = \emptyset,$

$Q_1 = \{P_1\}, \quad Q_2 = \{P_2, P_3\}, \quad Q_3 = \{P_4, P_5\},$

Fig. 4 Array describing
token *L*

$$
\begin{array}{l}
X\ .\ .\ .\\
X\ .\ .\ .\\
X\ .\ .\ .\\
X\ X\ X\ X
\end{array}
$$

$$P_1 = \{ S \longrightarrow \overset{U}{\underset{}{\boxed{a}}}\ R\ \},\ P_2 = \{ U \longrightarrow \overset{U}{\underset{}{\boxed{a}}}\ \}$$

$$P_3 = \{ R \longrightarrow \boxed{a}\ R\ \}, P_4 = \{ U \longrightarrow \boxed{a}\ \},\ P_5 = \{ R \longrightarrow \boxed{a}\ \}$$

The axiom array is initially in the region 1 and the other regions do not have objects. An application of the rules in the team Q_1 with the derivation mode t yields $\begin{smallmatrix}U\\a\ R\end{smallmatrix}$ and is sent to region 2. In region 2, if the first team Q_2 is applied with the derivation mode $*$, the array grows equal number of columns to the right of the array and equal number of rows upwards and it remains in region 2 and the process can be repeated. If the second team with the target indication *in* is applied with the derivation mode t, L-shaped angles with equal arms, the length of each arm being at least three are generated and sent to region 3. The picture language generated by Π, consists of rectangular arrays of all right angles (Token L)in the form of token *L*.

4.4 *Theorem*

The family $HP_4(CFPL)$ intersects with the family *(R:RIR)SML* [14].

Proof
We now compare our model with the model *(R:RIR)SML* introduced in [14], wherein Siromoney array grammars are studied endowed with the notions of indexed nonterminals and indexed production.

Let $G = (G_1, G_2)$ be the $(R : RIR)SMG$ where

$$G_1 = \{\{S, A\},\ \{S_1, S_2\},\ \{S \rightarrow S_1 A,\ A \rightarrow S_2, S_1\},\ S\}$$

generating strings of intermediates $S_1 S_2^n S_1$ for $n \geq 1$ and $G_2 = (G_{21}, G_{22})$ where

$$G_{21} = \{\{S_1, A_1, A_2, A_3\},\ \{x\},\ \{g_1, g_2\},\ P_{21},\ S_1\}\quad \text{with}$$

$$P_{21} = \left\{ \begin{array}{l} S_1 \rightarrow xA_1 g_2,\ A_1 \rightarrow A_2 g_1,\ A_1 \rightarrow xA_1 g_1,\\ A_2 g_1 \rightarrow xA_3,\ A_3 g_1 \rightarrow xA_3,\ A_3 g_2 \rightarrow x \end{array} \right\}$$

$$G_{22} = (\{S_2, B_1, B_2, B_3, B_4\},\ \{.,x\},\ \{f_1, f_2\},\ P_{22},\ S_2)$$

which generates the token H of x' s with the horizontal row of x's exactly in the middle which can also be generated by a $HP_4(CFPL)$ as in Example 3.2.

5 Conclusion

In this paper we have proposed a new generative model for picture arrays called Hybrid P system with $CFPG$. Specific patterns like characters are generated by the Hybrid P system with $CFPG$. The new model is compared with other generative models. It is worth examining further properties such as closure under set as well as language operations, of the system. Comparisons with other such generative models could also be done as future work.

References

1. Ceterchi, R., Mutyam, M., Paun, Gh, Subramanian, K.G.: Array rewriting P systems. Nat. Comput. **2**, 229–249 (2003)
2. Csuhaj-Varj, E., Dassow, J., Kelemen, J., Paun, Gh.: Grammar Systems: A grammatical approach to distribution and cooperation, Gordon and Breach Science Publishers, Topics in Computer Mathematics 5, Yverdon (1994)
3. Dassow, J., Freund, R., Paun, Gh: Cooperating array grammar system. Int. J. Pattern Recogn. Artif. Intell. **9**, 1–25 (1995)
4. Fernau, H. Freund, R.: Bounded Parallelism in Array Grammars Used for Character Recognition, In: Perner, P., Wang, P., Rosenfeld, A. (eds.), Advances in Structural and Syntactical Pattern Recognition (Proceedings of the SSPR'96), vol. 1121, pp. 40–49 Springer, Berlin (1996)
5. Helen Chandra, P., Saroja, K., Theerdus, S.M.: Array P Systems with Hybrid Teams. BIC-TA. In: Advances in Intelligent Systems and Computing 201, vol. 1, pp. 239–249(2013), Springer (2012)
6. Kari, L., Mateescu, A., Paun, Gh, Salomaa, A.: Teams in cooperating grammar systems. J. Exper. Th. AI **7**, 347–359 (1995)
7. Laroche, P., Nivat, M., Saoudi, A.: Context-sensitivity of puzzle grammars. Lect. Notes Comput. Sci. **654**, 195–212 (1992)
8. Nivat, M., Saoudi, A., Subramanian, K.G., Siromoney, R., Dare, V.R.: Puzzle grammars and context-free array grammars. Int. J. Pattern Recogn. Artif. Intell. **5**, 663–676 (1991)
9. Paun, Gh, Rosenbrg, G.: Prescribd teams of grammars. Acta Informatica **31**, 525–537 (1994)
10. Paun, Gh: Membrane Computing: An introduction. Springer, Berlin (2002)
11. ter Maurice, H.: Beek: Teams in grammar systems: hybridity and weak rewriting. Acta cybernetica **12**, 427–444 (1996)
12. Grammars, Hybrid Context-free Puzzle: In: Saroja Theerdus Kalavathy, S.M., Helen Chandra, P. (eds,) Proceedings of the International Conference On Mathematics in Engineering and Business Management, vol. II, pp. 33–38 (2012)
13. Siromoney, R., Huq, A., Chandrasekaran, M., Subramanian, K.G.: Stochstic Puzzle grammars. Int. J. Pattern Recogn. Artif. Intell. **6**, 257–273 (1992)
14. Subramanian, K.G., Revathi, L., Siromoney, R.: Siromoney array grammars and applications. Int. J. Pattern Recogn. Artif. Intell. **3**, 333–351 (1989)
15. Subramanian, K.G., Siromoney, R., Dare, V.R., Saoudi, A.: Basic puzzle languages. Int. J. Pattern Recogn. Artif. Intell. **5**, 763–775 (1995)

16. Subramanian, K.G., Thomas, D.G., Helen Chandra, P., Hoeberechts, M.: Basic puzzle grammars and generation of polygons. J. Automata Lang. Comb. **6**, 555–568 (2001)
17. Subramanian, K.G., Saravanan, R., Helen Chandra, P.: Cooperating basic puzzle grammar systems. Lecture Notes in Computer Science 4040, pp. 354–360. Springer (2006)
18. Subramanian, K.G., Saravanan, R., Geethalakshmi, M., Helen Chandra, P.: P systems with array objects and array rewriting rules. Prog. Nat. Sci. **17**, 479–485 (2007)

Enhanced Bee Colony Algorithm for Efficient Load Balancing and Scheduling in Cloud

K.R. Remesh Babu and Philip Samuel

Abstract Cloud computing is a promising paradigm which provides resources to customers on their request with minimum cost. Cost effective scheduling and load balancing are major challenges in adopting cloud computation. Efficient load balancing methods avoids under loaded and heavy loaded conditions in datacenters. When some VMs are overloaded with several number of tasks, these tasks are migrated to the under loaded VMs of the same datacenter in order to maintain Quality of Service (QoS). This paper proposes a modification in the bee colony algorithm for efficient and effective load balancing in cloud environment. The honey bees foraging behaviour is used to balance load across virtual machines. The tasks removed from over loaded VMs are treated as honeybees and under loaded VMs are the food sources. The method also tries to minimize makespan as well as number of VM migrations. The experimental result shows that there is significant improvement in the QoS delivered to the customers.

Keywords Cloud computing · Task scheduling · Bee colony algorithm · Load balancing · Qos

1 Introduction

Cloud computing is an emerging technology completely rely on internet, in which all the data and applications are hosted on a datacenters, which consists of thousands of computers interlinked together in a complex manner. The cloud providers adopt pay as you use model for their resource utilization. Over the Internet, the

K.R. Remesh Babu (✉)
Government Engineering College, Idukki Painavu, Kerala, India
e-mail: remeshbabu@yahoo.com

P. Samuel
Cochin University of Science and Technology, Kochi, India
e-mail: philipcusat@gmail.com

© Springer International Publishing Switzerland 2016
V. Snášel et al. (eds.), *Innovations in Bio-Inspired Computing and Applications*,
Advances in Intelligent Systems and Computing 424,
DOI 10.1007/978-3-319-28031-8_6

customers can use computation power, software resources, storage space, etc., by paying money only for the duration he has used the resource.

Besides Internet, customers, datacenters, and distributed servers are the main three components of a cloud eco system. Datacenter is a collection of servers hosting different applications and also provides storage facility. In order to sub-scribe for different applications, end user needs to connect to the datacenter. Usually a datacenter is situated far away from the end users. Distributed servers are the parts of a cloud environment which are present throughout the Internet hosting different applications.

In order to ensure QoS efficient scheduling and load balancing among nodes are required in the distributed cloud environment. In cloud computing ensuring QoS is crucial for customer satisfaction. An efficient load balancing mechanism tries to speed up the execution time of user requested applications. It also reduces system imbalance and gives a fair access to the users.

Better load balancing will result in good QoS metrics such as efficient resource utilization, scalability, response time, fault tolerance. Also migration time can be improved by better load balancing. The improvement in the above factors will ensure good QoS to the customers thereby less Service Level Agreement (SLA) violations.

The dynamic nature of cloud computing environment needs a dynamic algo-rithms for efficient and efficient scheduling and load balancing among nodes. Static load balancing algorithms will works only when small variation in the workloads. Cloud scheduling and load balancing problems are considered as NP hard problems.

Nature inspired algorithms plays a vital role in solving dynamic real time problems, which are hard to solve by normal methods. These NP hard problems are hard to solve within a time limit. Nature inspired algorithms produce optimal or near optimal solutions to these real time problems in polynomial time interval. The idea behind swarm intelligence algorithm is that local interaction of many simple agents to attain a simple objective.

The Bee Colony algorithm is a swarm intelligence algorithm [1] based on the foraging behavior of honey bee colonies to solve numerical function optimization problems. It mimics the foraging behavior of honey bees. It has advantages such as memory, multi-character, local search and solution improvement mechanism, so it is an excellent solution for optimization problems [2–4]. The algorithm consists of scout bees, forager bees and food source. In bee hives, scout bees forage for food sources. After finding a food source it returned to bee hive and performs a waggle dance. Based on waggle dance other bees in the hive get information about quantity of food and distance from the bee hive. Then forager bees follow the scout bees to the location of bee hive and begin to reap it. The positions of food sources are randomly selected by the bees.

In the proposed method, bee colony algorithm is modified and it is applied to efficiently schedule and balance the load among cloud nodes in the dynamic cloud environment. Here this method considers previous state of a node while distributing the load. For load balancing the bee colony algorithms parameters are mapped to cloud environment for achieving load balancing. The algorithm tries to achieve minimum response time and completion time. The remaining part of this paper is organized as follows. Section 2 describes about different kinds of load balancing methods in cloud. Enhanced bee colony algorithm and its architecture described in Sect. 3. The Sect. 4 gives experimental results and analysis. Finally this paper concludes in Sect. 5.

2 Related Works

Efficient scheduling and load balancing ensures better QoS to the customers and thereby reduces number of SLA violations. This section reviews some of the load balancing algorithms.

Modified throttled algorithm based load balancing is presented in [5]. While considering both availability of VMs for a given request and uniform load sharing among the VMs for number of requests served, it is an efficient approach to handle load at servers. It has an improved response time, compared to existing Round-Robin and throttled algorithms.

In [6], a load balancing approach was discussed, which manages load at server by considering the current status of all available VMs for assigning the incoming requests. This VM-assign load balancing technique mainly considers efficient utilization of the resources and VMs. By simulation, they proved that their algorithm distributes the load optimally and hence avoids under/over utilization of VMs. The comparison of this algorithm with active-VM load balance algorithm shows that their algorithm solves the problem of inefficient utilization of the VMs.

Response time based load balancing is presented in [7]. In order to decide the allocation of new incoming requests, proposed model considers current responses and its variations. The algorithm eliminates need of unnecessary communication of the Load Balancer. This model only considers response time which is easily available with the Load Balancer as each request and response passes through the Load Balancer, hence eliminates the need of collecting additional data from any other source thereby wasting the communication bandwidth.

In [8] a load balancing technique for cloud datacenter, Central Load Balancer (CLB) was proposed, which tried to avoid the situation of over loading and under loading of virtual machines. Based on priority and states, the Central Load Balancer manages load distribution among various VMs. CLB efficiently shares the load of user requests among various virtual machines.

Ant colony based load balancing in cloud computing was proposed in [9]. It works based on the deposition of pheromone. A node with minimum load is attracted by most of the ants. So maximum deposition of pheromone occurs at that node and performance is improved.

Cloud Light Weight (CLW) for balancing the cloud computing environment workload is presented in [10]. It uses two algorithms namely, receiver-initiated and sender-initiated approaches. VM Attribute Set is used to assure the QoS. CLW uses application migration (as the main solution) instead of using VM migration techniques in order to assure minimum migration time.

A resource weight based algorithm called Resource Intensity Aware Load balancing (RIAL) is proposed in paper [11]. In this method, VMs are migrated from overloaded Physical Machines (PM) to lightly loaded PMs. Based on resource intensity the resource weight is determined. A higher-intensive resource is assigned a higher weight and vice versa in each PM. The algorithm achieves lower-cost and faster convergence to the load balanced state, and minimizes the probability of future load imbalance, by considering the weights when selecting VMs to migrate out and selecting destination PMs.

A cloud partitioning based load balancing model for public cloud was proposed in [12]. This algorithm applies game theory to load balancing strategy in order to improve efficiency. Here a switch mechanism is used to choose different strategies for different situations.

Time and cost based performance analysis of different algorithms in cloud computing was given in [13]. A load balancing mechanism based on artificial bee colony algorithm was proposed in [14]. It optimizes the cloud throughput by mimicking the behavior of honey bees. Since bee colony algorithm arranges only a little link between requests in the same server queue, then maximization of the system throughput is suboptimal. Here, the increasing request does not leads to the increase of system throughput in certain servers.

An active clustering based load balancing technique is presented in paper [15]. It groups similar nodes together and works on these groups and produces better performance with high utilization of resources. Paper [16] proposes a Best-fit-Worst-fit strategy that efficiently places the virtual machines to the lesser number of active PMs. In this two level scheduling mechanism the tasks are scheduled using best-fit approach. Then the cloud broker uses worst-fit method for VM placement. They have considered cost and energy for the effective placement of VMs.

Weighted Signature based Load Balancing (WSLB), a new VM level load balancing algorithm is presented in [17]. This algorithm find the load assignment factor for each host in a datacenter and map the VMs according to that factor. Estimated finish time [18] based load balancing considers the current load of virtual machines in a datacenter and the estimation of processing finish time of a task before any allocation. This algorithm improves performance, availability and maximizes the use of virtual machines in their datacenters. In order to avoid a probable blocking of tasks in the queue, it permanently controls current load on the virtual machines and the characteristics of tasks during processing and allocation.

3 Modified Bee Colony Algorithm for Load Balancing

The proposed method uses the foraging behaviour of honey bees for effective load balancing across VMs and reschedules the cloudlets into under loaded VMs. For efficient implementation of honey bee algorithm, the foraging behaviour of honeybees is mapped into cloud environment in order to achieve load balancing. The mapping of bee colony parameters with cloud environment is given in Table 1.

In this proposed method the tasks are considered as honeybees. When the honeybees forage for food source, the cloudlets will be assigned in VMs for execution. Since the processing capacity varies for different VMs, sometimes VMs may be overloaded and others will be under loaded. In these circumstances in order to provide better performance and efficient load balancing mechanism is needed. When a particular VM is overloaded with multiple tasks then some tasks are need to be migrated and have to assign to an under loaded VM. In this case task to be migrated is chosen based on priority. In the proposed method tasks with lowest priority will be selected as a candidate for migration. This procedure is similar as honey is exhausted in a nectar and bees are ready to take off from the food source.

The architecture for the proposed load balancing method is given in Fig. 1. Cloud Information Service (CIS) is the repository that contains all the resources available in the cloud environment. It is a registry of datacenters. When a datacenter is created it has to register to the CIS. Datacenters are heterogeneous nature with specific characteristics. Usually a datacenter consists of several hosts. Hosts have number of processing elements (PEs) with RAM and bandwidth characteristics. In cloud environment these hosts are virtualized into different number of VMs based on user request. VMs may also have heterogeneous characteristics like hosts.

CIS collect information about the all resources in the datacenters. Based on this information the cloud broker submits these tasks to different VMs in a datacenter. In the proposed method the algorithm checks for overloaded conditions and it migrates task from overloaded VMs to under loaded VMs.

The proposed load balancing and scheduling mechanism works in four different steps. 1. VM Current Load Calculation. 2. Load Balancing and Scheduling Decision 3. VM Grouping 4. Task Scheduling.

Table 1 Mapping of enhanced bee colony parameters with cloud environment

Honey bee hive	Cloud environment
Honey bee	Task (Cloudlct)
Food source	VM
Honey bee foraging a food source	Loading of a task to a VM
Honey bee getting depleted at a food source	VM in overloaded condition
Foraging bee finding a new food source	Removed task will be rescheduling to an under loaded VM having highest capacity

Fig. 1 Load balancing architecture

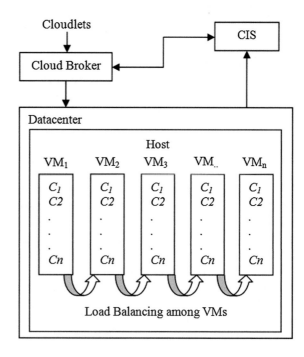

3.1 VM Current Load Calculation

The current load on a VM is measured based on the ratio between total lengths of the tasks submitted to that VM to the processing rate of that VM at a particular instance. Suppose N is the total numbers tasks assigned to a VM and Len is the length of single tasks and $MIPS$ is the Million Instruction Per Second rate of that VM, then using the Eq. (1) the current load can be calculated.

$$Load_{VM} = \frac{N*Len}{MIPS} \tag{1}$$

Then total load on a datacenter is the sum of load on each VMs. The equation for total load a datacenter $Load_{DC}$ is given by the Eq. (2).

$$Load_{DC} = \sum_{vm=1}^{n} Load_{VM} \tag{2}$$

The processing capacity of VM can be calculated using the Eq. (3) as given below.

$$Capacity_{VM} = PE_{num} * PE_{mips} + VM_{bw} \tag{3}$$

Here PE_{num} is the number of processing elements in a particular VM, PE_{mips} is the processing power of PE in MIPS rate and VM_{bw} is the band width associated for a VM.

A datacenter may have several VMS. So the total capacity of the entire data-center can be calculated from using the Eq. (4),

$$Capacity_{DC} = \sum_{vm=1}^{n} Capacity_{VM} \tag{4}$$

Then the proposed algorithm computes processing time of each task using Eq. (5).

$$PT = \frac{CurrentLoad}{Capacity} \tag{5}$$

Then the processing time required for datacenter to complete all the tasks in it can be calculated by the Eq. (6) given below,

$$PT_{DC} = \frac{Load_{DC}}{Capacity_{DC}} \tag{6}$$

Then the Standard Deviation (SD) is a good measure of deviations. The proposed method uses SD for measuring the deviations in the load on each VM. Equation (7) gives the SD of loads.

$$SD = \sqrt{\frac{1}{m} \sum_{i=1}^{m} (PT_i - PT)^2} \tag{7}$$

Then the load balancing decision is done based on the value of SD.

3.2   Load Balancing and Scheduling Decision

In this phase load balancing and rescheduling of tasks are decided. This decision depends on the SD value calculated using Eq. (7). In order to maintain system stability the load balancing and scheduling decision will take only when the capacity of the datacenter is greater than current load. Otherwise it will create imbalance in the datacenter. For finding the load a threshold value is set (value lies in 0–1) based on the SD calculated. The systems compare this value with calculated SD measure. The load balancing and scheduling done only if the calculated SD is greater than the threshold. This will improve the system stability.

3.3 VM Grouping

In order to increase the efficiency VMs are grouped into two groups: overloaded VMs and under loaded VMs. This will reduce the time required to find optimal VM for task migration. The overloaded VMs are the candidates for migration. In the proposed method these removed tasks are considers as honeybees and the under loaded VMs are their food sources. The VMs are grouped according to the SD and threshold value already calculated based on the load.

3.4 Task Scheduling

Before initiating load balancing the system have to find the demand to each overloaded VMs and supply to the under loaded VMs. Here the VMs are sorted based on the capacity in ascending order. The task migration is performed only when demand meets the supply. From the under loaded VM set, the proposed method selects a VM which has highest capacity as target VM. The method selects the task with lowest priority from an overloaded VM and it is rescheduled to an under loaded VM with maximum capacity.

Supply to a particular VM is the difference between its capacity and current load and it can be calculated using Eq. (8),

$$Supply_{VM} = Capacity - Load \tag{8}$$

Then the demand of a VM is calculated using the Eq. (9)

$$Demand_{VM} = Load - Capacity \tag{9}$$

The enhanced bee colony algorithm was given in Fig. 2. On submission of each task into the cloud, the VM will measure the current load status and calculates SD. If the SD of loads is greater than the threshold then load balancing process is initiated. During this load balancing process, VMs are classified into under loaded and overloaded VM sets. Then the submitted tasks are rescheduled to the VM having highest capacity.

4 Experimental Results

The performance analysis of the proposed method is carried out in a simulated environment. In this heterogeneous environment VMs having different specification are considered. Cloudlets with varying specifications are submitted into this cloud

1.	Start
2.	For each task do
3.	Calculate the load on VM and decide whether to do load balancing or not
4.	Group the VMs based on load as overloaded or under loaded.
5.	Find the supply of under loaded VMs and demand of overloaded VMs.
6.	Sort the overloaded and under loaded VM sets
7.	Sort the tasks in overloaded VMs based on priority.
8.	Find the capacity of VMs in the under loaded set.
9.	For each task in each overloaded VM find a suitable under loaded VM based on capacity.
10.	Update the overloaded and under loaded VM sets
11.	End of step 2.
12.	Stop

Fig. 2 Enhanced bee colony based load balancing algorithm

environment. The number of migrations and makespan are measured and compared with existing methods.

The makespan time of the proposed method with bee colony algorithm is shown in Table 2. The overall task completion time, i.e. makespan is graphically represented in Fig. 3. From these results it is clear that makespan is reduced into a significant amount while using enhanced bee colony algorithm. Then the users will get faster response than older methods. Response time is a good measure of QoS provided by the service provider. So here the provider can assure good QoS to their customers.

If number of task migrations is greater it will adversely affects the performance of the cloud and thereby reduces the QoS. A good load balancing and scheduling mechanism will reduce the number of task migrations. The proposed method is analyzed for number of task migrations. The results are tabulated in the Table 3.

The result in Table 3 shows that the number of task migrations is reduced while using enhanced bee colony algorithm. In most of the cases the algorithm out performs the existing bee colony algorithm. If frequent migration of tasks are happened it will adversely affect the performance of the entire cloud eco system and thereby its performance.

The above experimental results shows that how the proposed enhanced honey bee algorithm reduces the makespan as well as number of task migrations compared to the existing bee colony algorithm. Thus it helps efficient and effective use of

Table 2 Comparison of makespan

Number of cloudlets	Bee colony (s)	Enhanced bee colony (s)
10	50.1	43.85
15	70.1	68.85
20	80.1	78.85
25	110.1	100.1
30	120.1	118.85

Fig. 3 Comparison of
makespan

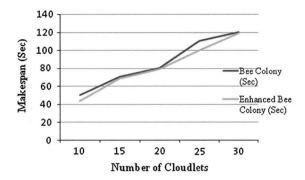

Table 3 Comparison of task
migration

Number of cloudlets	Bee colony	Enhanced bee colony
10	2	2
15	4	3
20	7	7
25	12	11

Fig. 4 Number of task
migrations

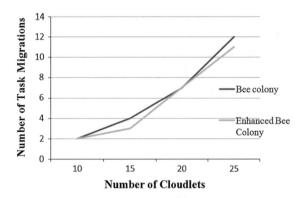

computational resource in the cloud environment. Since the algorithm minimizes the completion and reduces number of task migrations it will give better QoS to the end users. The number of task migration is given in the Fig. 4.

5   Conclusion

Nature inspired algorithms are good solution provider for real time dynamic optimization problems. This paper proposes an enhanced bee colony algorithm for efficient load balancing in cloud environment. Here the power of swarm intelligence algorithm is used to removes the tasks from overloaded VMs and submitted it to the most appropriate under loaded VMs. It not only balances the load, but also considers the priorities of tasks in the waiting queues of VMs. The task with least priority is selected for migration to reduce imbalance. So no tasks are needed to wait longer time in order to get processed. The experimental results show that, the proposed algorithm outperforms existing bee colony algorithm and minimize makespan and number of migrations and gives better QoS to end users.

In future the algorithm can be further enhanced with hybridization of other nature inspired algorithms like Ant Colony Optimization (ACO), Particle Swarm Optimization (PSO), etc.

References

1. Karaboga, D., Basturk, B.: A powerful and efficient algorithm for numerical function optimization: artificial bee colony (ABC) algorithm. Springer, J. Glob. Optim. **39**, 459–471 (2007)
2. Ajit, M., Vidya, G.: VM level load balancing in cloud environment.: In: IEEE Fourth International Conference on Computing, Communications and Networking Technologies (ICCCNT), pp. 1–5 (2013)
3. Fahim, Y., Ben Lahmar, E., El Labrlji, E.H., Eddaoui, A.: The load balancing based on the estimated finish time of tasks in cloud computing. In: 2nd World Conference on Complex Systems (WCCS), pp. 594–598 (2014)
4. Remesh Babu, K.R., Mathiyalagan, P., Sivanandam, S.N.: Pareto-Pareto based hybrid Meta heuristic ABC—ACO approach for task scheduling in computational grids. Int. J. Hybrid Intell. Syst. **11**(4/2014), 241–255 (2014)
5. Madivi, R., Kamath, S.S.: An hybrid bio-inspired task scheduling algorithm in cloud environment. In: International Conference on Computing, Communication and Networking Technologies (ICCCNT), pp. 1 –7 (2014)
6. Wang, L., Zhou, G., Xu, Y., Liu, M.: An enhanced Pareto-based artificial bee colony algorithm for the multi-objective flexible job-shop scheduling. Int. J. Adv. Manuf. Technol. **60** (Issue 9–12), 1111–1123. Springer (2012)
7. Domanal, S.G.R., Ram Mohana, G.: Load balancing in cloud computing using modified throttled algorithm. In: IEEE International Conference on Cloud Computing in Emerging Markets (CCEM), pp. 1–5 (2013)
8. Shridhar, G.D., Reddy, G.R.M.: Optimal load balancing in cloud computing by efficient utilization of virtual machines. In: IEEE Sixth International Conference on Communication Systems and Networks (COMSNETS), pp. 1–4 (2014)
9. Sharma, A., Peddoju, S.K.: Response time based load balancing in cloud computing. In: International Conference on Control, Instrumentation, Communication and Computational Technologies (ICCICCT), pp. 1287–1293 (2014)
10. Soni, G., Kalra, M.: A novel approach for load balancing in cloud data center. In: IEEE International Conference on Advance Computing Conference (IACC), pp. 807–812 (2014)
11. Dam, S., Mandal, G., Dasgupta, K., Dutta, P.: An ant colony based load balancing strategy in cloud computing. Springer Advanced Computing, Networking and Informatics, Vol. 28, pp. 403–413 (2014)
12. Mohammadreza, M., Amir, M.R., Anthony, T.C.: Cloud light weight: a new solution for load balancing in cloud computing. In: International Conference on Data Science and Engineering (ICDSE), pp. 44–50 (2014)
13. Chen, L., Shen, H., Sapra, K.: RIAL: resource intensity aware load balancing in clouds. In: IEEE Conference on Computer Communications (INFOCOM), pp. 1294–1302 (2014)

14. Xu, G., Pang, J., Fu, X.: A load balancing model based on cloud partitioning for the public cloud. In: IEEE Journal of Tsinghua Science and Technology, pp. 34–39 (2013)
15. Randles, M., Lamb, D., Taleb-Bendiab, A.: A comparative study into distributed load balancing algorithms for cloud computing. In: Proceedings of the IEEE 24th International Conference on Advanced Information Networking and Applications, Perth, Australia, pp. 551–556 (2010)
16. Yao, J., He, J.: Load balancing strategy of cloud computing based on artificial bee algorithm. In: IEEE 8th International Conference on Computing Technology and Information Management (ICCM), pp. 185–189 (2012)
17. Samal, P.: Analysis of variants in Round Robin Algorithms for load balancing in cloud computing. Int. J. Comput. Sci. Inf. Technol. **4**(3), 416–419 (2013)
18. Remesh Babu, K.R., Samuel, P.: Virtual machine placement for improved quality in IaaS cloud. In: IEEE Fourth International Conference on Advances in Computing and Communications (ICACC), pp. 190–194 (2014)

Robust Optimized Artificial Neural Network Based PEM Fuelcell Voltage Tracking

R. Vinu and Paul Varghese

Abstract Voltage control of Proton Exchange Membrane Fuel Cell (PEMFC) is necessary for any practical application. This paper considers a state space model for controller design and a Neural Network (NN) feed forward controller with an optimization technique called Harmony Search algorithm is considered to control the output voltage. This paper compares the results of the proposed controller with the NN feed forward controller. The comparison shows the proposed controller follows the reference voltage more closely than NN feed forward controller. Finally the performance of the controller is studied by evaluating Integral Squared Error (ISE), Integral Absolute Error (IAE) and Integral Time-weighted Absolute Error (ITAE) and the results are compared. The system error of the proposed controller is reduced to a least minimum value compared with the other.

Keywords Neural network · Feed forward · Harmony search

1 Introduction

Fuel cells are renewable energy sources that can directly generate electrical energy from chemical energy with zero emissions. So they are highly efficient and environment friendly [1, 2]. Now a days fuel cells has drawn more attention in the field of alternative energy generation. There are different types of fuel cells with power ranging from mW to MW which can be useful for mobile, portable and stationary applications. Out of that, Proton Exchange Membrane fuel cell (PEMFC) is considered very important because of its low operating temperature, quick start-up and high power energy ratio [3, 4]. The ideal standard voltage of a PEM fuel cell is 1.229 V. But it is very difficult for a PEMFC to maintain constant output voltage under varying load conditions [5, 6]. To achieve this an intelligent controller is required.

R. Vinu (✉) · P. Varghese
Anna University, Chennai, India
e-mail: vinur81@gmail.com

© Springer International Publishing Switzerland 2016
V. Snášel et al. (eds.), *Innovations in Bio-Inspired Computing and Applications*,
Advances in Intelligent Systems and Computing 424,
DOI 10.1007/978-3-319-28031-8_7

Hatti et al. [13] proposed a static model of PEM fuel cell using neural networks. The PEMFC model based on Artificial Neural Network (ANN) is represented using non parametric approach. This model is used to predict various parameters of PEMFC and to analyze their characteristics.

Keyhani et al. [7] proposed a neural network model of a 500 W Proton exchange membrane fuel cell. This model is developed by using recurrent neural networks and the performance of the model is validated. Keyhani et al. [9] developed a non linear state space model of 500 W proton exchange membrane fuel cell. The model is developed by considering open circuit voltage of PEM fuel cell, mass balance, thermodynamic balance, voltage losses and double layer effects. In this paper, this nonlinear state model is considered and Neural Network (NN) controller is designed to maintain the output voltage inspite of load variations. To improve the performance an optimization technique called Harmony Search is introduced.

Abhudhahir et al. [2] surveyed about the relevance of control systems for PEM fuel cells. Kamlakaran et al. [1] presented the different control strategies of a 1 kW PEM Fuel Cell. Niaki et al. [12] designed a Neural Network model based on Back Propagation Network to control the stack terminal voltage. This is achieved by controlling input air pressure signal. Less energy and simple control are the advantages of the proposed control algorithm.

Mohseni and Rezaei [5] presented a Predictive Neural Network Controller to control the voltage in the presence of temperature fluctuations. It is implemented on a dynamic electrochemical model of 5 kW PEM fuel cell. This controller reduces the effect of noise. Hatti et al. [4] implemented Quasi Newton Neural Networks based Controller to overcome the variations in load.

Yu et al. [8] developed controllers to avoid oxygen starvation by controlling the cell voltage. Controllers based on neural networks and fuzzy logic has been designed and their performance is compared by calculating Mean Absolute Error (MAE). Neural network controllers show better performance compared to other methods.

Liping Fan et al. [17] suggested fuzzy logic controllers for dynamic model based PEMFC to generate constant output voltage. Two fuzzy control schemes, one to control output voltage by adjusting hydrogen flow and another to control output voltage by adjusting oxygen flow are designed and their results are compared.

Askarzadeh et al. [15] proposed Radial Basis Function-based adaptive inverse control for voltage control by acting on the methane flow rate. This controller design provides reliability and efficiency to the system. Farhadi et al. [18] implemented a reinforcement Learning based controller to control the output voltage under load variations by varying zero and pole coefficients of the controller. This method provides fast transient response and zero steady state error. Zhu et al. [3] presented a detailed analysis about the different factors which influence the performance of PEMFC. A fuzzy logic controller has been designed to control the hydrogen and air flow rates. Experimental results are compared with the conventional controller which shows the intelligent controller can work better than the conventional one.

Zarabadipour and Khoeiniha [11] proposed a state space model for water management in a PEM fuel cell. An optimal PID controller based on Genetic algorithm is designed to control the cell voltage under load variations. Efficiency and life time of fuel cell is increased by adopting this method. Farhadi et al. [6] used PSO tuned PID Controller is designed for control system because of its functional simplicity and reliability. PID controller is used in maintaining constant voltage by taking input current as the control signal. The performance of this method is better compared with other techniques since the error value is reduced below 5 %.

Bilbao et al. [19] reviewed about the characteristics and applications of harmony search algorithm. Askarzadeh and Rezazadeh [20] proposed an innovative global Harmony search algorithm for identifying the model parameters of PEM Fuel cell. The proposed technique shows better results with other optimization techniques. This paper is organized as follows: The PEM Fuel Cell working and their characteristics are given in Sect. 2. The Neural Network controller and Harmony search algorithms are discussed in Sect. 3. Simulation Results and Discussions are presented in Sect. 4. Finally the Conclusion is presented in Sect. 5.

2 PEMFC Model

The Proton Exchange Membrane Fuel Cell consists of two electrodes (anode and cathode) separated by a proton exchange membrane which acts as an electrolyte as shown in Fig. 1. The inputs to the fuel cell are hydrogen and oxygen. Hydrogen flows through anode and oxygen flows through cathode. At anode hydrogen splits into positively charged hydrogen ions and negatively charged electrons. The electrolyte allows positive hydrogen ions to flow from anode to cathode. The negative electrons reach cathode through an external circuit thereby generating an electric current. At

Fig. 1   Working of fuel cell

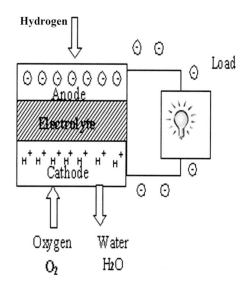

cathode, electrons and protons combine with oxygen by the use of catalyst to form water and heat [7, 8].

Fuel cell Modeling is very important to control the output voltage of a fuel cell. Many conventional electrochemical based models of PEM fuel cells are developed. These models provide certain understanding of the fuel cell but they cannot be used for controlling the fuel cells effectively [9]. The relationship between the parameters of the fuel cells is highly non linear. In this paper a non linear state space model is considered to control the output voltage.

The output voltage of PEM fuel cell is given by

$$V_{fc} = V_{ofc} - V_{losses} \tag{1}$$

where V_{ofc} is the fuel cell open circuit voltage and V_{losses} is the irreversible voltage losses existing in the fuel cell [10].

The open circuit output voltage of the PEM fuel cell can be given as follows

$$V_{ofc} = n_s E_0^{Cell} + \frac{n_s RT}{2F} \ln\left(\frac{P_{H_2}(P_{O_2})^{0.5}}{P_{H_2O}}\right) \tag{2}$$

where n_s is number of PEM fuel cell stacks, E_0^{Cell} is the reference potential at standard operating conditions (V), R is Universal gas constant [J/(mol K)], T is Stack Temperature (K), F is Faradays Constant (C/mol), P_{H_2} is the Partial Pressure of Hydrogen (atm), P_{O_2} is the Partial Pressure of Oxygen (atm), P_{H_2O} is the Partial Pressure of Water (atm) [11, 12].

Three types of voltage losses exist in PEM fuel cell. They are, Activation losses, Ohmic losses and Concentration losses. The irreversible voltage losses is given by

$$V_{losses} = V_{act} + V_{ohm} + V_{conc} \tag{3}$$

where V_{act} is the activation loss in PEM fuel cell due to sluggish electrode kinetics and it is given by

$$V_{act} = \frac{RT}{2F} \ln\left(\frac{I}{I_d}\right) \tag{4}$$

where I is stack current (A) and I_d is current density (A/m^2).

V_{ohm} is the ohmic loss associated with the conduction of protons through the electrolyte and electrons through internal electronic resistance and it is given by

$$V_{ohm} = V_A^O + V_C^O + V_M^O \tag{5}$$

where V_A^O is voltage across anode (volts), V_C^O is voltage across cathode (volts) and V_M^O is voltage across membrane (volts).

V_{conc} is the concentration losses exists due to the formation of concentration gradient of reactants at the surface of electrodes and it is given by

$$V_{conc} = \frac{RT}{eF} \ln(1 - \frac{I}{I_L})$$

(6)

where e is Number of electrons and I_L is Limiting current (A).

3   Controller Design

3.1   Neural Network Controller

An Artificial Neural Network (ANN) is a simplified model of the structure of biological Neural Network [13]. An ANN consists of several interconnected processing units called neurons, which helps to solve complex problems [14]. Neural networks can learn from the input characteristics and adapt themselves with the varying conditions in the environment [15]. In this paper, a Multi layer Feed forward controller is used to control the fuel cell output voltage. This feed forward controller employs Back Propagation (BP) algorithm for training the network. Back Propagation algorithm is a systematic method to train multilayer neural network which uses gradient descent based delta learning rule [16]. This method is used to minimize the total squared error of the output. Figure 2 shows the block diagram of Neural Network (NN) based controller system. The NN Controller based system consists of dual inputs V_{fc} and V_{ref} and error voltage e(t) is calculated to control the output voltage where $e(t) = V_{ref} - V_{fc}$ [17, 18].

3.2   Harmony Search Algorithm

Harmony search Algorithm is a random search optimization technique to find out the perfect state of harmony or solution [19]. Harmony Search Algorithm can be

Fig. 2 Neural network based controller system

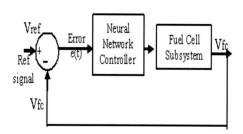

Fig. 3 Flow chart for
harmony search algorithm

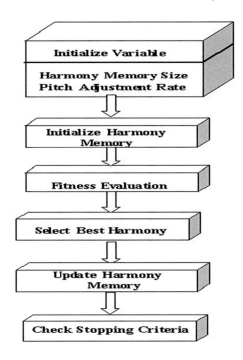

easily implemented with less adjustive parameters and has quick convergence. There
are three basic steps in this algorithm as shown in Fig. 3. They are as follows: Initialization of Harmony Memory (HM), Generation of new Harmony vectors and
Restructuring the HM.

Step 1: In this step, the problem and parameters of the algorithm are initialized.
Generally the optimization problem can be stated as given in (7)

$$minf(x) : x(j)\epsilon[l(j), u(j)], j = 1, 2, \ldots, n \qquad (7)$$

where f(x) is the objective function in which x is the set of design variables where
x varies from 1 to n, n is the number of design variables, and l(j) and u(j) are the
lower and upper limits for the design variable x(j) respectively. The parameters of
the algorithm are the Harmony Memory Range (HMR) i.e., the number of members
or vectors in harmony memory (HM), the Harmony-Memory Consideration Rate
(HMCR), the Pitch Adjustment Rate (PAR), the distance BandWidth (BW) and the
Number of Generations (NG) which gives the total number of iterations. It is clear
that the performance of HS is strongly determined by its parameters.

Step 2: Harmony Memory Formation
Initially the components at HM, i.e., HMR vectors, are determined.
   Let $x_i = x_i(1), x_i(2), \ldots, x_i(n)$ represent the *i*th randomly-generated harmony
vector:

$x_i(j) = l(j) + (u(j) - l(j)) * rand(0, 1)$ for $j = 1, 2, \ldots, n$ and $i = 1, 2, \ldots$, hms where $rand(0, 1)$ is a uniform random number between 0 and 1. Then, the HM matrix with the harmony vectors is given by

$$HM = \begin{bmatrix} X1 \\ X2 \\ \vdots \\ \vdots \\ Xhms \end{bmatrix} \tag{8}$$

Step 3: Generation of New Harmony vectors

A New Harmony vector x_{new} is generated by the following three operators: memory consideration, random re-initialization and pitch adjustment. By memory consideration, the value of the first decision variable $x_{new}(1)$ for the new vector is chosen randomly from any of the existing values in the current HM i.e., from the set $x_1(1)$, $x_2(1), \ldots, x_{hms}(1)$. To do this, a uniform random number r1 is generated within the range [0, 1]. If r1 is less than HMCR, the decision variable $x_{new}(1)$ is generated through memory considerations; other-wise, $x_{new}(1)$ is obtained from a random re-initialization between the search bounds [l(1), u(1)]. The values of the other decision variables $x_{new}(2), x_{new}(3), \ldots, x_{new}(n)$ are generated accordingly. Therefore, by applying memory consideration and random re-initialization, $x_{new}(j)$ can be given as

$$x_{new}(j) = \begin{cases} x_i(j) \in x_1(j), x_2(j), \ldots, x_{HMS}(j) & with\, probability\, HMCR \\ l(j) + ((u(j) - l(j)).rand(0, 1) & with\, probability 1 - HMCR \end{cases} \tag{9}$$

The values obtained by memory consideration is then checked again to find whether it should be pitch-adjusted. For this, the Pitch-Adjustment Rate (PAR) is assigned with the frequency for the adjustment and the Bandwidth factor (BW) which limit the search around the selected elements of the HM. The pitch adjusting decision is given by (10).

$$x_{new}(j) = \begin{cases} x_{new}(j) \pm rand(0, 1).BW & with\, probability\, PAR \\ x_{new}(j) & with\, probability\, 1 - PAR \end{cases} \tag{10}$$

Pitch adjustment process is accountable for the generation of new potential harmonies by doing slight modifications in the original variable positions. This process is similar to the mutation process in Genetic algorithms. Therefore, the decision variable is either modified by a random number between 0 to BW or left unchanged. In order to do the pitch adjustment operation, it is vital to ensure that points lying outside the feasible boundary [l, u] must be reassigned i.e., reduced to the maximum or minimum value of the interval.

Step 4: Restructuring the harmony memory

Whenever a New Harmony vector x_{new} is generated, the harmony memory is restructured based on the comparison between x_{new} and the worst harmony vector x_{worst} in the HM. Therefore x_{new} will reinstate x_{worst} and turn into a new member of the HM in case the fitness rate of x_{new} is better than the fitness rate of x_{worst} [20].

Step 5: Termination
Do steps 2 and 3 till the termination condition is satisfied.

4 Simulation Results and Discussions

Simulations are done using Matlab Simulink Platform [21, 22]. The main parameters used in the analysis are Area of a cell $A_s = 3.2 * 10^{-2} \text{m}^2$, Reference Potential $E_0^{cell} = 1.23$ V, Specific Heat Capacity of PEM Fuel Cell $C_{fc} = 500$ J/(molK), Faradays Constant F $= 96487$ C/mol, Total mass of PEM fuel cell stack, $M_{fc} = 44$ Kg, Number of PEM fuel cell stacks N_s, Volume of anode $V_a = 10^{-3} \text{m}^3$ and Volume of anode $V_a = 10^{-3}$ m.

Generally in a PEM fuel cell the voltage at the output varies with changes in load. The output voltage has to be controlled to use PEM fuel cell in any application. In this paper, controllers are designed to track the output voltage. In Neural Network based Controller in order to track the output voltage, a reference signal (step signal) is compared with the output voltage. An error signal is generated which is used to control the output voltage. Here a step signal with initial value $= 34$ V and Final value $= 38$ V is considered as reference signal. Figure 4 shows how the fuel cell output voltage of a neural network based controller follows the reference step signal.

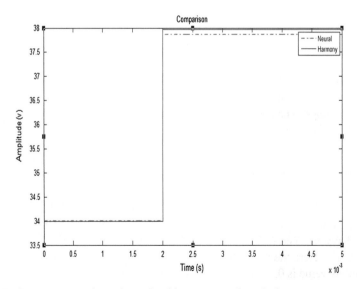

Fig. 4 Performance comparison of neural and harmony search methods

The performance of the above controller is analyzed by calculating the system error. In this paper, the performance analysis is done by calculating (1) Integral Squared Error (ISE), (2) Integral Absolute Error (IAE) and (3) Integral Time-weighted Absolute Error (ITAE).

(1) Integral Squared Error (ISE):
The system performance is measured by integrating the square of the system error over a fixed interval of time. ISE integrates the square of the error over time.

$$ISE = \int \varepsilon^2 dt \tag{11}$$

(2) Integral Absolute Error (IAE):
IAE integrates the absolute error over time. It does not add weight to any of the errors in a systems response

$$IAE = \int |\varepsilon| dt \tag{12}$$

(3) Integral Time-weighted Absolute Error (ITAE):
ITAE integrates the absolute error multiplied by the time over time. This adds weightage to errors which prolongs after a long time much more heavily than those at the initial period of the response.

$$ITAE = \int t|\varepsilon| dt \tag{13}$$

The error analysis plots based on NN Controller is as shown in Fig. 5. By applying Back Propagation Algorithm in a Multilayer Feed forward network the ISE is found to be 50.1418, IAE is 396.055 and ITAE is 1.349.

In this paper, an optimization method named harmony search method is proposed to minimize the system error compared with NN Controller, thereby to maintain the output voltage. The proposed method uses the following Parameters: Harmony Memory Size (HMS) = 5, Bandwidth (BW) = 0.2, Harmony Memory Consideration Rate (HMCR) = 0.95, Pitch Adjustment Rate (PAR) = 0.3. The Fuel Cell Output voltage using Harmony search optimization is given in Fig. 4. Here the Output Voltage follows the Reference step Voltage more closely compared with NN Controller. The Peak value of NN based Controller is 37.87 V and by applying harmony search optimization the peak value improves to 37.99 V. From this, we can infer that Harmony search based system follows the reference voltage more closely compared to NN based system. The Performance analysis based on Harmony Search method is given in Fig. 6. Here the ISE is found to be reduced compared with NN based Controller and its value is 0.1146. The IAE also reduces to 20.229 and ITAE to 0.0626.

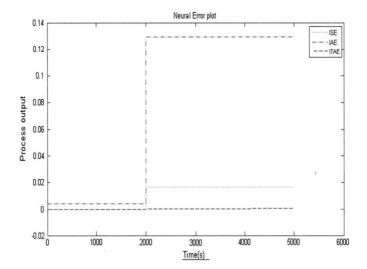

Fig. 5 Performance analysis based on NN controller

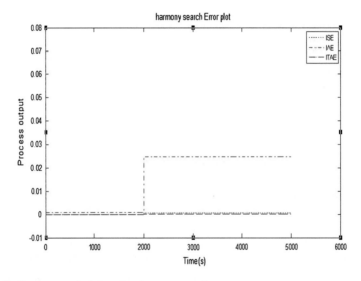

Fig. 6 Performance analysis based on harmony search

The error plot comparison of IAE and ITAE between NN based controller and Harmony search based controller is given in Figs. 7 and 8. The graph shows the reduction in the error values clearly, by applying Harmony Search optimization. By the optimization technique the error is reduced below 5 %.

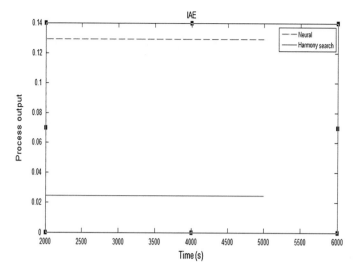

Fig. 7 Comparison of IAE values

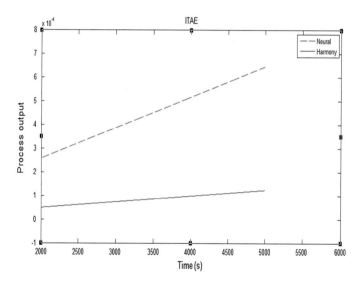

Fig. 8 Comparison of ITAE values

5   Conclusion

In this paper a NN feed forward controller with Harmony search optimization technique was proposed to control the output voltage. Harmony search Algorithm uses
less number of mathematical equations and generates the best solution only after
considering all the existing solutions. The simulation results shows that the proposed

controller output voltage follows very closely with the reference voltage and thereby reducing the error voltage to a least minimum value compared with NN feed forward controller. In this paper a step signal is considered but signals with different constraints can be considered and as a future improvement to this Neuro-fuzzy technique can be implemented.

References

1. Kamlakaran, M., Nandikesan, P., Choudhury, S.D., Shaneeth, K.P., Mohanty, S., Bhardwaj V.M.: Control strategy for PEM fuel cell power plant. In: IEEE 1st International Conference on Power and Energy in NERIST(ICPEN), pp. 1–3 (2012)
2. Abhudhahir, A., Mohamed Ali, E.A.: Survey of the relevance of control systems for PEM fuel cell. In: IEEE International Conference on Computer, Communication and Electrical Technology-ICCCET, pp. 322–326 (2011)
3. Zhu, J., Wang, H., Zhan, Y.: Performance analysis and improvement of a proton exchange membrane fuel cell using comprehensive intelligent control. In: International Conference on Electrical Machines and Systems ICEMS, pp. 2378–2383 (2008)
4. Hatti, M.: Neural network controller for fuel cells. In: IEEE International Symposium on Industrial Electronics ISIE, pp. 341–346 (2007)
5. Mohseni, M., Rezai, M.: A predictive control based on neural network for dynamic model of proton exchange membrane fuel cell. J. Fuel Cell Sci. Technol. 10(3) (2013)
6. Farhadi, P., Sojoudi, T.: PEMFC voltage control using PSO-tuned PID controller. In: Proceedings of IEEE NW Russia Young Researches in Electrical and Electronics Engineering Conference, pp. 32–35 (2014)
7. Keyhani, A., Khorrami, F., Puranik, S.V.: Neural network modelling of proton exchange membrane fuel cell. IEEE Trans. Energy Convers. 25(2), 474–482 (2010)
8. Yu, D.L., Gomm, J.B., Rgab, O.: Polymer electrolyte membrane fuel cell control with feed forward and feed back strategy. Int. J. Eng. Sci. Technol. 2(10), 56–66 (2010)
9. Keyhani, A., Khorrami, F., Puranik, S.V.: State space modelling of proton exchange membrane fuel cell. IEEE Trans. Energy Convers. 25(3), 804–813 (2010)
10. Fuel Cell HandBook. 7th ed. EG&G Technical Services. Inc., pp. 3.1–3.23 (2004)
11. Zarabadipour, H., Khoeiniha, M.: Optimal control design for proton exchange membrane fuel cell via genetic algorithm. In: Int. J. Electrochem. Sci. 7, 6302–6312 (2012)
12. Niaki, A.N., Fadaeian, T., Ghaderi, R., Rakthala, S.M., Ranjibar, A.: PEM fuel cell voltage tracking using artificial neural network. In: IEEE Electrical Power and Energy Conference, pp. 1–5 (2009)
13. Hatti, M., Nouibat, W., Tioursi, M.: Static modelling by neural networks of a PEM fuel cell. In: IEEE 32nd Annual Conference on Industrial Electronics. IECON, pp. 2121–2126 (2006)
14. Haykins, S.: Multi Layer Perceptrons in Neural Networks A comprehensive Foundation 2nd edn, Pearson Education, pp. 183–195 (2005)
15. Askarzadeh, A., Rezazadeh, A., Sedighizadeh, M.: Adaptive inverse control of proton exchange membrane fuel cell using RBF neural network. Int. J. Electrochem. Sci. 6, 3105–3117 (2011)
16. Deepa, S.N., Sivanandam, S.N.: Feed Forward Networks in Introduction to Neural Networks using MATLAB 6.0 2nd edn, TataMcGrawHill India (2006)
17. Liping, F., Yi, L., Chong, L.: Fuzzy logic based constant voltage control of fuel cells. TELKOMNIKA 10(4), 612–618 (2012)
18. Farhadi, P., Ghadimi, N., Imanzadeh, M., Karimi, M.: Voltage control of PEMFC using a new controller based on reinforcement learning. Int. J. Inf. Electron. Eng. 2(5), 752–756 (2012)
19. Bilbao, M.N., Del Ser, J., Geem, Z.W., Gil-Lopez, S., Landa-Torres, I., Manjarres, D., Salcedo-Sanz, S.: A survey on applications of the harmony search algorithm. Eng. Appl. Artif. Intell. 26(8), 1818–1831 (2013)

20. Askarzadeh, A., Rezazadeh, A.: An innovative global harmony search algorithm for parameter identification of a PEM fuel cell model. IEEE Trans. Ind. Electron. **59**(9), 3473–3480 (2012)
21. Deepa, S.N., Sivanandam, N.: MATLAB Environment for Soft Computing Techniques in Principles of Soft Computing, 2nd edn. Wiley, India (2011)
22. Neural Network Toolbox. MATLAB(R2011a) in 2011. http://www.mathworks.com

Gravitational Search Algorithm to Solve Open Vehicle Routing Problem

Ali Asghar Rahmani Hosseinabadi, Maryam Kardgar,
Mohammad Shojafar, Shahab Shamshirband and Ajith Abraham

Abstract Traditional Open Vehicle Routing Problem (OVRP) methods take account to definite responding to the all requests of customers whiles the main goal of proposed approach in OVRP is decreasing the vehicle numbers time and path traveled by vehicles. Therefore, in the present paper, a new optimization algorithm based on Gravity law and mass interactions is introduced to solve the problem. This algorithm being proposed based on random search concepts utilizes two of the four major parameters in physics including speed and Gravity and its researcher agents are a set of masses which are in connection with each other based on Newton's Gravity and motion laws. The proposed approach is compared with various algorithms and results approve its high effectiveness in solving the above problem.

A.A.R. Hosseinabadi (✉) · M. Kardgar (✉)
Young Research Club, Behshahr Branch, Islamic Azad University, Behshahr, Iran
e-mail: A.R.Hosseinabadi@iaubs.ac.ir

M. Kardgar
e-mail: kardgar_maryam@yahoo.com

M. Shojafar (✉)
Department of Information Engineering, Electronic and Telecommunication,
Sapienza University of Rome, via Eudossiana 18, 00184 Rome, Italy
e-mail: mohammad.shojafar@uniroma1.it; m.shojafar@hotmail.com

S. Shamshirband (✉)
Department of Computer System and Technology, University of Malaya,
Kuala Lumpur, Malaysia
e-mail: shamshirband@um.edu.my; shahab1396@gmail.com

A. Abraham (✉)
Machine Intelligence Research Labs, Scientific Network for Innovation
and Research Excellence, Auburn, WA, USA
e-mail: abraham.ajith@gmail.com

© Springer International Publishing Switzerland 2016
V. Snášel et al. (eds.), *Innovations in Bio-Inspired Computing and Applications*,
Advances in Intelligent Systems and Computing 424,
DOI 10.1007/978-3-319-28031-8_8

93

1 Introduction and Backgrounds

Open Vehicle Routing Problem (OVRP) includes finding the best route for a set of vehicles that are going to service a set of customers. In OVRP, each route includes a series of customers, begins from primary depot and ends in one of the customers [1, 2]. Some of the common limitations of OVRP are as follows: the capacity of all vehicles is the same, each customer must be visited just by one vehicle and his request must be satisfied by this visit, total requests of all customers in one route should not exceed the total capacity of vehicles. Authors in [3] implemented hybrid tabu search algorithm and improved tabu search algorithm [4, 5] to reduce the number of vehicles and cost of travel for OVRP [6]. According to simulation results, the proposed algorithm is able to reduce the number of vehicles required and costs related to travel. In [7], a Clonal Selection Algorithm has been presented for solving OVRP that contains a new definition for continuity of antibodies and Antibody Diversity algorithm. Authors in [8] proposed an algorithm to solve the open vehicle routing problem. Their algorithm was presented based on Genetic rules in order to enhance the performance of particle swarm optimization and differential evolution. In this algorithm, the dominant and recessive character includes any person. The particle swarm optimization and differential evolution will be done by dominant character and recessive character, respectively. The dominant character will be replaced by the recessive character if the proportion of dominant character is smaller than that of recessive character. Also, authors in [9] applied PSO to solve the OVRP. They have also presented a specific decoding method for implementation of PSO where a vector including customer positions in descending order is generated, each customer is assigned to a route based on his position, and finally a single mutation is applied to the generated routes. This method will be effective for solving the considered problem due to studying the possibility of routes and quality of responses. Besides, reference [10] used the procedure of variable neighborhood descent (VND) in Iterated local search (ILS) framework and implemented an approach called hybrid iterative local search for solving OVRP. To improve the obtained responses in VND procedure, they used four neighborhood structures including displacement, 2-opt* and 2-opt. the proposed method is able to obtain the best responses within the shortest possible time.

In the following, Sect. 2 will address the definition of the problem, the considered limitations, the gravitational search algorithm and the main idea, and description of Record-to-Record algorithm. The proposed algorithm is introduced in Sect. 3. The simulation results of the proposed algorithm are descried completely in Sect. 4 and finally Sect. 5 includes conclusion.

2   Problem Definition and Notation

The following assumptions are considered for solving OVRP. It is assumed that n is the number of customers, $N = \{1, 2, l \ldots, n\}$ is customer set, $V = \{0, 1, \ldots, n\}$ is the set of all customers and starting point so that zero will indicate the starting point, there are maximum K vehicles, the customer demand $i \in V - \{0\}$ and the capacity of each vehicle equals q_i and Q, respectively [11]. Furthermore, cost c_{ij} is considered as a scale equivalent to appropriate adjustments for a series of customers from i to j and cost w_k is equivalent to the performance of any vehicle k. Therefore, the OVRP includes the least number of required vehicles and a route for each vehicle, so that all customers' demand is satisfied, each customer is visited just by one vehicle, and the capacity of vehicles do not exceed the permissible limit [9]. For mathematical modeling of OVRP, we need two groups of variables. The first one is to model a series in which the customers are visited by vehicles and is defined as follows in (1):

$$x_{ij}^k = \begin{cases} 1 & \text{if } i \text{ precedes } j \text{ visited by vehicle } k \\ 0 & \text{otherwise,} \end{cases} \tag{1}$$

The second group shown as z_k is a binary variable and is defined as follows in (2):

$$z_k = \begin{cases} 1 & \text{if vehicle } k \text{ is active} \\ 0 & \text{otherwise} \end{cases} \tag{2}$$

If a vehicle services at least one customer, it would be regarded active [9]. With regard to considered parameters and variables, OVRP can be described as follows in (3):

$$\min \sum_{k=1}^{K} \sum_{i=0}^{n} \sum_{j=1}^{n} c_{ij} x_{ij}^k + \sum_{k=1}^{K} w_k z_k \tag{3a}$$

subjects to:

$$\sum_{k=1}^{K} \sum_{i=0}^{n} x_{ij}^k = 1, \qquad \forall j = 1, 2, \ldots, n \tag{3b}$$

$$\sum_{k=1}^{K} \sum_{j=1}^{n} x_{ij}^k = 1, \qquad \forall i = 1, 2, \ldots, n \tag{3c}$$

$$x_{ij}^k \leq z_k, : \forall k = 1, \ldots, K, \quad \forall i = 1, \ldots, n, \quad \forall j = 1, \ldots, n \tag{3d}$$

$$\sum_{i=0}^{n} x_{iu}^{k} - \sum_{j=1}^{n} x_{uj}^{k} = 0, \quad \forall k = 1, 2, \dots, K, \quad \forall u = 1, \dots, n \qquad (3e)$$

$$\sum_{(i,j) \in s \times s} x_{ij}^{k} \leq |S| - 1, \quad \forall S \subseteq V : 1 \leq\leq |S| \leq n,$$

$$\forall k = 1, 2, \dots, K \qquad (3f)$$

$$\sum_{j=1}^{n} q_j \left(\sum_{i=0}^{n} x_{ij}^{k} \right) \leq Q, \quad \forall k = 1, 2, \dots, K \qquad (3g)$$

$$\sum_{j=1}^{n} x_{0j}^{k} \leq 1, \quad \forall k = 1, 2, \dots, K \qquad (3h)$$

$$\sum_{i=1}^{n} x_{i0}^{k} \leq 0, \quad \forall k = 1, 2, \dots, K \qquad (3i)$$

$$x_{ij}^{k} \in \{0, 1\}, \quad \forall k = 1, 2, \dots, K,$$
$$\forall i = 1, 2, \dots, n, \quad \forall j = 1, 2, \dots, n, \qquad (3j)$$

$$z_k \in \{0, 1\}, \quad \forall k = 1, 2, \dots, K \qquad (3k)$$

Creating balance between route and vehicle is the objective function. Equation (3a) shows cost of all vehicles' routes after their moving from initial point and also cost of the first part of each route. Equation (3b) shows the costs of assigning machine or setup cost. To be precise, Eqs. (3b) and (3c) guarantee the entry and exit of just one from each customer. Equation (3d) is related to variables x and z and shows that all customers are serviced by active machines. Equations (3e) and (3f) represent the continuity of each machine and deletion of sub-tours, respectively. Equation (3g) indicates the maximum capacity of vehicle. Equations (3h) and (3i) show that only one vehicle must start from initial point in order to service customers and no vehicle would return to the initial point. Finally, Eqs. (3j) and (3k) are equal to a definition of variables x and z for each vehicle k [9, 12].

3   The Proposed Algorithm

In the proposed algorithm, the gravitational search algorithm (GSA) is used to solve the open vehicle routing problem (OVRP). This algorithm is intended to reduce the time and traveled distance by vehicle and to minimize the number of required vehicles for OVRP. With regard to the complexity and difficulty of OVRP, its solving would be difficult even with small number of customers and vehicles (see [13]). Therefore, due to the features of gravitational search algorithm, using it is suggested to solve OVRP. In this paper, the nature of gravitational search algorithm (GSA) is regarded as an appropriate strategy for solving OVRP because of important factors of the problem and the number of vehicles. Unlike other algorithms, gravitational search algorithm (GSA) does not progress with random intelligent solution; however, it works by trying existing solutions and has final distinctive conditions which can complete the counting of a special repeater. Thus, although GSA has some random-based elements within itself, it does not progress by mere probability; although it use local search neighborhood to solve the problem, it does not always move in a manner among them; and although it has specific behavior like that of the greedy algorithms, it does not always find the best way to search. The GSA applies the law that controls objects' motion in physical space to direct the motion from a complex search space.

GSA starts to work with an initial or current response that means it is a single chromosome approach. In the next step, the second or candidate response will be created using the first response, based on the considered neighbor, customers and vehicle that all of them can remove problem limitations. In this step, the gravity force between these two bodies is calculated and regarded as a solution for the problem. After that, the speed and gravitational mass are updated. One of the reasons of the above algorithm's superiority toward other algorithm is due to its speed parameter that cause this algorithm always searches the best and most optimal responses. This action will increase the mass of an object in Gravity force. As we know, the heavier the object is, the higher would be its Gravity force; as a result, it will attract the optimal responses toward itself. This process will be repeated until the object has more weight by attracting optimal responses at each time; therefore, the algorithm will restore the most optimal response. The initial responses of an OVRP can be obtained based on an algorithm such as sweep or randomly. The proposed system, as a combination of gravitational search algorithm and the initial response which is assigned to it using sweep algorithm, solves OVRP as follows: The hybrid gravitational search algorithm begins the conditions of OVRP by defining three parameters. At first, an initial response is generated using sweep algorithm, then GSA algorithm will be run on the generated response, and finally Record-to-record algorithm will be used and run on the response obtained from GSA in order to optimize the response. In the following, this procedure will be described clearly. At first, the sweep algorithm due to its nature gives an initial response including routes and the number of required vehicles for traveling these routes. The time of traveling routes having been specified using the number of vehicles is also important. Furthermore, reduction of time will be our desirable goal. Using the following procedure, the gravitational search algorithm will improve time, distance, and reduce the number of vehicles.

At first, we consider *three* matrices of distance, initial speed and time. The distance matrix will be calculated for each sample separately based on tested problems which are derived from [14] in standard form. According to the gravitational search algorithm, the initial speed at first has initial constant response. In this problem, the initial speed is intended 10 for all bodies.

The initial speed is updated in each level after displacement of responses by gravity. As mentioned above, in the initial speed matrix, the initial speed 10 will be assigned to each customer who is regarded as a body. Thus, the time (mass) matrix in terms of distance and speed is obtained using the following in Eq. (4):

$$T = \frac{\sqrt{(y_B - y_A)^2 + (x_B - x_A)^2}}{V_{in}\{A, B\}} \tag{4}$$

where T indicates customer's mass, $(y_B - y_A)^2 + (x_B - x_A)^2$ is the distance between two customers A and B, and $V_{in}\{A, B\}$ shows the speed between *two* customers A and B.

Using GSA with these conditions, a proper reserve factor must be defined. The reserve factor is equal to the number of reserved customers for vehicles in the future. The set of customers assigned to vehicles is a solution itself. As two matrices $(n * n)$ that is equal to number of customers and vehicles, this solution can be used to display a response. Each row and column of mentioned matrix indicated one of the customers of group. The value held in each row and column also shows the number of vehicles to which customers belong. When the algorithm is complete, reserved vehicles for customers as well as their reserved factor will be displayed. In the next step, record-to-record algorithm will be run on it as one-point and two-point and the responses will be saved. Then gravitational search algorithm would be implemented on obtained response, and thus the results are initial responses of gravitational search algorithm. Using GSA, one-point and two-point operations will be again performed on obtained response that is different from normal displacements. The value of speed and time matrices would be updated by each time of running algorithm if they were optimized more. In each step, the maximum value is selected in each row and column of speed matrix and then the assigned vehicles to the customer are displaced in case of change. After that, the cost of current response will be calculated due to the changes. If the cost of current response were higher than the best cost of previous response, it would be replaced otherwise it would not be updated in table. Therefore, a new gravitational search algorithm will be again implemented on previous response, a new response will be tested on other candidate bodies, and the speed and mass values will be updated. The previous response is called current response and the new obtained response that is compared with current one is called candidate response. The algorithm will stop when the assigned speed is zero or the number of iterations of algorithm will reach the maximum value determined by the problem.

4 Performance Evaluation

The proposed algorithm has been tested on well-known data set of Christofide et al. [15], Fisher [16] and Li et al. [1]. In Fisher model, there are 14 C1-C14 problems [16]. Compared with VRP model, the maximum length of considered route is a multiple of 0.9 because considered length of VRP is not appropriate. The last column indicates low limit taking Kmin (the sum of customers demands that are divided by vehicle capacity). In addition, the large-scale data set of Li et al. [1] is also taken in to account. The second case includes eight problems O1-O8 with 200-480 customers and has no limitation about the length of route. Each problem has geometric symmetry and customers are around depot as a circle.

4.1 *Computational Results*

In this section, you can see the results of simulation done to solve OVRP. Implementation has been done with programming language *C#*.NET and programs were run on a computer with processor 2.06 GHz Pentium-IV and RAM 6 GB. The results of simulating the proposed algorithm have been compared with 11 samples of different algorithms including (Clonal selection Algorithm [7], ITS [3], ITS [4], DHPD [8], ORTR [1], HES [17], BLSA [18], BBMOOVRP [19], IVND [10]). With regard to steps mentioned in previous section, the output results can be seen in Figs. 1 and 2.

According to the compared algorithms and searching the Internet, it is worth noting that solving the open vehicle routing problem using gravitational search algorithm is completely new and without any similar. It is also among initial simulations using rage above algorithms. With regard to the comparison, the results show time superiority, minimum number of vehicle and minimum distance traveled by vehicle for the proposed algorithm to the compared algorithms.

Figures 1 and 2 show the results of using the proposed algorithm with other algorithms to solve OVRP in a variety of easy and hard problems. As you can see, the proposed algorithm was able to achieve more acceptable and optimal responses to the compared algorithms. Furthermore, it has been able to reduce the number of vehicles very well in each hard problem. Due to the Time column of attached tables, reaching the final response in a short time-that is one of the most important factors in solving OVRP so that the proposed algorithm was able to solve this factor well-is one of the significant and unique superiorities of the proposed algorithm compared to other algorithms. Followed the simulation results, you can see the output forms of some samples of various problems using proposed algorithm. These forms are divided in to two parts, easy and hard problems. Easy problems include (C1, C2, C3, C4, C14, F11, F12) and hard ones include (O1–O8). As it can be seen, the proposed algorithm has been able to result appropriate outputs. Note that, F11 and F12 numbers of vehicles are 30,000 and 2210, K_{min} are 4 and 6, and the numbers of considered costumers which have request are 71 and 134, simultaneously.

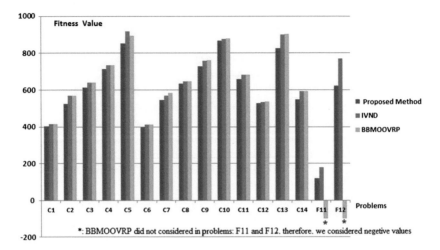

Fig. 1 The results of comparing three methods (proposed approach, IVND [10] and BBMOOVRP [19]) to minimize the total traveled distance and the number of vehicles

In Fig. 1, the distance traveled (fitness) and the number of vehicles for 16 independent problems have compared in our approach, IVND [10], and BBMOOVRP [19]. From the Fig. 1, we conclude that the fitness value for both approaches are increased by increasing Q, N and k_{min}, but the rate of fitness in proposed method is less than BBMOOVRP in most cases, which means our is able to find proper results in less travels of vehicles and it is confirmed that by increasing the number of the vehicles (value of k_{min}) and numbers of the customers in OVRP the fitness should be risen. Although BBMOOVRP did not considered last two numerous vehicles in the network but our approach considered these two cases and compared to IVND showed that the distance traveled for each vehicles to response to the incoming requests is less than IVND according to faster converge into proper value. On the other hand, in Fig. 1, it is shown that the duration for solving the problem in our approach is significantly better than BBMOOVRP approach. due to the huge duration time for IVND, we drop it from comparing.

Figure 2 represents a comparison of proposed method with seven other evolutionary optimization algorithms: two ITSs algorithms [3, 4], a ORTR algorithm [1], a BLSA [18], Colonal Selection algorithm [7], and HES algorithm [17]. We have partitioned the problems into two groups of instances (C1-C5) which is represents in Fig. 2a, b time (exponential results compared to proposed method) and fitness values, simultaneously. The demonstrated results for each of the instances present the best cost over ten runs and the quality of the solutions produced by the each algorithm. Figure 2c plots the distance traveled by vehicles or fitness value for the second partitioned of the instances which are C6-C14 and also two specific huge vehicles and normal requests (i.e., F11 and F12) for HES and ORTR evolutionary algorithms, because, the rest did not provide any results for these partition of instances. Also, we do not consider the solving problem duration for these problems among these

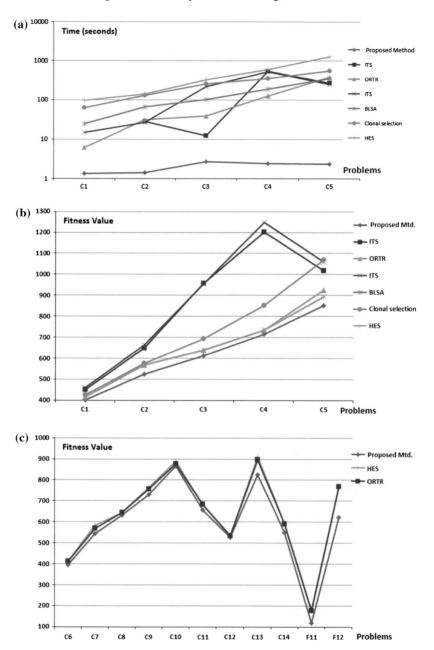

Fig. 2 Comparison of the proposed algorithm with other evolutionary algorithms from the literature. **a** Solving problem time (i.e., duration in seconds) for C1-C6 independent easy types problems. **b** Fitness value (distance traveled by vehicles) for C1-C6 independent easy types problems. **c** Fitness value (distance traveled by vehicles) for C6 to C12 and F11 and F12 independent easy types problems

algorithms, whiles these values are significantly larger than proposed method. For example, The F12 problem time for HES and ORTR is approximately more than 350 times (quasi 753 seconds), and 75 times (quasi 158.2 seconds) higher than our approach, simultaneously.

For example, in Fig. 2b, The average quality of the obtained fitness solutions in the proposed method among these problems is 620.6 %, 650 for the HES, 656 for the ORTR, 722.5 for the Colonel selection, 877.1, 855.6 for ITS [3], ITS [4], and 649.8 for the BLSA. Compared to the all approaches the proposed algorithm finds better solutions in all instances, by increasing N and Q the fitness for all methods increased by the rate of increment is lowest in our approach due to fast converging to the optimum result. On the other hand, in Fig. 2a, The average solution finding duration in the proposed method is 2.05, the HES is 491.2, the is 117.2, the Colonel selection is 274.6, ITS [3] is 208.9, ITS [4] is 213, and 148.2 for the BLSA. So, it is obvious that for the solving time our approach is the best among these algorithm.

5   Conclusion

In this paper, a gravitational search algorithm is presented to solve the above problem. The advantages of this algorithm include its speed, short running time, and much low assessment values. Reduction of running time, finding the shortest route among customers and reduction the number of required vehicles are the objectives of this algorithm. The efficiency of this algorithm has been compared with 11 samples of various algorithms. The results showed that proposed method has improved significantly than compared algorithms. The minimum value of this improvement will be 118.327 in small and medium systems to solve F11 problem and the maximum value will be 866.479 to solve C10 problem.

References

1. Li, F., Golden, B., Wasil, E.: The open vehicle routing problem: algorithms, large-scale test problems, and computational results. Comput. Oper. Res. **34**(10), 2918–2930 (2007)
2. Shamshirband, S., Shojafar, M., Hosseinabadi, A.A.R., Abraham, A.: Ovrp_ica: an imperialist-based optimization algorithm for the open vehicle routing problem. In: Hybrid Artificial Intelligent Systems, pp. 221–233. Springer (2015)
3. Huang, F., Liu, C.: A hybrid tabu search for open vehicle routing problem. In: 2010 International Conference On Computer and Communication Technologies in Agriculture Engineering (CCTAE), vol. 1, pp. 132–134. IEEE (2010)
4. Huang, F., Liu, C.: An improved tabu search for open vehicle routing problem. In: 2010 International Conference on Management and Service Science (MASS), pp. 1–4. IEEE (2010)
5. Hosseinabadi, A.A.R., Kardgar, M., Shojafar, M., Shamshirband, S., Abraham, A.: Gels-ga: hybrid metaheuristic algorithm for solving multiple travelling salesman problem. In: 2014 14th International Conference on Intelligent Systems Design and Applications (ISDA), pp. 76–81, Nov 2014

6. Pooranian, Z., Harounabadi, A., Shojafar, M., Hedayat, N.: New hybrid algorithm for task scheduling in grid computing to decrease missed task. World Acad. Sci. Eng. Technol. **55**, 924–928 (2011)
7. Pan, L., Fu, Z.: A clonal selection algorithm for open vehicle routing problem. In: Proceedings of the 3rd International Conference on Genetic and Evolutionary Computing (WGEC09), pp. 786–790 (2009)
8. Hu, F., Wu, F.: Diploid hybrid particle swarm optimization with differential evolution for open vehicle routing problem. In: 2010 8th World Congress on Intelligent Control and Automation (WCICA), p. 2692–2697. IEEE (2010)
9. MirHassani, S.A., Abolghasemi, N.: A particle swarm optimization algorithm for open vehicle routing problem. Expert Syst. Appl. **38**(9), 11547–11551 (2011)
10. Chen, P., Qu, Y., Huang, H., Dong, X.: A new hybrid iteratedlocal search for the open vehicle routing problem. In: Pacific-Asia Workshop on Computational Intelligenceand Industrial Application, 2008. PACIIA'08, vol. 1, pp. 891–895. IEEE (2008)
11. Rababah, A.: High accuracy hermite approximation for space curves in rd. J. Math. Anal. Appl. **325**(2), 920–931 (2007)
12. Rababah, A.: Distances with rational triangular bézier surfaces. Appl. Math. Comput. **160**(2), 379–386 (2005)
13. Shojafar, M., Abolfazli, S., Mostafaei, H., Singhal, M.: Improving channel assignment in multi-radio wireless mesh networks with learning automata. Wirel. Pers. Commun. **82**(1), 61–80 (2015)
14. Chen, A-l, Yang, G-k, Zhi-ming, W.: Hybrid discrete particle swarm optimization algorithm for capacitated vehicle routing problem. J. Zhejiang Univ. Sci. A **7**(4), 607–614 (2006)
15. Toth, P., Vigo, D.: The vehicle routing problem. Soc. Ind. Appl. Math. (2001)
16. Fisher, M.L.: Optimal solution of vehicle routing problems using minimum k-trees. Oper. Res. **42**(4), 626–642 (1994)
17. Repoussis, P.P., Tarantilis, C.D., Bräysy, O., Ioannou, G.: A hybrid evolution strategy for the open vehicle routing problem. Comput. Oper. Res. **37**(3), 443–455 (2010)
18. Zachariadis, E.E., Kiranoudis, C.T.: An open vehicle routing problem metaheuristic for examining wide solution neighborhoods. Comput. Oper. Res. **37**(4), 712–723 (2010)
19. Marinakis, Y., Marinaki, M.: A bumble bees mating optimization algorithm for the open vehicle routing problem. Swarm Evol. Comput. **15**, 80–94 (2014)

PIRIDS: A Model on Intrusion Response System Based on Biologically Inspired Response Mechanism in Plants

Rupam Kumar Sharma, Hemanta Kr Kalita and Biju Issac

Abstract Intrusion Detection Systems (IDS) are one of the primary components in keeping a network secure. They are classified into different forms based on the nature of their functionality such as Host based IDS, Network based IDS and Anomaly based IDS. However, Literature survey portrays different evasion techniques of IDS. Thus it is always important to study the responsive behavior of IDS after such failures. The state of the art shows that much work have been done on IDS on contrary to little on Intrusion Response System (IRS). In this paper we propose a model of IRS based on the inspiration derived from the functioning of defense and response mechanism in plants such Systemic Acquired Resistance (SAR). The proposed model is the first attempt of its kind with the objective to develop an efficient response mechanism in a network subsequent to the failure of IDS, adopting plants as a source of inspiration.

Keywords Intrusion detection system · Intrusion response system · Bio-inspired · Nature · Biologically inspired · Learning · KDD99 · Anomaly detection · Host intrusion system · Network security · Bot nets · Bot · SAR · Plants defense

1 Introduction

Biological Thinking has been a source of inspiration to engineers and researchers across the globe to explore possible solutions to the complex problems. In computer network biological inspirations have been used to design strategies both for attack and

R.K. Sharma (✉) · H.K. Kalita (✉)
North Eastern Hills University, Shillong, India
e-mail: sun1_rupam1@yahoo.com

H.K. Kalita
e-mail: hemanta91@yahoo.co.in

B. Issac
Teesside University, Middlesbrough, UK

© Springer International Publishing Switzerland 2016
V. Snášel et al. (eds.), *Innovations in Bio-Inspired Computing and Applications*,
Advances in Intelligent Systems and Computing 424,
DOI 10.1007/978-3-319-28031-8_9

defense. Human Immune System (HIS)/Artificial Immune System (AIS) has been extensively used as inspiration by researchers to model a robust intrusion detection system because of high level of protection exhibited even to most of the unseen pathogens. Arisytis is an example of Artificial Immune System Toolkits [1]. Current AIS research includes Negative Selection, clonal selection and immune network theory as the most popular underlying theories. The defense life cycle can be demonstrated as follows. Prevention phase consists of training phase whereupon a classifier is built using machine learning algorithm such as ANN (Artificial Neural Network), Bayesian Network, decision trees etc. Network traffic are then monitored for probable anomaly. If any traffic successfully by pass the preventive rules, there presence are detected on the network by different host based and network based intrusion detection techniques. The last phase mitigation complements the entire life cycle by responding to already performed attacks on the network, such as triggering a response mechanism to slow down or eradicate the malicious activity from further proliferation in the network [2]. However, much research have been done on IDS (Intrusion Detection System) in comparison to limited work on Intrusion response systems owing to the complexity of developing and developing an automated response. Till today much of the existing IDS have the response in form of an alert generated to bring into notice the administrator of certain activity otherwise restricted in the network. A generic taxonomy of intrusion response system is presented by Stakhonova et al. [3]. The authors have broadly classified the intrusion response system into two types, by (a) degree of automation and by (b) activity of triggered response. (a) is further classified into Notification systems, Manual Response systems and Automatic Response systems. (b) is classified into passive response and active response. Automatic Response systems again classified as per ability to adjust (static, adaptive), by time of response (proactive, delayed), by cooperation ability (autonomous, cooperative), by response selection method (static, dynamic, cost-sensitive map-ping). Malware programs such as worms, virus, bots etc. can cause considerable impact on a network rendering the network vulnerable. In most cases a particular host is compromised and a zombie is created on the network. Computer worms are programs that self propagate in a network exploiting vulnerabilities with limited or no human intervention whereas, virus are programs that need human intervention to abet their propagation. The mode of propagation of worms sometimes can render the detection mechanisms impossible. The contagion strategy is an example of passive worm that uses embedded propagation. In such cases the worm appends or replaces a normal message. Today one of the biggest concerns in security is the rising bots and bot-net in networks. Bots are computer programs designed to perform predefined functions remotely, automatically and repeatedly once they are initiated by a victim's system or by an end user of the network [4]. In this paper we propose a bio-inspired method of response to intrusion in a Network. The portrayed model is designed taking inspiration from the defense model in plants. The outline of the paper is as follows. Section 2 gives an overview of the defense and response mechanism in plants. Section 3 gives an insight into the proposed bio-inspired model and also the mathematical foundations underlying the response time in plants and infection time by a malicious program in a Network. Section 4 marks the conclusion of the paper (Fig. 1).

Fig. 1 Defense life cycle

2 Overview of Defense and Response Mechanism in Plants

Plants are constantly exposed to various pathogens all the time. However, the strong defense mechanisms that plants constantly expose against these pathogens have been significant in keeping plants alive. The immune system in plants can be broadly classified into two types; one uses transmembrane pattern recognition receptors (PRR). PRR responds to microbial or pathogen-associated molecular patterns (MAMPs or PAMPS) such as flagellin. The second acts inside the cell using the polymorphic NB-LRR protein products encoded by most R-genes [5] (Fig. 2).

The first layer of defense plants exhibits is the plasma membrane. Microbes must first breach this cell wall in order to intrude inside host cell. The plasma membrane of plants has undergone a regular evolution both in mechanical properties and receptors capable of sensing cellular damage. Such breaches of cell wall should alarm the host about possible invasions. Pathogens which overcome these defensive

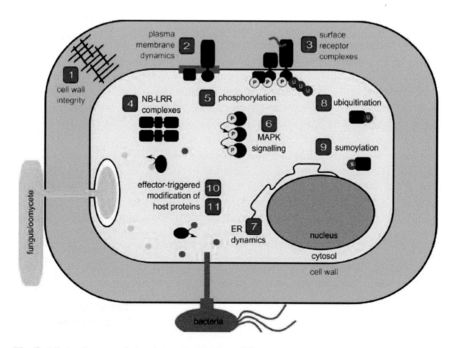

Fig. 2 Mechanisms regulating immunity in plants [6]

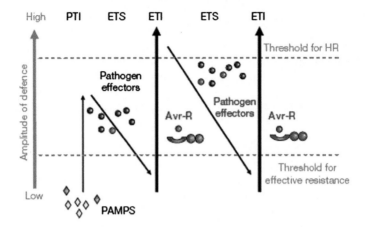

Fig. 3 A zig-zag model illustrates the quantitative output of the plant immune system [5]

layers are counterfeited by two response mechanisms in plants, namely; microbial-associated molecular patterns (MAMP/PAMP) triggered immunity (MTI/PTI) and effector-triggered immunity (ETI) (Fig. 3).

Whenever PAMPSs (or MAMPs) are recognized by PRR, results in PAMP-triggered immunity (PTI). However, successful pathogens that could breach PTI deploy huge number of effectors to render pathogen virulence. Such effectors change the usual functionality of PTI resulting in effector-triggered susceptibility (ETS) [7–9]. A given effector is recognized by plants specifically by one of the NB-LRR proteins, resulting in effector-triggered immunity (ETI) [5]. The consequence of ETI sometimes is also in form of hypersensitive response resulting in programmed cell death (PCD) of the infected cells and production of antimicrobial molecules such as—1,3-glucanase in the surrounding tissues resulting in local resistance. Early MAMP response triggers ROS, NO, ethylene and a later deposition of callose and synthesis of antimicrobial components. Modification of self proteins of plants by the effectors triggers activation of R proteins in plants. Most commonly R protein RPM1 or RPS2 guards RIN4 (RPM1-INTERACTING PROTEIN 4). RIN4 are targeted and modified by three distinct pathogen effectors from P. syringae (AvrRpm1, AvrB and AvrRpt2). RIPK (RPM1-INDUCED PROTEIN KI-NASE), was shown to phosphorylate RIN4 in response to pathogen effectors AvrRpm1 and AvrB. Phosphorylation of RIN4 is important for activation of R proteins. Such activity of pathogens in fact has portrayed the substantial decrease in pattern triggered immunity (PTI). The regulation of stomata after interaction of RIN4 with plasma membrane-associated H^+-ATPases is also discovered, which are primary site of pathogen entry. Pathogenic bacteria swim towards open stomata. To prevent such activity stomata close to Pst and to Escherichia coli [10]. Literature study indicates that plant immune system uses R proteins to monitor effector-triggered modification of self-molecules, rather than to monitor presence of non-self molecules. The regulation of R proteins under normal conditions should be controlled manner such that it don't trigger false regulation in absence of pathogens. This phenomenon is known as hybrid necrosis in plants. The mobile signal generated in the infected tissues should travel to distal parts carrying vital information about the primary pathogen infection. The onset of SAR is accompanied by increased accumulation of the signaling hormone salicylic acid in the phloem. Salicylic acid methyltransferase activity, which converts salicylic acid into methylsalicylic acid (MeSA) is required in the tissue that generates the immune signal. Conversely, MeSA esterase activity, which converts MeSA back into salicylic acid, is required for signal perception in systemic tissues. Experiments also demonstrates that defective in induced resistance 1-1 (dir1-1) gene transports a lipid-based immune signal to systemic tissues. Organophosphate compound glycerol-3-phosphate (G3P) is a signal generated in the infection site and transmitted to distal tissues to induce systemic immunity. Likewise Azelaic acid induced 1 (AZI1) is also involved in production and/or translocation of a mobile immune signal [11, 12] (Fig. 4).

Indirect recognition of effectors indicates that some R-proteins might not directly bind avirulence effectors but monitor host targets and observe their perturbation. This phenomenon is described as the guard hypothesis. Loss or perturbation of the guardee by effectors leads to R-protein dependent HR based resistance. Flg22 recognition leads to several plant defense reactions, such as production of reaction oxygen species (ROS), activation of mitogen-activated protein kinases (MAPK), ethylene production, callose deposition at the cell wall and expression of defense related genes leading to enhanced immunity as well as growth arrest [8]. Recent Studies [9] indicate that at least some NB-LRR proteins enter nucleus to activate defenses, probably as a consequence of effector recognition. Most NB-LRR proteins detect effector proteins indirectly by associating with host proteins that are targeted by effectors. AvrRpm1 and AvrB trigger RPM1 resistance probably by inducing the phosphorylation of RIN4. The p. syringae effector HopAO1 modifies host chloroplast, suppress the production of defense hormone salicylic acid etc. Taken together, pathogenic bacteria use effectors to modulate diverse host response

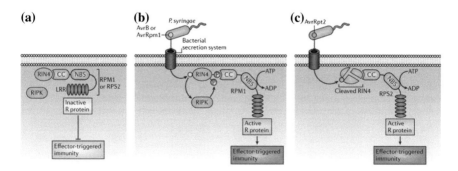

Fig. 4 The guard model, surveillane of the host immune regulator by RIN4 by the R proteins RPM1 and RPS2

to their advantage. AvrPto directly interacts with several receptor kinases, including FLS2 and EFR in Arabidopsis and LeFLS2 in tomato plants to block PAMP/MAMP induced defenses and enhances bacterial virulence [13, 14].

3   Pirids (Plant Based Inspiration of Response in Intrusion Detection System)

Section 2 outlines in brief the mechanism in plants as response to pathogen attacks. In this section we try to derive an analogy between response mechanism in plants and a similar derived automatic response mechanism in a computer network whenever an end system in the trusted network is target for compromise.

Figure 5 shows the architecture of an extended bus topology. Each end system in the topology behaves like a leaf in a plant. Figure 6 shows the sequence of events in SAR. Nodes implementing signature based intrusion detection might subject to failure to previously unseen signature [15]. Such system is proved to fail detection against traffic framed intelligently simulating behaviour similar like camouflage [15] in plants. Malicious programs breaching such security measures might succeed in creating havoc in the network. We therefore, propose a multi-layered defense mechanisms based on the inspiration derived from plants. The structure of the proposed model can be described by the diagram below (Fig. 7).

The proposed model have three layer of defense. Level 0, level 1 and level 2 respectively. Level 0 Defense behaves similar like the PRR (Pattern Recognition Receptors) in plants. It is the most generic defense response mechanism of the model. The incoming packets are continuously sniffed on the fly and the signatures

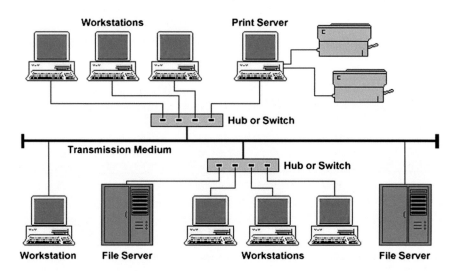

Fig. 5 Structural similarity between an extended bus topology and plant

Fig. 6 Sequence of events in SAR

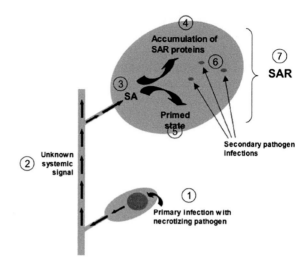

Fig. 7 Three level defense mechanism of a node

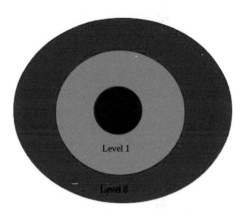

of the packets are extracted. The extracted signatures are then send to two threads for signature matching. The first thread checks the perfect matching of extracted signature using Boyer Moore String searching algorithm [16] with an existing signature database. The time complexity of the algorithm is $O(n + m)$, but usually sub-linear. However, attackers most of the time write programs that are capable of self-mutation [17]. In such cases perfect matching often fails to meet the detection process. Hence, second thread is spawned for signature matching using longest common subsequence string matching algorithm. This ensures that signature even though spread in discrete statements or altered statements will still be detected. If any of the thread returns successful detection, following actions are triggered.

- PAMP/MAMP triggers response mechanism such as regulation of stomata and triggers downstream signalling. Deriving the same inspiration the communication will be stripped down to a narrow line from a broad spectrum. As for example if signature extracted from HTTP packet match signatures from the database such as sql injections, cross-site-scripting etc., that particular source ip can be blocked next time from accessing the resource in the destination and thereby mitigating further attack attempts from the same source.

• Signal the second layer defense mechanism, layer 1, about intrusion attempt detected by layer 0 and subsequently activate layer 1 defense mechanism.

Level 1 Defense behaves similar like guard proteins in plants. The guard programs or guard agents monitors the signature of the critical programs being guarded. If the signature of the critical program guarded is changed by another non-self programs undetected by level 0, guard program is activated. The memory agents corresponding to a guard agent will record the information and generate a memory vector. The memory vector is the new signature generated from the infected program. Once the activation rate of guard programs exceeds certain given threshold value, the following responses will be instantiated. Distributes the memory vector to neighbours for embedding the new information into the level 0 signature database of the neighbours (SAR in plants).

Update the level 0 signature database.

Triggers the level 3 defense mechanism.

Level 3 defense is the extreme response of the system. The analogy is like that of HR (Hyper Sensitive Response) in plants. The activation of this level will trigger the following response

Initiate a cooling time. Cooling time is an approximate time for which the system is going off from the network.

Trigger self destructive programs which shuts the system off the networks such that further propagation of malicious programs is restricted in the network from the infected system.

Once the cooling time is out, the system waits for its recovery by behaviour of collaborative effort from the neighbours. The infected program will be deleted and the new one's received automatically from the neighbours will be re-installed.

A. Mathematical Model and analogy of Plant response time to infection and Malicious Program spread in a Network

The transition from one system to another with with $u(t)$ as control variable and $x(t)$ as state variable can be defined as [18]

$dx/dt = f(t, x(t), u(t))$, $0 \leq t \leq T$, (1), with initial condition $x(0) = x_1$ where, $f(x, t, u)$ is the state function and T is the terminal time (assuming T is finite). Then there is an objective functional $J(u) = \ln(g(t, x(t), u(t))dt$, where $g(x, t, u)$ is a given continuously differentiable function and $x(t)$ follows as a response to $u(t)$. The fundamental problem is to determine $u(t)$ that maximizes $J(u)$. Precise choice of $u(t)$ will give very good control to stop spread infection from one system to another in the network. Under environmental conditions susceptible plants (S) might turn diseased (D), upon subjected to pathogen or are able to withstand the infection (R) via host defense mechanism. In such cases the sum total of the probability distribution is

$$S + R + D = 1$$

Likewise if systems susceptible to attack in a network is (S') from which (D') systems are infected in a network and (R') systems could withstand the attack then the above equation can be modified into

$$S' + R' + D' = 1$$

The model equation for the treated plants are as follows [18].

Pre-inoculation:

$$dR/dt = (e(t) - \gamma R)(1 - R) \quad For \quad 0 \leq t \leq t_p; R(0) = R_i$$

e(t) is the effectiveness of the elicitor used.

Post-inoculation:

$$dR/dt = (e(t) - \gamma R)(1 - R - D) \text{ For } t_p \leq t \leq T; R(T_p) = R_p$$
$$dD/dt = \beta D(1 - R - D); D(t_p) = D_i$$

and the elicitor effectiveness in plant is defined to be

$$e(t) = kt/(t^2 + L^2)$$

where, γ is the specific rate the resistant tissue becomes susceptible and β is the specific rate at which the disease spreads. The value R_p represents the degree of the resistance at the time of the pathogen inoculation. The equation for untreated plants can be written as follows:

$$dR/dt = -\alpha R(1 - R - D); R(0) = R_i$$
$$dD/dt = \beta D(1 - R - D); D(0) = D_i$$

where α is the rate at which untreated plants lose their resistance due to pathogen attack. Giuseppe et al. [19] have derived the equation for the n of machines that will be compromised in the interval of time dt ('a' being constant) as

$$n = (N\,a) \cdot K(1-a)dt$$

where K = average initial compromise rate i.e. the number of vulnerable hosts that an infected host can compromise per unit time at the beginning of the outbreak. a (t) = is the proportion of vulnerable machines which have been compromised at the instant t. $N \cdot a(t)$ = is the number of infected hosts, each of which scans other vulnerable machines at a rate K per unit time. But since a portion of $a(t)$ of the vulnerable machines is already infected, only $K \cdot (1 - a(t))$ new infections will be generated by each infected host, per unit of time. If 'N' is constant in a network,

$$N\,da = (N\,a) \cdot K(1-a)dt$$
$$\text{thus, } da/dt = Ka(1-a) \text{ or}$$
$$a = e^k(t-T)/1 + e^k(t-T),$$

where the above equation is a form of logistic curve and T is a time parameter representing the point of maximum increase in the growth. The guard agents used in level 1 defence can be a set of detectors. The overview of detectors as from [20] is a string that do not have any match with any of the protected data. The protected data with progress of time are monitored by detectors. If ever a change is known to occur the corresponding detector is activated. The protected strings are the guarded proteins in plants and the detector set is the set of RP proteins in plants, i.e. the set of all those strings which are not matching with any of the self strings. The change could be modification of the existing file, append of new contents to the existing file or deletion of contents or change in the permission rights of the file considered protected. However, it is important in such perspective that protected strings do not change frequently over time and are nearly stable strings (we do not often change the statements of a program frequently). Forrest et al. [20] have stated that the probability P M that two random strings match at least r contiguous locations if: m = the number of alphabet symbols. 1 = the number of symbols in a string (length of the string) r = the number of contiguous matches required for a match.

$$P_M = m^r[(1-r)(m-1)/m + 1]$$

The approximation holds good if only $m^{-r} \ll 1$. Theirs results also concludes that as the length of the string increases P_M have a linear rise. However, there is an exponential decrease in P_M as r increases. The number and size of detector string is also estimated such that any random change to the protected string could be detected with certain fixed probability. if N_{R0} = Number of initial detector strings before censoring* N_R = Number of detector strings after censoring.

N_S = The number of self strings. P_M = The probability of match between two random strings. f = The probability of a random string not matching any of the N S self strings $= (1 - P\ M)^{NS}$, $P_f =$ The probability that N_R detectors fail to detect an intrusion. If P_M is small and N_S is large, then f approximate to

$$e\ pow\ (-P_M N_S)\ and$$
$$N_R = N_{Ro} \times f\ and$$
$$P_f = (1 - P_M)^{Nr}$$

If P_M is small and N_R is large, then $P_f = e^{-PmNr}$
Thus,
$N_R = N_{Ro} \times f = -\ln P_f/P_M$ solving above equations we get the following

$$N_{R0} = -\ln P_f /(P_M \times (1 - P_M)N_s)$$

The above formula predicts the number of initial strings that will be required to detect a random change.

4 Conclusion

The paper portrays a defense model based on the inspiration of the defence mechanism in plants. The implementation of the above model and the analysis of experimental results is undertaken and will be discussed in the near future. However, challenges such as effective detector set generation, proper signal transmission with memory from the infected node to other nodes, nature of cooling time and collaborative effort by the neighbours will be considered in more depth. The automatic recovery mechanism will also be explored to incorporate minimum human intervention in the recovery mechanism.

References

1. Ma, W., Tran, D., Sharma, D.: Negative selection with antigen feedback in intrusion detection. In: Artificial Immune Systems, pp. 200–209. Springer (2008)
2. Shameli-Sendi, A., Cheriet, M., Hamou-Lhadj, A.: Taxonomy of intrusion risk assessment and response system. Comput. Secur. 45, 1–16 (2014)
3. Stakhanova, N., Basu, S., Wong, J.: A taxonomy of intrusion response systems. Int. J. Inf. Comput. Secur. 1(1), 169–184 (2007)
4. Weaver, N., Paxson, V., Staniford, S.,. Cunningham, R.: A taxonomy of computer worms. In: Proceedings of the 2003 ACM workshop on Rapid malcode, pp. 11–18. ACM (2003)
5. Jones, J.D., Dangl, J.L.: The plant immune system. Nature 444(7117), 323–329 (2006)
6. Jones, A.M., Monaghan, J., Ntoukakis, V.: Editorial: mechanisms regulating immunity in plants. Front. Plant Sci. 4 (2013)

7. Nicaise, V., Roux, M., Zipfel, C.: Recent advances in pamp-triggered immunity against bacteria: pattern recognition receptors watch over and raise the alarm. Plant Physiol. **150**(4), 1638–1647 (2009)
8. Zipfel, C.: Pattern-recognition receptors in plant innate immunity. Curr. Opin. Immunol. **20**(1), 10–16 (2008)
9. Zhou, J.-M., Chai, J.: Plant pathogenic bacterial type iii effectors subdue host responses. Curr. Opin. Microbiol. **11**(2), 179–185 (2008)
10. Melotto, M., Underwood, W., Koczan, J., Nomura, K., He, S.Y.: Plant stomata function in innate immunity against bacterial invasion. Cell **126**(5), 969–980 (2006)
11. Spoel, S.H., Dong, X.: How do plants achieve immunity? Defence without specialized immune cells. Nat. Rev. Immunol. **12**(2), 89–100 (2012)
12. Vlot, A.C., Klessig, D.F., Park, S.-W.: Systemic acquired resistance: the elusive signal(s). Curr. Opin. Plant Biol. **11**(4), 436–442 (2008)
13. Chisholm, S.T., Coaker, G., Day, B., Staskawicz, B.J.: Host-microbe interactions: shaping the evolution of the plant immune response. Cell **124**(4), 803–814 (2006)
14. Muthamilarasan, M., Prasad, M.: Plant innate immunity: an updated insight into defense mechanism. J. Biosci. **38**(2), 433–449 (2013)
15. De Boer, P., Pels, M.: Host-based Intrusion Detection Systems. Amsterdam University, Amsterdam (2005)
16. Boyer, R.S., Moore, J.S.: A fast string searching algorithm. Commun. ACM **20**(10), 762–772 (1977)
17. Sharif, M.I., Lanzi, A. Giffin, J.T., Lee, W.: Impeding malware analysis using conditional code obfuscation. In: NDSS (2008)
18. Latif, A., Syaza, N.: Mathematical modelling of induced resistance to plant disease: a thesis presented in partial fulfilment of the requirements for the degree of doctor of philosophy in mathematics at Massey University, Albany campus, New Zealand. Ph.D. dissertation, The author (2014)
19. Serazzi, G., Zanero, S.: Computer virus propagation models. In: Performance Tools and Applications to Networked Systems, pp. 26–50. Springer (2004)
20. Forrest, S.., Perelson, A.S., Allen, L., Cherukuri, R.: Self-nonself discrimination in a computer. In: 2012 IEEE Symposium on Security and Privacy, pp. 202–202. IEEE Computer Society (1994)

Particle Swarm Optimization Method Based Consistency Checking in UML Class and Activity Diagrams

Renu George and Philip Samuel

Abstract Unified Modeling Language models are the de facto industry standard for object-oriented modeling of the static and dynamic aspects of software systems. To ensure software quality, it is essential to maintain consistency between the models. Inconsistencies among the diagrams of a model may result in serious faults which are hard to detect and may lead to project failure. Complex systems require large number of diagrams and hence detection of inconsistencies among the diagrams has a significant role during the design phase of software development. In this paper we describe a method for detection of inconsistencies among the class and activity diagrams using particle swarm optimization technique. Particle Swarm Optimization (PSO) is a soft computing technique that provides solutions to optimization problems by maximizing certain objectives in a complex search space. The PSO algorithm is applied to detect inconsistency and to optimize the consistency value of the attributes to ensure consistency. The application of PSO algorithm provides consistent, optimized diagrams that result in the generation of more accurate code.

Keywords UML model · Consistency · Particle swarm optimization

1 Introduction

Large complex systems involve different models and maintaining consistency among the models is difficult. The refinement of the models may also introduce inconsistencies. Inconsistency can be referred to conflicting requirements, violation of constraints and any design model possessing these properties is termed

R. George (✉)
College of Engineering, Chengannur, Kerala, India
e-mail: renugeorge@ceconline.edu

P. Samuel
Information Technology, SOE, Cochin University of Science and Technology,
Kochi, Kerala, India
e-mail: philips@cusat.ac.in

© Springer International Publishing Switzerland 2016
V. Snášel et al. (eds.), *Innovations in Bio-Inspired Computing and Applications*,
Advances in Intelligent Systems and Computing 424,
DOI 10.1007/978-3-319-28031-8_10

117

inconsistent. There are two types of consistency issues: intra-consistency and inter-consistency problems [23]. Consistency issues between different models is termed intra-consistency problem and between different diagrams of the same model is termed inter-consistency problem. Identification of inconsistencies during the design phase results in the development of accurate code. Manually detecting the inconsistencies is a tedious and time consuming task.

Unified modeling language (UML) has become the de facto standard for requirements modeling. UML provides diagrams to model the static and dynamic aspects of the system. Class diagrams represent the static aspects of the system, the classes required for implementation of the system, the relationships between classes and the attributes and methods of each class. UML also provides different diagrams such as activity diagram, sequence diagram, state chart diagram and collaboration diagram to model the dynamic aspects of the system.

A recent and emerging paradigm in bio-inspired computing for implementing adaptive systems is Swarm Intelligence (SI). The behavior of real world insect swarms is considered as the basis of Swarm Intelligence and the behavior is used as a problem-solving tool in SI. A population of interactive elements that performs a collaborative search of space to optimize some global objective is called swarm [18].

Particle swarm optimization (PSO) proposed by Kennedy and Eberhart in 1995 is inspired by the social behavior of bird flocking and fish schooling searching for food. PSO is a computational intelligence oriented, stochastic, population-based global optimization technique [2]. In terms of memory requirement and speed, it is computationally inexpensive and requires only primitive operators. PSO has a simple concept with a unique searching mechanism, computational efficiency, and easy implementation and has been extensively applied to many engineering optimization areas.

The process of finding the maximum or minimum value of a function or process satisfying a set of constraints is termed optimization. The goal of optimization is to maximize efficiency. In this paper, we focus on detection of inter-consistency problems between diagrams of the same model by applying the principle of PSO. The software system is modeled using class diagram and activity diagram. The methods of a class are represented using activity diagrams. Consistency checking can be treated as a form of optimization that checks each component of the UML class and activity diagram and verifies that the attributes involved in the implementation of the activity is consistent. If inconsistency is detected, the consistency value is optimized and consistent diagrams result in the generation of accurate code. Generation of accurate code results in a saving in cost and time.

2 Related Work

Detecting inconsistency in UML diagrams is very crucial to the development of accurate software. The paper rule-based detection of inconsistency in UML models [3] defines a production system language and rules specific to software designs modeled using UML class diagram and sequence diagram. Inconsistencies can be detected, users can be notified, solutions can be recommended and inconsistencies can be automatically fixed using this approach.

Inconsistency between different versions of a UML model expressed as a collection of class, sequence and state diagrams is detected and resolved in [6] using description logic. UML model is specified as a collection of concepts and roles using a knowledge representation tool, Loom. Detection and resolution of inconsistencies are performed using logic rules.

A metamodel independent method for checking model consistency is proposed in [8]. Here models are represented by sequence of elementary construction operations and consistency rules are then expressed as logical constraints on such sequences. The approach mainly deals with class diagram and use case diagrams.

A method on specifying consistency rules on different aspect models expressed in UML is specified in [9]. Consistency checking and consistency rules are also proposed. The paper instant consistency checking for UML [10] presents an approach for dealing with model changes and quickly, correctly, and automatically deciding what consistency rules to evaluate. A method of checking inconsistencies in UML models using description logic is specified in [11].

The application of swarm optimization technique in software development environment is specified in [12]. An algorithm to provide decision-making support for class responsibility assignment in a class diagram by reassigning the methods and attributes to classes using Particle Swarm Optimization are presented in the paper. The building blocks of swarm intelligence and how they are used to solve the routing problem is discussed in [13]. A general framework called Ant Colony Routing in which most swarm intelligence routing algorithms can be placed is presented in the paper.

A possible solution for the optimal operation of distribution networks which takes into account the impact of Distributed Generations is discussed in [14]. The distribution problem planning is done using PSO. Detection of concurrency problems such as deadlock and starvation using Particle Swarm Optimization algorithm is proposed in [15]. Thread execution interleaving that have a high probability of revealing deadlock and starvation faults is optimized by PSO which results in reduced complexity and increased accuracy.

A framework to test the consistency of data coming from different types of data sources is described in [16]. Consistency requirements are modeled using OCL and other modeling elements and model instances should be represented in XMI format for consistency check. An analysis of the PSO algorithm is performed in [20]. PSO is a simple algorithm with fewer parameters.

Consistency issues in behavioral models are dealt with in [17] and a methodology for dealing with consistency problems is presented. Consistency tests are formulated by mapping the relevant aspects of the model to a semantic domain and the methodology is applied to concurrent models in UML-RT.

An adaptive PSO algorithm to perform automatic tracking of the changes in a dynamic system by adding two new changes to PSO: environment detection and response is described in [21]. Functional test case selection based on multi-objective PSO is discussed in [22]. Selection of functional test cases considering coverage criteria and requirements effort using multi-objective PSO is investigated in the paper.

3 Particle Swarm Optimization

PSO is a machine learning technique based on the social behavior of bird flocking or fish schooling. It is based on the collective movement of a number of particles or birds in search global optimum. Each particle has a position and performance. A group of random particles (solutions) is used to initialize the system and search for optima is performed by updating generations. Each particle communicates with its neighbors about its performance, records its best performance so far and knows the position of the highest performing neighbor. The position of each particle is updated based on the displacement at the previous time step in the same direction it was following; the displacement in the direction towards the position where the highest performance of the particle so far was recorded; and displacement in the direction towards the position of the highest performing neighbor at that moment [1].

At every iteration, the position of the particle relative to the goal is evaluated and the best position of the particles in the neighborhood is shared with this particle and this information is used by the particle to update its position and velocity [19].

Each particle is updated by following two best values in every iteration: pbest and gbest. Fitness or the best value it has achieved so far is called pbest and gbest or global best is the best value obtained so far by any particle in the population [5]. The particle updates its velocity and positions after finding the two best values, with the following Eqs. (1) and (2).

$$v[] = v[] + c1 * \text{rand}() * (pbest[] - present[]) + c2*\text{rand}()*(gbest[] - present[]) \tag{1}$$

$$present[] = present[] + v[] \tag{2}$$

v[]	is the particle velocity,
present[]	is the current particle (solution),
pbest[] and gbest[]	are defined as stated before,
rand()	is a random number between (0, 1),
c1, c2	are learning factors

The PSO algorithm consists of an initialization part that initializes the particles in the search space, computation of fitness value and updating the velocity and position of the particles. The steps are iterated until minimum error criteria or the maximum number of iterations is reached.

4 Consistency Checking

Consistency checking consists of analyzing the models to identify unwanted configurations defined by inconsistency rules. The model is inconsistent if such configurations are found [4]. The entire model is analyzed to check for model consistency. The class and activity diagrams are analyzed to detect inconsistencies if any. Class diagrams depict what classes there are, the attributes and methods of each class and the relationship between different classes. Class diagrams provide detailed information about the attributes and methods in a class, the data type and visibility information of each attribute. The visibility of a method, its return type and the parameters of the method can also be specified in the class.

Activity diagrams depict the dynamic behavior. The implementation of each method in a class diagram is specified using an activity diagram. There is a one to one correspondence between a method in the class diagram and an activity diagram.

Consistency checking is a form of optimization that analyzes each activity in the activity diagram and verifies that the attributes involved in the implementation of the activity are defined in the corresponding class. The aim of our work is to optimize attribute specification to ensure consistency. Inconsistency rules are used to detect the inconsistencies that may occur among the different diagrams of a model. Our method checks two types of inconsistencies: (a) Attribute missing inconsistency and (b) Attribute specification inconsistency.

4.1 Attribute Missing Inconsistency

An activity diagram corresponds to the implementation of a method in a class. The attributes involved in the implementation of the activities specified in the nodes of the activity diagram should be defined in the corresponding class. If the attribute is not defined in the class, the information related to the attribute such as the class in which the attribute is to be defined, type and visibility shall be missing and this leads to inconsistency and generation of code with errors.

4.2 Attribute Specification Inconsistency

Attribute specification inconsistency refers to the situation where the attribute is mentioned in the class, but the relevant information related to the attribute such as type and visibility is missing.

Particle swarm optimization method is used to detect and fix the inconsistencies. A fitness function is applied to each particle. The aim of our work is to optimize the fitness value to achieve consistency. Attribute specification index, ASI is computed from the fitness value of the particles. The value is used to determine the stopping criteria of the PSO algorithm. Detection and correction of inconsistencies during the design stage of software development life cycle results in generation of more accurate code.

5 Consistency Checking Using Particle Swarm Optimization Algorithm

We have implemented a method to check consistency between class and activity diagram by applying the principle of PSO and generate optimized diagrams by maximizing the fitness value of particles.

5.1 Algorithm

The algorithm for implementing the PSO method for consistency checking is outlined below.

```
Design UML class diagram and activity diagrams.
    Parse the diagram and obtain diagram specification
    Identify the attributes in the implementation of
    activity in the nodes of the activity diagram.
    Compute attribute missing index of each node in the
    activity diagram
    Initialize the search space with particles.
Set initial velocity = 0, c1 = c2 = 1, inertia weight = 1.
Repeat
    Apply fitness function to each particle.
    Record pbest and gbest
    Compute velocity of the particle using Eq. (1)
    Update position using Eq. (2)
    Check ASI using Eq. (3)
Until ASI = 1 or maximum number of iterations is attained
```

5.2 Particle Initialization

To apply the PSO algorithm, an initial population of particles called swarm is to be considered. The type of particles depends on the problem we are trying to optimize [5, 7]. The attributes specified in the class diagram and activity diagram are treated as particles. A particle is represented as a six tuple consisting of id, name, type, visibility, method_id and class_id. Name refers to the name of the attribute, type represents the data type, visibility refers to the access specifier, method_id represents the methods that access the attribute and class_id represents the id of the class in which the attribute is defined. Attribute missing inconsistency check is performed during particle initialization to identify missing attributes. If the attribute is missing, a new particle with the name of the attribute is created.

5.3 Fitness Function

The fitness value, computed by applying a fitness function to each particle, is a measure of consistency of the particle. In our method, the fitness value is computed as a function of the parameters in the representation of a particle. Each parameter is assigned a weight according to its significance and the fitness function computes the weighted sum of the parameters. The goal of our algorithm is to maximize the fitness function. A high value of fitness function implies that the particle is consistent.

The fitness function is applied to all particles and the fitness value is computed in each iteration. For each particle, we compute attribute specification index, ASI which is the ratio of the fitness value of the particle to the fitness value of the particle with gbest value. ASI provides a measure of how consistent the attribute specification is. A specification index value less than one indicates inconsistency.

$$ASI = f(Particle)/f(gbestParticle) \tag{3}$$

5.4 Position/Velocity Adjustment

If the consistency check for attribute specification inconsistency returns a value less than one, the particle is inconsistent. The velocity and position of particles are updated based on the values of present, pbest and gbest. Consistency is achieved by moving the parameters of the particles in the search space to their respective positions according to PSO Eqs. (1) and (2). To apply the PSO algorithm for solving model consistency problem, we define velocity as the number of parameters to be added to a particle in each iteration and position as the total number of parameters added to the particle. The process is repeated until an optimum solution that satisfies the consistency criteria is achieved or maximum number of iterations is reached.

5.5   Case Study

To perform consistency check on class and activity diagram, we have implemented
a project Library. The project consists of a class diagram with four classes Book,
Member, Librarian and Issue and an activity diagram for the method issueBook() in
the class Librarian. Each class has its own attributes and methods (Fig. 1).

The activity diagram for the method issueBook() method in the class Librarian is
depicted in Fig. 2. The activity diagram checks the status of the book. If the book is
not available, the issue process ends. If the book is available, the book limit of the
member is checked. If the book limit has been reached, the book is not issued.
Otherwise the book is issued to the member and the memberid, bookid and
issuedate are updated in the Issue class.

The issue class contains only the attributes book id and issue date. There is no
attribute to store the id of the member to whom the book is issued. If found
unnoticed during coding, an erroneous software with critical error will be created.

Figures 1 and 2 are inconsistent since the method issueBook refers to an attribute
not defined in the class Issue.

The class and activity diagrams are parsed to identify attributes and nodes in the
diagram. Particles are created corresponding to each attribute defined in the class
diagram. The nodes in the activity diagram and the attributes involved in the
computation are identified. For each attribute we check for attribute missing
inconsistency. If the attribute is missing, new particle with the attribute name is

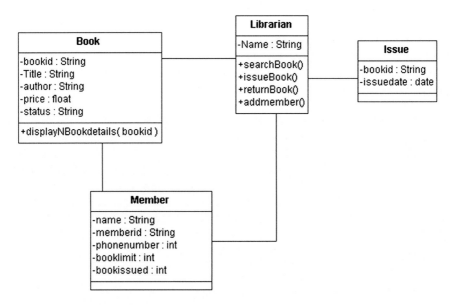

Fig. 1 Class diagram Library

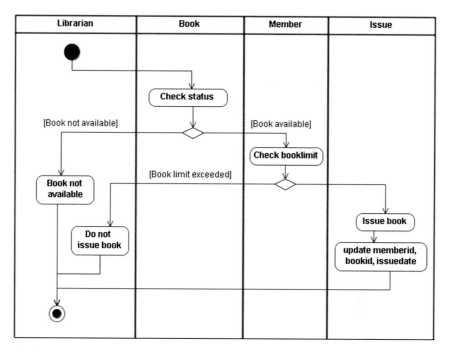

Fig. 2 Activity diagram for issueBook()

added to the set of particles. The PSO algorithm is applied iteratively until the ASI value of all the particles are one or the maximum number of iterations is achieved.

The node 'update memberid, bookid, issuedate' refers to three attributes memberid, bookid and issuedate whereas only two attributes bookid and issuedate are defined in the class Issue. Attribute missing inconsistency check detects the missing attribute and a new particle is created. During iteration of the PSO algorithm, ASI value of the memberid particle has a value less than one since the parameters are not defined. In each iteration, depending on the velocity value, the parameters are updated resulting in a change in the position. The algorithm is applied repetitively until the termination condition is reached.

6 Conclusion

We have developed a method to perform inter-consistency check on UML diagrams. Particle swarm optimization algorithm is employed to perform the consistency check to produce optimized diagrams without any inconsistencies. The attributes are treated as particles and a fitness function is defined to perform the consistency check and an ASI value is defined to check for the stopping criteria. Each particle is defined as a tuple consisting of different parameters. The fitness function evaluates the parameters of each particle and a numerical value is computed. The position and velocity of the particles are adjusted based on the result of consistency check. Detection and correction of inconsistency results in optimizing the value of fitness function. Large complex systems require many models and inconsistencies if undetected results in the development of software with errors. Correction of errors in the software may turn out to be costly and time consuming. Our method helps to detect inconsistencies during the design stage by employing PSO thereby saving cost, time and effort and results in the production of more accurate code.

References

1. Floreano., D, Mattiussi, C.: Bio-inspired Artificial Intelligence: Theories, Methods, and Technologies. MIT Press, Cambridge (2008)
2. Kennedy., J, Eberhart, R.: Particle swarm optimization. In: Proceedings of IEEE International Conference on Neural Networks, vol. IV, pp. 1942–1948 (1995)
3. Liu, Wn., Easterbrook, S., Mylopoulos, J.: Rule-based detection of inconsistency in uml models. In: Workshop on Consistency Problems in UML-Based Software Development, vol. 5 (2002)
4. Blanc, X., Mougenot, A., Mounier, I., Mens, T.: Incremental detection of model inconsistencies based on model operations. In: CAiSE, vol. 9, pp. 32–46 (2009)
5. Particle Swarm Optimization.: http://www.swarmintelligence.org
6. Van Der Straeten., R., Mens, T., Simmonds, J., Jonckers, V.: Using description logic to maintain consistency between UML models. In: "UML" 2003-The Unified Modeling Language. Modeling Languages and Applications, pp. 326–340. Springer Berlin Heidelberg (2003)
7. O'Keeffe, M., O, Cinneide, M.: Towards automated design improvement through combinatorial optimization. In: Proceedings of Workshop on Directions in Software Engineering Environments (2004)
8. Blanc., X., Mounier, I., Mougenot, A., Mens, T.: Detecting model inconsistency through operation-based model construction. In: ACM/IEEE 30th International Conference on Software Engineering, 2008. ICSE'08, pp. 511–520. IEEE (2008)
9. Dubauskaite, R., Vasilecas, O.: Method on specifying consistency rules among different aspect models, expressed in UML. Elektronika ir Elektrotechnika 19(3), 77–81 (2013)
10. Egyed., A.: Instant consistency checking for the UML. In: Proceedings of the 28th International Conference on Software Engineering, pp. 381–390. ACM (2006)
11. Van Der Straeten, R., Simmonds, J., Mens, T.: Detecting inconsistencies between UML models using description logic. Description Logics 81 (2003)
12. Saini, D.K., Sharma, Y.: Soft computing particle swarm optimization based approach for class responsibility assignment problem. Soft Comput. 40(12) (2012)
13. Ducatelle., F, Di Caro, G.A., Gambardella, L.M.: Principles and applications of swarm intelligence for adaptive routing in telecommunications networks. Swarm Intell. 4(3) 173–198 (2010)
14. Shamshiri, M., Gan, C.K., Mariana, Y., Ruddin, M., Ghani, A: Using particle swarm optimization algorithm in the distribution system planning. Aust. J. Basic Appl Sci 7(3), 85–92 (2013)
15. Revathi, C., Mythily, M.: A Uml/Marte detection of starvation and deadlocks at the design level in concurrent system. Int.J. Comput. Technol. Appl. 4(2), 279–285 (2013)

16. Nytun, J.P., Jensen, C.S.: Modeling and testing legacy data consistency requirements. In: "UML" 2003-The Unified Modeling Language. Modeling Languages and Applications, pp. 341–355. Springer, Berlin Heidelberg (2003)
17. Engels, G., Küster, J.M., Heckel, R., Groenewegen, L.: A methodology for specifying and analyzing consistency of object-oriented behavioral models. In: ACM SIGSOFT Software Engineering Notes, vol. 26, no. 5, pp. 186–195. ACM (2001)
18. Kennedy, J., Kennedy, J.F., Eberhart, R.C., Shi, Y.: Swarm Intelligence. Morgan Kaufmann (2001)
19. Peram, T., Veeramachaneni, K., Mohan, C.K.: Fitness-distance-ratio based particle swarm optimization. In: Swarm Intelligence Symposium, SIS'03, Proceedings of the 2003 IEEE, pp. 174–181. IEEE (2003)
20. Bai., Qinghai.: Analysis of particle swarm optimization algorithm. Computer and information science 3, no. 1, p. 180 (2010)
21. Hu, X., Eberhart, R.C.: Adaptive particle swarm optimization: detection and response to dynamic systems. In: WCCI, pp. 1666–1670. IEEE (2002)
22. De Souza, L.S., de Miranda, P.B.C., Prudencio, R.B.C., de Barros, F.: A multi-objective particle swarm optimization for test case selection based on functional requirements coverage and execution effort. In: 23rd IEEE International Conference on Tools with Artificial Intelligence (ICTAI), pp. 245–252. IEEE (2011)
23. Huzar, Z., Kuzniarz, L., Reggio, G., Sourrouille, J.L.: Consistency problems in UML-based software development. In: UML Modeling Languages and Applications, pp. 1–12. Springer Berlin, Heidelberg (2005)

Data Centric Text Processing Using MapReduce

N. Sandhya and Philip Samuel

Abstract Processing huge volume of data opened new opportunities in ecommerce, engineering, business and large computing applications. MapReduce programming model is a parallel data processing approach for execution on computer clusters. This model provides an abstraction to design scalable computing algorithm for big data processing. For batch processing types of data processing, MapReduce model provides faster computation. The key/value pair generation of MapReduce program creates memory overhead and deserialization overhead due to data redundancy. Redundancy of data is one of the most important factors that consumes space and affect system performance while using large set of data. This overhead can be avoided considerably by using a novel approach that we developed named Data Triggered Multithreaded Programming (DTMP) model. In this paper, we demonstrate the use of DTMP model using a large dataset with author details and his publications. The Data Triggered Multithreaded Programming can dynamically allocate the resources and can identify the data repetition occurring during computation. DTMP model when applied to the MapReduce programming model brings performance improvement to the system. The major contributions of this work are a simple, scalable and powerful processing of text data that enables automatic parallelization and distribution of large-scale computations.

Keywords Big data computing · Data centric architectures · Data parallelism · MapReduce · Scalable · Data triggered multithreading

N. Sandhya (✉) · P. Samuel
Information Technology, SOE, Cochin University of Science and Technology,
Kochi 682022, India
e-mail: nairsands@gmail.com

P. Samuel
e-mail: philips@cusat.ac.in

© Springer International Publishing Switzerland 2016
V. Snášel et al. (eds.), *Innovations in Bio-Inspired Computing and Applications*,
Advances in Intelligent Systems and Computing 424,
DOI 10.1007/978-3-319-28031-8_11

1   Introduction

The large volume of computing devices, social network sites, ecommerce services, online banking transactions and mobile data has led us to an era of data explosion. Currently, we create new data in the order of exabytes every day [1]. In traditional high-performance computer applications (e.g., for weather forecasting), it is common place for a high-performance computer to have processing nodes and storage nodes connected together by a high-capacity interconnect. MapReduce assumes an architecture where processors and storage (disk) are co-located. In such a setup, we can take advantage of data locality by running code on the processor directly attached to the block of data we need [2].

In the distributed file system, frameworks have been developed for large volume of data processing. One of the most important programming model is MapReduce [3, 4]. It is an emerging programming model for data-intensive applications. The idea behind MapReduce is from functional programming, where the programmer defines Map and Reduce functions to process large sets of distributed data.

The Data Triggered Multithreaded Programming (DTMP) Model inherits the power of eliminating redundant computation from data triggered thread but enhances the design of the programming model and runtime system to demonstrate the ability to support massive data-level parallelism [2, 3]. The DTMP model provides a new type of data trigger declaration that allows programmers to trigger computation more efficiently. The DTMP model also allows programmers to describe the ordering of triggered computation. The DTMP model supports many threads running at the same time and executes threads in and out-of-order fashion [4].

In this paper, we propose a Data Triggered Multithreaded Programming together with MapReduce programming model [5]. This DTMP model is a data centric programming model obtained from the data-triggered threads (DTT) model to better understand the need for data-centric computing. DTMP model initiates parallel computation when the application changes memory content.

This paper is organized as follows. Section 2 discusses Related work. In Sect. 3 we provide the details of Multithreaded MapReduce Model. Section 4 is Case study of the Threaded MapReduce Model using an Author dataset. Section 5 is the Conclusion of the paper.

2   Related Works

As the data grows, the availability of high performance and relatively low-cost hardware database systems are parallelized to run on multiple hardware platforms to manage scalability [6]. MapReduce model, whose basic idea is to simplify the distributed computing platform that offers two main functions Map and Reduce. The MapReduce programming model can be used to solve parallel problems [6].

This model can be used in applications such as data mining, machine learning and scientific computations. Hadoop, a popular big data processing framework implements this MapReduce programming model [7].

Dataflow architectures are similar to data triggered programming which tries to achieve parallelism by triggering computation. The data flow architectures require hardware support which is more costly and complex [8]. The data triggered model does not require any hardware support and provides composing applications. The data triggered programming shares the same functionalities as Cilk [9] and CEAL [10] that extends dataflow like programming and execution models on conventional architectures. Cilk exploits dataflow parallelism like functional programming language. CEAL encourages programmers to use incremental algorithms on changing data to avoid redundant computation. The DTMP model extends Data-Triggered Threads, which is designed to exploit both dataflow-like parallelism and reduce redundant computation [6].

The programming model of DTMP is similar to Habanero [7] in triggering multithreaded computation asynchronously, in allocating tasks and load balancing. The DTMP model, the program can avoid redundant computation that Habanero does not address.

3   Multithreaded MapReduce Model

The proposed model implements both Data Triggered Multithreaded Programming model followed by the MapReduce model which can improve the system performance in a cost effective manner.

As shown in Fig. 1. We have data trigger and a trigger point for the DTMP model and this process outputs the tasks to a queue. We can modify the memory content of the data through data trigger process. Trigger point that waits for the completion of all outstanding events of a certain support thread functions.

The queued tasks are the input to the MapReduce model. In the Map Reduce Model the data splits into different groups and these are stored in the distributed file system. In this model the main two phases are Map and Reduce. Data from the distributed file system are taken and fed to the Map phase. Here dataset is represented in the form of key/value pairs. These key/value pairs are sorted and shuffled. Thus generate an intermediate key/value pair. This intermediate key/value pairs are input to the Reduce function. The output of the Reduce function is the final result which is stored in the distributed file system.

3.1   MapReduce Programming Model

MapReduce programming has made large complex text data processing easier to understand, efficient and tolerant of hardware failures during computation [11]. This

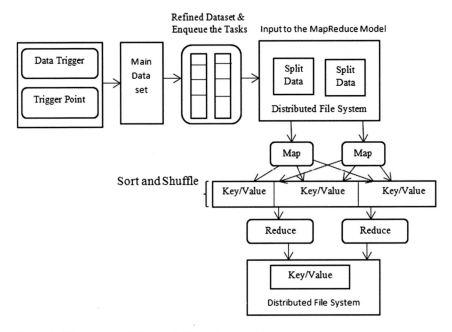

Fig. 1 Architecture of multithreaded MapReduce model

technology is applied for batch processing of large volumes of data, and it is not suitable for recent demands like real-time processing [6]. MapReduce [12] is a parallel data processing approach for execution on computer cluster [2]. For text file processing MapReduce is one of the important programming model. The key strengths of the MapReduce programming framework are parallelism combined with its simple programming framework and it can be applicable to a large variety of application domains. This requires dividing the workload across more number of machines. The degree of parallelism depends on the input data size [2].

MapReduce programming model defines *Map* function and *Reduce* function. The *Map* function performs sorting and filtering of the input data. The input data to *Map* function is in the form of key/value pair [12]. The *Map* function in the MapReduce programming is fed with data stored on the distributed file system are split across the nodes in the cluster. The Map tasks are started on the compute nodes and Map function is applicable to each Key/Value pair and the output of the Map function is intermediate Key/Value pairs. This intermediate Key/Value pairs are stored in the local file system and sorted by the keys [2].

The Reduce() function performs the addition operation. The Reduce function takes the intermediate Key/Value pair as the input and a list of intermediate values with that key as its input. All key-value pairs for the same key are fed to the same function [13]. A *Reduce* function is applied to all values grouped for one key and in turn generates Key/Value pairs. Key/value pairs from each reducer are written on the distributed file system. The final output will be set of values.

3.2 Data Triggered Multithreaded Programming

The DTMP programming eliminates redundant computation but improves the design of the programming model and runtime system to describe the capability to support massive data-level parallelism [12]. The DTMP programming provides a triggering of data for declaration that allows programmers to more efficiently trigger computation [8]. The DTMP programming also allows programmers to describe the ordering of triggered computation. The DTMP programming supports multiple threads running at the same time and executes threads in and out of order manner. Based on the changes happen in the memory address locations that triggers the data computation, the runtime system of DTMP can dynamically schedule computation to the most appropriate computing resource to reduce the amount of cache misses and data synchronization traffic. The runtime system can also balance the workload among processing units [8]. The user of the DTMP model achieves parallelism by declaring data triggers and an associated thread function.

The Data Triggered Multithreaded Programming model, the user can declare a variable, a data structure and a triggering thread [8]. The triggering thread can be assigned to the variable. The fields in the data structure contain the attributes of data for processing. The supporting thread function describes the computation, when the program changes the value of a data trigger [14]. In our DTMP model we are using data trigger and trigger point. The data trigger operation can modify the data trigger's memory content, and trigger point waits for the completion of all occurrences of all the supporting thread functions [14].

When this application executes data trigger operation, then these following steps are executed [14].

- The DTMP system checks for any changes in the current version. If the system detects any changes, then it will create new thread function event containing the changed address and the supporting thread function connected with the triggering data. DTMP system will increase the trigger point to the next value associated with the thread function. If the system does not detect any changes then it will not trigger any data.
- The changed address is analysed and then enqueue the appropriate task.
- When supporting thread function finishes its execution and the polling thread notifies the system and release the completed tasks in the queue. Then the trigger point will decreases its counter value associated with the supporting thread function.
- When the program reaches the trigger point of the supporting thread function, then it stops its operation and checks for any incomplete tasks associated with the trigger point. Thus the program resumes, if pending task does not exist with the trigger point.

Data repetition creates large amount of unwanted threads results in performance degradation [8]. The data triggered is defined with continuously changing element. In the DTMP model the destination value of the assignment changes and the data thread triggers when a same data fields appears in the data set. If all the attributes or fields of one data match then the triggering of data occurred. This application can be used to trace any particular event or element of large dataset to reduce redundancy.

4 Case Study of Authors Dataset

The proposed model uses a large dataset, which consists of author and his publication details, i.e., authorId, author name, publications details. So for the processing of dataset we are taking authorid and his number of publications in this example.

We process this author dataset using data triggered multithreading first and then this dataset will be input to the MapReduce programming model. The data triggered programming requires declaration of the data triggers to achieve parallelism and associated with each trigger needs a supporting thread function. The following steps are required for the first processing of the data triggered programming. For each author, a data structure field is created.

i. We can declare a variable to the author name.
ii. Value is assigned for each author id as the data trigger which is associated with author name.

As shown in Table 1. Data structure is defined with name 'author' and contains fields such as authorid, papercount, and citations. These are the details of the dataset which we are processing. Next we defined the data triggered thread function which contains thread pointer and filetype pointers for the fields in the data structure.

This triggering model allows the user to declare a data trigger after an assignment statement. The data triggers multithreaded execution when the destination value of the assignment changes. I.e., when an author id with same author name and same publications arrives this triggers and redundancy recognized and repeated

Table 1 Data triggered programming model for author dataset	Author declaration
	Typedef struct author {
	authorid;
	papercount;
	citations; ....}
	author;
	Data triggered function
	int author_thread(void *aid_ptr) {
	int i, j, k;
	fptype authorid;
	fptype publishno;
	fptype citation;
	return 0;# trigger_point data author_thread Inner loop

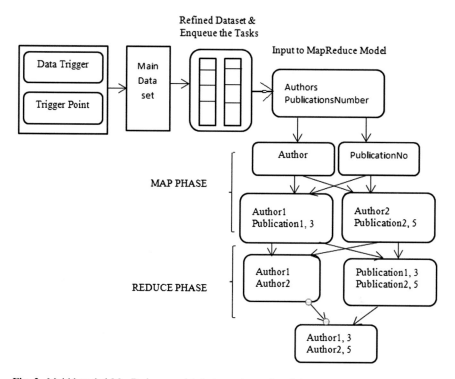

Fig. 2 Multithreaded MapReduce model design using author dataset

authorid are not entered to the refined dataset. If both the fields are same then only it is considered as redundant data. Redundant threads can be eliminated to an extent.

As shown in Fig. 2. completing redundancy check by the DTMP model, this is entered to a dataset. From main dataset which is of refined form with no redundant data, here we enqueue the tasks and the input is fed to the MapReduce programming model. The author name, publications and citations are available in the dataset. In the Map phase the sorting and shuffling takes place and author names are taken as keys and publication numbers are taken as values. When a MapReduce task is submitted to the system then the map task applies the Map function to every Key/Value pair (author1, publicationno). Zero or more intermediate Key/Value pairs (list (authors, publications)) are generated for the same input Key/Value pair. These are stored in the distributed file system and sorted by the keys (Table 2).

Completion of the map tasks notifies the reducer by the MapReduce engine. In the reduce phase, the output files from the map tasks are taken in parallel and sort the files to combine the Key/Value pairs into a set of new Key/Value pairs i.e., (authors, list(publications)), where all publications with the same author are listed. The reduce function is applied and the final list total number of authors and publications are generated. Based on this author and publication list journal ratings are calculated.

Table 2 Algorithm Map and Reduce functions	Algorithm 1: Map function
	Input: String Authorid, publicationno
	Key: Messageid, value: document value
	String[] twittermessage = value.split("l")
	Intermediatekey(article[0], ParseFloat(twittermessage [2]))
	Algorithm 2: Reduce function
	Input: String key, Iterator values
	Float totalmessage = 0;
	While values.hasNext() do
	Totalmessage + = values.next();
	End

If the above dataset is processing only through MapReduce programming model then redundant data will be more. Since we are taking this dataset of author details from different sites, blogs or databases, repetition of data will be there. Data triggered threads usage avoids this to a minimum. In this data triggered model we are assigning values to some variable. If repetition occurs then the variable changes its value and data triggering occurs.

5   Conclusion

We are in the world of big data where, scientific computing needs to handle more than terabytes of data efficiently. While using large set of data, redundancy is one of the important factors which consumes space and affect the system performance. By processing the large dataset using multithreaded data triggering method redundant data can be reduced.

Repetition of data required more thread generation during processing and it affects the system performance. MapReduce model has limited capability of controlling redundant data. For batch processing types of data processing MapReduce model provides faster computation. Due to large volume of data the key-value pair generation of MapReduce program creates memory overhead and deserialization overhead. This overhead can be avoided to the maximum by using Data Triggering Programming Model. Processing large volume of data using this DTMP model and then sending to MapReduce Model helps to improve the performance of the system. Thus the data triggered programming together with MapReduce programming brings performance improvement. Thus our system has the power to reduce redundancy and improve performance.

This proposed system can also achieve better scalability than normal MapReduce programming model. This type of model can used for social networking sites such as twitter and facebook. As we use data triggering model, tracing particular member or particular issue becomes simpler. This proposed system strengthens the MapReduce technology with our DTMP model and serves as a basis for further experimentation.

References

1. Arvind, Nikhil, R.S.: Executing a program on the mit tagged-token dataflow architecture. IEEE Trans. Comput. 300–318 (1990)
2. Li, F., Ooi, B.C., Tamer Ozsu, M., Wu, S.: Distributed data management using MapReduce. In: ACM Computing Surveys (CSUR), **46**(3) (2014)
3. Bhatotia, P., Wieder, A., Rodrigues, R., Acar, U.A., Pasquin, R.: Incoop, MapReduce for incremental computations. In: ACM SOCC '11 (2011)
4. Tseng, H.-W., Tullsen, D.M.: Data-triggered threads: eliminating redundant computation. In: 17th International Symposium on High Performance Computer Architecture, pp. 181–192 (2011)
5. Dean, J., Ghemawat, S.: Mapreduce: simplified data processing on large clusters. In: ACM Proceedings, pp. 107–113, Jan 2008
6. Arvind, Nikhil, R.S.: Executing a program on the mit tagged-token dataflow architecture. IEEE Trans. Comput. 300–318 (1990)
7. Cave, V., Zhao, J., Shirako, J., and Sarkar, V.: Habanero-java: the new adventures of old x10. In: Proceedings of the 9th International Conference on Principles and Practice of Programming in Java, PPPJ '11, pp. 51–61 (2011)
8. Hong, S., Kim, H.: An analytical model for a GPU architecture with memory-level and thread-level parallelism awareness. In: ACM SIGARCH Computer Architecture News, pp. 152–163 (2009)
9. Brunett, S., Thornley, J., Ellenbecker, M.: An initial evaluation of the tera multithreaded architecture and programming system using the the c3i parallel benchmark suite. In: Proceedings of the 1998 ACM/IEEE Conference on Supercomputing (SC 1998), pp. 1–19 (1998)
10. Lewis, B., Berg, D.J.: Multithreaded Programming with Pthreads. Prentice Hall (1998)
11. Hammer, M.A., Acar, U.A., Chen, Y.: CEAL: A C-based language for self-adjusting computation. In: ACM SIGPLAN 2009 Conference on Programming Language Design and Implementation, pp. 25–37 (2009)
12. Steffan, J., Colohan, C., Zhai A., Mowry, T.: A scalable approach to thread-level speculation. In: 27th Annual International Symposium on Computer Architecture, pp. 1–12 (2000)
13. Lin, J., Chris, D.: Data-intensive text processing with MapReduce. Synth. Lect. Hum. Lang. Technol. **3**, 1–177 (2010)
14. Tseng, H.-W., Tullsen, D.M.: Data-triggered multithreading for near-data processing. In: 1st Workshop on Near-Data Processing (WoNDP) (2013)

Request Reply Detection Mechanism for Malicious MANETs

S. Sreelakshmi and K.G. Preetha

Abstract Mobile Ad hoc Network is a collection of self-organizing mobile nodes that can communicate either directly or indirectly. The open medium and wide distribution of nodes make MANET highly vulnerable to different kinds of attacks. Most of the common routing protocols assume that all participating nodes are well behaving. The presence of malicious nodes cannot be neglected as it affects the network resources like bandwidth, throughput, transmission delay, etc. and packet transmission adversely. This paper presents an acknowledgement based routing scheme that detects selfish nodes. The observations shows that the system behaves well when compared with the most commonly used routing protocol AODV in terms of packet drops, throughput, PDR and normalized routing overhead.

Keywords Selfish nodes · Reliability · MANETs · Routing overhead · Packet drop

1 Introduction

A Mobile Ad hoc NETwork or MANET is defined as a system of mobile nodes that are connected wirelessly. These nodes can communicate either directly or indirectly in a multi-hop fashion without the support of any fixed infrastructure. All the nodes in the network has equal status and can associate with any network as they wish. The term ad hoc means for this purpose, hence MANETs are maintained for specific applications [1, 17]. Figure 1 shows the structure of a typical MANET.

The application of MANET is required in such environments where there is no pre-existing fixed infrastructure for communication. There is no centralised facility to observe or control the communication, hence the whole process is much complex

S. Sreelakshmi (✉) · K.G. Preetha
Rajagiri School of Engineering & Technology, Kochi, Kerala, India
e-mail: sls.savidham@gmail.com
URL: http://rajagiritech.ac.in

K.G. Preetha
e-mail: preetha_kg@rajagiritech.ac.in

© Springer International Publishing Switzerland 2016
V. Snášel et al. (eds.), *Innovations in Bio-Inspired Computing and Applications*,
Advances in Intelligent Systems and Computing 424,
DOI 10.1007/978-3-319-28031-8_12

139

Fig. 1 MANET scenario

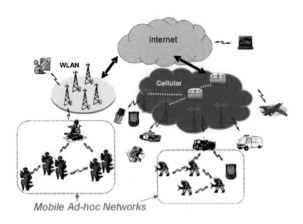

Mobile Ad-hoc Networks

than a fixed one. MANETs are widely known for their self configuring and self maintaining properties. Any node can be added in any network if it is within the range. It also has the freedom to leave any network as per wish. Establishing communication among a group of soldiers in a battlefield is a good example. In such situations a fixed network may not be available for communication. Hence MANETs can be used. The proposed system ensures packet delivery even in the presence of misbehaving nodes using acknowledgements. It is an on demand routing scheme so as to ensure both security and reliability. The system considers mainly three category of nodes: selfish nodes, incapacitated nodes and packet dropping nodes. Such nodes drops each and every packet that it receives and it will not forward any packet to any other node. The reception of acknowledgement confirms successful packet delivery to the destination. Otherwise a malicious path is suspected and it goes to Request-Reply Detection (RRD) phase. In RRD, it detects the misbehaving nodes and excludes them from further participation. An alternate path is discovered and the data packet is transmitted successfully to the destined node.

The paper is structured as follows. In Sect. 2, the background information related to the routing and security attacks of MANETs are described. The proposed system is presented in detail in Sect. 3. Section 4 mainly discusses the simulation details with corresponding results. The conclusion of the paper is given in Sect. 5. Finally Sect. 6 describes the future work that can be done.

2   Current Research Directions

There are many issues to be considered in case of MANETs. Some of the routing issues are discussed in [1, 3, 18, 19]. Error prone channel makes the medium dynamic. Another key challenge of MANET is hidden terminal problem. Inorder to prevent this, we can use RTS-CTS mechanism [2]. Bandwidth constrained variable capacity links is another problem. The wireless links are subjected to different types of interferences. This may lead to issues like congestion. Constraints on energy and security of data are other issues.

Nitin Nagar evaluates the features of routing protocols for wireless sensor network [4, 5]. These papers also describes the issues of multipath routing, mobility, self configuring, scalability. In [6], the advantages of multipath routing and enhancements with load balancing, fault tolerance and higher aggregate bandwidth are discussed. It also describes the major components in multipath routing; i.e., route discovery, route maintenance and traffic allocation.

Another key research area is security of MANETs. Conventional networks use dedicated nodes to carry out basic functions like packet forwarding, routing, and network management [20–23]. In ad hoc networks these are carried out collaboratively by all available nodes. A variety of new protocols have been developed for finding/updating routes for providing end to end communication. But no proposed protocol has been accepted as standard yet. However these new routing protocols, based on cooperation between nodes, are vulnerable to different attacks. Unfortunately, many proposed routing protocols for MANETs do not consider security. Moreover their specific features like the lack of central points, the dynamic topology, the existence of highly-constrained nodes, are challenges for security.

A brief description of different types of attacks that can occur in MANET is given in [7, 9]. Another typical attack nowadays found in MANET is collusion attack. Farah Kandah has proposed a collusion injected attack and its causes in [8]. This paper shows how a group of malicious nodes can work together to prevent a well behaving node from its communication with the network.

In 2003, Y.Hu proposed a secure routing protocol for MANETs, called SEAD [10, 11]. SEAD is a secure and efficient distance vector routing protocol for mobile wireless ad hoc networks. In [12], a model based on the Sequential Probability Ratio Test was developed to characterize to distinguish malicious nodes from well behaving nodes. [13–15] discusses some other mechanisms to detect malicious nodes in MANETs.

3 Proposed Methodology

The paper proposes mechanisms to distinguish selfish nodes, incapacitated nodes and malicious nodes from well-behaving nodes. The following subsections will describe how to handle selfish nodes, incapacitated nodes, packet dropping nodes and also how to detect routing problems.

3.1 Handling Selfish Nodes

Selfish nodes should be avoided from the network as they affects the resource utilisation adversely and cause problems like delay, data missing and so on. But these nodes need not behave selfishly forever. Even though these nodes are blacklisted from packet transmission there should be some mechanism to check whether they releive their selfishness afterwards. If so, then they can be used for packet transmission later. The proposed technique Request Reply Detection(RRD) can be used to detect selfish nodes in a network. It is an improvement of the widely known TWOACK mechanism [16] to detect malicious nodes in MANETs. The source sends the packet whenever it want. If the reply from destination is not received within a pre-defined time, then it suspects a selfishness. Then RRD is executed and confirms the selfishness. In RRD, each node has to send a reply two hops back. Consider the scenario shown in Fig. 2. There are seven nodes. Here node 3 is supposed to give back a reply message to node 1, node 4 to node 2 and so on. If any node does not receive its corresponding reply, it confirms its suspection. Assume node 3 is selfish node. Then it will not transmit the packet to 4 and hence 5 will not receive any request. So corresponding reply is also not given back. Now node 2 suspects either node 4 or node 3 as malicious as it does not receive the reply from node 4. Hence it asks node 1 whether it has received reply from node 3. If yes, that implies node 4 is the adversary; else node 3 (See Fig. 3). Hence RRD algorithm can confirm that the nodes are selfish. fter confirming the presence of such nodes, all other nodes in the network are made aware about this attack so that it can be avoided for further transmission (Fig. 4).

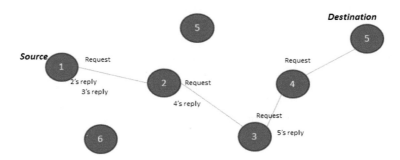

Fig. 2 RRD in case of malicious free networks

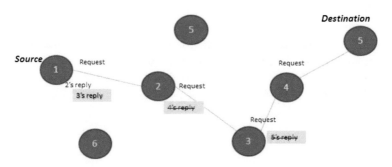

Fig. 3 RRD in case of malicious networks

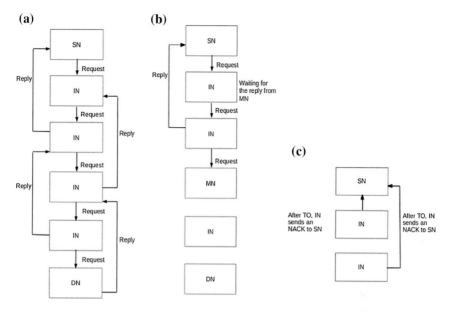

Fig. 4 **a** Algorithm flow in case of malicious free networks; **b** Algorithm flow in case of malicious networks; **c** Detection of nodes *SN=Source Node, DN=Destination Node, IN-Intermediate Node, MN-Malicious Node*

Selfish nodes need not necessarily be always selfish. Hence after a particular time interval these nodes can be checked for selfishness by pinging them using simple Hello messages. These nodes can be confirmed as 'always selfish' if it does not reply for a particular number of Hello messages. But this final confirmation not a necessity in case of highly dynamic MANETs. The detailed explanation of RRD is illustrated in Fig. 5.

3.2   Handling Incapacitated Nodes

Incapacitated nodes are such nodes that are not intentionally malicious but they are unable to perform transmission due to the lack of resources. Such nodes can be detected using the same RRD mechanism. But these need not be blacklisted from the network forever. Hence these nodes are considered in further transmission after checking for its capability (Figs. 6 and 7).

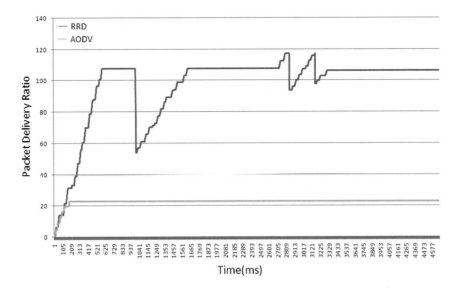

Fig. 5 Packet delivery ratio

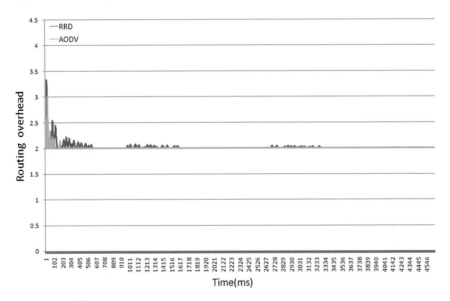

Fig. 6 Routing overhead

Fig. 7 Throughput

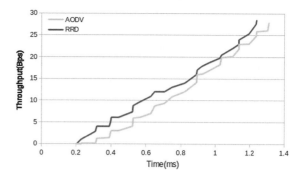

3.3 Handling Packet Dropping Nodes

These nodes are category of malicious nodes that drops each and every packet that they receive. They wont participate in packet transmission and does not allow other nodes to behave well. Hence these nodes should be detected and blacklisted from the network forever. The detection of these nodes can also be performed using the proposed RRD mechanism.

3.4 Handling Routing Issues

The acknowledgement from the destination may not reach source not only due to selfishness but also due to some other routing issues like collison, congestion, so on. This should not be mistaken as malicious attack. But in RRD phase we can clearly distinguish such issues from malicious attacks. Hence this problem does not arise in the proposed scheme.

4 Performance Evaluation

The simulation scenarios and methodology is described in this section. The performance of the proposed system is compared with existing AODV [13, 24] protocol, based on the packet drops and routing overhead.

4.1 Scenarios for Simulation

The performance of our system can be evaluated under mainly two scenarios; in presence of selfish nodes and then in the absence of selfish nodes.

4.1.1 Malicious Free Networks

In case of non-malicious networks, the system discovers the most optimal route to the destination node dynamically. The destination node sends back an acknowledgement when it receives the packet. This ensures the packet reception at the destination.

4.1.2 Malicious Networks

In this section, situations like a path from source to destination may contain some malicious nodes are considered. In such scenarios, the data packet will not reach the destination. The presence of malicious node is confirmed by non reception of acknowledgement. Then the source finds an alternate path to destination. Hence reliability can be ensured in either cases.

4.2 Simulation Parameters

The simulation is conducted on Network Simulator [26], NS2.35 environment on Ubuntu 12.04. The simulation parameters are provided in Table 1. In order to measure and compare the performances of the proposed scheme, it adopts the following parameters:

1. *Packet delivery ratio(PDR)*: The packet delivery ratio defines the ratio of the number of packets received by the destination node to the number of packets sent by the source node.
2. *Packet Drops*: The lower value of the packet lost means the better performance of the protocol. See Fig. 8.
3. *Normalized overhead*: It can be defined as the ratio of routing packets to received packets by destination.
4. *Throughput*: Throughput refers to how much data can be transferred from one location to another in a given amount of time [23]. See Fig. 7.

4.3 Simulation Results

The simulation results of packet delivery ratio with AODV is shown in Fig. 5. Since the packet drop is minimal in RRD, the PDR is high. AODV does not ensure reliabilty too. Hence RRD is better. The routing overhead for both the protocols are same initially. In case of malicious MANETs, the routing overhead is slightly high because of the discovery of alternate route. But it always ensures reliability, so that this overhead can be neglected. The simulation results of normalized overhead is shown in Fig. 6.

Table 1 Simulation parameters

Parameter	Value
Simulator	Network simulator 2.35
Number of nodes	20
Topology	Random
Interface type	Phy/wireless PHY
MAC type	802.11
Queue type	Droptail/priority queue
Queue length	200 packets
Antenna type	Omni antenna
Propagation type	Two ray ground
Transport agent	UDP
Application agent	CBR
Simulation time	50 s

Fig. 8 Packet drop

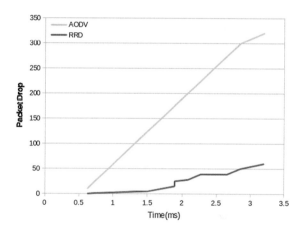

5 Conclusion

MANETs are highly prone to different types of attacks. Packet delivery should be ensured even in the presence of malicious nodes. Thus the proposed work can ensure security and reliability in packet transmission. An acknowledgement based methodology is used to guarantee packet delivery. If the source node does not receive any acknowledgement within a threshold time, then the source suspects a malicious path and goes to RRD phase. The selfish nodes can be detected using the RRD mechansim. Then it reinitiates a routing algorithm to find an alternative route to destination by excluding the malicious route.Thus it ensures packet transmission even in the presence of malicious nodes.

6 Future Work

Research based on RRD can be extended in the following interesting ways.

- The data packets transmitted are not secure. It can be captured by any malicious nodes. To ensure proper confidence and integrity, validation mechanisms can be included.
- The algorithm can be used not only to malicious nodes with packet dropping capability, but also to other adversaries.

References

1. Singh, G., Singh, J.: MANET: issues and behaviour analysis of routing protocols. In: International Journal of Advanced Research in Computer Science and Software Engineering, Feb 2004
2. Ju, H.J., Rubin, I., Kuan, Y.C.: An adaptive RTS/CTS control mechanism for IEEE 802.11 MAC protocol. In: National Science Foundations (2002)
3. Nagar, N., Biagioni, E.S.: Open issues in routing techniques in Ad Hoc wireless sensor networks. In: Proceedings of the International Conference on Parallel and Distributed Processing Techniques and Applications, vol. 4 (2002)
4. Suri, A., Iyengar, S.S., Cho, E.: Ecoinformatics using wireless sensor networks: an overview. In: Proceedings of the Ecological Informatics, Elsevier (2006)
5. Biagioni, E., Bridges, K., Chee, B.J.S.: PODS: a remote ecological Micro sensor network. http://www.botany.hawaii.edu/pods/overview.htm
6. Mueller, S., Tsang, R.P., Ghosal, D.: Multipath routing in mobile Ad Hoc networks: issues and challenges. In: Performance Tools and Applications to Networked systems, Lecture Notes in Computer Science, Springer (2004)
7. Shrivastava, S., Jain, S.: A brief introduction of different type of security attacks found in mobile Ad-hoc network. In: ACM/Kluwer Wireless Networks Journal (ACM WINET), Vol. 9(5), Sep 2003
8. Kandah, F., Singh, Y., Wang, C.: Colluding Injected attack in mobile Ad-hoc networks. In: IEEE INFOCOM 2011 Workshop on M2MCN-2011 (2011)
9. Tiranuch, A., Wu, J.: A survey on intrusion detection in mobile Ad hoc networks. In: Proceedings of the Wireless/Mobile Network Security, Springer (2006)
10. Hu, Y., Johnson, D., Perrig, A.: SEAD secure efficient distance vector routing for mobile wireless ad hoc networks. In: Proceeding of the 4th IEEE Workshop Mobile Computing Systems and Applications, p. 313 (2002)
11. Lai, W.-S., Lin, C.-H., Liu, J.-C., Huang, Y.-L., Chou, M.-C.: I-SEAD: a secure routing protocol for mobile Ad hoc networks. In: International Journal of Multimedia Ubiquitous Engineering, vol. 3(4) Oct 2008
12. Santhi, K., Bakeyalakshmi, P.: Principal component analysis in routing to identify the intrusion by sequential hypothesis testing. In: International Journal of Engineering and Science ISBN: 2319-6483, ISSN: 2278-4721, vol. 1(8), pp. 13-19, Nov 2012
13. Abdalla, A.M., Almazeed, A.H., Dr. Zewail, A.: Detection and isolation of packet dropping attacker in MANETs. In: International Journal of Advanced Computer Science and Applications, vol. 4(4) (2013)
14. Kalman, G., Parag, M., Matthias, H., Ralf, S.: Detection of colluding misbehaving nodes in mobile Ad hoc and wireless mesh networks. In: IEEE Global Communication conference 2007, Nov 2007

15. Hu, Y., Perrig, A., Johnson, D.: ARIADNE: a secure on-demand routing protocol for ad hoc networks. In: Proceedings of the 8th ACM International Conference on Mobile Computing, pp. 12–23. Atlanta, GA (2002)
16. Kang, N., Shakshuki, E., Sheltami, T.: EAACK-A Secure intrusion-detection system for MANETs. In: IEEE Journal on selected areas in communications, vol. 30(2), Feb 2013
17. Hoebeke, J., Moerman, I., Dhoedt, B., Demeester, P.: An overview of mobile Ad hoc networks: applications and challenges. In: Department of Infor-mation Technology (INTEC), Ghent University IMEC vzw (2005)
18. Jayakumar, J.G., Gopinath, G.: Ad hoc mobile wireless networks rout-ing protocol a review. J. Comput. Sci. 3(8), 574–582 (2007)
19. Perkins, C.E., Royer, .: Ad-hoc on-demand distance vector routing. In: Proceedings of the 2th IEEE Workshop on Mobile Computing Systems and Applications, pp. 90–100. New Orleans, LA, Feb 1999
20. Lundberg, J.: Routing security in ad hoc networks. In: Proceedings of the Helsinki University of Technology, HUT TML (2000)
21. Deng, H., Li, W., Agrawal, D.P.: Routing security in wireless Ad hoc network. In: IEEE Communications Magzine, vol. 40(10) (2002)
22. Papadimitratos, P., Haas, Z.J.: Secure routing for mobile ad hoc networks. In: SCS Communication Networks and Distributed Systems, Modeling and Simulation Conference (CNDS 2002), Jan (2002)
23. Karpijoki, V.: Security in ad hoc networks. In: Proceedings of the Helsinki University of Technology, Seminars on Network Security (2000)
24. Mistry, N., Jinwala, D.C., IAENG, M., Zaveri, M.: Improving AODV protocol against black hole attacks. In: IMECS2010 (2010)
25. Kannan, S., Kalaikumaran, T., Karthik, S., Arunachalam, V.P.: A review on attack prevention methods in MANET. J. Mod. Math. Stat. 5(1), 37–42 (2011)
26. NS-2.35 simulator. http://nsnam.org/
27. http://ns2tutor.weebly.com/

Ensemble of Flexible Neural Trees for Predicting Risk in Grid Computing Environment

Sara Abdelwahab, Varun Kumar Ojha and Ajith Abraham

Abstract Risk assessment in grid computing is an important issue as grid is a shared environment with diverse resources spread across several administrative domains. Therefore, by assessing risk in grid computing, we can analyze possible risks for the growing consumption of computational resources of an organization and thus we can improve the organization's computation effectiveness. In this paper, we used a function approximation tool, namely, flexible neural tree for risk prediction and risk (factors) identification. Flexible neural tree is a feed forward neural network model, where network architecture was evolved like a tree. Our comprehensive experiment finds score for each risk factor in grid computing together with a general tree-based model for predicting risk. We used an ensemble of prediction models to achieve generalization.

Keywords Risk assessment · Flexible neural tree · Feature selection · Grid computing

S. Abdelwahab (✉)
Faculty of Computer Science and Information Technology,
Sudan University of Science and Technology, Khartoum, Sudan
e-mail: saabdelghani@pnu.edu.sa

S. Abdelwahab
Computer Science and Information College,
Princess Norah Bint Abddulrahman University, Riyadh, Saudi Arabia

V.K. Ojha (✉) · A. Abraham (✉)
IT4Innovations, VSB Technical University of Ostrava, Ostrava, Czech Republic
e-mail: varun.kumar.ojha@vsb.cz

A. Abraham
e-mail: ajith.abraham@ieee.org

A. Abraham
Machine Intelligence Research Labs (MIR Labs), Washington, USA

© Springer International Publishing Switzerland 2016
V. Snášel et al. (eds.), *Innovations in Bio-Inspired Computing and Applications*,
Advances in Intelligent Systems and Computing 424,
DOI 10.1007/978-3-319-28031-8_13

1 Introduction

Risk assessment is an important issue and challenge in grid computing that enables users to use resources spread across a grid by an administrator of an organization. Hence, organization needs an effective mechanism for monitoring or assessing risks involved in distributing computational resources. Therefore, a prediction model that predicts the risk involved in grid computing shall improve the effectiveness of resource distribution. On the other hand, we also need to identify important factors that influence the prediction of risk. Abdelwahab et al. [1] have reported several risk factors associated with grid computing that threaten security measures.

In this work, we used flexible neural tree for predicting and identifying risks in a grid-computing environment. Two approaches were adopted to gird-computing risk assessment problem. The first approach involved developing an effective prediction model that can predict risk in grid computing environment. To do this, we take care of generalization. Therefore, we constructed an ensemble system using the developed prediction models. In the second approach, we determined the score of the risk factors involved in grid computing. We assigned score to each risk factor, score one (highest) to the risk factor that contributes most in predicting risk, and score zero (lowest) to the one that has no influence on risk prediction. In other words, we determined the risk factors that a grid-computing administrator needs to consider for improving resource distribution. In this work, we proposed to use cross-validation, where we used flexible neural tree (FNT) model for prediction. Additionally, we compared our feature selection with the methods presented in [2]. FNT-based feature selection shows a methodological process for computing predictability score of individual features. Our objective was to obtain best model that may predict risk with high accuracy.

Rest of the paper is organized as follows. In Sect. 2, related works in the literature are discussed followed by the methodology used, and associated risk factors in Sect. 3. Experimental set-up and the obtained results are presented in Sect. 4 followed by conclusions in Sect. 5.

2 Literature Review

Birkenheuer et al. [3] introduced the concept of risk assessment in grid computing. They investigated "AssessGrid" project for the assessment and management of risk into grid infrastructures for the grid actors: end-user, broker, and resource provider in context of service-level agreement (SLA). Many researchers have addressed risk assessment (RA) in grid computing [4–6]. The probability of resource failure plays a significant role in risk assessment process in grid computing. However, the main drawback of probability-based models is: all models were built on unrealistic assumption that the resource failure represents Poisson Process [5]. Carlsson et al. [4] developed a framework for resource management in grid computing by using predictive probabilistic approach. They introduced an upper limit failures number and approximated the likelihood of success of a specific computing task. They used a fuzzy non-parametric regression technique to estimate possibility distribution of future number of node failures. On the other hand, Alsoghayer et al. [5] proposed a mathematical model by using historical and discrete time analytical model (Markov model) to predict risk of resource failure in grid environment. Although a significant number of researchers have proposed RA methods, the risk information in grid computing is limited, due to the dependability of risk assessment efforts on the node or machine level [7]. However most of the proposed methods addressed risk assessment in grid with the aspect of resource failure. In our work, we addressed the risk assessment in grid computing in the context of security aspect.

3 Research Methodology

3.1 Risk Assessment in Grid-Computing Environment

Risk assessment is an important issue and challenge in grid computing because grid computing is a collection of diverse computers and resources spread across several administrative domains with the purpose of resource sharing. Therefore, we need to avoid security breaches by reducing risk. Hence, to offer reliable grid computing services, we need a mechanism to assess risks and take necessary precautions to avoid them. We collected many risk factors associated with grid computing from the literature. Then we conducted an online survey with international experts to evaluate the collected risk factors. We asked the experts to determine the influence of these factors by categorizing it under three levels: severe, moderate, and marginal. At the next step, we assigned a numeric range to each included factor depending on its concept and chance of occurrence. Based on expert knowledge and some statistical approaches, we then simulated 1951 instances based on a generic grid environment. The original dataset consists of 20 input attributes (risk factors), and the output (risk value). The dataset is used to develop a prediction model that predicts the risk involved in grid computing. The risk factors (attributes) are summarized in Table 1.

3.2 Flexible Neural Tree

In a prediction problem, where we have dataset with n many independent variables X and a dependent variable Y, an approximation model finds relationship between independent variables X and a dependent variable Y. Moreover, it tries to find unknown parameter θ such that the root mean square error (RMSE) between the model output $\hat{Y}$ and actual output Y is zero. A wide range of applications accept

Table 1 Risk factors in grid-computing environment

Risk factor		Definition	Refs.
Services level agreement violation	SLAV	SLAV represents an agreement between a service user and a provider in the context of a particular service provision	[11]
Cross domain attack	CDA	In CDA, attacker compromises one site to spread attack to other federated sites	[12]
Job starvation	JS	In JS, stranger job scheduled on the host use local (host) resources	[13]
Resource failure	RF	It is a failure if: (1) resource stops because of resource crash; (2) available resources does not meet the minimum levels of QoS	[14]
Resource attacks	RA	It is illegal use of host resources by attacker	[13]
Privilege attack	PA	User may gain excess privilege to accessing command shell. If grid computing allows access to command shell using a predefined scripts	[15]
Confidentiality breaches	CB	Unauthorized, unanticipated, or unintentional disclosure could result in loss of public confidence, or legal action against the organization	[13]
Integrity violation	IV	Integrity refers to the trustworthiness of data or resources. It is act of preventing improper or unauthorized change	[13]
Distributed denial of services	DDoS	DoS attacks involve sending large number of packets to a destination to prevent legitimate users from accessing information or services	[13]
Data attack	DA	In grid security, DA is a scheme in which malicious code is embedded into innocuous-looking data, which lead to destructive results.	[15]
Data exposure	DE	DE is other side of widespread connectivity in which (while improving productivity) makes it easier to obtain unauthorized access to sensitive data	[15]
Credential violation	CV	Credentials are tickets or tokens used to identify, authorize, or authenticate a user. Comprise CV causes theft of user credentials	[13]
Man in the middle attack	MMA	MMA is an attack, where the attacker secretly relays and possibly alters the communication between two parties	[13]
Privacy violation	PV	PV is the interference of a person's right to privacy by various means such as showing photos in public	[15]
Sybil attack	SA	In Sybil attacks, few entities fakes multiple identities. So, it is concern for the systems that rely upon implicit certification	[13]
Hosting illegal content	HIC	HIC is done by exploiting the leased nodes	[15]
Stealing input or output	SIO	It is a way to steal the data received by the system or to steal data sent from it	[15]

(continued)

Table 1 (continued)

Risk factor		Definition	Refs.
Shared use threat	ShUTh	Incompatibility between the attributes of grid users and conventional users causes ShUTh. Hence, no strict separation between participants	[15]
Stealing or altering the software	SS	SS caused by unauthorized means entering altered data, false data, unauthorized data, or unauthorized instruction to a system	[15]
Policy mapping	PM	Multiple administrative domains with multiple policies causes difficulty to users to map different policies across the grid	[13]

artificial neural network (ANN) as the most convenient tool for the approximation [8]. Thus, it becomes a universal approximator. ANN performance heavily relies on its structure, parameters, and activation-functions (squashing function) [8] optimization. Researchers have investigated various ways in the past to optimize the individual components of ANN using evolutionary procedure [9]. Chen et al. [10] proposed a model called flexible neural tree (FNT) that addressed ANN optimization in all of its components, including structure and parameters and does automatic feature selection. FNT was conceptualized around a multi-layered feed-forward neural network to build a tree-based model, where network structure and parameters were optimized by using meta-heuristic optimization algorithms (the nature inspired stochastic algorithms for function optimization).

We define FNT as a set of function-nodes and terminals, where the function-node indicates a computational node and terminals indicate a set of all input features. The function instruction set F and a terminal instruction set T for generating FNT model are described as:

$$S = F \cup T = \{ +_2, +_3, +_4, \cdots, +_N \} \cup \{ x_1, x_2, \ldots, x_n \} \qquad (1)$$

where $+_i (i = 2, 3, \ldots, n)$ indicates that a function-node can take i arguments, whereas, the leaf node (terminal node) receives no arguments. Figure 1 illustrates a function-node/computational-node of an FNT.

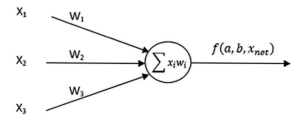

Fig. 1 A computational node of a flexible neural tree

In Fig. 1, the computational node $+_i$ receives i inputs through i connection weights (random real values) and two adjustable parameters/arguments a_i and b_i of the squashing (transfer) function, that limits the total output of the function-node within a certain range. A transfer-function used at the function-node is:

$$f(a_i, b_i, net_n) = e^{-\left(\frac{(net_n - a_i)}{b_i}\right)}, \tag{2}$$

where net_n is the net input to the ith function-node also known as excitation of the node. It is computed as:

$$net_n = \sum_{j=1}^{n} w_j x_j, \tag{3}$$

where $j = 1, 2, 3 \ldots$ is the input to the ith node. Therefore, the output of the ith node is given as:

$$out_n = f(a_i, b_i, net_n) = e^{-\left(\frac{(net_n - a_i)}{b_i}\right)}, \tag{4}$$

Figure 2 illustrates an example of a typical FNT. The root node of the FNT given in Fig. 2 indicates the output of the entire tree-based model. The leaf nodes of the tree indicate the selected input feature and the edges of the tree indicate the underlying parameters (or the weights) of the model.

Meta-heuristics are the stochastic algorithms that uses the exploration and exploitation of a given search space to find a global optimum solution for an optimization problem. Two different classes of meta-heuristic were used for the optimization for two different parts of the FNT: (a) genetic programming was used for the optimization of the structure [9]; and (b) swarm based meta-heuristics was used for the optimization of the parameters [16].

Fig. 2 A typical FNT with instruction set $F = \{+_2, +_3,\}$ and $T = \{x_1, x_2, x_3\}$

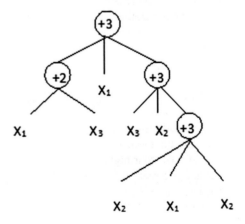

3.3 Ensemble of FNT

A collective decision with consensus of many members is better than a decision of an Individual. Hence, ensemble of many models (predictors) may offer the most general solution to a problem [17]. There are two components in ensemble system [18] construction: (1) Construction of as diverse and as accurate models as possible. (2) Combining the models using a combination rules. To construct diverse and accurate models, we use the following techniques: (a) Training models with different sets of data, the algorithm Bagging is an example [19]; (b) Training models with a different set of input features, the algorithm Random Sub-space is an example; and (c) Training models with different set of parameters. Once many models are constructed with high diversity and accuracy, then we need to combine them for a collective decision. We used weighted mean combination method, where the weights for the models were computed by using meta-heuristic algorithm. In this work, we used, genetic algorithm for searching weights of the predictors (FNTs). Hence, ensemble output was computed as:

$$RMSE^{F'}(w_1, w_2, \ldots, w_k) = \sqrt{\frac{1}{N} \sum_i^N \sum_j^k w_j f_j (x_i - y_i)^2} \tag{5}$$

where x_i, and y_i denote the ith input-target pair in the learning set that consists of total of N samples and w_j is the weight of jth predictor.

4 Experimental Set-up and Results

We conducted our experiments using a platform independent software tools that realize the mentioned methodology. We processed our dataset using the developed software tool for constructing a predictive model and for understating the significance of input feature selection. We mentioned in Table 1 that our collected dataset have 20 input variables. Hence, we need to figure out which of these variables influence the grid-computing environment.

We conducted several experiments with the parameter settings as per Table 2. Since, the computation model mentioned is stochastic in nature, each instance of experiment offers distinct results in terms of accuracy and feature selection. We used RMSE to measure the accuracy, in other words, fitness of approximation model. Additionally, we use correlation coefficient to measure the correlation that tells the relationship between two variables (here, the two variables: the actual output, and the models' output) reveals the quality of the constructed model.

We selected four highly accurate and divers FNT models for making ensemble. In Table 3, we present FNT model results over tenfold cross validation dataset. We constructed an ensemble of FNT model that shows significant improvement over using individual FNT model.

Table 2 Parameter settings of the HFNT tool

	Parameter name	Parameter utility	Values
1	Tree height	Maximum number of levels that a tree model can acquire during evolution	5
2	Tree arity	Maximum number of siblings a function-node can acquire during evolution	4
3	Tree node type	Indicates the type of transfer-function a node can acquire during evolution	Gaussian
4	GP population	Number of candidates taking a part in the process of the evolution	30
5	Mutation probability	Probability that a candidate will take part in the mutation process to form a new candidate	0.4
6	Crossover probability	Probability that a candidate will take part in the crossover process to form a new candidate	0.5
7	Elitism	Probability that a fittest candidate will survive/propagate to the next generation	0.1
8	Tournament size	It indicates the size of the pool used for the selection of the candidates that will take part in evolutionary process	15
9	MH algorithm population	The initial size of the swarm (population)	50
10	MH algorithm node range	Defines search-space of transfer-function	[0,1]
11	MH algorithm edge range	Defines the search-space for the edges	[−1.0,1.0]
13	Structure iteration	Iteration of structure optimization	100,000
14	Parameter iteration	Iteration of parameter optimization	10,000

Table 3 FNT results based on tenfolds cross validation

Exp.	Training		Test		Ensemble weights
	RMSE	r	RMSE	r	
1	0.03648	0.999	0.05861	0.998	0.589584
2	0.04546	0.999	0.04952	0.998	0.359202
3	0.04609	0.999	0.07277	0.931	0.053049
4	0.05292	0.998	0.0835	0.907	0.000001
Ensemble	–	–	0.0311	0.999	–

In the second phase of experiment, 25 FNT models were constructed. Our objective is to find significant input features. In other words, we determined significant risk factors. We calculated the score of risk factors A_j, that is score of jth as follows:

$$Score(A_j) = \frac{\sum_i^M \left(fm_i \times \mathbb{I}(A_j)\right)}{M}, \tag{6}$$

where fm_i is the fitness of model i, M is total number of models (here it is 25), and function $\mathbb{I}(A_j)$ is an identity function that returns 1 if attribute (risk factor) A_j is selected by model m_i, otherwise it returns 0. Once we calculate the score of all attributes, i.e., all N attributes (here N is 20), we calculated the final score by normalizing their values as follows:

$$Score(A_j) = \frac{Score(A_j)}{\max\limits_{j=1\,to\,N}(A_j)} \tag{7}$$

where max is a function that returns maximum value. Hence, our calculated score is given in Fig. 3.

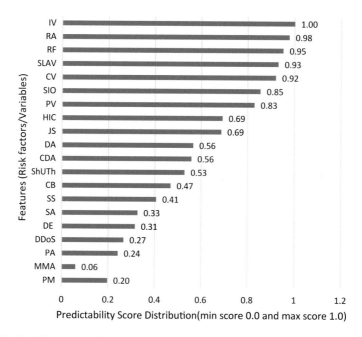

Fig. 3 Predictablityscore (influence of individual variables in risk assessment)

5   Discussions

The objective of this study is to construct the best prediction model, for assessing risk in computational grid as well as determine the significant input features. To avoid high computational cost and enhance the prediction accuracy, irrelevant input features were reduced from the dataset before constructing the prediction models. We used a correlation-based feature selection that evaluates the worth of a subset of attributes by considering the individual predictive ability of each feature along with the degree of redundancy between them. We applied different search methods, such as best first search and exhaustive search. Exhaustive search guarantees that all reachable nodes are visited in the same level before proceeding to the next level of the tree, so the possible moves in the search space are examined regularly) for feature selection. The attributes subset obtained after applying best first search (Backward) had three attributes: *CV, SIO* and *HIC* while *RA, CV*, and *HIC* were obtained when using exhaustive search.

In this work, we used FNT for feature selection, which gave more accurate and clear features selection because it assigned each feature a score according to (7) that determined the significance level of the feature (input variable) to the prediction of risk. A clear understanding of risk variables significance level served the objective of giving priority to managing risk variables by the administrator who manages grid computing environment. FNT feature selection was based on evolutionary process and the selection is automatic. We obtained best features that attained predict ability score above 0.8: *SLVA, RF, RA, IV, CV, PV* and *SIO* (Fig. 3). On the other hand, the feature selection presented in [2] was RA, CV, and HIC, which were redundant and have poor impact during construction of ensemble since there is loss of diversity. We use FNT model to validate using tenfold cross-validation. We finally construct an ensemble of accurate and diverse FNT models that offer better generalization than an individual FNT model.

6   Conclusions

In this paper, we proposed an adaptive feature selection and function approximation tool for prediction of risk in grid computing environment. We also developed a method to calculate predictability of input feature using empirical results of FNT models. As a result, we were able to identify the significant risk factors in grid computing environment. Furthermore, we used tenfold cross-validation for validating the constructed FNT models. Finally, we used ensemble of four highly accurate and diverse FNT models for constructing ensemble of FNTs, which offered a general predictive model for assessing risk in grid computing environment.

Acknowledgments This work was supported by the IPROCOM Marie Curie Initial Training Network, funded through the People Programme (Marie Curie Actions) of the European Union's Seventh Framework Programme FP7/2007–2013/, under REA grant agreement number 316555.

References

1. Abdelwahab, S., Abraham, A.: A review of the risk factors in computational grid. J. Inf. Assur. Secur. **8**(6), 270–278 (2013)
2. Abdelwahab, S., Ojha, V.K., Abraham, A.: Neuro-fuzzy risk prediction model for computational grids. In: The Second International Afro-European Conference for Industrial Advancement. Springer (2015)
3. Djemame, K., Gourlay, I., Padgett, J., Birkenheuer, G., Hovestadt, M., Kerstin, Kao, O.V.: Introducing risk management into the grid. In: Second IEEE International Conference on e-Science and Grid Computing, e-Science'06, pp. 28 (2006)
4. Carlsson, C., Fullér, R.: Probabilistic versus possibilistic risk assessment models for optimal service level agreements in grid computing. IseB **11**(1), 13–28 (2013)
5. Alsoghayer, R., Djemame, K.: Resource failures risk assessment modelling in distributed environments. J. Syst. Softw. **88**, 42–53 (2014)
6. Carlsson, C., Fullér, R.: Risk assessment of SLAs in grid computing with predictive probabilistic and possibilistic models. In: Greco, S. et al. (eds.) Preferences and Decisions, pp. 11–29. Springer, Berlin (2010)
7. Sangrasi, A., Djemame, K.: Component level risk assessment in grids: a probablistic risk model and experimentation. In: IEEE International Conference on Digital Ecosystems and Technologies Conference (DEST). IEEE (2011)
8. Haykin, S.: Neural Networks: A Comprehensive Foundation. Macmillan College Publishing Company (1994)
9. Golberg, D.E.: Genetic Algorithms in Search, Optimization, and Machine Learning, pp. 95–99. Addion Wesley (1989)
10. Chen, Y., Yang, B., Dong, J., Abraham, A.: Time-series forecasting using flexible neural tree model. Inf. Sci. **174**(3-4), 219–235 (2005)
11. Rana, O.F., Warnier, M., Quillinan, T.B., Brazier, F., Cojocarasu, D.: Managing violations in service level agreements. In: Grid Middleware and Services, pp. 349–358. Springer (2008)
12. Syed, R.H., Syrame, M., Bourgeois, J.: Protecting grids from cross-domain attacks using security alert sharing mechanisms. Future Gener. Comput. Syst. **29**(2), 536–547 (2013)
13. Chakrabarti, A., Damodaran, A., Sengupta, S.: Grid computing security: a taxonomy. IEEE Secur. Priv. **6**(1), 44–51 (2008)
14. Lee, H.M., Chung, K.S., Jin, S.H., Lee, D.-W., Lee, W.G., Jung, S.Y.Y., Chang, H.: A fault tolerance service for QoS in grid computing. In: Computational Science—ICCS 2003, pp. 286–296. Springer (2003)
15. Smith, M., Friese, T., Engel, M., Freisleben, B.: Countering security threats in service-oriented on-demand grid computing using sandboxing and trusted computing techniques. J. Parallel Distrib. Comput. **66**, 1189–1204 (2006)
16. Kennedy, J.: Particle swarm optimization. In: Encyclopedia of Machine Learning, pp. 760–766. Springer (2010)
17. Dietterich, T.G.: Ensemble methods in machine learning. In: Multiple Classifier Systems, pp. 1–15. Springer (2000)
18. Mendes-Moreira, J., Soares, C., Jorge, A.M., Sousa, J.F.D.: Ensemble approaches for regression: a survey. In: ACM Computing Surveys (CSUR), vol. 45, p. 10 (2012)
19. Breiman, L.: Bagging predictors. Mach. Learn. **24**(2), 123–140 (1996)

A Novel Approach for Malicious Node Detection in MANET

K.P. Anjana and K.G. Preetha

Abstract Mobile Ad-hoc Network (MANET) is a collection of wireless nodes without any centralized administration thereby forming an ad-hoc wireless network. MANET has limited battery power, memory, bandwidth, etc. In order to save these resources some nodes stop cooperating in the routing function. These nodes, known as malicious nodes refuse to forward data packets or drop them, thereby degrading the performance of the network. To detect such nodes, an Intrusion Detection System (IDS) is used. Several existing such systems includes Watchdog, Nuggets-based System, TWO-ACK system, etc. However, they fail to detect these nodes in the presences of ambiguous collisions, receiver collisions, limited transmission power, false misbehaviour, and partial dropping. In this paper, a new method is proposed to overcome all the previously specified limitations. The proposed system works under two phases: Detection and Broadcast phase. The detection mode finds out the malicious nodes in the network and determines the truthfulness of the misbehaviour report generated and the broadcast mode distributes the information about malicious nodes throughout the network.

Keywords MANET · Malicious nodes · Intrusion detection system

1 Introduction

As the technology is advancing, newer wireless devices are coming into the markets, and several research works are going on in the field of wireless networks [7]. Technically speaking any wireless network can be divided into two main categories, namely Infrastructure based wireless network and Ad-Hoc Wireless network. The former type of wireless network is the one in which there is a fixed infrastructure

K.P. Anjana (✉) · K.G. Preetha
Rajagiri School of Engineering and Technology, Cochin, Kerala, India
e-mail: anjanakovuckal@gmail.com

K.G. Preetha
e-mail: preetha_kg@rajagiritech.ac.in

© Springer International Publishing Switzerland 2016
V. Snášel et al. (eds.), *Innovations in Bio-Inspired Computing and Applications*,
Advances in Intelligent Systems and Computing 424,
DOI 10.1007/978-3-319-28031-8_14

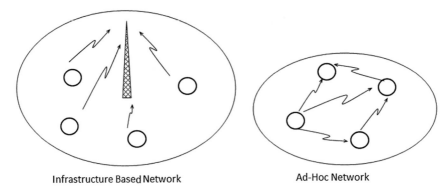

Infrastructure Based Network   Ad-Hoc Network

Fig. 1 Types of wireless network

along with a centralized management system, whereas in the later type there is no fixed infrastructure and everything is completely decentralized (Fig. 1).

The ad-hoc type wireless networks are coming into the scenario. Cost and time efficiency are the interesting characteristics of ad-hoc networks. Because of its open infrastructure, setting up an ad-hoc network is an easy and fast process. Other characteristics of an ad-hoc network includes multi-hop routing, self-creation, self-administration, device heterogeneity, etc. Mobile Ad-hoc Network shortly known as MANET is a type of wireless ad-hoc networks. MANET is mainly composed of heterogeneous mobile devices like mobile phones, PDAs, laptops, etc. and there is direct communication between the devices. MANET includes all the features of any ad-hoc network along with an exclusive feature that all the nodes in MANET act as both a normal host as well as a router. Since MANETs are easy and fast to deploy and have fewer maintenance requirements, they find usage in a wide range of fields [3]. Few areas where these are employed includes military, collaborative and distributed computing, emergency and disaster management operations, wireless sensor networks, Vehicular Ad-hoc networks, etc.

Despite so many advantages, there are several limitations imposed on MANET because of its extremely dynamic topology, power constraint, less computational power and so on. MANET also face security issues in which they are prone to attacks. These attacks are classified into passive and active attacks depending upon the degree of modifications made to the data exchanged. Active attacks are again classified into internal and external attacks on the basis of who is performing the attacks. Internal attacks are those in which a node belonging to the network itself is performing the attack. Most of the limitations of MANET contribute to the misbehaviour of different nodes in the network. The commonly seen misbehaviour is simply dropping the packets of other nodes and not actively participating in the routing and forwarding process. Nodes might exhibit misbehaviour with an aim to save its own resources, like battery power. As the number of misbehaving nodes increase, the network becomes weaker and less efficient.

There are several techniques currently available that help to detect the misbehaving nodes of the network and thereby restrict their activities in the network. Few examples of such techniques are Watchdog, Nuggets-Based System, TWOACK, EAACK, etc. But these techniques cannot handle issues like limited transmission, receiver collisions, etc. The technique proposed in this paper overcomes most of the limitations of existing system and hence improves the overall efficiency of the intrusion detection system. Here source node detects the malicious node, generates a report based on this and informs the remaining nodes in the network about this.

2 Literature Survey

2.1 Security Issues

The development process of a security protocol for MANET is a difficult task because of reasons like shared broadcast radio channel, lack of centralized authority, lack of association, limited resource availability etc.

There are two types of attacks possible on MANETs, they are passive attacks and active attacks. In a passive attack only the data been exchanged within the network is only extracted without any alteration, and the network operation is also not disrupted. In an active attack, the data is either altered or destroyed, which in turn affects the operations of networks [8]. In active attacks, the attacker node has to utilizes its own resources, however in passive attack, the attacker node misbehaves in order to save their own resources. Misbehaviour among the nodes often results in lack of cooperation within the network. Such nodes are termed as malicious nodes or selfish nodes. Selfish nodes themselves do not get involved in providing any services to other nodes however, they utilize the services provided by other nodes in the network [2]. Different types of attacks caused due to lack of cooperation are blackhole attack, wormhole attack, rushing attack, byzantine attack, etc.

2.2 Intrusion Detection System

Intrusion detection system is a software running inside every node, which allows the nodes to detect the source of any misbehaviour in the network. There are two types of IDS, namely network-based and host-based IDS. Network-based IDS is employed at the boundaries of the network, where they analyze the packets crossing the boundary of the network. The host-based IDS is employed on the host itself, monitoring the process running on the host. Most of the existing system are mainly classified into three categories:

- Credit-Based IDS: Based on give and take policy. Every node gets paid by a source node for forwarding it's packet. Every credit gained is used to transmit it's own packet.
- Reputation-Based IDS: Every node maintains a reputation metric for every other node in the network. This metric could be based on the forwarding behaviour pattern of the nodes. The metric value is then used to determine whether a node is malicious or not.
- Acknowledgement-Based IDS: This system finds out the malicious nodes in the network purely on the basis of reception of acknowledgement for a packet sent.

Several existing IDS systems are based on one of above category. Few examples are given below:

Watchguard and Pathrator. Marti et al. proposed an idea, in order to improve the performance of ad-hoc network in presence of compromising or malfunctioning nodes who agree to forward but in reality fail to do so [4]. It is a reputation based IDS system, working on network employed with Dynamic Source Routing (DSR) protocol. It has two modules running in it. One is the Watchdog module, checking whether the downstream nodes of a particular node is forwarding the packet or not. If the node is not forwarding the packet then it is considered as a malicious node and added to a blacklist. While the other module called the Pathrater chooses a path from source to destination with no malicious nodes.

Nuggets System. It is a credit-based IDS system. In the nuggets system, the payment is in the form of nuggets also called as beans. The sender puts a certain number of nuggets on the data packet to be sent. Each intermediate node earns nuggets in return for forwarding the packet. If the packet exhausts its nuggets before reaching its destination, then it is dropped [5].

Two-ACK System. Two-Ack system was proposed by Liu et al., which is an acknowledgement based IDS system [1]. It detects misbehaving links by acknowledging every data packet transmitted over every three consecutive nodes along the path from the source to the destination.

EAACK System (Enhanced Adaptive Acknowledgement Scheme). Proposed by Shakshuki et al., is a system that is mainly composed of three phases, first is ACK phase followed by S-ACK and finally ended with the MRA phase [6]. During the first phase the system checks whether the send packet was received at the destination node or not, by waiting for an acknowledgement. The S-ACK mode is activated to determine which node is a malicious one. This phase is ends with a misbehaviour report generated by a node. In the MRA phase, the system verifies whether the misbehaviour report generated in the previous step was a valid one or not.

2.3 Limitation of Existing IDS

All the above mentioned intrusion detection systems are subjected to one or more limitations specified below,

Fig. 2 Ambiguous collision

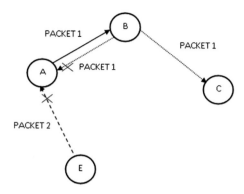

Ambiguous Collision. Due to this type of collision a node fails to determine whether its neighbouring node has actually forwarded the packet or not. This type of collision is not handled by the Watchdog System.

In Fig. 2, an ambiguous collision occurs at node A. Thus node A is unable to determine whether the collision was caused due to a packet 2 send from another neighbouring node or by node B's forwarding of the packet 1 to all it neighbours including node A. Hence node A is not clearly able to determine whether node B has actually forwarded the packet or not.

Receiver Collision. Under this category, a node fails to distinguish between the two situation of the node forwarding a packet and whether the packet is actually received by the expected node. Watchdog and Pathrater fail to handle to this type of problem.

In Fig. 3, a receiver collision occurs at node C. Here A has overheard the B's transmission of packet 1 to C, thus it believes that C has successfully received packet 1. But unfortunately packet 1 collided with packet 2 at C, thereby making C lose both the packets.

Fig. 3 Receiver collision

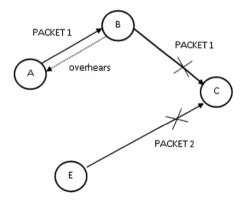

Fig. 4 Limited transmission
power

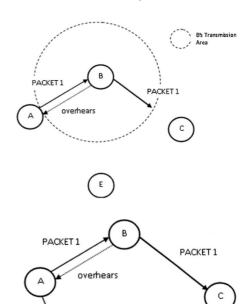

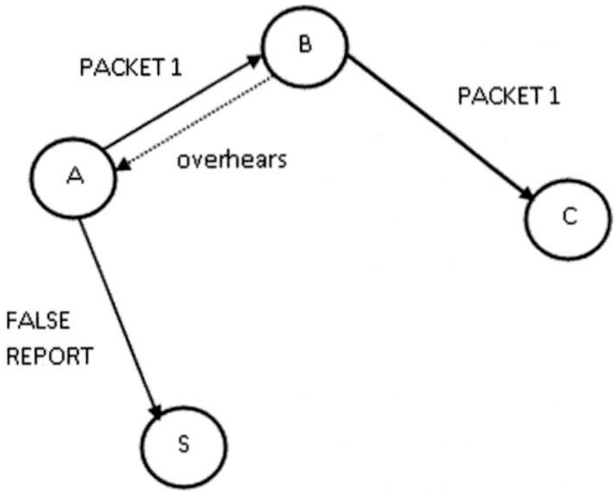

Fig. 5 False misbehaviour
report

Limited Transmission Power. In this a node purposefully limits it's transmission
power so that it cannot be overheard by few of it's neighbouring nodes or so that the
packet doesn't reach it's next-hop node. In Fig. 4 node B has limited it's transmission
power to node A only. When it forwards packet 1 to C, the packet won't reach C, since
C is not within B's transmission power but A overhears B's transmission and believes
that B is a normal node.

False Misbehaviour Report
Under this attack a node purposefully sends a report stating a perfectly working normal node as malicious node [9]. In Fig. 5 node A forwarded packet 1 to node B. It
overheard B's transmission of forwarding the packet to node C, but still it generates
a Misbehaviour report stating node B as malicious and sends it to the source S.

3 Proposed System

The proposed system tries to overcome ambiguous collision, receiver collision and
false misbehaviour report generation. The entire working of the system is divided
into two phases, namely detection phase and broadcast phase. In the detection phase,

the malicious nodes are discovered, while in the broadcast phase the news of detection is spread throughout the network.

The detection phase is further divided into three sub-modes, acknowledgement mode, conviction mode and confirmation mode. Acknowledgement phase to ensure that data packets have been send and received without loss. In case any packets are being dropped in the mid-way, the conviction phase finds out which node(s) is dropping the packets. The Confirmation phase determines the accuracy of the previous phase's outcome.

Steps of execution

First the detection phase is executed followed by the broadcast phase.

Detection Phase

1. Initially all nodes are set to Acknowledgement Phase.
2. Source node S transmits the data packet and waits for an acknowledgement from the receiver node R.
3. If acknowledgement is received then S moves on to transmit the next data packet.
4. If an acknowledgement is not received, S retransmit the same packet. Retransmission takes place for a maximum of three times. After which the Conviction Mode would be activated.
5. A one-hop request packet is forwarded to the next-hop node in the route.
6. If a reply is received then the next-pair of nodes repeat this same step.
7. Else a Misbehaviour report is generated and Confirmation Mode is activated.
8. Confirmation Mode verifies the truthfulness of the report. If found valid the source node starts the broadcast phase.

Broadcast Phase

1. The node found to be malicious is added to MAL (Malicious Node List) at.
2. If the mistrust value associated with this node is higher than a threshold value then, a broadcast message is generated.
3. This message is flooded throughout the network and the timer value associated with the recently discovered malicious node is reset to MAXVALUE.

3.1 Acknowledgement Mode

Consider Fig. 6, node S is the source node and node R is the receiver node. S is transmitting data to R. Node A, B, C and D are all intermediate nodes. S transmits data packet DP1 to R and waits for an acknowledgement ACK1 from R.

Path taken by DP1: S-A-B-C-D-R
Path taken by ACK1: R-D-C-B-A-S

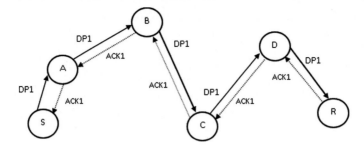

Fig. 6 Acknowledgement mode

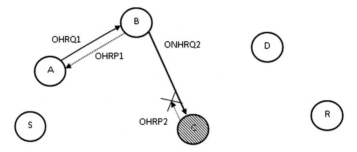

Fig. 7 Conviction mode

3.2 Conviction Mode

Main objective is to find the misbehaving node in the network. Under this mode a pair of two consecutive nodes are considered. Consider the pair of A and B shown in Fig. 7. Let C be malicious node. A sends a one-hop request to B. Whenever a one-hop packet is received a node is required to immediately send a one-hop reply back to the source of request message. A malicious node would simply drop all the packets coming to it, hence no reply would be sent. Here B sends one-hop reply to A.

Next the pair of B and C are considered. B sends a one-hop request to C, but since C is malicious it won't reply. Thus B generates a misbehaviour report stating C as malicious and sends it to S.

3.3 Confirmation Mode

The Confirmation Mode is responsible to verify the truthfulness of the received report. Consider Fig. 8, where node S receives a report stating that node C is malicious. Thus S checks for an alternative route to node R in its route table. This new route should not contain both the malicious node as well as the node that had generated that report. Then S sends that same data packet (marked as MRA packet) to

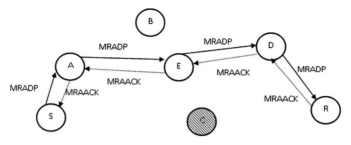

Fig. 8 Confirmation mode

R via this new route. On receiving this packet, node R checks whether it has already received the same packet or not. If it has, then that indicates that the report was false, and if not then the report was true. If the report were termed as false then the node which generated it would be considered as malicious.

Path taken by MRA-DP: S-A-E-D-R
Path taken by MRA-ACK: R-D-E-A-S

3.4 Broadcast Mode

Once a node receives a true outcome after the Confirmation mode, the malicious node is added to MAL (Malicious Node List) and timer value set to MAX. If the trust value associated with the recently added node is higher than a threshold value then a notification message is generated and flooded throughout the network.

The source as well as the receiver nodes of this notification message are supposed to delete all the routes passing through the nodes specified in the message.

3.5 Malicious Node List

Every node in the network maintains a special list called the Malicious Node List (MAL). Whenever a malicious node has been detected, that node is immediately added to this list. Each entry in this list is associated with a timer value and a trust value. The trust value is initially set to a minimum value. Each time a Misbehaviour report of a particular node is received from a unique node, its value is incremented. Whenever an entry is added to the list the timer value is set to a maximum value i.e. MAX. This timer value is decremented by 1 at regular intervals and it is reset to MAX whenever a misbehaviour report is received from some other source node.

As soon as the timer value reduces to zero, that entry is immediately deleted from the list. In some case malicious nodes can act as a normal node in the future. When the timer value reaches 0, each malicious node gets a second chance to again participate in routing. Also the notification message would be generated if and only if the trust value of a node is higher than a threshold value. The trust value would be checked either when a mis-behaviour report is received or when a node has a data packet to transmit.

The malicious node list has the following fields:

1. NODE ADDRESS: The IP address.
2. SOURCE NODE: The IP address of the recent node that reported the malicious node.
3. TIMER: The timer value.
4. TRUST: The trust value of the node, maintained by every other node in network.

In this system, whenever a source node identifies that its packets are being dropped in the mid-way, it immediately activates the detection phase to determine the possible malicious node(s). Finally, in the broadcast phase the information related to malicious node is spread across all the nodes in the network. The proposed system thus offers advantages by detecting the malicious nodes efficiently and accurately. It also overcomes several limitations of currently existing intrusion detection systems.

4 Conclusion

The system proposed in this paper is an acknowledgement based intrusion detection system for MANET. It overcomes all the limitations of currently used IDS like ambiguous collisions, receiver collisions, limited transmission power, etc. It works in two phases namely detection phase and broadcast phase. The detection determines the malicious nodes in the network, where the broadcast mode is used to spread the news of a malicious node throughout the network. The source node uses a special list called the MAL list, which stores all malicious nodes IDs and the respective trust values. This list is used for preventing the malicious nodes from participating in further communications.

References

1. Liu, K., Deng, J., Varshney, P.K., Balakrishnan, K.: An acknowledgment-based approach for the detection of routing misbehaviour in MANETs. IEEE Transaction on Mobile Computing 6(5), 536–550 (2007). IEEE Press, New York
2. Akbhani, R.H., Patel, S., Jinwala, D.C.: DoS attacks in mobile ad-hoc networks: a survey. Proceedings of Advanced Computing & Communicaions Technologies. 2nd International Meeting, 2012, pp. 535–541. IEEE Press, New York (2012)
3. Aarti and Tyagi, S.S: Study of MANET: characteristics, challenges, application and security attacks. In: Proceedings of International Journal of Advanced Research on Computer Science and Software Engineering, vol. 3, no. 5, pp. 252–257. IEEE Press, New York (2013). http://www.ijarcsse.com/docs/papers/Volume_3/5_May2013/V3I5-0267.pdf

4. Marti, S., Giuli, T.J., Lai, K., Baker, M.: Mitigating routing misbehavior in mobile ad hoc networks. In: 6th International Conference on Mobile Computing and Networking. MOBICOM'00, pp. 255–265. IEEE Press, New York (2000)
5. Buttyan, L., Hubaux, J.: Stimulating cooperation in self-organizing mobile ad-hoc networks. In: Mobile Networks and Applications, vol. 8 no. 5, pp. 579–592. IEEE Press, New York (2003)
6. Shakshuki, M., Sheltami, R.: EAACK-A secure intrusion detection system for MANEI. IEEE Trans. Ind. Electron. **60**(3), 1089–1097 (2013). IEEE Press, New York
7. IETF MANET working group. http://www.ietf.org/html.charters/manet-charter.html
8. Toh, C.K.: Ad Hoc Mobile Wireless Networks Protocols and Systems. Prentice Hall PTR, New Jersey (2001)
9. Kang, N., Shakshuki, E., Sheltami, T.: Detecting forged acknowledgements in MANETs. In: Proceedings of IEEE 25th International Conferences on Advanced Information and Networking Application, Biopolis, Singapore, pp. 488–494. IEEE Press, New York (2011)

Improving the Productivity in Global Software Development

D. Manoj Ray and Philip Samuel

Abstract Globalization has led to the expansion of information technology and distributed software development. Most of the software development companies face various challenges in distributing the project. As a consequence of the dispersed nature of global software development projects, communication, coordination, and control become more difficult which adversely influence effort estimation of the software development. The major ingredients that impact software development productivity of globally distributed projects are project delivery rate, team size and communication complexity. The paper analyses the factors affecting the productivity of the globally distributed projects. The project distribution can be effectively done depending on the estimated productivity for the different sites. The project distribution to the multiple sites can be done in the order of decreasing productivity factor.

Keywords Global software development · Function point analysis · Productivity · Communication complexity

1 Introduction

The application software became an essential and mandatory support for the business of many organizations satisfying the perceived need of potential users and for personal use. Thus the software development becomes a major role in infor-

D. Manoj Ray (✉)
Department of Computer Science, Cochin University of Science
and Technology and College of Engineering Karunagappally,
Kochi 682022, India
e-mail: manojray@yahoo.com

P. Samuel (✉)
Division of Information Technology, SOE, Cochin University
of Science and Technology, Kochi 682022, India
e-mail: philips@cusat.ac.in

© Springer International Publishing Switzerland 2016   175
V. Snášel et al. (eds.), *Innovations in Bio-Inspired Computing and Applications*,
Advances in Intelligent Systems and Computing 424,
DOI 10.1007/978-3-319-28031-8_15

mation technology fields and thousands of companies emerged all over the world. The globalization in the innovations and markets has impacted on software development dramatically. Global Software Development (GSD) is an outsourcing procedure in which the internationally distributed developers from various social foundations and remote areas take an interest in the product improvement endeavours which can bring about new imaginative thoughts and practices which may be helpful for the organization. The sole explanation behind picking this method is because of its expense adequacy and the accessibility of expansive experienced work pool [1, 2]. The software development time for the GSD likewise less because of time zone variability. Time zone adequacy can be accomplished by selecting sites in a manner that advancement can occur the 24 h a day [3]. Accordingly, so as to spare time, cost and assets much organization, hence focused on Global Software Development (GSD) to convey the software product. Modularizing the improvement transform in GSD gives preference of development of diverse elements and part in parallel [4].

The globalization had a few blemishes for the most part the separation and time zone contrast that anticipated human communication between clients and developers and massive occupation exchange. The researchers demonstrate that forty percent of software development ventures are floundering due to lack of institutionalized and practicable routines for risk management because of innate complexities [5]. A noteworthy trademark that unsuccessful tasks had in like manner was inappropriate planning, which has a vast impact on the group brought about an offshore custom software development venture [6, 7]. The GSD acquaints the difficulties in connection with communication, coordination and control of development processes. In this way influencing project organization, project control, and item quality the new procedures and instruments are therefore vital [8].

As per the attributes of the GSD communication, coordination and control are the significant difficulties, hence the effort estimation technique incorporates these complexities for its accuracy. The project performance measures such as productivity, profit and quality are less studied in terms of GSD. The paper analyses the factors affecting the productivity, effort and cost of the GSD projects. The project distribution depends on the productivity factor of the sites. The paper is divided into four sections where this section discussed the nature of GSD projects. The next section discusses about the need of effort estimation methods for the GSD. The third section, explains the challenges of GSD in detail which followed by the software productivity.

2 Project Size Estimation

Effort estimation is the procedure of assessing the sensible effort, the aggregate man hours required to build up the software product. The successful project is delivered in time and budget plan fulfilling the client necessities viably, which can be evaluated before the development.. Since the effort needed to create software is a key segment of the expense of development, the right utilization of effort estimation procedures is significant to the success of the projects [9].

The effort estimation approaches can be categorized as expert estimation, formal estimation model and combinational based estimation model. Some estimation methods can be used to estimate the required amount of effort to successfully deliver a software product are Constructive Cost MOdel (COCOMO), Software LIfe-cycle Model (SLIM) etc. [10].

An estimation technique is a set of procedures, supported by suitable experimental formulae and supported with historical reference data that help derive the predictable result inside of a better than average level of exactness [11]. There are a number of software project estimation methods available to project managers that help arrive at estimated size, effort, schedule, resource loading and other parameters. One of the most popular estimation methods, even now, is the measure of the software application delivered in Lines-Of-Code. The estimation could be based on LOC or person months as effort. COCOMOII computes software development effort as a function of program size, which is estimated thousands of source lines of code (SLOC). Function Points [12] is the well known method giving the quantity of Function Points (FP) count based on functionalities being conveyed by the application. Another technique is the use case point model where use cases are utilized to find the functional requirements of the product [13]. The work in [14] clarifies the capability of fruitful use of the use case point system for assessing the size and effort of software development projects.

Function point Analysis (FPA) method is an algorithmic method for the software size estimation using the metric function point invented by Albrecht at IBM in 1979, then revised by International Function Point User Group (IFPUG) [12]. Function Point Estimation is a standard system for measuring software development from the user's perspective by evaluating the functionality. The generic FPA model is to expand usability, on the premise of the objective application's qualities and system architecture. Some models have been proposed for estimating software development effort and cost, using FPA [15]. The number of function points for a project is calculated by first counting the number of external inputs, external outputs, internal logical, external interfaces, and external inquiries to be used by the software. Each parameter is assigned a complexity rating of simple average, or complex.

Many research works are carried out in the area of effort estimation of the individual development site; little research has been performed in the effort estimation of GSD. The effort estimation of GSD projects is much tedious tasks than the traditional software due to the differences in language, cultural barriers, time zone between the sites. The productivity rate changes among the development sites for each variable in the characteristics.

3 Challenges in GSD

The effort estimation methods for the GSD projects considering all its complexities were rarely found in the researches. But the characteristics of the GSD applications differ in some aspects from other applications. The GSD companies highlighted the impact of communication, cooperation and control within the distributed development life cycle activities [16, 17]. These life cycle activities of GSD projects are affected by the geographical, temporal and socio-cultural distances [18–20]. Temporal difference is a directional measure of the separation in time experienced by two actors wishing to interact.

3.1 Communication

Communication is the exchange of complete and unambiguous information that is, the sender and receiver can reach a common understanding [3]. Communication issues lead to coordination problems as the quantity of sites expanding, the communication results in a possibilty of data over-burden in distributed groups. The team members face redundancy of information [21]. The quality and the recurrence of the communication is less in GSD when it is contrasted with that of non distributed groups [22, 23]. As a result of the scattered way of global software development, communication between the team members turns out to be more troublesome [24] which antagonistically impact undertaking project cost and effort estimation of the product development. The physical separation, time shifts which are the components recognized as having the effect on the communication issue causes the expense overhead for GSD projects, having an effect on productivity. For enhancing the quality of effort estimation which is one of the difficulties in the software project management, the communication element ought to consider for cost and effort estimation of GSD tasks [25]. The study in [26] proposed a system produced for variables, for example, geographical, socio-cultural and temporal separation influencing communication in GSD.

The categories of the communication in the software development range are the contrast the type of communication, for example, synchronous and asynchronous, the communication channel, for example, formal and casual, the communication direction as upward, descending and/or level [27, 28]. The complexity of the communication varies as per the categories which rely on upon the cost and development time which ought to be assessed for the effort estimation techniques for the GSD ventures.

An obvious disadvantage of being separated by temporal difference is that the number of overlapping hours during a workday is reduced between sites, which results in difficulty for synchronous team meetings. Team members might have to work flexible hours in order to coordinate with their remote colleagues through real-time teleconferences, increasing the cost and effort of coordinating regularly [29].

The face to face gatherings because of the other communication embeds the travelling costs for the GSD projects causes the expanded overhead in the development cost [30]. The communication in the classification of asynchronous, informal with generally geographically distributed is more unpredictable. Temporal separation can be brought on by time zone distinction or time moving work designs [29, 31].

Social contrasts intensify communication in GSD process, and society and language affect the comprehension of communication sides [32]. Socio-cultural separation is a degree to which individuals from gathering vary in measurements of dialect, societal position, religion, governmental issues, financial conditions and essential presumptions [26]. Because of high socio-cultural separation among distributed team members, different communication issues can occur.

3.2 Coordination

Coordination is the demonstration of incorporating every undertaking with each hierarchical unit, so the unit adds to the general objective [3]. The coordination breakdown issues are more noteworthy in GSD even in the vicinity of gathering of procedures, organizational mechanism and communication tools set up to expand the capacity of teams to perform the tasks [33]. Because of the time zone distinction between living up to expectations hours no immediate coordination is conceivable. In any case, the coordination cost which identifies with repairing outcomes of misconception, modifying may increment [34]. As colleagues are from diverse socio-cultural backgrounds, the language and cultural training may be obliged to set up a compromised culture [19]. The commonality between the team members may be expanded to be at the dubious information, abilities and skills of developers from different sites [27].

3.3 Control

Control is the procedure of holding fast to objectives, arrangements, measures, or quality levels [3]. The geographical separation of GSD tasks causes trouble to pass on vision and technique for the development. Despite the fact that the time zone viability can be used, the management of project artefact may subject to defers [17]. The apparent dangers from ease choices influence the collaboration and can bargain the advantages of GSD [35]. The diverse view of authority or hierarchy between developers likewise influences the control life cycle of global development [18, 36].

4 Productivity

Software Productivity is the proportion between the measure of programming created to the work and cost of delivering it. Unlike lines of code and pages of documentation that are easy to count, program functionality does not have a characteristic unit of measure and is in this manner harder to evaluate. Software metrics that we can use as a quantifiable measure of different attributes of a software development process should be built up [37]. Software productivity gives the software managers and experts an arrangement of valuable, tangible data points for sizing, estimating, managing, and controlling software projects with meticulousness and exactness [38]. Software productivity can be measured customarily utilizing the lines of code or capacity focuses and the productivity is the LOC or FP created every hour by the software engineer [39].

The real ingredients that affect software development productivity are software product (an application), the development resources and processes, and environment [11]. In the GSD, productivity relies on upon different components, such as project delivery rate, team size and communication complexity. These characters will influence the productivity adversely or positively. Delivery rate, can be seen as the proportion of output/input. In software development, output is the software product and input is the exertion spent by the developers amid software production stages. The environment under which the production site controls the variability in data endeavours. The general effort of the whole activities incorporates requirement analysis, architecture and design, coding and unit testing, system and integration and cost of quality. Productivity of the group is the proportion of application size and aggregate effort. Productivity of the project based on the competency of engineers, related knowledge in the given innovation and the software development environment [11].

$$Effort = Project\ Size * Productivity \tag{1}$$

The project size can be measured utilizing Function Points. Productivity in number of hours per FP. So the effort computed in hours or person days. Higher delivery rate will build the productivity. The productivity of a project relies on tools, the experience of the team, and stability of the requirements. So the productivity measuring group ought to have the capacity to characterize and measure every one of the fixings so as to have the capacity to compute the delivery rate figures. Improvement of delivery rate measures the normal rate at which development project delivers functions to the client, which is measured in capacity focuses per elapsed month. These rate measurements are helpful in the choice of better sites to distribute the projects.

5 Discussion

In the Function Point methods Project delivery rate means how many Function Points can develop in a man hour. Function Point per man hour value varies depending on the nature of the company. The project should be distributed to the site having higher productivity. Productivity factor gets reduced with the increase in team size. The productivity factor is inversely proportional to the team size.

Design issues, coding or build issues, changes in codes and version changes are the common reasons occurrence in communication among team members. As the team size increases the communication channels between the team members exponentially increases, so communication complexity increases which adversely affect the productivity. This overhead should account, while estimating final effort for the project. Figure 1 shows the team size versus communication paths. For projects, the productivity factors depend on the technology platform on which the application is being developed, languages and database type whose values changes from product to product.

For the effort and duration estimation of the project, the functional size of the project will be combined with the estimated delivery rate and speed of delivery. Effort can be also calculated as size times the project delivery rate. The duration in months can be found by dividing the size by the speed of delivery. The size of the project can be estimated using the functional points. The speed of delivery varies for the sites depending on the developing application type, language and platform. The sites with higher speed of delivery should be selected for distributing the GSD projects so to increase efficiency. Figure 2 shows the comparison between delivery rate and productivity. If the delivery rate increases the productivity will also increase. The figure is plotted using the data set obtained from the guidelines to software measurement published by the management reporting committee of IFPUG.

The productivity factor varies from site to site for GSD projects. Productivity in sites varies with number of function point handle, communication complexity. The communication complexity varies based on the factors such as time zone differences, language barrier, and cultural diversity. Each factor is assigned with a complexity rating of low, medium or complex according to the characteristics of sites to be selected for distributing the project is proposed as shown in Table 1.

Fig. 1 Team size versus communication paths

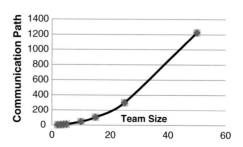

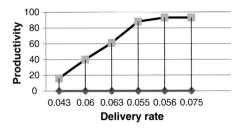

Fig. 2 Delivery rate versus productivity

Table 1 Rate factors for finding communication complexity of each site

Rate	Communication factors		
	Time zone difference	Language barrier	Cultural diversity
Low	Less than 2 h	Whole team members are native speakers	Domain knowledge is very high
Medium	Between 2 and 6 h	More than 50 % team members are native speakers	High socio-cultural distance, but share a common organizational culture
Complex	Greater than 6 h	More than 50 % team members are non-native speakers	High socio-cultural distance and different organizational cultures

The low, medium and high ratings take the numerical values 0, 1 and 2 respectively. The communication complexity of each site is defined by the sum of rate of factors as it increases with these factors.

$$\text{Communication Complexity}, C = \sum_{0}^{n} ai \qquad (2)$$

where a_i is the rate of factors such as time zone differences, language barrier, cultural diversity, etc. The a_i varies from 0 to n according to the factors affecting the communication complexity. The time zone difference becomes more complex as the geographical distance between the sites increases. The language barrier factor is complex when more than 50 % of team members are native speakers. The cultural diversity increases with high socio-cultural distance and different organizational culture. As communication complexity increase the productivity of each site decreases.

The selection of site for distributing the project depends on the productivity. The project distributor should assess the productivity index factor for each site depending on the team size, delivery rate and communication complexity. The project can be assigned to the site with greater productivity index factor. If the multiple distribution of the project is required select the sites in the order of decreasing productivity index factor.

6 Conclusion

The efficient distribution of projects in the global software development is crucial for the success of the projects. The challenges faced by the global software development environment affect the effort and productivity. The estimation of the project size also affects the accurate estimation of productivity. Globally distributed projects are affected by the productivity of the distributed sites which is influenced by the various factors such as communication complexity, team size and speed of delivery. The project distributes among the sites done depending on the productivity estimation of the sites.

References

1. Daman, D., Moitra, D.: Global software development how far we have come? IEEE Softw. 740–745 (2006)
2. Haq, S., Raza, M., Zia, A., et al.: Issues in global software development: a critical review. IJ. Softw. Eng. Appl. (20i, 4) 590–595
3. Carmel, E., Agarwal, R.: Tactical approaches for alleviating distance in global software development. IEEE Softw. **18**(2), 22–29 (2001)
4. Iqbal, A., Abbas, S.S.: Communication Risks and Best Practices in global software development. Master Thesis. MSE-2011-5406. Blekinge Institute of Technology, Sweden (2011)
5. Betz, S., Mäkiö, J., Stephan, R.: Offshoring of software development—methods and tools for risk management. In: International Conference on Global Software Engineering, IEEE (2007)
6. Fabriek, M., van den Brand, M., Brinkkemper, S.: Reasons for success and failure in offshore software development projects. In: Proceedings of ECIS, pp. 446–457 (2008)
7. Khan, S.U., Niazi, M., Ahmad, R.: Critical success factors for offshore software development outsourcing vend, ors: a systematic literature review. In: Fourth IEEE International Conference on Global Software Engineering, pp. 307–316 (2009)
8. Jim´enez, M., Piattini, M., Vizca´ıno, A.: Challenges and improvements in distributed software development: a systematic review. Hindawi Publishing Corporation Advances in Software Engineering, vol. 2009, Article ID 710971
9. Peixoto, C.E.L., Audy, J.L.N., Prikladnicki, R..: Effort estimation in global software development projects preliminary results from a survey. In: International Conference on Global Software Engineering, pp. 123–127. IEEE (2010)
10. Muhairat, M., Aldaajeh, S., Al-Qutaish, R.E.: the impact of global software development factors on effort estimation methods. Eur. J. Sci. Res. 46(2) 221–232 (2010). ISSN:1450-216X
11. Partthasarathy, M.A.: Practical Software Estimation -Function Point methods for Insourced and Outsourced Projects. Pearson (2007)
12. Marthaler, V.: International Function Point Users Group (IFPUG)—Function Point Counting Practices Manual. Release 4.1.1 (April 2000)
13. Kusumoto, S., Matukawa, F., Inoue, K.: estimating effort by use case points: method, tool and case study. In: Proceedings of the 10th International Symposium on Software Metrics (METRICS'04). IEEE (2000)
14. Damodaran, M., Washington, A.N.E.: Estimation using use case points. In: The Proceedings of the Information Systems Education Conference, vol. 19 (2002)
15. Pressman R.S.: Software Engineering: A Practitioner's Approach, 4th edn. The McGraw-Hill Companies, Inc. (1997)

16. Damian D., Lanubile F., Oppenheimer H.L.: Addressing the challenges of software industry globalization: the workshop on global software development. In: Proceedings 25th International Conference on Software Engineering, pp. 793–794. IEEE Computer Society, Los Alamitos (2003)
17. Usman, M., Dr. Azam, F., Hashmi, N.: Analysing and reducing risk factor in 3-C's model communication phase used in global software development. In: International Conference on Information Science and Applications (ICISA), pp. 1–4 (2014)
18. Ågerfalk, P.J., Fitzgerald, B., Holmström1, H., et al.: A framework for considering opportunities and threats in distributed software development. In: Proceedings of the International Workshop on Distributed Software Development Paris, 29 Aug 2005
19. Holmstrom, H., Fitzgerald, B., Agerfalk, P., Conchuir, E.O.: Agile practices reduce the distance in global software development. Inform. Syst. Manage. 23 (2006)
20. Holmstrom, H., Conchúir, E.O., Ågerfalk, P.J., et al.: Global software development challenges: a case study on temporal, geographical and socio-cultural distance. In: International Conference on Global Software Engineering (ICGSE2006), Costãodo Santinho, Florianópolis, Brazil (2006)
21. Franz, H.: The impact of computer mediated communication on information overload in distributed teams. System Sciences. HICSS-32. In: Proceedings of the 32nd Annual Hawaii International Conference, p. 15 (1999)
22. Bass, M.: Monitoring GSD projects via shared mental models: a suggested approach. In: Proceedings of the International Workshop on Global Software Development for the Practitioner, pp. 34–37 (2006)
23. Yasir Hassan, S., Mushtaq, R., Sani, U.: Communication issues in GSD. Int. J. Adv. Sci. Technol. 40, 69–76 (2012)
24. Eykelhoff, M.: Communication in global software development. A pilot study, In: 7th Twente Student Conference on IT, Enshedede, 2005
25. Manoj Ray, D.: Communication as an essential factor for effort estimation in GSD. In: Online Proceedings of Trends in Innovative Computing 2012—Intelligent Systems Design, pp. 131–135 (2012)
26. Khan, A.A., Basri, S., Fazel-e-Amen.: A survey based study on factors affecting communication in GSD. Res. J. Appl. Sci. Eng. Technol. 7(7) 1309–1317 (2014)
27. Battin, R.D., Crocker, R., Kreidler, J., et al.: Leveraging resources in global software development. IEEE Softw. 18(2), 70–77 (2001)
28. Carmel, E.: Global software teams: collaborating across borders and time zones. Prentice Hall PTR, Upper Saddle River (1999)
29. Herbsleb, J.D., Mockus, A.: An empirical study of speed and communication in globally distributed software development. IEEE Trans. Softw. Eng. pp. 481–494 (2003)
30. Grinter R.E., Herbsleb J.D. Perry D.E.: The geography of coordination: dealing with distance in R&D work. In: Proceedings on the ACM SIGGROUP Conference on International Conference on Supporting Group Work, pp. 306–315. ACM Press, New York (1999)
31. Sarker, S., Sahay, S.: Implication of time and space for distributed work.: an interpretive study of US–Norwegian systems development teams. Eur. J. Inform. Syst. 3–20 (2004)
32. Wu, S.: Overview of communication in global software development process. In: International Conference on Service Operations and Logistics, and Informatics (SOLI), pp. 474–478. IEEE (2012)
33. Cataldo, M., Bass, M. Herbsleb, J.D., et al.: On Coordination mechanisms in global software development. In: Proceedings of the International Conference on Global Software Engineering (2007)
34. Espinosa, J.A., Carmel, E.: The effect of time separation on coordination costs in global software teams: a dyad model. In: Proceedings of the 37th Annual Hawaii International Conference on System Sciences (HICSS'05), pp. 1–10. IEEE Computer Society (2004)
35. Kiel, L.: Experiences in distributed development a case study. In: International Workshop on Global Software Development: GSD 2003 (co-located with ICSE 2003) pp. 44–47

36. Krishna, S., Sahay, S., Walsham, G.: Managing cross-cultural issues in global software outsourcing. Commun. ACM **47**(4), 62–66 (2004)
37. Bijoy, B., Luchetski, J.: Software Metrics: Quantifying and Analysing Software for Total Quality Management. Systems Development Handbook (P. Tinnirello, Ed). 4th edn. CRC Press LLC (2000)
38. Jones. C.: Applied Software Measurement: Assuring Productivity and Quality. 2nd edn. McGraw Hill (1996)
39. Purna Sudhakara, G., Farooqband, A., Patnaik, S.: Measuring productivity of software development teams. Serbian. J. Manag. **7**(1), 65–75 (2012)

Prediction of Heart Disease Using Random Forest and Feature Subset Selection

M.A. Jabbar, B.L. Deekshatulu and Priti Chandra

Abstract Heart disease is a leading cause of death in the world. Heart disease is the number one killer in both urban and rural areas. Predicting the outcome of disease is the challenging task. Data mining can be can be used to automatically infer diagnostic rules and help specialists to make diagnosis process more reliable. Several data mining techniques are used by researchers to help health care professionals to predict the heart disease. Random forest is an ensemble and most accurate learning algorithm, suitable for medical applications. Chi square feature selection measure is used to evaluate between variables and determines whether they are correlated or not. In this paper, we propose a classification model which uses random forest and chi square to predict heart disease. We evaluate our approach on heart disease data sets. The experimental results demonstarte that our approach improve classification accuracy compared to other classification approaches, and the presented model can help health care professional for predicting heart disease.

Keywords Heart disease · Random forest · Data mining · Feature selection · Chi square

M.A. Jabbar (✉)
Muffakham Jah College of Engineering and Technology, Hyderabad, India
e-mail: jabbar.meerja@gmail.com

B.L. Deekshatulu
IDRBT, RBI, Hyderabad, India
e-mail: deekshatulu@hotmail.com

P. Chandra
ASL, DRDO, Hyderabad, India
e-mail: priti_murali@gmail.com

© Springer International Publishing Switzerland 2016
V. Snášel et al. (eds.), *Innovations in Bio-Inspired Computing and Applications*,
Advances in Intelligent Systems and Computing 424,
DOI 10.1007/978-3-319-28031-8_16

187

1   Introduction

Heart disease also called as coronary artery disease is a condition that affects the heart. Heart disease is a leading cause of death worldwide. Physicians generally make decisions by evaluating current test results of the patients. Previous decisions taken by other patients with the same conditions are also examined. So diagnosing heart disease requires experience and highly skilled physicians. Heart disease diagnosis is an important yet complicated task. The automation or decision support system would be extremely advantageous. Data mining can be used to automatically infer diagnostic rules and help specialists to make diagnosis process more reliable. Data mining have shown a good result in prediction of heart disease and is widely applied for prediction of heart disease.

Random forest is an ensemble classifier which combines bagging and random selection of features. Random forest can handle data without preprocessing. Random forest algorithm has been used in prediction and probability estimation.

Feature selection is a process of identifying and removing redundant and irrelevant features and increasing accuracy. Feature subset selection methods are classified into four types. (1) Embedded method (2) Wrapper method (3) Filter method (4) Hybrid method. Chi square test is used as a feature selection measure to determine the difference between expected frequency and observed frequency.

Major contributions of our paper are summarized as

(1) Propose a new method which employs the random forest algorithm for prediction of heart disease.
(2) Apply chi square feature selection algorithm to identify and remove irrelevant attributes.
(3) Apply feature selection algorithm (chi-square) to improve the accuracy in predicting heart disease.

The rest of the paper is organized as follows. Section 2 presents related work. We will review various articles related to heart disease. Section 3 deals with literature review. Section 4 presents proposed approach. Experimental results are discussed in Sect. 5. Finally Sect. 6 summarizes our paper.

2   Related Work

In this section, we will review some articles related to heart disease.

Polat et al. proposed hybrid method which uses fuzzy weighted preprocessing and artificial immune system [1]. Their proposed medical decision making method consists of two phases. In the first phase fuzzy weighted preprocessing is applied to heart disease data set to weight the input data. Artificial immune system is applied to classify the weighted input. They applied their methodology on Cleveland heart disease data set which consists of 13 attributes. The method uses 10 fold cross validation.

Diagnosis of heart disease through neural network ensembles was proposed by Das et al. [2]. Their method creates a new model by combining posterior probabilities from multiple predecessor models. They implemented the method with SAS base software on Cleveland heart disease data set and obtained 89.01 % accuracy.

P.K. Anooj developed a clinical decision support system to predict heart disease using fuzzy weighted approach. The method consists of two phases. First phase consists of generation of weighted fuzzy rules, and in second phase fuzzy rule based decision support system is developed. Author used attribute selection and attribute weight method to generate fuzzy weighted rules. Experiments were carried out on UCI repository and obtained accuracy of 57.85 % [3].

Robert Detrano et al. proposed probability algorithm for the diagnosis of coronary artery disease. The probabilities that resulted from the application of the Cleveland algorithm were compared with Bayesian algorithm. Their method obtained an accuracy of 77 % [4].

Decision tree for diagnosing heart disease patients was proposed by Shouman et al. [5]. Different types of decision trees are used for classification. The research involves data discretization, decision tree selection and reduced error pruning. Their method outperforms bagging and j48 decision tree. Their approach achieved 79.1 % accuracy.

Diagnosis of heart disease through bagging approach was proposed by Tu et al. [6]. The proposed bagging algorithm to identify warning signs of heart disease. They made a comparison with decision tree. Their approach claimed an accuracy of 81.4 %.

Andreeva used C4.5 decision tree for the diagnosis of heart disease. Feature extraction and specific rule inferring from heart disease data set considered. Their proposed approach achieved an accuracy of 75.73 % [7].

3 Literature Review

This section reviews literature used in this paper.

3.1 Heart Disease

Heart disease is a range of conditions that affect our heart. Heart disease also called as coronary heart disease (CHD), is a deposition of fats inside the tubes which supplies blood to the heart muscles. Heart disease actually starts as early as 18 years and patients only came to know about heart disease when the blockage exceeds about 70 %. Theses blockages develop over the years and lead to rupture of the membrane covering the blockage due to pressure increases. If the chemicals released by broken membrane mixed with blood and lead to a blood clot, results to heart disease [8].

The reasons which increase blockage are called as risk factors. These risk factors are classified as modifiable and non modifiable risk factors. Non modifiable risk factors are age, gender, and heredity. These risk factors can't be modified and they will always keep causing heart disease. Risk factors which can be changed by our efforts are called as modifiable risk factors. Some modifiable risk factors are (1) Food related (2) Habit related (3) Stress related (4) Bio chemical and miscellaneous risk factors.

Effective decision support system should be developed to help in tackling the menace of heart disease.

3.2 Random Forest (RF)

Random forest algorithm is one of the most effective ensemble classification approach. The RF algorithm has been used in prediction and probability estimation. RF consists of many decision trees. Each decision tree gives a vote that indicate the decision about class of the object. Random forest item was first proposed by Tin kam HO of bell labs in 1995. RF method combines bagging and random selection of features. There are three important tuning parameters in random forest (1) No. of trees (n tree) (2) Minimum node size (3) No. of features employed in splitting each node (4) No. of features employed in splitting each node for each tree (m try). Random forest algorithm advantages are listed below.

(1) Random forest algorithm is accurate ensemble learning algorithm.
(2) It produces highly accurate classifier for many data sets.
(3) Random forest runs efficiently for large data sets.
(4) It can handle hundreds of input variables.
(5) Random forest estimates which variables are important in classification.
(6) It can handle missing data.
(7) Random forest has methods for balancing error for class unbalanced data sets.
(8) Generated forests in this method can be saved for future reference [9]. Following algorithm illustrates random forest method.

Algorithm Random forest

　　Step 1: From the training set, select a new bootstrap sample.
　　Step 2: Grow on a un pruned tree on this bootstrap sample.
　　Step 3: Randomly select (m try) at each internal node and determine best split.
　　Step 4: if each tree is fully grown. Do not perform pruning.
　　Step 5: Output overall prediction as the majority vote from all the trees.

3.3 Chi-Square Method

Chi square is a statistical test that is used to measure divergence from the distribution of feature occurrence which is independent of the class value [10]. Chi square requires the following conditions to be satisfied

(1) Data must be quantitative
(2) One or more categories of data required
(3) Independent observations
(4) Sample size should be adequate and simple
(5) Data must be in frequency form
(6) All observations must be read.

Chi square formula used for the data sets is

$$X^2 = \sum \frac{(O - e)^2}{e}$$

where O is observed frequency and e is expected frequency. Thus chi square represents a summed normalized square deviation of the observed values from the corresponding expected values.

4 Proposed Method

The literature survey represents various techniques for prediction of heart disease. Each method has its own advantages and their short comings. The proposed technique uses random forest algorithm and chi square feature selection for prediction of heart disease. Feature subset selection is a process that selects a subset of original attributes and reduces feature space [11].

We applied Random forest with chi square as feature selection on heart disease data set collected from various corporate hospitals in Hyderabad (Heart disease data set T.S) and also on heart stalog data set. In our proposed work we used chi square to reduce number of attributes and keep only attributes which contribute more towards the diagnosis of heart disease.

Confusion matrix is used to compare actual classification of heart disease data set, with number of correct and incorrect predictions made by the model. The traditional classification matrix is shown below.

	Disease present	Disease absent
Test positive	10	7
Test negative	2	11

To evaluate the performance of our proposed model, we used following classification measures [12].

(1) Specificity = TN/(FP + TN)
(2) Sensitivity = TP/(TP + FN)
(3) Disease prevalence = (TP + FN)/(TP + FP + TN + FN)
(4) Positive predictive value: TP/(TP + FP)
(5) Negative Predictive value: TN/(FN + TN)
(6) Accuracy = (TP + TN)/(TP + FP + TN + FN)

where
TP => Positive tuples that are correctly labeled by the classifier
TN => Negative tuples that are correctly labeled by classifier
FN => Positive tuples that are incorrectly labeled by classifier
FP => Negative tuples that are incorrectly labeled by classifier

Attributes for our heart disease data set T.S are listed in Table 1.

Proposed algorithm:

Step 1: Load the heart disease data set
Step 2: Rank the features in descending order based on chi square value. A high value of chi square indicates feature is more related to class. Apply backward elimination algorithm. Back ward elimination algorithm starts from the full feature set, and iteratively removes one by one feature with low chi value. In each iteration only one feature is removed, which mostly affects overall model accuracy, as long as the accuracy stops increasing. Least rank feature will be pruned. Chi square is used to select high ranked features.
Step 3: Select the feature with highest value.
Step 4: Apply Random forest algorithm on the remaining features (features with high chi square) of the data set that maximizes the classification accuracy.
Steps 5: Find the accuracy of the classifier.

Table 1 Heart disease data set attributes

1	Age	Numeric
2	Gender	Nominal
3	BP	Numeric
4	Diabetic	Nominal
5	Height	Numeric
6	Weight	Numeric
7	BMI	Numeric
8	Hypertension	Nominal
9	Rural	Nominal
10	Urban	Nominal
11	Disease class	Nominal

Steps 1 to 4 deals with feature selection. High ranked features are selected for classification using chi square approach. From Step 3 to 4 RF classification will be applied to the selected feature subset. After applying classification, accuracy of the classifier will be calculated.

5 Experimental Results

To evaluate the performance of our approach, we used the measures listed in Sect. 4. Accuracy comparison of Heart Disease data set-Cleveland [13] is shown in Table 2 and Fig. 1. Naïve bayes approach obtained an accuracy of 78.56 %, whereas decision table obtained an accuracy of 82.43 %. The results are obtained using 10 cross validation. Our approach obtained 7.97 % improvement over C4.5 algorithm. Accuracy comparison for Heart Disease data set T.S is compared with Decision tree (DT) is shown in Table 3 and Fig. 2. Our approach obtained 100 % accuracy, where as DT obtained an accuracy of 98.66 %. Comparision of various parameters for heart disease data set T.S is listed Table 4 and Fig. 3. Specificity shows that the probability of testing the result of heart disease will be negative when the heart disease is not present. Positive predictive value (PPV) is the probability that the

Table 2 Accuracy comparison for heart disease data set (Cleveland data set)

Sl. no	Approach	Accuracy
1	PART C4.5	75.73
2	Naïve bayes	78.56
3	Decision table	82.43
4	Neural nets	82.77
5	Our approach	83.70

Fig. 1 Accuracy comparision of heart stalog data set by various approaches

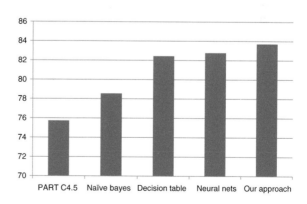

Table 3 Accuracy comparison for heart disease data set (TS Data set)

Sl. no	Approach	Accuracy
1	Decision Trees (DT)	98.66
2	Our approach	100

Fig. 2 Accuracy comparision
of heart disease data set-T.S

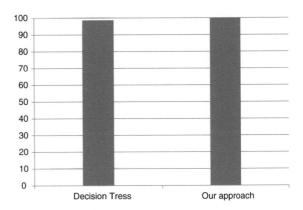

Table 4 Comparison of various parameters for heart disease data set T.S

Sl. no	Parameter	Our approach	Decision tree
1	Sensitivity	100	100
2	Specificity	100	92.86
3	Disease prevalence	82.67	81.33
4	Positive Predictive Value (PPV)	100	98.39
5	Negative Predictive Value (NPV)	100	100

Fig. 3 Comparision of
various parameters for heart
disease data set-T.S

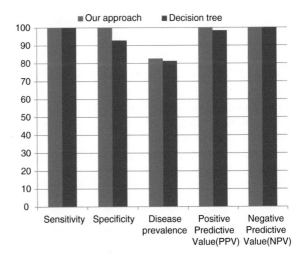

heart disease is present when the diagnosis test is positive. PPV value for DT is 98.39 % where as our approach records 100 %. Negative predictive value (NPV) recorded by our approach is 90 % and positive predictive value is 75.8 for heart stalog data set which is shown in Table 5 and Fig. 4. Clinically, the disease prevalence(DP) is the same as the probability of disease being present before the

Table 5 Comparison of various parameters for heart disease data set—HEART STALOG

Sl. no	Parameter	Our approach	Decision tree
1	Sensitivity	85.8	80.18
2	Specificity	82.3	80.50
3	Disease prevalence	39.2	41.11
4	Positive Predictive Value (PPV)	75.8	74.17
5	Negative Predictive Value (NPV)	90.0	85.33

Fig. 4 Comparision of various parameters for heart stalog data set

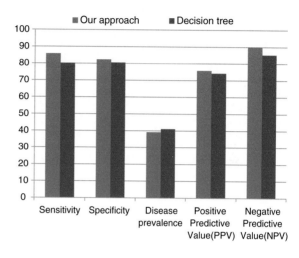

test is performed (prior probability of disease). The above experimental results suggests that our proposed approach efficiently achieve high degree of dimensionality reduction and improve accuracy with predominate features. Overall our approach outperforms other approaches. This indirectly helps patient's no. of diagnosis tests to be taken for prediction of heart disease.

Specificity of heart disease data set T.S obtained by our approach is 100 % where as it is 92.86 % by decision tree. Our approach obtained 1.8 % improvement over decision tree for heart stalog data set, which is shown in Figs. 3 and 4.

6 Conclusion

In this research paper, we presented an efficient approach for prediction of heart disease using Random forest. We adopted backward elimination method for feature selection using chi square measure for heart disease classification. Feature selection measure improves the classification accuracy. Our proposed approach (Random forest and Chi square) achieved an accuracy of 83.70 % for heart stalog data set. Applying Random forest has shown improved accuracy in prediction of heart disease. This research systematically tested using 10 fold cross validation to identify most accurate method. We compared our approach with other classification algorithms. Our approach outperforms other classification approaches for effective classification of heart disease. This type of research will play an important role to help health care professionals for prediction of heart disease.

References

1. Polat, K., Gunes, S., Tosun, S.: Diagnosis of heart disease using artificial immune recognition system and fuzzy weight preprocessing. Pattern Recognit. **39**, 2186–193 (2006)
2. Das, R., Turkoglu, I., Sengur, A.: Effective diagnosis of heart disease through network ensembles. Expert Syst. Appl. **36**, 7675–7680 (2009)
3. Anooj, P.K.: Clinical decision support system: risk level prediction of heart disease using weighted fuzzy rules. J. King Saud Univ. CIS, **24**, 27–40 (2012)
4. Detrano, R., Janosi, A., Stein burn, W., et al.: International application of new probability algorithm for the diagnosis of CAD. Am. J. Cardiol. **64**(5), 304–310 (1989)
5. Shouman, M., Turner, T., Stocker, R.: Using decision tree for diagnosing heart disease patients. In: 9th Australian Data Mining Conference, Australia, vol 121. ACM (2011)
6. Tu, M.C. et al.: Effective diagnosis of heart disease through bagging approach. In: Biomedical Engineering and Approach, pp. 1–4, BMEI 2009, IEEE (2009)
7. Andreeva: Data modeling and specific rule generation via data mining techniques. In: International Conference on Computer System and Technologies, Comsystech 2006, pp. 1–6 (2006)
8. Saaol times, Monthly magazine, Modifiable risk factors of heart disease, pp. 6–10, July (2015)
9. home.etf.rs/~vm/os/dmsw/Random%20Forest.pptx. Last Accessed 10 Aug 2015
10. Forman, G.: An extensive empirical study of feature selection metrics for text classification. J. Mach. Learn. Res. **3**, 1289–1305 (2003)
11. Sonwang, P., et al.: Computer network security based on SVM approach. In: 11th International Conference on Control, Automation, and Systems
12. Med Calc: www.medcalc.org. Last Accessed 5 Aug 2015
13. UCI machine learning repository: archive.ics.uci.edu/ml. Last Accessed 15 Aug 2015

Software Requirement Elicitation Using Natural Language Processing

Murali Mohanan and Philip Samuel

Abstract Software requirements are usually written in natural language or speech language which is asymmetric and irregular. This paper presents a suitable method for transforming user software requirement specifications (SRS) and business designs written in natural language into useful object oriented models. For sentence detection, tokenization, parts of speech tagging and parsing of requirement specifications we incorporate an open natural language processing (OpenNLP)tool. It provides very relevant parts of speech (POS) tags. This parts of speech tagging of the SRS is quite useful for further identification of object oriented elements like classes, objects, attributes, relationships etc. After obtaining the required and relative information, Semantic Business Vocabulary and Rules (SBVR) are applied to identify and to extract the object oriented elements from the requirement specification.

Keywords Requirement elicitation · Software requirement specification · OpenNLP · SBVR · Class model generation

1 Introduction

The major challenge in software design is the ability to comprehend tedious, long-drawn-out user requirements as outlined by the clients. Software analysis if done precisely saves a lot of time of the system analyst and the software design

M. Mohanan (✉)
Department of Computer Science, SOE,
Cochin University of Science and Technology, Kochi, India
e-mail: mohananmurali@gmail.com

P. Samuel
Division of Information Technology, SOE,
Cochin University of Science and Technology, Kochi, India
e-mail: philips@cusat.ac.in

© Springer International Publishing Switzerland 2016
V. Snášel et al. (eds.), *Innovations in Bio-Inspired Computing and Applications*,
Advances in Intelligent Systems and Computing 424,
DOI 10.1007/978-3-319-28031-8_17

197

phase can be started right away. In the field of information technology, there have been innumerable changes in the way this problem has been tackled. Though there are many traditional approaches which aim at recognising the functionalities of the information system, the modern object-oriented approach based on Natural Language Processing has garnered maximum popularity because of its strong role in object oriented modeling.

The natural language processing is a research area in which many researchers proposed several methods for analyzing the natural language (NL) requirements [1, 2]. Nan Zhou and Xiaohua Zhou (2004) proposed a methodology to generate object oriented model from the user requirement document. This approach used natural language processing to analyze the written requirements and domain based ontology is used to improve the class identification performance. The author used a linguistic pattern to differentiate the class and attributes, numeric pattern to analyze the relationship and parallel structure pattern to found more classes and its attributes. But it was not good enough in automatically identifying object oriented elements.

In this paper we have addressed the problem related to the software analysis and development phase. Here we use open natural language processing (OpenNLP) to process software requirement specification (SRS). The OpenNLP [3, 4] is used to produce parts of speech (POS) tag from the SRS which contains natural language statements. The POS tag captures the required details such as noun, verb, adverb, etc. of the natural language statements. Sentence splitting, tokenization and Pos tagging [5] are the phases of OpenNLP. These phases help to process the requirement specifications in NL which are easy to understand by both the user and the machine [6].

The recent trends of the software engineering largely depends on object oriented paradigm that widely use unified models. Unified Modelling Language (UML) is commonly used for modelling the user software requirements, documents the software assets, development and redevelopment of software [7]. Our research work proposes a methodology which is used to extract object oriented elements from NL processed SRS. Object oriented analysis applies the object oriented paradigm to model software information systems by defining classes, objects and relationships between them.

The UML model is an important component for Object Oriented analysis and design. The existing tools such as ReBuilder [8], CM-Builder [9], GOOAL [10], NL-OOML [11], UML-Generator [12] generates the UML class diagram automatically from the natural languages. The problem with these tools are they generate the object oriented models with lower accuracy due to the informal nature of NL and its ambiguity [13, 14].

This paper is focused on reducing the complexity in designing the object oriented models from the user requirements. User specifies their requirements in natural languages such as English. Using OpenNLP, user requirement statements are processed first. It tokenizes the input and then syntactically and semantically processes the input. Since the natural language processing is such a difficult task, our research work is divided into two phases to provide to efficient work. The initial phase is OpenNLP tool process and second phase is Semantic Business Vocabulary and Rules (SBVR) process [15].

2 Proposed Methodology

Aiming to give a suitable support for software developers as well as software engineers we have proposed a neoteric approach for natural language processing and object oriented modeling. This work is focused on natural language processing and then to extract useful object oriented elements. Software requirements are usually written in the natural language or speech language which is asymmetric and irregular. In our work the open natural language processing (NLP) analyzes the user requirements and provides the parts of speech (Pos) tagging. The SBVR process implemented here is used to extract object oriented elements like classes, objects, attributes, relationships etc. from the NL processed SRS. The SBVR process includes SBVR vocabulary extraction and rule generation. This can be further refined to form Unified Models which depict the major functionalities of a software system (Fig. 1).

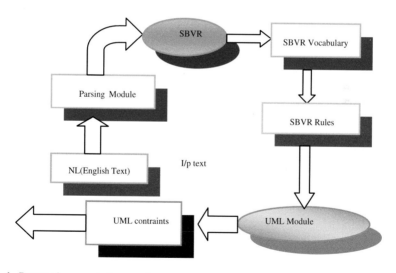

Fig. 1 Proposed approach framework

3   Open Natural Language Processing (OpenNLP)

The OpenNLP is a research area that aims to obtain how computer understands and process the natural language. The OpenNLP tool used in the proposed system understands the natural language as suggested by Deeptimahanti and Babar [16]. The OpenNLP starts to execute by extracting tokens from user requirement statements and then it proceeds to the syntax [3] and semantic analyzes [17] by parsing each and every sentence. The parser removes the stop words which has no information and also removes the function words such as *on, over, between, has, do* and generates the content word which has noun, adjectives, adverb and verb. Thus a Pos tagger [18] is generated.

The Apache OpenNLP library is a machine learning based toolkit for the processing of natural language text. The common NLP tasks, such as tokenization, sentence segmentation, part-of-speech tagging, named entity extraction, chunking, parsing, and co-reference resolution are done using this tool. OpenNLP project is developed to create a mature toolkit for the above mentioned NLP tasks.

3.1   General Library Structure

The Apache OpenNLP library contains several components, enabling one to build a full natural language processing pipeline. These components contain parts which can be enabled to execute the respective natural language processing tasks and to train them as a model and also to evaluate the model. Each of these facilities is accessible via its application program interface (API) [13].

3.2   Methods Used in Our Concept Are

Sentence Detection
The OpenNLP Sentence Detector can detect that a punctuation character marks the end of a sentence or not. In this sense a sentence is defined as the longest white space trimmed character sequence between two punctuation marks. The first and last sentence makes an exception to this rule. The first no whitespace character is assumed to be the begining of a sentence, and the last non whitespace character is assumed to be a sentence end.

Tokenization
The OpenNLP Tokenizer segments an input character sequence into tokens. Some of the tokens generated are punctuation, words, numbers, etc.

Input text

E.g. `Robert Clive,50yearsold,will join the Company as an executive chairman Jan.15.`
`Mr. Harry is President of Ford B.V., the German automobile company.`

Output is shown as individual tokens in a whitespace separated representation.

`Robert Clive, 50 years old, will join the Company as an executive chairman Jan. 15.`
`Mr. Harry is President of Ford B.V., the German automobile company.`

Tagging
The Part of Speech Tagger marks tokens with their corresponding word type based on the token itself and the context of the token. A token might have multiple pos tags depending on the token and the context. The OpenNLP POS Tagger uses a probability model to predict the correct pos tag out of the tag set. To limit the possible tags for a token a tag dictionary can be used which increases the tagging and runtime performance of the tagger.

`Robert Clive, 50 years old, will join the company as an executive chairman Jan. 15.`
`Mr. Harry is President of Ford B.V., the German automobile company.`

POS Tagger generates the following:

`Robert_NNPClive_NNP,_,  50_CD  years_NNSold_JJ,_,  will_`
`MDjoin_VBthe_DTCompany_NNas_INan_DTexecutive_JJchairman_N`
`NJan._NNP 15_CD._.`
`  Mr._NNP   Harry_NNPis_VBZPresident_NNof_INFord_NNPB.V.`
`_NNP,_, the_DTGerman_NNPautomobile_VBGcompany_NN`

Parsing
OpenNLP parsing can be done by training the API.

Input text E.g. `The slow white cat jumps over the quick rat.`
The parser output is.

`  (TOP (NP (NP (DT The) (JJ quick) (JJ white) (NN cat) (NNS`
`jumps)) (PP (IN over) (NP (DT the)`
`  (JJ slow) (NN rat))) (..)))`

4   SBVR

Semantic Business Vocabulary and Rules [19] is introduced by the Object Management Group (OMG) in 2008 for software and business people. It describes the desired vocabulary and rules for providing the semantic documentation of vocabulary, facts and rules of business. It provides a multilingual, unambiguous and rich capability of languages that are used by software designers and business people in various domains.

4.1   SBVR Vocabulary and Rules

Semantic Business Vocabulary and Rules generates the vocabulary and rules for a particular business domain. In our research work it helps in identifying various object oriented elements from natural language processed requirements specification. Thus SBVR provides a suitable way to capture the object oriented items from the requirement specification in NL [20]. In a software model SBVR vocabulary describes the specific software domain and SBVR rules describe the specific logic. The elements of the SBVR vocabulary are concepts and fact types. The concepts represent an entity of a specific domain. Object types, individual concept, verb concept, etc., are the types of concepts. The common nouns are referred to the noun

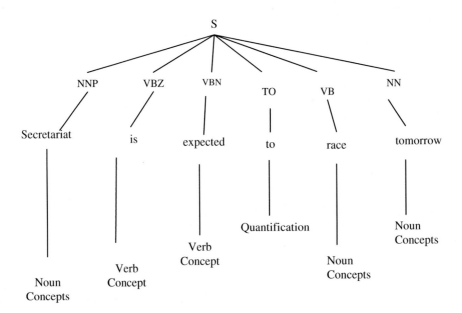

Fig. 2 Parse tree representation for SBVR vocabulary extraction

concepts, the proper nouns are considered as individual concepts, the auxiliary verbs and action verbs are the verb concepts.

The combination of noun concepts and verb concepts are the fact types in SBVR Vocabulary. The fact type which is represented in *is-property-of* relationship is considered as characteristics fact type which is extracted as suggested in [19]. Plural nouns (prefixed with s), articles (a, an, the) and cardinal numbers (2 or two) are considered as Quantification (Fig. 2).

The associative fact types are identified by the binary fact types in parts of speech (POS) tagging. *"The belt conveys the parts"* is an example sentence taken to understand the binary fact type. An association is there in the mentioned sentence between the words *belts* and *parts*. In SBVR model, association is mapped to associative fact types, aggregation is considered as partitive fact types and generalization is denoted as categorization fact types. SBVR elements are extracted from the output of the Open NLP.

5 Model Design

In this paper the UML model is considered as the business domain. In the UML class model, the noun concepts are class names and their respective attribute names and object names are denoted as individual concepts. Also operation names are considered as action verbs and the fact types are referred to as associations & generalizations.

To generate the UML model SBVR rule has to be extracted to analyze the specific software logic. The SBVR rule is based on any one of the fact types of SBVR vocabulary. Definitional rule and behavioral rule are the two types [21] of SBVR Rules. The definitional rule defines the organizational setup whereas behavioral rule describes the conduct of an entity. Semantic formulation, logical formulation, quantification and model formulation are processes to be performed to generate the SBVR rules from the fact type. The SBVR rule is constructed by applying the semantic formulation to each fact type.

6 SBVR Rules Generation

With regard to the scope of our project the SBVR rules are generated by the basic semantic formulations proposed in SBVR version 1.0 [19]. The semantic formulations considered here are logical formulation, quantification and modal formulation and are explained as follows. Figure 3 represents logical formulation.

From the extracted vocabulary the required tokens are identified to map the logical operators. Tokens such as *not, no* are considered as the logical operator negation ($\ulcorner$). That, and is denoted as conjunction ($\wedge$) similarly or is disjunction ($\vee$)

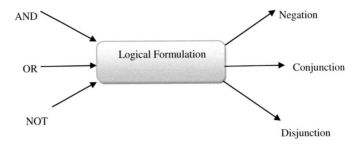

Fig. 3 Logical Formulation

and tokens like *infer, imply, indicate, suggest* are considered as the logical operator implication ($\rightarrow$).

Quantification mentions the scope of the concept and it is applied in this work by mapping the tokens as given below:-

More than, greater than $\rightarrow$ atleast n quantification
Less than $\rightarrow$ atmost n quantification and
Equal to, positive statement $\rightarrow$ n quantification

Figure 4 shows a sample quantification.
The modal formulation describes seriousness of a constraint. In SBVR two modal formulations are there, one is possibility formulation (PF) and the second is obligation formulation (OBF). The structural requirement is represented by the PF and the behavioral requirement is represented by the OBF. The model verbs mapping to these formulations are as shown below:-

Can, may – PF
Verb concept, should – OBF

Figure 5 shows modal formulation.

Fig. 4 Quantification

Fig. 5 Modal formulation

Structured English Notation is the final step in rule generation and it is performed as in SBVR 1.0 document, Annex C [19]. In this phase, notations are provided for generated tokens. For example the noun concepts are underlined e.g. Employee; the verb concepts are italicized e.g. *could be*; the SBVR keywords are bolded e.g. **at most**; the individual concepts are double underlined. Attributes are also italicized but with different colour.

7 Object Oriented Analysis of SBVR

The final step of the proposed work is the object oriented analysis from the SBVR rule to extract the object oriented elements such as classes, its attributes, objects, methods, generalization, aggregation and associations. This extraction procedure is as narrated in the following sentences. In the SBVR rule the generic entity is represented by the noun concept. On this basis the noun concept is mapped to classes. Similarly the particular entity is obtained from the individual concept so it is mapped to objects. The attributes of a class are obtained by all the characteristics of the noun concepts without action verb. The verb concepts (issue(), order()) are mapped with methods. Association is extracted by the unary relationship, binary relationship and multiplicity. In the SBVR rule following are some relations:

$$\text{Unary fact type} \rightarrow \text{Unary relationship}$$
$$\text{Associative fact type} \rightarrow \text{binary relationship}$$
$$\text{Quantification (noun concept)} \rightarrow \text{multiplicity.}$$

In extracting the generalization the partitive fact type is divided into subject-part and object part, where the subject-part is main class and the object-part is sub class in generalization. The categorization fact type in SBVR rule is considered as aggregation. Similar to the generalization extraction the categorization fact types are divided as subject part and object part and are maintained as main class and sub class respectively. SBVR rules describe the specific logic in the software domain.

8 Experiment and Result

A simple case study is taken for the proposed work and it is captured from the domain of a robot system. The following problem statement is first processed with the openNLP which is further processed by the SBVR process to identify object oriented items.

An assembly unit consists of a user, a belt, a vision system, a robot with two arms, and a tray for assembly. The user puts two kinds of parts, dish and cup, onto the belt. The belt conveys the parts towards the vision system. Whenever a part enters the sensor zone, the vision system senses it and informs the belt to stop immediately. The vision system then

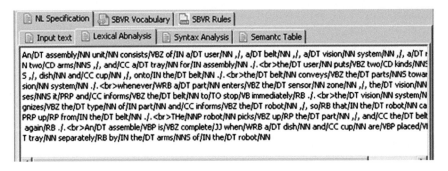

Fig. 6 Generated Pos Tagger

recognizes the type of part and informs the robot, so that the robot can pick it up from the belt. The robot picks up the part, and the belt moves again. An assembly is complete when a dish and cup are placed on the tray separately by the arms of the robot.

The problem statement is given as the input to our tool. Initially the openNLP processes these statements and provides the parts of speech. This output of the openNLP is shown in Fig. 6.

This tagger is further processed by the SBVR to identify object oriented items from the generated vocabulary.

9 Conclusion

This research work is carried out for providing a robust solution to reduce the ambiguity in natural language and to extract object oriented information from the user requirements. This is a neoteric approach which largely supports machine processing. An open natural language processing tool is implemented to analyze and to parse the SRS which provides the essential parts of speech (POS). After obtaining the required and relative information in the form of POS tags. With the support of Semantic Business Vocabulary and Rules relevant vocabulary and rules are formed. This provides an automatic method to capture the object oriented elements in requirements specification. The automatic extraction of object oriented items from the natural language processed user requirements is a novel concept. Software designers can further refine the gathered information and can develop solid object oriented models.

References

1. Arora, C., Sabetzadeh, M., Briand, L., Zimmer, F.: Automated checking of conformance to requirements templates using natural language processing. In: IEEE Transactions on Software Engineering. doi:10.1109/TSE.2015.2428709
2. Falessi, D., Cantone, G., Canfora, G.: Empirical principles and an industrial case study in retrieving equivalent requirements via natural language processing techniques. In: Software Engineering, IEEE Transactions on, vol. 39, no: 1, pp. 18–44 (2013)
3. Fernandez, P.M., Garcia-Serrano, A.M.: The role of knowledge-based technology in language applications development. In: Expert Systems with Applications, vol. 19, pp. 31–44 (2000)
4. Kok, S., Domingos, P.: Learning the structure of markov logic networks. In: Proceedings of the ICML-05, pp. 441–448. Bonn, Germany, ACM Press (2005)
5. Lane, P.C.R., Henderson, J.B.: Incremental syntactic parsing of natural language corpora with simple synchrony networks. IEEE Trans. Knowl. Data Eng. 13(2), 219–231 (2001)
6. Bajwa, I.S., Lee, M.G., Bordbar, B.: SBVR business rules generation from natural language specification. In: Artificial Intelligence for Business Agility—Spring Symposium (SS-11-03), pp. 2–8 (2011)
7. Perez-Gonzalez, H.G.: automatically generating object models from natural language analysis. In: 17th Annual ACM SIGPLAN Conference on Object-oriented Programming, Systems, Languages, and Applications, ACM, New York, USA, pp. 86–87 (2002)
8. Oliveira, A., Seco, N., Gomes, P.A.: CBR approach to text to class diagram translation. TCBR Workshop at the 8th European Conference on Case-Based Reasoning, Turkey (2006)
9. Harmain, H.M., Gaizauskas, R.: CM-Builder: A natural language-based CASE tool for object-oriented analysis. Autom. Softw. Eng. 10(2), 157–181 (2003)
10. Perez-Gonzalez, H.G., Kalita, J.K.: GOOAL: a graphic object oriented analysis laboratory. In: 17th Annual ACM SIGPLAN Conference on Object-oriented Programming, Systems, Languages, and Applications (OOPSLA'02), NY, USA, pp. 38–39 (2002)
11. Anandha, G.S., Uma, G.V.: Automatic construction of object oriented design models [UML Diagrams] from natural language requirements specification. PRICAI 2006: trends in artificial intelligence, LNCS 4099, pp. 1155–1159 (2006)
12. Bajwa, I.S., Samad, A., Mumtaz, S.: Object oriented software modeling using NLP based knowledge extraction. Eur. J. Sci. Res. 35(01), 22–33 (2009)
13. Li, K., Dewar, R.G., Pooley, R.J. Object-oriented analysis using natural language processing, linguistic analysis, pp. 75–76 (2005)
14. Mich, L.: Ambiguity identification and resolution in software development: a linguistic approach to improve the quality of systems. In: Proceedings of the 17th IEEE Workshop on Empirical Studies of Software Maintenance, Florence, Italy, pp. 75–76 (2001)
15. Feuto, P.B, Cardey, S, Greenfield, P.: Domain specific language based on the SBVR standard for expressing business rules. In: Enterprise Distributed Object Computing Conference Workshops (EDOCW), 17th IEEE International, pp. 31–38 (2013)
16. Deeptimahanti, D.K., Babar, M.A., An automated tool for generating UML models from natural language requirements. In: 24th IEEE/ACM International Conference on Automated Software Engineering ASE'09, pp. 680–682 (2009)
17. Dinarelli, M., Moschitti, A., Riccardi, G.: Discriminative reranking for spoken language understanding. In: Audio, Speech, and Language Processing, IEEE Transactions on, vol. 20, No: 2, pp. 526–539 (2012)
18. Toutanova, K., Manning, C. D.: Enriching the knowledge sources used in a maximum entropy part-of-speech tagger. In: Joint SIGDAT Conference on Empirical Methods in Natural Language Processing and Very Large Corpora, pp. 63–70 (2000)
19. OMG.: Semantics of Business vocabulary and Rules. (SBVR) Standard v.1.0. Object Management Group, (2008). http://www.omg.org/spec/SBVR/1.0

20. Bajwa, I.S., AsifNaeem, M.: On specifying requirements using a semantically controlled representation. In: 16th International Conference on Applications of Natural Languages to Information Systems. Springer, Alicante, Spain, pp. 217–220 (NLDB 2011)
21. Kleiner, M., Albert, P., Bézivin, J.: Parsing SBVR Based Controlled Languages. Model Driven Engineering Languages and Systems, pp. 122–136 (2009)
22. Zhou, N., Zhou, X.: Automatic acquisition of linguistic patterns for conceptual modeling. Course INFO 629: Concepts in Artificial Intelligence, Drexel University. Fall (2004)
23. Ambriola, V., Gervasi, V.: On the systematic analysis of natural language requirements with CIRCE. In: Automated Software Engineering, vol. 13, No. 1, pp. 107–167 (2006)
24. Priyanka, M., Rashmi, P.: Article: Published by Foundation of Computer Science, New York, USA. Generating UML diagrams from natural language specifications. Int. J. Appl. Inf. Syst. **1** (8), 19–23 (2012)

Application of Hexagonal Coordinate Systems for Searching the K-NN in 2D Space

Vojtěch Uher, Petr Gajdoš, Tomáš Ježowicz and Václav Snášel

Abstract Efficient searching of the k-nearest neighbors (k-NN) is a widely discussed problem. Most of the known 2D methods is based on division of a space to some quads or rectangular clusters. It is convenient for simple orthogonal querying of the space. However, a radius of neighbourhood is circular, thus the non complying quads have to be eliminated. This paper describes a novel approach of searching k-NN using hexagonal clustering of the 2D unordered point clouds. The hexagonal grid fully fills the 2D space as well. The shape of a hexagon is closer to the circular one and hexagonal coordinate systems are efficiently used to simply address the surrounding hexagons intersected by neighbourhood of a point. The paper contains performance tests of the proposed algorithm.

Keywords k-nearest neighbors · k-NN · Hexagon · Hexagonal grid · Hexagonal coordinate system · Point cloud

1 Related Work

The k-NN problem solves the searching of the k closest points from a query point in an unordered point cloud (PC). A condition of a maximum radius of the neighborhood can be added. The k-NN is an important part of many algorithms like clustering, classification and generally machine learning algorithms. It can be applied

V. Uher (✉) · P. Gajdoš (✉) · T. Ježowicz · V. Snášel
Department of Computer Science and National Supercomputing Center, VŠB-Technical University of Ostrava—Ostrava, Ostrava, Czech Republic
e-mail: vojtech.uher@vsb.cz

P. Gajdoš
e-mail: petr.gajdos@vsb.cz

T. Ježowicz
e-mail: tomas.jezowicz@vsb.cz

V. Snášel
e-mail: vaclav.snasel@vsb.cz

© Springer International Publishing Switzerland 2016
V. Snášel et al. (eds.), *Innovations in Bio-Inspired Computing and Applications*,
Advances in Intelligent Systems and Computing 424,
DOI 10.1007/978-3-319-28031-8_18

for data preprocessing or fitness evaluation for selected bio-inspired algorithms, triangulations or N-body simulations.s However, naive k-NN implementation is very time consuming, so a more efficient approach is required. There are many articles about searching k-NN in 2D, or generally in n-dimensional spaces. Many of them describe some space division based on an orthogonal grid. The orthogonal quads are then usually structured into some kind of trees [1] like e.g. Quad-tree [2, 3], BVH [4] or trees with different space partitioning like BSP-tree [2, 5] and k-d Tree [6, 7]. Voronoi diagram [1, 2] can be also used especially in 2D space. The problem of these methods is that the objective structures are slow in practice. Trees are sparse and thus unbalanced, so they have to be dynamically recalculated. A traversing of the structure leads to wasteful memory jumps and node checks. Another algorithms use some space filling curve (SFC) for efficient hashing of the points. SFC just connects the points which are close to each other [8]. The points are linearly stored in this order in a memory. The properties of SFCs are used in many k-NN methods and range queries [9–11].

The mentioned methods usually need a time consuming preprocessing and they are based on hierarchical structure traversal. The aim of this paper was to find a straightforward method for searching the k-NN to avoid unnecessary computations and memory accesses. We presented a similar method in [12] based on a level indexing of surrounding quads and used it for a parallel computation of neighborhood. The orthogonal grid can be easily computed and stored in a memory, but a neighborhood is circular, thus many quads have to be checked and eliminated. Instead of quads a hexagonal grid is used in this paper. Hexagonal grid can completely fill the 2D space just like quads (see Fig. 1). Each hexagon represents one cluster of points. Hexagonal grids are widely used in computer games, but it seems that there are not many articles about their use for solving k-NN in unordered point clouds. That is why many sources for our research were online guides from game developers, especially from Patel's websites [13, 14]. Grünbaum and Shephard wrote a exhausting summary of many different tiling patterns in [15] including hexagonal tilings. The paper [16]

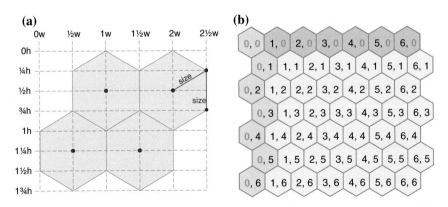

Fig. 1 **a** Pointy topped hexagonal grid and its measurements **b** Offset coordinates

presents kinetic data structure for the all nearest neighbors problem of a set of moving points in the plane based on hexagonal Delaunay graph. Hexagonal grids are also used often for clustering of network devices to get better signal coverage or power savings, for example [17, 18]. A basic hexagonal binding was described in [19] as a scatterplot matrix technique. The theory about hexagonal grids is described in the next section.

2 Background

2.1 Hexagonal Grids

There are many variants of hexagonal grids. The following paragraphs define the grid used for our k-NN method. Hexagonal grids and their properties were nicely described by Patel in [13, 14], thus the terminology declared in his articles will be used here. The schemas of hexagonal grids were inspired by Patel's visualizations as well.

The diagram in the Fig. 1a shows basic properties of the applied grid. The pointy topped variant of grid was used (see [13]). The picture describes width (w), height (h) and edge *size* of regular hexagons. Principal is the edge *size*. Whole grid is then computed according to this value. The height h and width w of hexagon are defined as

$$h = size * 2, \quad w = \frac{\sqrt{3}}{2} * h \tag{1}$$

The vertical (*vert*) and horizontal (*horiz*) distances between two adjacent hexes are

$$vert = h * \frac{3}{4}, \quad horiz = w \tag{2}$$

2.2 Hexagonal Coordinate Systems

There are multiple variants of hexagonal coordinates (see [13, 14] for more details and visualizations). The three basic ones are used in this paper and they are briefly introduced in the following subsections.

2.2.1 Offset Coordinates

The simplest coordinate system is the offset system shown in the Fig. 1b. It is the *odd-r horizontal* type according to [13]. Every second row is intended to make a complete hexagonal grid. The hexagons are addressed by row number l and col number c just like 2D array. However, addressing of surrounding hexagons is not straightforward.

(a)   **(b)**

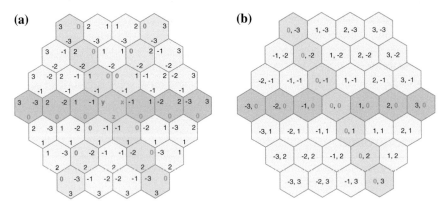

Fig. 2   **a** Cube coordinates **b** Axial coordinates

2.2.2   Cube Coordinates

The cube coordinates are adapted for simple moving to all six directions (see Fig. 2a). It is based on three primary axes x, y and z. Obviously, one axis is redundant here just to make addressing of surrounding hexagons easier. Cube coordinates are represented by traditional cube grid axes for a diagonal plane at $x + y + z = 0$. It means that each direction on the hexagonal grid is a combination of two directions on the cube grid. The cube coordinates are the reference coordinates in this paper. Most of hexagonal operations is computed over it. Other systems can be converted to this one and vice versa.

2.2.3   Axial Coordinates

The last system is very similar to the previous one. Axial coordinates practically use only two dimensions from the cube coordinates. If the condition $x + y + z = 0$ is met the third redundant dimension is simply neglected. From comparison of grids in Fig. 2 it is obvious that the axis y is the redundant one in this case. So, according to cube coordinates there are relations $r = z$ and $q = x$, where r is a row and q is a column addressing a hexagon in axial coordinate system.

2.2.4   Conversion of Coordinates

All three coordinate systems can be simply converted to each other. The equations are mentioned here for completeness just like they were defined in [13]. The x, y, z are cube coordinates, q, r are axial coordinates and l, c are offset coordinates defined before.

Cube to axial:

$$q = x, \quad r = z \tag{3}$$

Axial to cube:

$$x = q, \quad z = r, \quad y = -x - z \tag{4}$$

Cube to offset (odd-r):

$$c = x + \frac{z - z \& 1}{2}, \quad l = z \tag{5}$$

Offset (odd-r) to cube:

$$x = c - \frac{l - l \& 1}{2}, \quad z = l, \quad y = -x - z \tag{6}$$

2.2.5 Point Localization in Hexagonal Grid

The crucial task is computation of corresponding hexagon which contains a specified point. It is possible to compute a center point of hexagon according to its coordinates and properties of the hexagonal grid described in Sect. 2.1. A conversion from 2D point to hexagonal coordinates is then inversion of this process. This conversion to cube coordinates can also be found in [13] and the corresponding equations are

$$x = \frac{x_p * \frac{\sqrt{3}}{3} - y_p * \frac{1}{3}}{size}, \quad z = \frac{y_p * \frac{2}{3}}{size}, \quad y = -x - z, \tag{7}$$

where x_p and y_p are coordinates of a point from 2D point cloud and $size$ is the edge size defined in Sect. 2.1. However, the resulting cube coordinates are doubles, thus they have to be rounded then to the nearest integers with respect to the constraint $x + y + z = 0$ (see [13]).

3 Our Approach

The presented algorithm searches maximally the k nearest neighbors within a defined radius. It was designed to be maximally straightforward and to be potentially parallelized. A point cloud is usually a set of unordered points. This set has to be divided to clusters, which represent a generalized location of contained points. Hence the described hexagonal grids are involved to address the clusters. The reference system uses the cube coordinates, because they offer an easy walk over surrounding hexagons (see Fig. 2a). All coordinate systems can be converted to each other if necessary. Clustering of the space based on a regular hexagonal grid may lead to empty clusters, thus a reasonable searching structure has to be defined. The convenient point hashing and a way of searching neighbors are the main contributions of this paper. The parts of algorithm are described in the following subsections.

3.1 Initialization Phase

A point cloud of N points has to be loaded to memory. It is necessary to compute a bounding box (BB) of the whole point cloud by a simple search of min/max coordinates. Then the sizes of the bounding box ($xSize, ySize$) are found out. Next, a regular hexagonal grid has to be defined. From the essence of regular grid, only the edge *size* of hexagon (Sect. 2.1) has to be declared. The other measurements are computed from this size. The exact *size* is obtained as a rate of BB width, thus the computations over grid are independent on a specific point cloud. The point clouds may have many different arrangements, so the bounding box can also have arbitrary sizes. However, the hexagonal grid can fill a rectangular box too. Unlike in the case of Octrees, the regular grid is not deformed by an aspect ration of BB. The origin of the coordinate system is at top left corner of the grid just like it was symbolized in Fig. 1b. The number of hexagons for each row (*colNum*) and the number of rows (*rowNum*) of the grid are computed as

$$colNum = round(xSize/w) + 1, \quad rowNum = round(ySize/vert) + 1, \qquad (8)$$

where *vert* is the vertical distance between two adjacent hexes and w is the width of hexagon (Eqs. (1) and (2)).

The radius r defines the circle area, which has to be checked for nearby points. The explicit length should be defined relatively to BB as well. The hexagonal neighborhood consists of R hexagonal rings as it is shown in the Fig. 3a. The R is defined as a number of hexagonal rings hit by radius r and it can be simply computed according to hexagon width. An advantage of the hexagon is that it is geometrically close to a circle. A ring has a hexagonal shape as well. Hence, it is possible to simply decide whether a hexagon or a whole ring lies within the radius or not. The points of inner hexes can be considered as neighbors automatically. Then only outer rings have to be checked hex by hex to find the inner points. The pattern of hexagonal neighborhood is the same for all query points, so the number of inner rings can be computed in advance and reused many times.

3.2 Building of Searching Structure

The structure represents a searching core for the presented k-NN. As it was mentioned in the Sect. 1, some space trees are generally used for searching k-NN. However, the problem consists in unbalanced sparse trees and memory alignment. Moreover, the proposed algorithm is supposed to be designed for further parallelization.

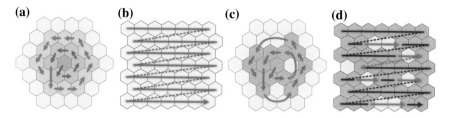

(a)   **(b)**   **(c)**   **(d)**

Fig. 3 **a** Indexation of hexagonal neighborhood **b** Cluster memory alignment **c** Sparse neighborhood **d** Sparse grid memory alignment

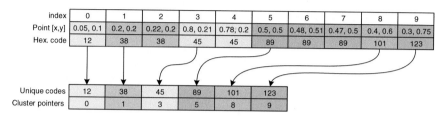

index	0	1	2	3	4	5	6	7	8	9
Point [x,y]	0.05, 0.1	0.2, 0.2	0.22, 0.2	0.8, 0.21	0.78, 0.2	0.5, 0.5	0.48, 0.51	0.47, 0.5	0.4, 0.6	0.3, 0.75
Hex. code	12	38	38	45	45	89	89	89	101	123

Unique codes	12	38	45	89	101	123
Cluster pointers	0	1	3	5	8	9

Fig. 4 Data representation of point cloud and searching structure

This structure consists of all non empty clusters linearly stored in a memory in the order defined by rows of grid (see Fig. 3b). Each hexagon represents a space cluster of points. However, some clusters are empty and thus they corrupt the ideal indexation of 2D grid. The simplified sparse structure is based on the idea of space filling curves (e.g. [12, 20]). The hexagonal coordinates of clusters are converted to hash codes defining their cluster order on the SFC (Fig. 4). This could be simply done by computation of hexagon indexes in the linearized 2D array representing the grid (see Fig. 3d). A SFC hash *code* is computed from 2D offset coordinates by

$$code = l * colNum + c. \tag{9}$$

The edge *size* of hexagon is defined. The structure is build as follows:

1. Cube coordinates are computed for all points of the point cloud (Eq. 7) and then converted to offset ones (Eq. 5)
2. A hash code is computed (Eq. 9) for all offset coordinates from previous step and stored to memory
3. The points and their hash codes are then sorted according to the codes, thus the points from the same cluster are stored one next to each
4. The algorithm goes through all generated hash codes, detects their differences and then stores unique codes with corresponding cluster pointers to the structure

The Fig. 4 shows the memory representation of exemplary normalized point cloud. The structure now contains only existing clusters ordered according to their hash codes. Each cluster is represented by a pair of code and pointer to the first cluster point in the point array. Both, the simplified structure and the point array are perfectly aligned in the memory. Then a specific cluster can be found by binary search of the corresponding hash code.

3.3 Searching the Nearest Neighbors

The algorithm uses the searching structure to address hexagons of neighborhood. The neighborhood is defined by precomputed pattern of R hexagonal rings as it was explained in Sect. 3.1. A query point p has to be localized in hexagonal grid (Eq. 7). Hexagons of neighborhood are then checked one by one in the manner shown in the Fig. 3a. Existence of a hexagon in the structure is verified by binary search of the corresponding hash code. Points of found clusters are then checked sequentially. The structure contains only non empty clusters, thus the empty ones are skipped. The illustration of sparse neighborhood is in the Fig. 3c. The shifts on the hexagonal grid are simply done by recursive incrementation or decrementation of hexagonal cube coordinates as it is illustrated in the Fig. 2a. The structure solves problem with addressing of the sparse grid.

The number of neighbors k, query point p and neighborhood pattern of R rings are declared. The searching algorithm goes as follows:

1. Cube coordinates of query point p are computed (Eq. 7), they represent a central hexagon of the neighborhood
2. All hexagons of neighborhood are browsed in the order explained in Fig. 3a or c
3. A hash code is computed (Eqs. 5 and 9) for each hexagon of the neighborhood and then it is binary searched in the prepared structure. The existing clusters are returned
4. Points of found clusters are returned according to a pointer addressing the first cluster point in the point cloud
5. A number of cluster points is obtained from the searching structure as a pointer difference of two adjacent clusters
6. Points from clusters corresponding to the inner hexagons of the query radius are considered to be neighbors automatically. Points from other clusters have to be checked sequentially
7. This process is repeated until all hexagons of R neighboring rings are checked
8. The complete set of found neighbors is sorted according to their distances from p and the k best ones within the radius r are selected as result

The described algorithm searches maximally the k nearest neighbors of query point p within the defined radius. If the radius is big enough, precisely the k best will be returned. This method can be applied for estimation of neighbors as well as for exact results. A faster estimate can be obtained e.g. if the first k points found within the radius r are straightforwardly returned as result (sorting is omitted).

The whole process is repeated for all query points from the point cloud, but the searching structure is build only once and used many times. This algorithm is very fast and it is prepared for parallelization on GPU as it is presented in Sect. 4. It filters out all wasteful clusters of points and it browses only relevant points.

4 Experiments and Discussion

This section describes some performance tests of presented k-NN algorithm. The performance was measured on 4 types of randomized point clouds shown in Fig. 5. The measured times are in seconds and they represent exact k-NN computation times for all points of a point cloud including structure building time and sorting of neighboring points. Every computation was done ten times and the final computation times were averaged. The graph in the Fig. 6 shows computation times for point cloud sizes from 10^4 to 10^7 points and $k = 32$. It seems that there are no significant differences

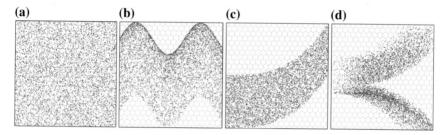

Fig. 5 Four types of randomized point distributions

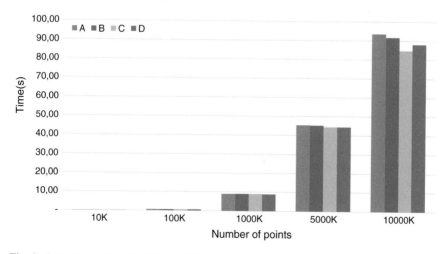

Fig. 6 Computation times for different PC sizes and datasets (**A–D**)

Fig. 7 Numbers of
computed distances for
different k

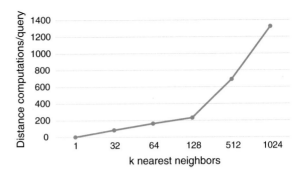

between point distributions A–D. The graph also shows that thousands of points can be solved in less than one second, the neighbors of 10^7 points were solved in about 1.5 min. For comparison, the brute force method computes k-NN of 10^5 points about 17 min. The searching structure building takes usually about $\frac{1}{30}$ of total computation time. The edge size BB rate of the hexagonal grid was set from 0.001 to 0.05 according to a number of points. The greater the point number is the gentler the grid should be.

The second test was focused on the number of point to point distance computations. The Fig. 7 represents average numbers of distances computed per one k-NN query for different k. The distance test was performed in an uniformly distributed point cloud (Fig. 5a) with $2 * 10^6$ points. The grid *size* was uniformly set to 0.001 of BB width. The graph shows that the number of sequentially checked points was strongly reduced. However, this number varies according to the grid density, because the algorithm computes distances of all points found in hexagonal clusters of defined R rings. Even better results can be obtained for a convenient grid settings of a specific point cloud. A study about effects of the algorithm input parameters is a matter of our next research.

The paper [3] monitors a performance of Quadtree and R*-Tree. The 2D k-NN problem is also discussed. These two algorithms are very common for searching the nearest neighbors. The proposed algorithm described in this paper is able to compete them in the computation times as well as in the number of required distance computations and it is even faster. In comparison to them, this method is very straightforward. A reasonable clustering leads to balanced query times without any sophisticated objective structure. The searching structure can be build very quickly and its memory is well aligned. The memory utilized by the algorithm consists of N 2D points, their codes and structure array of code-pointer pairs representing the existing clusters (Fig. 4). The found neighbors can be organized in many ways in memory. The worst case is allocation of $k * N$ point indices to handle all neighborhoods, but also buffer of k point indices is sufficient if the found neighbors are involved to related computations immediately. Unlike methods described in [3], this algorithm is well prepared for parallelization on the GPU using e.g. CUDA architecture. Every thread can process one neighborhood and store the found cluster pointers to a queue. The points of found clusters are then checked in parallel.

All experiments run on the following hardware: Intel Core i5 760 @ 2.8 GHz, 16 GB RAM, Windows 7 64-bit.

5   Conclusion

A novel approach for k-nearest neighbors in 2D unordered point clouds was introduced. A clustering based on the regular hexagonal grid was used for building the searching structure. This structure conveniently solves the difficulties with sparse spaces. The algorithm utilizes a straightforward logic of hexagonal grid to address neighboring clusters. A query to the searching structure is managed by binary search. The properties and performance of the algorithm were discussed and it showed that the exact neighbors can be obtained very quickly. The method was compared with the most common algorithms (Quadtree and R*-Tree) too. The novel approach can compete them in the computation times and it is even faster. Its main advantage consists in a convenient design for further parallelization. Next research will be focused on the parallelization on the GPU and testing of the algorithm on real datasets. A study about algorithm settings for different point cloud should be done as well.

Acknowledgments This work was supported by the IT4Innovations Centre of Excellence project (CZ.1.05/1.1.00/02.0070), funded by the European Regional Development Fund and the national budget of the Czech Republic via the Research and Development for Innovations Operational Programme and by Project SP2015/146 "Parallel processing of Big data 2" of the Student Grand System, VŠB—Technical University of Ostrava.

References

1. Tsaparas, P.: Nearest neighbor search in multidimensional spaces, Technical Report (1999)
2. Berg, M.D., Cheong, O., Kreveld, M.V., Overmars, M.: Computational Geometry: Algorithms and Applications, 3rd ed. Santa Clara, CA, USA: Springer TELOS (2008)
3. Kim, Y.J., Patel, J.: Performance comparison of the r^*-tree and the quadtree for knn and distance join queries. IEEE Trans. Knowl. Data Eng. **22**(7), 1014–1027 (2010)
4. Yin, M., Li, S.: Fast bvh construction and refit for ray tracing of dynamic scenes. Multimed. Tools Appl. **72**(2), 1823–1839 (2014)
5. Havran, V., Kopal, T., Bittner, J., Žára, J.: Fast robust bsp tree traversal algorithm for ray tracing. J. Graph. Tools **2**(4), 15–23 (1998)
6. Sample, N., Haines, M., Arnold, M., Purcell, T., Purcell, T.: Optimizing search strategies in k-d trees (2001)
7. Kybic, J., Vnučko, I.: Approximate all nearest neighbor search for high dimensional entropy estimation for image registration. Signal Process. **92**(5), 1302–1316 (2012)
8. Lawder, J.: The application of space-filling curves to the storage and retrieval of multidimensional data (2000)

9. Gajdos, P., Jezowicz, T., Uher, V., Dohnalek, P.: A parallel fruchtermanreingold algorithm optimized for fast visualization of large graphs and swarms of data. Swarm Evolutionary Comput. (2015)
10. Skopal, T., Krátký, M., Pokorný, J., Snášel, V.: A new range query algorithm for universal b-trees. Inf. Syst. **31**(6), 489–511 (2006)
11. Connor, M., Kumar, P.: Fast construction of k-nearest neighbor graphs for point clouds. IEEE Trans. Vis. Comput. Graph. **16**(4), 599–608 (2010)
12. Uher, V., Gajdos, P., Jezowicz, T.: Solving nearest neighbors problem on gpu to speed up the fruchterman-reingold graph layout algorithm. In: 2015 IEEE 2nd International Conference on Cybernetics (CYBCONF), pp. 305–310, June 2015
13. Patel, A.J.: Red blob games—hexagonal grids. http://www.redblobgames.com/grids/hexagons/. Accessed Sep 2015
14. Patel, A.J.: Amit's thoughts on grids. http://www-cs-students.stanford.edu/amitp/game-programming/grids/. Accessed Sep 2015
15. Grünbaum, B., Shephard, G.: Tilings and Patterns. ser. Dover Books on Mathematics Series. Dover Publications, New York (2013)
16. Rahmati, Z., King, V., Whitesides, S.: Kinetic data structures for all nearest neighbors and closest pair in the plane. In: Proceedings of the Twenty-ninth Annual Symposium on Computational Geometry, ser. SoCG '13, pp. 137–144. ACM, New York, NY, USA (2013)
17. Wang, D., Xu, L., Peng, J., Robila, S.: Subdividing hexagon-clustered wireless sensor networks for power-efficiency. In: Proceedings of the 2009 WRI International Conference on Communications and Mobile Computing, ser. CMC '09, vol. 02, pp. 454–458. IEEE Computer Society, Washington, DC, USA (2009)
18. Salzmann, J., Behnke, R., Timmermann, D.: Hex-mascle x2013; hexagon based clustering with self healing abilities. In: Wireless Communications and Networking Conference (WCNC), 2011 IEEE, pp. 528–533, March 2011
19. Carr, D.B., Littlefield, R.J., Nichloson, W.L.: Scatterplot matrix techniques for large n. In: Proceedings of the Seventeenth Symposium on the Interface of Computer Sciences and Statistics on Computer Science and Statistics, pp. 297–306. Elsevier North-Holland Inc, New York, NY, USA (1986)
20. Chen, H.-L., Chang, Y.-I.: Neighbor-finding based on space-filling curves. Inf. Syst. **30**(3), 205–226 (2005)

Locality Aware MapReduce

Reema Rhine and Nikhila T. Bhuvan

Abstract The large amount of data produced need to be processed properly. This can be done using Apache Hadoop, which is an open source software library. HDFS and MapReduce are the two core components of Apache Hadoop. The overall performance can be increased by improving the performance of MapReduce. A Locality Aware MapReduce idea is introduced here. It includes an input splitting strategy and also a MapReduce scheduling algorithm. Both of them are based on the locality of data. The input splitting, called the Improved Input Splitting, which works based on locality, clusters data blocks from a same node into the same single split, so that it is processed by one map task. In the scheduling algorithm, to assign tasks to a node, local map tasks are always preferred over non-local map tasks, no matter which job a task belongs to. That is, here the algorithm performs scheduling by checking for a local data when a free slot is available. Non-local data is always given a second preference. Since the scheduling is done based on locality it is called Locality Aware Scheduling. Each of these methods, when executed separately and combined showed a better performance than the one without any modification.

Keywords Locality · MapReduce · Scheduling

1 Introduction

Big data [1] is a buzzword used to describe a massive volume of both structured and unstructured data that is so large it is difficult to process using traditional database and software techniques. It refers to the huge amount of data that keeps on growing exponentially with time. The source of this big data includes social networking sites,

R. Rhine (✉) · N.T. Bhuvan
Department of Information Technology, Rajagiri School of Engineering
and Technology, Cochin, Kerala, India
e-mail: reema092@gmail.com

N.T. Bhuvan
e-mail: nikhilatb@rajagiritech.ac.in

© Springer International Publishing Switzerland 2016
V. Snášel et al. (eds.), *Innovations in Bio-Inspired Computing and Applications*,
Advances in Intelligent Systems and Computing 424,
DOI 10.1007/978-3-319-28031-8_19

ecommerce data, photos, videos etc. This large amount of data needs to be processed properly so that all the important information can be extracted from it for later use. Earlier, organisations lacked a proper system which could handle the processing of this huge data. Now, they use Hadoop to process this huge amount of data.

Apache Hadoop is an open source software library. It is a framework which is capable of distributed processing of data. It consists of two main components, HDFS and MapReduce. The HDFS handles the storage of data in Hadoop and the MapReduce deals with the processing of the data.

The performance of MapReduce improves as data transmissions in the MapReduce cluster decreases. The reduced data transmissions reduces the network traffic and the execution time. There are several studies going on for locality optimisation in MapReduce. LEEN [2] is a locality-aware partition to reduce the data transfers at the shuffle phase. Delay scheduling [3] also gave importance to locality while scheduling.

This paper introduces a technique to improve the performance of MapReduce. It is called the Locality Aware MapReduce. It includes an input data splitting method called the Improved Input Data Splitting and a scheduling algorithm based on locality called the Locality Aware Scheduling algorithm.

The Improved Input Splitting divides the input to map phase in such a way that the data belonging to a node is partitioned into the same split. It avoids the non-local reading of data. The Locality Aware Scheduling performs the scheduling of different tasks based on locality. This scheduling algorithm gives priority to local tasks first. The non-local task is given second preference. In this paper, Sect. 2 describes the different components of Hadoop, Sect. 3 explains about the implementation details of the Locality Aware MapReduce, Sect. 4 gives the evaluation results.

2   Background

Hadoop [4] consist of two main components, the Hadoop Distributed File System (HDFS) [5] and the MapReduce [6].

2.1   HDFS

Hadoop Distributed File System handles the storage of data in hadoop. It has a master slave architecture. It has a single master called the NameNode and several slaves called the Data Node. The data to be stored is divided and stored in different slaves. The master holds the metadata information of the data stored in the slaves. The metadata has the storage details. So in order to access a data, a client needs to get the metadata information first. So the NameNode is accessed first to get the metadata of the concerned data. Once the client acquires this information, it can directly connect the DataNode to access the data stored. The data to be stored in the DataNodes are divided into several blocks. They are also replicated according to the replication factor specified for fault tolerance.

2.2 MapReduce

MapReduce handles the programming aspects of Hadoop. MapReduce consist of two main phases, the Map phase and the Reduce phase. The Map phase is responsible for the mapping of data which includes the sorting and filtering of data. The Reduce phase performs the reduction operation and gives the final output. In between the Map and Reduce phase lies the Shuffle phase. The output of the Map phase is given to the Shuffle phase. The Shuffle phase shuffles the data and forms the input for the Reduce phase. The input to the Reduce phase is a key and all its associated records. JobTracker is the central coordinator or MapReduce process. It is present in the Master. All the DataNodes contain a TaskTracker which performs the processing of data as per the instructions from the JobTracker.

2.3 Input Splitting

A split [7] is the input processed by a single map task. So the splitting of data occurs before the map phase. The splits are further divided into records which are key value pairs. The length of the split is measured in bytes. Every input split has a storage location. A spilt has no input data but a reference to the data. The inputs to the MapReduce is handled by the InputFormat class. The getSplits function in the InputFormat class performs the splitting of data.

2.4 Scheduling

There are several scheduling [8] techniques for MapReduce. A good MapReduce scheduler [9] must avoid unnecessary data transmissions. By default, the First In First Out scheduling is used. Here when the slave nodes with empty map slots send heartbeat messages, the JobTracker searches for local task in the first job in the job queue. If there is no local task available in the first job, a non-local task is assigned to the empty map slot. Here, the strict job order is followed. The second job in the job queue is considered only after the completion of all the tasks of the first job even though there is no local map task available for the first job and plenty of them in the second job. So even though it takes locality into account, the strict job order affects the performance.

3 Implementation

The Locality Aware MapReduce implementation consists of an Improved Input Splitting method and a Locality Aware Scheduling algorithm. Both these methods function based on locality.

Fig. 1 The process

The process is shown in Fig. 1. The input first goes through the Improved Input Splitting phase. Then in the map phase the Locality Aware Scheduling occurs. After the map phase comes the reduce phase to give the final output.

3.1 Improved Input Splitting method

The splitting of data occurs before the Map phase. The splits [10] form the input to the Map phase. Using the default procedure of the getSplits function, the splitting of data may occur in such a way that the data belonging to different nodes may be partitioned into a single split. If partitioned in this manner, in order to do a map task, the split formed requires the non-local reading of data. This increases the network traffic and also the overall execution time.

The Improved Input Splitting method performs the splitting by considering the locality factor. The splits are forms in such a way that the data belonging to the same node forms the single split. Thus, when this split is accessed for the map task, only the local reading of data is required. This avoids the unnecessary data transmissions and hence, decreases the network traffic and the execution time. The algorithm for the Improved Input Splitting is given below.

Algorithm 1 Improved Input Splitting Algorithm

1: BlockList = Read the File Meta Data;
2: BLOCKNUMBER = Get the Block Number;
3: SLOTNUM = Get the Slots Number;
4: SplitMemNum = BLOCKNUMBER/SLOTNUM;
5: **for** each block in BlockList **do**
6:     ReplicaList = Find the Replica;
7: **end for**
8: **for** each replica in ReplicaList **do**
9:     **for** every Array i in SplitList **do**
10:         **if** replica.location == SplitList[i].location **then**
11:             Add the List as ArrayMember;
12:         **end if**
13:     **end for**
14:     NewArray(SplitList);
15: **end for**
16: Return(SplitList);

The DataNodes always send heartbeat messages to the NameNode. These messages helps to identify the status of the DataNodes. The node utilization details provided by the JobTracker helps to choose the appropriate data replica. The location of the data replica helps to choose the appropriate input data for the map task. If a particular replica is chosen, then, all the data of that node is given to the local map task. This avoids the non-local reading of data.

Metadata information is obtained from the NameNode. SplitList is an array list which contains the blocks of the same partition and SplitMemNum is the arrays in this list. Proper replica for each block is chosen. After this they are clustered in partitions based on locality. This is done by checking the location of the replica. If it exists in the SplitList, it is added to the location partition otherwise it is inserted in a new list into SplitList of this new location.

3.2 Locality Aware Scheduling

The Locality Aware scheduling algorithm gives priority to local tasks first irrespective of the job it belongs to. Non-local task is only a second preference. This algorithm gives every slave node a fair chance to grab local tasks before any non-local tasks are assigned to any slave node. The Locality Aware Scheduling Algorithm is given below.

Algorithm 2 Locality Aware Scheduling Algorithm

1: **for** each node i of the N slave nodes **do**
2:     set LocalityMarker[i]=null;
3: **end for**
4: Upon receiving a heartbat from node i:
5: **while** node i has free slots, i.e., its free slot count s is greater than 0 **do**
6:     set previousMarker=LocalityMarker[i]
7:     **for** each job j in the JobQueue **do**
8:         **if** job j has an unassigned local task t **then**
9:             assign t to node i
10:             s=s-1;
11:             **if** LocalityMarker[i]==null **then**
12:                 LocalityMarker[i]=1
13:             **else** LocalityMarker[i]+=1
14:             **end if**
15:             Break For
16:         **else** Continue
17:         **end if**
18:     **end for**
19:     **if** previousMarker==LocalityMarker[i] **then**
20:         LocalityMarker[i]=0;
21:         Break While
22:     **else** LocalityMarker[i]==0
23:         assign node i a non-local task t from the first job in the JobQueue
24:         s=s-1;
25:         Break While
26:     **end if**
27: **end while**

When a slave node has empty slot, it sends heartbeat messages. On getting the heartbeat message, the JobTracker checks the jobs in the job queue. If any job has a map task whose input is stored in that slave node, it is assigned to it. This algorithm relaxes the job order for finding the tasks. If local map task is not found in the first job, the scheduler checks the succeeding jobs for local task.

In this algorithm, if a node fails to get a local task for the first time, no task is assigned to it. Since it is likely that all the slaves with empty slots have given the heartbeat message and once been considered for local task assignment, if a node fails to find a local task for the second time, a non- local task is assigned to it. This algorithm also makes sure that at most one non-local task is assigned during a heartbeat interval.

Here, in this algorithm, a locality marker [11] marks the status of each node. If no job in the queue has local map task available, the algorithm decides whether or not to assign non-local map task depending on the marked value. Also the locality markers are cleared when a new job is added to the queue.

4   Results

The platform used for evaluation was Hadoop 0.21 version. A multinode cluster was setup with one master and three slaves. Eclipse indigo was used as the programming tool. WordCount and WordMean programs were used for evaluation.

Inputs of different sizes were used for the evaluation. Comparison was done to see the performance of simple method without any improvement, using the Improved Input Splitting method, using the Locality Aware Scheduling algorithm and using the Improved Input Splitting method and the Locality Aware Scheduling method combined.

The execution of the simple WordCount and WordMean programs are shown in Figs. 2 and 3 respectively. Here the time taken for the execution of the simple Word-Count program is 24,285 ms and for the simple WordMean program is 22,239 ms.

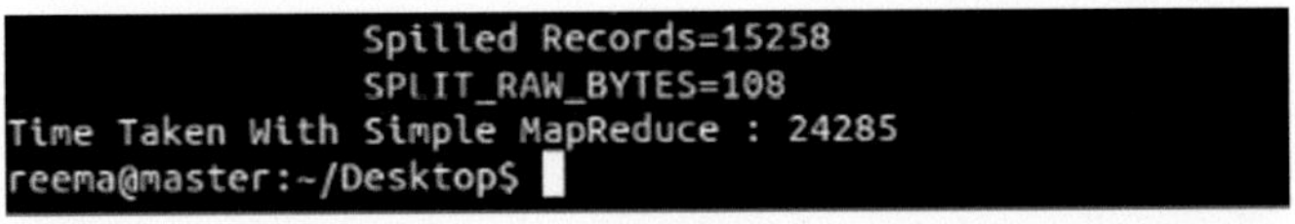

Fig. 2   Execution of the simple WordCount program

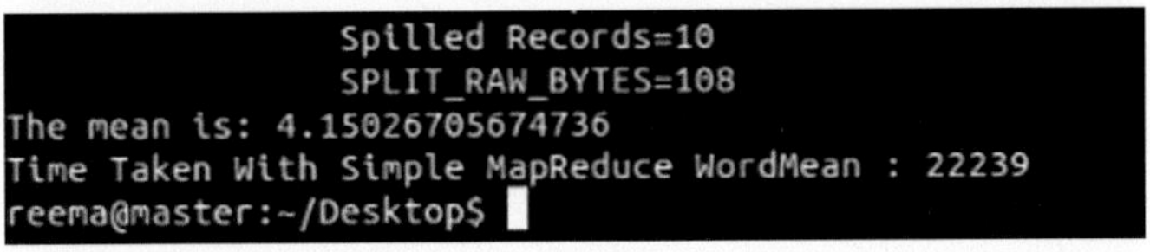

Fig. 3   Execution of the simple WordMean program

```
            Spilled Records=15258
            SPLIT_RAW_BYTES=108
Time Taken with Data Partitioning algorithm and scheduling algorithm together: 22932
The Replication factor for the system is 2
 will be able to split data between :2 nodes

reema@master:~/Desktop$
```

Fig. 4 Execution of the WordCount program with the Improved Input Splitting method and the Locality Aware Scheduling method combined

```
            Spilled Records=10
            SPLIT_RAW_BYTES=108
The mean is: 4.15026705674736
Time Taken with Data Partitioning algorithm and scheduling algorithm together : 19137
The Replication factor for the system is 2
 will be able to split data between :2 nodes

reema@master:~/Desktop$
```

Fig. 5 Execution of the WordMean program with the Improved Input Splitting method and the Locality Aware Scheduling method combined

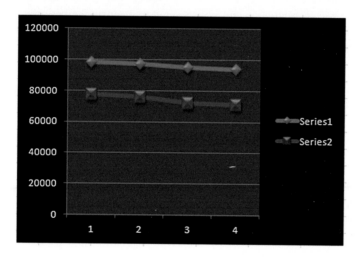

Fig. 6 Comparison with input size 227 MB

Figures 4 and 5 respectively shows the execution of the WordCount and Word-Mean programs using the Improved Input Splitting method and the Locality Aware Scheduling method combined. Here the WordCount program completes its execution in 22,932 ms and the WordMean program in 19,137 ms.

Given below is the comparison graph showing the performance of WordCount and WordMean with input size 227 MB. The x axis have values 1, 2, 3 and 4 which represent simple method without any improvement, using Improved Input Splitting method, using the Locality Aware Scheduling algorithm and using the Improved Input Splitting method and the Locality Aware Scheduling method combined respectively. The y axis represents time in ms (Fig. 6).

From the comparison graph it is seen that both the Locality Aware Scheduling method and the Improved Input Splitting Method performs better than the simple method without any improvement. Also it is found that the Locality Aware Scheduling combined with the Improved input splitting mechanism shows the best performance as it took the least amount of time.

5 Conclusion

MapReduce is a powerful platform for large-scale data processing. Here a Locality Aware MapReduce is introduced. It consists of two methods, Improved Input Splitting Method and the Locality Aware Scheduling method. Improved Input Splitting,which works based on locality, clusters data blocks from a same node into the same single partition, so that it is processed by one map task. Locality Aware Scheduling method performs scheduling by checking for a local data when a free slot is available. A multinode cluster was setup to evaluate the performance using the WordCount and WordMean programs. It was found that the Locality Aware Map Reduce method performs better than the default method.

References

1. Big Data, http://www.webopedia.com/TERM/B/big-data.html
2. Ibrahim, S., Jin, H., Lu, L., et al.: LEEN: Locality/fairness-aware key partitioning for mapreduce in the cloud. In: IEEE Second International Conference on Cloud Computing Technology and Science (CloudCom), pp. 17–24 (2010)
3. Zaharia, M., Borthakur, D., Sen Sarma, J., et al.: Delay scheduling: a simple technique for achieving locality and fairness in cluster scheduling. In: Proceedings of the 5th European Conference on Computer Systems, 2010, pp. 265–278 (2010)
4. Apache Hadoop, http://hadoop.apache.org
5. HDFS, http://hadoop.apache.org/hdfs/
6. Dean, J., Ghemawat, S.: MapReduce: simplified data processing on large clusters. Commun. ACM **51**(1), 107–113 (2008)
7. Understanding MapReduce Input Split Sizes and MapR-FS Chunk Sizes, https://www.mapr.com/blog/understanding-mapreduce-input-split-sizes-and-mapr-fs-chunk-sizes
8. Thomas, L., Syama R.: Survey on MapReduce scheduling algorithms. Int. J. Comput. Appl. **95**(23), 9–13 (2014)
9. Kc, K., Anyanwu, K.: Scheduling hadoop jobs to meet deadlines. In: 2nd IEEE International Conference on Cloud Computing Technology and Science (CloudCom), pp. 388–392 (2010)
10. Wang, C., Wu, Q., Tan, Y., Wang, W., Wu, Q.: Locality based data partitioning in MapReduce,: Computational Science and Engineering (CSE), 2013. In: 16th IEEE International Conference, pp. 1310–1317, 3–5 Dec 2013
11. He, C., Lu, Y., Swanson, D.: Matchmaking: a new mapreduce scheduling technique. In: IEEE Third International Conference on Cloud Computing Technology and Science (CloudCom), pp. 40–47, 29 Nov–1 Dec 2011

Hybrid Feature Selection Using Correlation Coefficient and Particle Swarm Optimization on Microarray Gene Expression Data

Arunkumar Chinnaswamy and Ramakrishnan Srinivasan

Abstract Diagnosis of cancer is one of the most emerging clinical applications in microarray gene expression data. However, cancer classification on microarray gene expression data still remains a difficult problem. The main reason for this is the significantly large number of genes present relatively compared to the number of available training samples. In this paper, we propose a hybrid feature selection approach that combines the correlation coefficient with particle swarm optimization. The process of feature selection and classification is performed on three multi-class datasets namely Lymphoma, MLL and SRBCT. After the process of feature selection is performed, the selected genes are subjected to Extreme Learning Machines Classifier. Experimental results show that the proposed hybrid approach reduces the number of effective levels of gene expression and obtains higher classification accuracy and uses fewer features compared to the same experiment performed using the traditional tree-based classifiers like J48, random forest, random trees, decision stump and genetic algorithm as well.

Keywords Feature selection · Correlation coefficient · Particle swarm optimization · Extreme learning machine

A. Chinnaswamy (✉)
Department of Computer Science and Engineering, Amrita School of Engineering,
Amrita Nagar, Coimbatore 641112, Tamilnadu, India
e-mail: arunkumar.chinnaswamy@gmail.com

R. Srinivasan
Department of Information Technology, Dr. Mahalingam College of Engineering
and Technology, Pollachi 642003, Tamilnadu, India
e-mail: ram_f77@yahoo.com

© Springer International Publishing Switzerland 2016
V. Snášel et al. (eds.), *Innovations in Bio-Inspired Computing and Applications*,
Advances in Intelligent Systems and Computing 424,
DOI 10.1007/978-3-319-28031-8_20

1 Introduction

Hybridization method is used to generate DNA samples in microarray gene expression data. This process can be done in two ways. In the first method, during the hybridization process, messenger RNA (mRNA) is stained using matrices sample taken from tissue or blood stream becomes cDNA. RNA profiling can be noisy and may not be sampled unevenly over time [1]. The second method is the Affymetrix chips are hybridized using oligonucleotides on the surface of the array chip. It is possible to monitor and simultaneous measure thousands of activation levels of gene expression in a single experiment. This is considered as the key advantage of DNA microarray technology. Protein production helps identify the different types of memberships. This is achieved because the gene expression level refers to a specific protein production gene [2]. The clinical medicine progress is only possible because of valuable results produced by microarray experiments performed on a variety of issues of gene expression profile. Microarray data can be applied to the problems of classification of cancer also. This was a recent development in the field of clinical research. Microarray data on cancer DNA is combined with statistical techniques for analyzing gene expression profiles to identify potential biomarkers for diagnosis and treatment of various types of cancer [3].

Statistical analysis of differentially expressed genes helps to assign them to different classes. This process improves the understanding of basic biological processes in the system. Using the concept of technology of microarray gene expression, it is possible to study the simultaneous activity of thousands of genes. The relative abundance of mRNA in the gene can be found by using gene expression profiles [4]. Results obtained represent the state of the cell. Discriminant analysis of microarray data is an excellent tool for medical diagnostics of diseases, treatment and prevention. The main purpose of the classification is to build an effective model that can identify differentially expressed genes and could also be used to identify classes in the unknown samples [5]. Some of the challenges in the microarray data are the smallest number of training and testing data available, the higher dimensionality of the data and the variations that could sneak in experiments performed to estimate the levels of gene expression. The two main tasks in the analysis of gene expression microarray are feature Selection and classification. To perform the classification process with an acceptable level of accuracy, the process of feature selection becomes crucial. Microarray gene expression data contains hundreds of thousands of genes or feature information. Only a small subset of genes exhibit strong correlation between them. Feature selection is a process that effectively selects differentially expressed genes in the dataset and forms a new subset for efficient classification. There may be situations in which a low-ranked gene could perform well in the rankings and a critical gene could be left out in the selection of functions [6]. The prediction accuracy would increase only with the best method of feature selection which otherwise would be impossible to understand. Another important measure is to avoid overfitting and build faster and cost effective models [7]. In this study, we use a hybrid approach that combines the benefits of a filter and a wrapper to perform feature selection. They are easy to use, simple and computationally efficient [8].

2 Materials and Methods

There are three different categories of Feature Selection. They are Supervised, Unsupervised and semi-supervised [9]. Also Feature subset selection methods in machine learning algorithms can be classified into four types namely filter approach, wrapper approach, embedded approach and hybrid approach [10]. There are several methods available in literature to perform the process of Feature Selection namely Filter Approaches that do not require a learning algorithm for a predictive model [11], Wrapper Approaches that are tailor-made for a particular classifier [12, 13], regularized least squares, branch and bound algorithms and Support Vector Machines. The Filter approach is independent of the learning algorithm and hence it is easy to implement and performs faster computation. Once the feature selection is done, it can be used as input to the different classifiers. Wrapper approaches are computationally intensive as it needs to evaluate each of the features to perform the classification. Various feature ranking and feature selection techniques have been proposed such as Correlation-based Feature Selection (CFS), Principal Component Analysis (PCA), Gain Ratio (GR) attribute evaluation, Chi-square Feature Evaluation, Fast Correlation-based Feature selection (FCBF), Information gain, Euclidean distance, i-test and Markov blanket filter. Some of the above filter methods do not perform feature selection. Instead they perform feature ranking and hence they need to be combined with suitable search methods to estimate the number of attributes required for classification. Such filters are often used with forward selection, which considers only additions to the feature subset, backward elimination, bi-directional search, best-first search, genetic search and other methods [14].

2.1 Correlation Coefficient

A correlation based heuristic evaluation function is used to compute the correlation coefficient using the Correlation-based Feature Selection (CFS). It overcomes the disadvantage of univariate filter approaches that does not take into account the interaction between features [15, 16]. The identification ability of each of the attributes is used to evaluate a subset of attributes. A multivariate approach is effective in identifying the correlation that exists among the different genes in the dataset [17]. Pearsons correlation coefficient is very sensitive to the presence of outliers and noise [18]. The relationship between variables (Genes) can be measured by the process of correlation [2]. The linear relationship between two variables in best described in statistics using correlation coefficient or Pearson Product Moment Correlation (PPMC).

Formula for calculating Pearson correlation between features x_i and y_i is given in (1)

$$Correlation = \sum (x_i - mean(x_i) * y_{i\_} mean(y_i) / n * SD(x_i) * SD(y_i)) \qquad (1)$$

Pearson correlation coefficient between attributes is found out. Attributes having low inter-correlation are selected [19]. The WEKA tool is used to implement CFS. The selected genes were used to study the different types of cancer. The attributes exhibit high correlation if the value of correlation coefficient lies between 0.5 and 1 and is said to be less correlated if its value lies between 0.3 and 0.5 [27].

2.2 Particle Swarm Optimization

Particle Swarm Optimization (PSO) is a stochastic evolutionary algorithm based on a population, on the basis of adequate socio-psychological principles to solve engineering problems based on several variables. Swarm Intelligence is embedded in this method. It involves the concept of sharing of information that is simulated by bio inspired behavior. It allows particles to acquire benefits based on discoveries and previous experience, while looking for food. For applications based on PSO, each particle that flies through the search space represents a candidate solution.

The pseudo code for Particle Swarm Optimization [20] is given as below:

```
for each particle i ∈ 0.1,....,s do
Initialize position xᵢ
Initialize position vᵢ
Set pᵢ = xᵢ
end for
repeat
for each particle i ∈ 0.1,....,s do
evaluate the fitness function for each particle i, f(xᵢ)
evaluate personal best (pbest) and global best (gbest)
for each dimension j ∈ 1, . . ., d do
calculate new velocity vᵢ,ⱼ(Δt + 1) using Eq. (4)
calculate new position xᵢ(Δt + 1) using Eq. (5)
end loop
end loop
until some convergence criteria is satisfied.
```

The position of a particle is biased by the best position visited using their own knowledge and position of the particle considered by a better knowledge of neighboring particles. When the neighborhood is a swarm of particles, the particle is said to be the world's best particles. The global optimum is measured by a fitness function which varies as a function of the optimization problem [20]. Each particle in the swarm is represented by the following uniqueness:

x_i: current position of the ith particle,
v_i: current velocity of the ith particle,
p_i: best previous position of the ith particle,
$gbest$: global best particle in its neighborhood.

The personal best position of particle i is the best position experienced by the particle so far. If f is the objective function, the personal best of a particle, at time step Δt is calculated as:

$$p_i(\Delta t + 1) = \begin{cases} p_i(\Delta t) & if\, f(x_i(\Delta t + 1)) > = f(p_i(\Delta t)) \\ x_i(\Delta t + 1) & if\, f(x_i(\Delta t + 1)) < f(p_i(\Delta t)) \end{cases} \qquad (2)$$

If gbest denotes the global best particle, it is given as:

$$gbest(\Delta t)\epsilon\{p_0, p_1, \dots p_s\} = min\{f(p_0(\Delta t)), f(p_1(\Delta t)), \dots f(p_s(\Delta t)) \qquad (3)$$

where s is the size of the entire swarm.

The numbers of particles are initialized at random locations that correspond to feature subsets and then swarm towards promising areas via the global best solution so far and each particle's local best. The smallest subset with maximum quality is returned.

3   Experimental Framework

3.1   Preprocessing

Microarray gene expression data suffers from the problems of missing values due to several experimental reasons. The lymphoma dataset used for our study suffers from this problem. In order to solve this issue, preprocessing is performed on the raw dataset using the impute method. In this case, the missing values are treated using the 'mode' statistical operation wherein the missing values are filled with the value that occurs more often in the dataset. This imputed data is then subjected to feature selection and classification to achieve better classifier accuracy.

3.2   Hybrid Approach to Feature Selection

The most common challenge in bioinformatics is in the process of selecting relevant and non redundant genes from the dataset. Complex biological problems can only be solved by predicting and classifying the genes in the most efficient way.

Microarray data finds its applications in the areas of cancer classification, disease diagnosis, prediction and treatment and most importantly in the area of gene identification that would be used in drug development at later stages. However, the problem of classification is time consuming because of the fact that the sample size is very small and the dimensionality of the data is very large. The process of feature selection performed before classification reduces the running time and also increases the accuracy of prediction. Lot of research is carried out in predicting the essential features (feature selection) before the classification process and therefore increases the accuracy of prediction. The process of gene selection is based on two key aspects namely reduction in the total number of features without reduction in predictive accuracy and to select key genes or features that exhibit close relationship among one another. The predictive accuracy is improved since Feature selection uses relatively fewer features since only selective features need to be used in disease diagnosis and treatment.

In this study, we used a hybrid approach to select feature genes in microarrays, and used three different feature selection algorithms to evaluate the performance of the proposed method. The filter model part correlation-based feature selection (CFS) is used to evaluate the ability of each feature which differentiates between different categories. The reasoning behind this method is that it can calculate the importance of each feature with respect to the class. The value of c_1 and c_2 are initialized to 1 and 2 respectively. The initial number of particles is assumed to be 100. The maximum number of generations is set to 50. Correlation based feature selection is used as an attribute evaluator and the particle swarm optimization is used as a search strategy. We used Weka software package to implement the proposed hybrid method to select suitable gene subset and then classification is performed on the selected gene subset.

3.3 Extreme Learning Machines Classifier

A feedforward neural network has slower learning speed compared to a conventional support vector machine. This becomes a bottleneck when feedforward neural networks are deployed. The two key reasons are the iterative tuning of all the parameters of the networks by using such learning algorithms and the usage of slow gradient based learning algorithms to train the neural networks. On the contrary, Support Vector Machines are most widely used to solve binary classification problems because of their outstanding classification capability. The most common form of Support Vector Machines deployed are the Least Squares—Support Vector Machines (LS-SVM) [21, 27].

Extreme Learning Machines (ELM) utilizes the concept of generalized single hidden layer feedforward neural networks (SHFFN). The hidden layer in this network does not require any tuning. This concept is employed in Radial Basis Function networks, polynomial functions, Support Vector Machines and conventional single hidden layer and multi hidden layer feedforward neural networks. The unique feature that distinguishes ELM from the other methods is that ELM does not require the tuning of the generalized single hidden layer except predefined network architecture and the neurons are generated randomly whereas the other techniques does require the tuning of the hidden layers [22, 27].

The uniqueness of ELM is that the target functions and the training and testing dataset are independent of the hidden node parameters. In theory, all the parameters of the ELM are determined using analytical approaches rather than tuning. In practice, to increase the efficiency of the real world applications, the output weights are determined using approaches such as iterativeness, non-iterativeness, incremental or non-incremental implementations [23]. The hidden node parameters possess universal approximation and separation capability. This is possible because of the independent nature of the hidden node that is not only independent of the training set but also among them. These hidden nodes and their relative mappings are called as ELM Random nodes/neurons/features. The weights of the ELM based classifiers can be computed to determine the decision boundary and it does not require the storage of the training dataset [24]. ELM stands out from the conventional classification algorithms since ELM generates the hidden node parameters without any knowledge of the training set whereas the conventional algorithms require the knowledge of the training dataset in advance. A drawback of the ELM is that a single ELM is unstable. Hence researchers are working on to create an ELM ensemble in order to improve the stability of the classifier [27].

An ELM employs the Moore-Penrose generalized inverse technique of a weight matrix and then estimates by simple leastsquares method. The simple ELM training algorithm is as follows:

If there are N samples (x_j, t_j), where $x_j = [x_{i1}, x_{i2}, ..., x_{in}]^T \in R^n$ and $t_j = [t_{i1}, t_{i2}, ..., t_{in}]^T \in R^n$, then the standard SLFN with N hidden neurons and activation function $g(x)$ is defined in (4) as

$$\sum \beta_i g(w_i \cdot x_j + b_i) = 0_j, \quad j = 1, \ldots, N \tag{4}$$

where $w_j = [w_{i1}, w_{i2}, ..., w_{in}]T$ is the weight vector that connects the ith hidden neuron and the input neurons, $\beta_j = [\beta_{i1}, \beta_{i2}, ..., \beta_{in}]T$ is the weight vector that connects the ith neuron and the output neurons, and b_i is the threshold of the ith hidden neuron. The "." in the $w_i \cdot x_j$ is the inner product of w_i and x_j.

Equation (6) can be mathematically expressed as in (5) below

$$\sum \beta_i g(w_i \cdot x_j + b_i) = t_j, \quad j = 1, \ldots, N, \ i = 1, \ldots, N \tag{5}$$

Equation 7 can also be expressed in matrix format as $H\beta = T$. This is depicted as shown below in a matrix format.

$$H(w_1, \ldots, w_l, b_1, \ldots, b_l, x_1, \ldots, x_N) = \begin{pmatrix} w_1 \cdot x_1 + b_1 \ldots \ldots w_1 \cdot x_1 + b_l \\ w_1 \cdot x_N + b_1 \ldots \ldots w_1 \cdot x_N + b_l \end{pmatrix}$$

The matrix H is called the hidden layer output matrix of the neural network. If the number of neurons in the hidden layer is equal to the number of samples, then H is square and invertible. If the number of neurons in the hidden layer does not match with the total number of samples, the system of equation needs to be solved numerically by using the formula shown in (6).

$$\min_{\beta} \|\mathbf{H}\beta - \mathbf{T}\| \tag{6}$$

The result that minimizes the norm of this least squares is given in (7) below

$$\beta^{1} = \mathbf{H}^{1}\mathbf{T} \tag{7}$$

where $\mathbf{H}^{1}$ is the Moore-Penrose generalized inverse of the matrix. By using this method, it is possible to achieve minimum training error, smallest norm of weights and best generalized performance and a unique solution for $\mathbf{H}\beta = \mathbf{T}$ [25–27].

The algorithm for the Extreme Learning Machines can be expressed in 3 simple steps [25]

Step 1: Define a hidden layer node N, randomly assign input weights a_i and hidden layer biases b_i (i = 1......N)
Step 2: Calculate the hidden layer Output Matrix H
Step 3: According to Eq. 5, calculate the output weight β^{1}

4 Results and Discussion

The dataset used is downloaded from [28]. This database is considered as the benchmark dataset for microarray data. The dataset consists of non-overlapping, disjoint data designated as training dataset and testing dataset. They include Lymphoma, MLL and SRBCT samples. The data format is shown in Table 1 and includes the name of the disease, the number of sample genes and total number of disease classes. The above three diseases fall under the category of multi class problems.

The process of feature selection is performed using the hybrid approach. The total number of genes ranges from 2000–12500. The number of relevant genes selected by using the hybrid approach is shown in Table 2.

Table 1 Description of the dataset used for the study

Name of the dataset	Number of genes in the raw dataset	Number of classes
SRBCT	2308	4
Lymphoma	4026	3
MLL	12582	3

Table 2 Number of genes selected using proposed hybrid method

Name of the dataset	Number of genes in raw dataset	Number of genes selected by proposed method
SRBCT	2308	63
Lymphoma	4026	306
MLL	12582	1058

Table 3 Classifier accuracy—ELM versus traditional tree based classifiers

Dataset	ELM	J48	Random forest	Random tree	Decision stump	Genetic programming
SRBCT	93.7	81.9	93.9	72.3	59.0	93.9
Lymphoma	96.8	63.2	89.5	78.9	65.8	92.1
MLL	85.6	77.7	80.6	65.3	65.3	81.9

It is evident from the above table that the proposed hybrid method selects 2–8 % of the relevant and informative genes from the raw dataset.

The selected genes are then subjected to classification using Extreme Learning Machines Classifier. Also they are subjected to classification using traditional classifiers. The classification accuracy is comparatively higher when feature selected genes using the hybrid approach are classified using Extreme Learning Machines as evident in Table 3.

The classifier accuracy in Table 3 above shows that ELM produces the highest accuracy for SRBCT dataset with 93.7 % and MLL with 85.6 %. Also it produces comparatively high accuracy of 96.8 % for the lymphoma dataset.

5 Conclusion

In this paper, we have discussed the classifier accuracy of the proposed hybrid approach that combines the correlation coefficient with particle swarm optimization. This is compared with the traditional tree based classifiers like J48, Random Forest, Random Tree, Decision Stump and Genetic Algorithm as well. It is evident that the extreme learning machines classifier produces more or comparatively better accuracy than the other tree based classifiers available in literature. The proposed hybrid method that has higher potential in aiding further research in the area of feature selection simplified the process of gene selection which is evident from the experimental results. The proposed method significantly reduces the number of genes needed for classification and has also contributed to the improvement in classifier accuracy. The proposed method has greater scope of application to problems in other domains in future.

References

1. Mitra, S., Das, R., Hayashi, Y.: Genetic networks and soft computing. IEEE/ACM Trans. Comput. Biol. Bioinform. **8**(1) (2011)
2. Yang, C.-S., Chuang, L.-Y., Ke, C.-H., Yang, C.-H.: A hybrid feature selection method for microarray classification. IAENG Int. J. Comput. Sci. **21** (2008)
3. Yang, C.-S., Chuang, L.-Y., Yang, C.-H., IG-GA: a hybrid filter/wrapper method for feature selection of microarray data. J. Med. Biol. Eng. **30**(1), 23–28
4. Maji, P., Das, C.: Relevant and significant supervised gene clusters for microarray cancer classification. IEEE Trans. Nano Biosci. **11**(2) (2012)
5. Chuang, L.-Y., Chang, H.-W., Tu, C.-J., Yang, C.-H.: Improved binary PSO for feature selection using gene expression data. Comput. Biol. Chem. **32**(1), 29–38 (2008)
6. Sharma, A., Imoto, S., Miyano, S.: A top-r feature selection algorithm for microarray gene expression data. IEEE/ACM Trans. Comput. Biol. Bioinform. **9**(3) (2012)
7. Sakellariou, A., Sanoudou, D., Spyrou, G.: Investigating the minimum required number of genes for the classification of neuromuscular disease microarray data. IEEE Trans. Inform. Technol. Biomed. **15**(3) (2011)
8. Rajapakse, J.C., Mundra, P.A.: Multiclass Gene selection using pareto-fronts. IEEE/ACM Trans. Comput. Biol. Bioinform. **10**(1) (2013)
9. Wang, J., Zhao, P., Hoi, S.C.H., Jin, R.: Online feature selection and its applications. IEEE Trans. Knowl. Data Eng. **26**(3) (2014)
10. Song, Q., Ni, J., Wang, G.: A fast clustering-based feature subset selection algorithm for high-dimensional data. IEEE Trans. Knowl. Data Eng. **25**(1) (2013)
11. Liu, S., Patel, R.Y., Daga, P.R., Liu, H., Fu, G., Doerksen, R.J., Chen, Y., Wilkins, D.E.: Combined rule extraction and feature elimination in supervised classification. IEEE Trans. Nano Biosci. **11**(3) (2012)
12. Leung, Y., Hung, Y.: A Multiple-filter-multiple-wrapper approach to gene selection and microarray data classification. IEEE/ACM Trans. Comput. Biol. Bioinform. **7**(1) (2010)
13. Ji, G., Yang, Z., You, W.: PLS-based gene selection and identification of tumor-specific Genes. IEEE Trans. Syst. Man Cybern.—Part C: Appl. Rev. **41**(6) (2011)
14. Karegowda, A.G., Manjunath, A.S., Jayaram, M.A.: Comparative study of attribute selection using gain ratio and correlation based feature selection. Int. J. Inform. Technol. Knowl. Manage. **2**(2), 271–277 (2010)
15. Hall, M.A.: Correlation-based Feature Selection for Machine Learning. University of Waikato (1999)
16. Lazar, C., Taminau, J., Meganck, S., Steenhoff, D., Coletta, A., Molter, C., de Schaetzen, V., Duque, R., Bersini, H., Nowe, A.: A survey on filter techniques for feature selection in gene expression microarray analysis. IEEE/ACM Trans. Comput. Biol. Bioinform. **9**(4) (2012)
17. Fu, L.M., Youn, E.S.: Improving reliability of gene selection from microarray functional genomics data. IEEE Trans. Inform. Technol. Biomed. **7**(3) (2003)
18. da Costa, J.F.P., Alonso, H., Roque, L.: A weighted principal component analysis and its application to gene expression data. IEEE/ACM Trans. Comput. Biol. Bioinform. **8**(1) (2011)
19. Kumar, A.P., Valsala, P.: Bioinformation **9**(16), 824–828 (2013)
20. Kar, S., Sharma, K.D., Maitra, M.: Gene selection from microarray gene expression data for classification of cancer subgroups employing PSO and adaptive K-nearest neighborhood technique. Expert Syst. Appl. 612–627 (2015)
21. Huang, G.-B., Zhu, Q.-Y., Siew, C.-K.: Extreme learning machine: a new learning scheme of feedforward neural networks. In: Proceedings of IEEE International Joint Conference on Neural Networks, vol. 2 (2004)
22. Wang, Y., Cao, F., Yuan, Y.: A study on effectiveness of extreme learning machine. Neurocomputing, Elsevier

23. Zhang, R., Huang, G.-B., Sundararajan, N., Saratchandran, P.: Multicategory classification using an extreme learning machine for microarray gene expression cancer diagnosis. IEEE/ACM Trans. Comput. Biol. Bioinform. **4**(3) (2007)
24. Lu, H.-J., An, C.-L., Zheng, E.-H., Lu, Y.: Dissimilarity based ensemble of extreme learning machine for gene expression data classification. Neurocomputing (2014)
25. Yoon, H., Park, C.-S., Kim, J.S., Baek, J.-G.: Algorithm learning based neural network integrating feature selection and classification. Expert Syst. Appl. (2013)
26. Chandrasekar, C., Meena, P.S.: Microarray Gene expression for cancer classification using fast extreme learning machine with ANP. Int. J. Eng. Res. Appl. **2**(2), 229–235 (2012)
27. Arunkumar, C., Ramakrishnan, S.: Binary Classification of cancer microarray gene expression data using extreme learning machines. In: 2014 IEEE International Conference on Computational Intelligence and Computing Research, pp. 1–4 (2014)
28. http://www.biolab.si/en/

Resource Aware Adaptive Scheduler for Heterogeneous Workload with Task Based Job Sampling

Athira V. Panicker and G. Jisha

Abstract Resource aware adaptive scheduling for Mapreduce jobs aims at improving resource utilization across machines. Mapreduce schedulers mainly have fixed number of execution slot on each tasktracker that represents the capacity of cluster. Here a method of dynamically adjusting the number of slots on tasktracker based on task completion gaol is implemented to maximize the resource utilization. A method of task based job sampling is used to get job profile information that inturn used to adjust the slots dynamically. Accuracy of our estimations where assessed based on completion time goal and actual execution time.

Keywords Hadoop · Resource awareness · Job scheduling · Dynamic slot

1 Introduction

With the growth of Internet, the amount of information grows exponentially over time, which in turn leads to an ever-growing demand for efficient and scalable data processing frameworks and platforms. MapReduce [1] is a programming model aimed at providing a simple data processing scheme on large data set stored distributed across large clusters of computers. It allows the user to submit each job with a map function and a reduce function specified. The input data to be processed are firstly split into smaller tasks before being passed to the map function. The map function then transforms the input data into key-value pairs. Lastly the reduce function is responsible for aggregating and further transforming the key-value pairs and eventually writing the output into a distributed file system. Hadoop have a master slave architecture. The master node contains the NameNode and JobTracker. It is in

A.V. Panicker · G. Jisha (✉)
Department of Information Technology, Rajagiri School of Engineering
and Technology, Cochin, Kerala, India
e-mail: jishag@rajagiritech.ac.in

A.V. Panicker
e-mail: athiravpanicker@gmail.com

© Springer International Publishing Switzerland 2016
V. Snášel et al. (eds.), *Innovations in Bio-Inspired Computing and Applications*,
Advances in Intelligent Systems and Computing 424,
DOI 10.1007/978-3-319-28031-8_21

charge of receiving job submissions, splitting a job into smaller tasks, and assigning the tasks to slave nodes based on the heartbeat, which contains the status message of a slave, received from each slave every few seconds. The slave nodes are also known as the DataNodes and the TaskTrackers. Each slave has a fixed number of map slots and reduce slots for processing tasks.

The work presented in this paper concentrates on designing a novel resource aware Adaptive scheduler for Hadoop that schedules Map Reduce jobs based on individual jobs resource demand. The scheduler dynamically adjust the slots based on task completion time goal if it is not given it assumes that job should be completed at the earliest. The different jobs will have different resource requirement. There are two types of workloads I/O bound workloads that use more I/O operations and CPU bound workloads that use more CPU operations. Heterogeneity can occur both in jobs as well as in the various nodes of the cluster like CPU, disk etc.

The slot is said to be a schedulable unit and in hadoop resources are abstracted into two types of slots they are map and reduce slots. Here the slot is associated with a particular job. Every job needs a number of map slots and reduce slots. The scheduler will dyanamically allocate the resources to job according to the resource demands. The resource demands and resource utilization of machine can be calculated periodically and can be used to allocate resources inorder to increase the performance.

The main idea of our work is to design a more effective and efficient scheduler by exploiting real time resource utilizations. This done by dynamically adjusting the slots according to resource demands.

2   Scheduling In Hadoop: Related Works

In hadoop Map Reduce the submitted jobs are divided into many tasks. The process of assigning this task to different task tracker is called as scheduling. The scheduling decisions are taken by Job tracker whereas Task trackers are responsible for executing tasks. This task assignment is done by sending heartbeat message by Task Tracker to job Tracker by a task assignment flag. There are different kinds of schedulers used in hadoop they are:

FIFO: FIFO scheduler is said to be the original MapReduce scheduler. In this scheduler the jobs are arranged based on submission time and their priorities. The master assigns the mapslots of first job in queue whenever it receives a heartbeat from slave indicating their availability for processing more map task. Here the task assignment is based on Data locality that is if input data of task resides on the given slave it is given priority for assignment over those task with input data resides on same rack but not the same machine. If both this condition fails the hadoop assigns task with input data outside rack.

Hadoop on Demand: Hadoop on Demand [2] is a scheduler which is introduced to address one of the drawback of FIFO scheduler. In FIFO scheduler the first job in queue is assigned when the slot become free so in that case there will be starvation of other job in presence of a long running job. So HOD addresses this problem by

Table 1 Existing scheduler comparison

Scheduler	Advantage	Disadvantage	Resource awareness
HOD	Private cluster	under utilization	–
FIFO	High throughput	Short job starves on long running job	Data locality
Fair scheduler	Guaranteed resource availability	Complexity in configuration	Copy compute splitting
Capacity scheduler	Guaranteed resource availability	Complexity in configuration	Memory intensive jobs

providing private hadoop cluster for each user through node allocation using torque resource manager.

Fair Scheduler: The main concept of Fair scheduler [3] are pools. It gives an illusion of providing private cluster for each user. Each user is given a pool that specifies minimum share in units of task slots. The minimum share is always guaranteed by the scheduler. The allocation algorithm of fair scheduler can be summarized into three phases they are: (1) Satisfy the pool whose minimum share is greater than or equals to the demand (2) Allocate resources to the other pools up to its minimum share (3) Residual given to the unfilled, starting with the least fulfilled.

Capacity Scheduler: The capacity scheduler [4] uses multiple queue instead of pools in fair scheduler. The user can specify the high memory requirement of job so that the scheduler ensures that slot with enough memory only can accept task of that job (Table 1).

3   Resource Aware Adaptive Scheduler

The Resource aware adaptive scheduler contains four phases they are 1. Placement algorithm and utility calculator 2. Tasktracker Profiler 3. Taskbased job sampling and 4. Task Scheduler. The tasktracker profiler and task based job sampling collect the tasktracker utilization and resource demand of job respectively. This profiling information is used by the placement algorithm and utility calculator to decide the placement of task in slots based on completion time goal and the transition of task among the slots are done by the task scheduler.

3.1   Scheduler Architecture

The scheduler described here takes into account the resource requirement of jobs before scheduling the task. The resources includes CPU, Disk I/O and network I/O. The resource demands are calculated by task based job sampling and task tracker

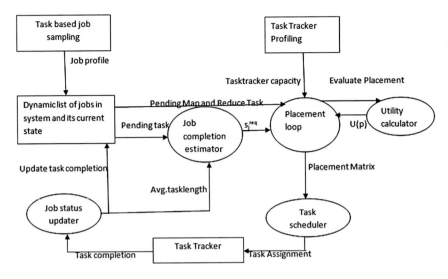

Fig. 1 Scheduler Architecture

profiling calculate tasktracker utilization. The Placement algorithm [5] and ultility calculation help us to dynamically adjust the slots (Fig. 1).

3.2 Placement Algorithm and Utility Calculator

The Placement algorithm [5] dynamically adjusts the number of map and reduce tasks allocated per tasktracker by considering the resource constraints, Task completion time and utility value [5]. Placement algorithm runs at each control cycle. The algorithm is utility driven it produces a placement matrix that balances between the jobs to increase the performance in heterogeneous environment. For each placement matrix Job Utility Calculator calculates a utility value and it is given to the placement algorithm to choose better placement matrix. In each cycle the algorithm tries to increase the utility value and it removes allocated task of higher utility jobs and replace it with task of jobs with lowest utility value. The jobtracker have a list of active jobs and list of tasktracker with its description that contains the number of completed task and pending task. Upon completion of task the jobstatus updater updates the number of pending map and reduce task in the job descriptor. Based on this information the job completion estimator estimates the number of map task that should be allocated simultaneously (s_{req}^j) to complete the task according to job completion time goal.

3.3 Task Tracker Profiling

The Task Tracker profiler profiles the information from the task trackers in the cluster. It profiles both static information as well as dynamic information about resources from task trackers. The main three resource types used here are CPU, disk I/O and Network I/O. The static information can be obtained from linux configuration files. The runtime CPU capacity of a Task Tracker is calculated by the scheduler [6] as

$$c_{cpu} = v_{cpu} \times n_{cpu} \times 1 - u_{cpu} \qquad (1)$$

Here v_{cpu} and n_{cpu} denotes clock rate and number of cores of cpu in each the Task Tracker respectively u_{cpu} is CPU usage in percentage. Disk I/O and network I/O are calculated seperately for both read and write [6].

$$c_{dread} = v_{dread} - w_{dread} \qquad (2)$$

$$c_{dwrite} = v_{dwrite} - w_{dwrite} \qquad (3)$$

$$c_{nread} = v_{ndownload} - w_{ndownload} \qquad (4)$$

$$c_{nwrite} = v_{nupload} - w_{nupload} \qquad (5)$$

Here v_{dread} and v_{dwrite} are the maximum sequential disk read and write speeds in KB/s respectively. w_{dread} and w_{dwrite} are the current disk read and write speed measured. $v_{ndownload}$ and $v_{nupload}$ is the maximum read and write bandwidth of the network, and $w_{nupload}$ and $w_{ndownload}$ are current measured outgoing and incoming network traffic in KB/s.

Each Task Tracker are profiled periodically and master keeps only the latest version of resource scores. Inorder to avoid outliers in predictions we use the worst score in past few seconds.

3.4 Job Sampling in Task

Inorder to calculate the resource demand of a job a method of job sampling based on task is used. A sample task of a job is run in a task tracker and resource requirement like CPU usage, Disk I/O score and Network I/O score can be calculated.

3.4.1 Data Size

For a given task, the total input data size is the sum of disk input and network input.

$$s_{in} = s_{disk-in} + s_{network-in} \qquad (6)$$

For sample task,

$$s_{out} = s_{s-disk-in} + s_{s-network-in} \qquad (7)$$

3.4.2 Network I/O Scores

Network I/O profiling is done by the following equations where vnetwork-download and vnetwork-upload are static data obtained from configuration file; and wnetwork-download and wnetwork-upload are measured and calculated with the following formulas:

Measured current network Download bandwidth:

$$w_{nd} = \frac{s_{n-accumulative-in} - s_{pre-n-accumulative-in}}{t_{now} - t_{pre}} \qquad (8)$$

Measured current network upload bandwidth:

$$w_{nu} = \frac{s_{n-accumulative-out} - s_{pre-n-accumulative-out}}{t_{now} - t_{pre}} \qquad (9)$$

The network I/O scores cnetwork-read and cnetwork-write are both in KB/s.

3.4.3 Disk I/O Scores

Similar to network I/O score calculation, we obtain disk I/O scores in KB/s with following equations:

Measured current disk read speed:

$$w_{d-read} = \frac{s_{d-accumulative-in} - s_{pre-d-accumulative-in}}{t_{now} - t_{pre}} \qquad (10)$$

Measured current disk write speed:

$$w_{d-write} = \frac{s_{d-accumulative-out} - s_{pre-d-accumulative-out}}{t_{now} - t_{pre}} \qquad (11)$$

From resource requirement of sample task we can calculate resource requirement of job on assumption that task of same job has same resource requirements.

3.5 Task Scheduler

The task scheduler is the one which implements placement matrix while ensuring that there is enough tasktracker capacity. Inorder to avoid the overloading the new placement matrix is not immediately enforced while the previous cycle is still running. The new task are started only if previous is complete and needed resources are available.

4 Evaluation and Discussion

4.1 Experiment Environment

We establish a Hadoop cluster with 1 master, 3 slave nodes. Each node has i7 processor with 3.4 GHZ and 8 GB RAM. All machine were running the ubuntu 14.04 operating system.

WorkLoad Details:

- CPU Workload: Pi Estimation-Approximate the value of pi by counting the number of points that falls within the unit quarter circles.
- I/O Workload: Word Count counts the occurrence of each word in given input files.

4.2 Experimental Results

Table 2 shows the comparison of user given completion time goal and actual execution time (Figs. 2 and 3).

Table 3 shows the comparison of User Estimated slots and Actual slots used (Fig. 4).

Table 2 Comparison of user given time goal and actually execution time

–	Word count1	Pi estimator	Word count2	Word count3	All jobs
User expected completion time (in ms)	23967	22345	21977	22345	24956
Actual job completion time (in ms)	17534	18467	17348	18345	18926

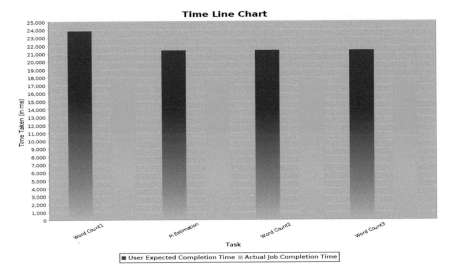

Fig. 2 Comparison of user given time goal and actually execution time

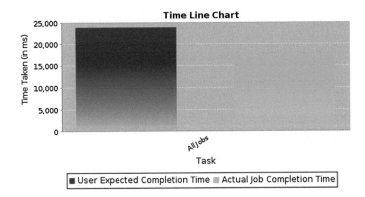

Fig. 3 Comparison of user given time goal and actually execution time for All jobs

Table 3 Comparison of User estimated slots and Actual slots used

–	Word count1	Pi estimator	Word count2	Word count3
User estimated map and reduce slots	2 map, 3 reduce	4 map, 4 reduce	2 map, 3 reduce	2 map, 3 reduce
Actual map and reduce slots used	2 map, 3 reduce	3 map, 4 reduce	2 map, 3 reduce	2 map, 2 reduce

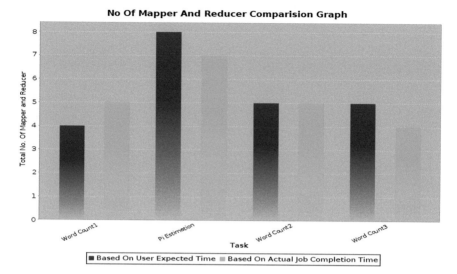

Fig. 4 Comparison of User Estimated slots and Actual slots used

5 Conclusion

In this paper, we proposed the resource aware adaptive scheduler for heterogeneous workload which decides the assignment based on the resource demand of the job as well as the resource supply capacity of the Task Tracker. Here dynamicaly the slots are adjusted to utilize Hadoop cluster more effectively with respect to CPU and I/O. Simulation results have shown how the slots are adjusted to reach the user given completion time goal.

References

1. Apache Hadoop. http://hadoop.apache.org
2. Isard, M., Prabhakaran, V., Currey, J., Wieder, U., Talwar, K., Goldberg, A.: Quincy: fair scheduling for distributed computing clusters. In: SOSP (2009)
3. The Hadoop Capacity Scheduler. http://hadoop.apache.org/common/docs/r0.19.2/capacitynscheduler.html
4. Polo, J., Carrera, D., Becerra, Y., Torres, J., Ayguade, E., Steinder, M., Whalley, I.: Performance Management of Accelerated MapReduce Workloads in Heterogeneous Clusters
5. Polo, J., Castillo, C., Carrera, D., Becerra, Y., Whalley, I., Steinder, M., Torres, J., Ayguade, E.: Resource-aware adaptive scheduling for MapReduce clusters. In: Middleware, ser. Lecture Notes in Computer Science, vol. 7049, pp. 187–207. Springer (2011)
6. Wei, L.: Resource Aware Scheduling for Hadoop, unpublished

Generating Picture Arrays Based on Grammar Systems with Flat Splicing Operation

K.G. Subramanian, G. Samdanielthompson, N. Gnanamalar David and Atulya K. Nagar

Abstract The operation of splicing on strings was introduced in the study of recombinant behaviour of DNA molecules. A particular class of splicing, known as flat splicing on strings was recently considered and this operation was extended to picture arrays. Making use of this operation on arrays, we propose a grammar system, called flat splicing array grammar system (FSAGS), as a new model of picture generation. The components of a FSAGS generate picture arrays working in parallel using the rules of a two-phase grammar called 2RLG with two different components communicating using the array flat splicing operation. We exhibit the power of FSAGS in generating certain "floor designs", besides establishing some comparison results bringing out the generative power of FSAGS.

Keywords Flat splicing · Picture array · Picture language

1 Introduction

The bio-inspired operation of splicing on strings of symbols, which involves the idea of "cutting" and "pasting", was introduced by Head [5] in his seminal paper on a formal model of the recombinant behaviour of DNA molecules under restriction enzymes and a ligase. This operation has been extensively investigated by many

G. Samdanielthompson · N. Gnanamalar David
Department of Mathematics, Madras Christian College, Tambaram 600059, India
e-mail: samdanielthompson@gmail.com

N. Gnanamalar David
e-mail: ngdmcc@gmail.com

K.G. Subramanian (✉) · A.K. Nagar
Department of Mathematics and Computer Science, Faculty of Science,
Liverpool Hope University, Liverpool, L16 9JD, UK
e-mail: kgsmani1948@gmail.com

A.K. Nagar
e-mail: nagara@hope.ac.uk

© Springer International Publishing Switzerland 2016
V. Snášel et al. (eds.), *Innovations in Bio-Inspired Computing and Applications*,
Advances in Intelligent Systems and Computing 424,
DOI 10.1007/978-3-319-28031-8_22

251

researchers, establishing several language theoretic results. Recently, a specific kind of splicing, called flat splicing on words has been introduced in [1]. The operation of flat splicing on a pair of words (α, β) involves "cutting" α and "inserting" β into it, as directed by the rule. On the other hand, inspired by problems in image processing, several generative models of two-dimensional picture arrays based on language theory, have been proposed and studied (see, for example, [4, 10]). Recently, the operation of flat splicing on words has been extended to picture arrays [6] and an array flat splicing system (*AFS*) has been defined to describe picture array languages.

Motivated by the idea of modelling distributed complex systems, the notion of grammar system consisting of several grammars that cooperate according to some well-defined protocol, was introduced [2]. Making use of the operation of splicing, a grammar system, called splicing grammar system was introduced in [3] with communication between its components being done by splicing of strings. This system was extended to arrays in [9] by introducing splicing array grammar system. Here we make use of the array flat splicing operation and introduce a grammar system called flat splicing array grammar system (*FSAGS*) and examine the power of this system in generating picture array languages. The components of the *FSAGS* involve the rules of the two-phase grammar 2*RLG* [4, 8] working in parallel and the array flat splicing operation [6] is used for communication among two different components in terms of "insertion" of an array generated in a component into an array generated in a different component. We also exhibit the use of (*FSAGS*) in generating "floor designs", besides establishing some comparison results bringing out the power of (*FSAGS*).

2   Preliminaries

For notions and results related to formal one-dimensional string languages and two-dimensional array languages, the reader is referred to [4, 7].

A word (also called a linear word or string) w is a finite sequence $w_1 w_2 \ldots w_n$ of symbols w_i, $1 \le i \le n$, taken from a finite alphabet Σ. The set of all words over Σ, including the empty word λ with no symbols, is denoted by Σ^*. We denote by $|w|$, the length of a word w which is the number of symbols in the word. For any word w, we denote by ${}^t w$ the word w written vertically. For example, if $w = aab$ over $\{a, b\}$, then

$$
{}^t w = \begin{matrix} a \\ a \\ b \end{matrix} .
$$

A picture array (also called an array or a picture) M of dimension $m \times n$ over an alphabet Σ is a rectangular array with m rows and n columns and is of the form

$$M = \begin{array}{ccc} p_{11} & \cdots & p_{1n} \\ \vdots & \ddots & \vdots \\ p_{m1} & \cdots & p_{mn} \end{array}$$

where each symbol $p_{ij} \in \Sigma, 1 \leq i \leq m, 1 \leq j \leq n$. We denote by $|M|_r$ and $|M|_c$, the number of rows and the number of columns of M respectively. The set of all rectangular arrays over Σ is denoted by Σ^{**}, which contains the empty array λ with no symbols. $\Sigma^{++} = \Sigma^{**} - \{\lambda\}$. A picture language is a subset of Σ^{**}.

For an array A of dimension $m \times n$ and an array B of dimension $p \times q$, the column catenation $A \circ B$ is defined only when $m = p$ and the row catenation $A \diamond B$ is defined only when $n = q$. Informally speaking, in column catenation $A \circ B$, B is attached to the right of A. In row catenation $A \diamond B$, B is attached below A.

Furthermore, $X \circ \lambda = \lambda \circ X = X \diamond \lambda = \lambda \diamond X = X$, for every array X.

An array grammar model of the non-isometric variety generating rectangular arrays of symbols is the two-dimensional right-linear grammar (Originally called 2D matrix grammar [8]), which we call here as a two-dimensional right-linear grammar, consistent with the terminology used in [4].

Definition 1 A two-dimensional right-linear grammar (2RLG) [4, 8] is $G = (V_h, V_v, V_i, T, S, R_h, R_v)$ where V_h, V_v, V_i are finite sets of symbols, respectively called as horizontal nonterminals, vertical nonterminals and intermediate symbols; $V_i \subset V_v$; T is a finite set of terminals; $S \in V_h$ is the start symbol; R_h is a finite set of horizontal rules of the form $X \rightarrow AY, X \rightarrow A X, Y \in V_h, A \in V_i$; R_v is a finite set of vertical rules of the form $X \rightarrow aY$ or $X \rightarrow a, X, Y \in V_i, a \in T \cup \lambda$.

There are two phases of derivation in a 2RLG. In the first horizontal phase, starting with S the horizontal rules are applied (as in a regular grammar) generating strings over intermediates, which act as terminals of this phase. In the second vertical phase each intermediate in such a string serves as the start symbol for the second phase. The vertical rules are applied in parallel in this phase for generating the columns of the rectangular arrays over terminals. In this phase at a derivation step, either all the rules applied are of the form $X \rightarrow aY, a \in T$ or all the rules are of the form $X \rightarrow Y$ or all the rules are the terminating vertical rules of the form $X \rightarrow a, a \in T$ or all of the form $X \rightarrow \lambda$. When the vertical generation halts, the array obtained over T, is collected in the picture language generated by the 2RLG. Note that the picture language generated by a 2RLG consists of rectangular arrays of symbols. We denote by $L(2RLG)$ the family of array languages generated by two-dimensional right-linear grammars.

We illustrate the working of 2RLG with an example.

Example 1 Consider the 2RLG G_L with regular horizontal rules $S \rightarrow AX, X \rightarrow BX$, $X \rightarrow B$ where A, B are the intermediate symbols and the vertical rules $A \rightarrow aA, A \rightarrow a, B \rightarrow bB, B \rightarrow a$ where a, b are the terminal symbols. In the first phase, the horizontal rules generate words of the form AB^n, $n \geq 1$. In the second phase every symbol in such a word AB^n is rewritten in parallel either by using the rules $A \rightarrow aA, B \rightarrow bB$ in which case the derivation can be likewise continued adding rows of the form $ab \ldots b$

Fig. 1 A picture array
generated in Example 1

a b b b b

a b b b b

a b b b b

a a a a a

or by using the rules $A \rightarrow a, B \rightarrow a$ in which case the derivation terminates adding
a row of the form $aa \dots a$, thus generating picture arrays, one member of which is
shown in Fig. 1.

An operation, called flat splicing on linear words, is considered by Berstel et
al. [1]. A flat splicing rule r is of the form $(\alpha|\gamma - \delta|\beta)$, where $\alpha, \beta, \gamma, \delta$ are words
over a given alphabet Σ. For two words $u = x\alpha\beta y$, $v = \gamma z\delta$, an application of the flat
splicing rule $r = (\alpha|\gamma - \delta|\beta)$ to the pair (u, v) yields the word $w = x\alpha\gamma z\delta\beta y$. In other
words, the second word v is inserted between α and β in the first word u as a result
of applying the rule r.

The notion of flat splicing on words [1] has been extended to arrays in [6], by
introducing two kinds of flat splicing rules, called column flat splicing rule and row
flat splicing rule and thus a new model of picture generation, called array flat splicing
system is introduced in [6].

Definition 2 [6] Let V be an alphabet.

(i) A column flat splicing rule is of the form $({}^t(a_1a_2)|{}^t(x_1x_2) - {}^t(y_1y_2)|{}^t(b_1b_2))$
where $a_1, a_2, b_1, b_2 \in \Sigma \cup \{\lambda\}$ with $|a_1| = |a_2|$ and $|b_1| = |b_2|, x_1, x_2, y_1, y_2 \in \Sigma \cup \{\lambda\}$ with $|x_1| = |x_2|$ and $|y_1| = |y_2|$.

(ii) A row flat splicing rule is of the form $(c_1c_2|u_1u_2 - v_1v_2|d_1d_2)$ where $c_1, c_2, d_1, d_2 \in \Sigma \cup \{\lambda\}$ with $|c_1| = |c_2|$ and $|d_1| = |d_2|, u_1, u_2, v_1, v_2 \in \Sigma \cup \{\lambda\}$ with $|u_1| = |u_2|$ and $|v_1| = |v_2|$.

(iii) Let $r_1, r_2, \dots, r_{m-1}$ be a sequence of $(m-1)$ column flat splicing rules given
by

$$r_i = ({}^t(\alpha_i\alpha_{i+1})|{}^t(\gamma_i\gamma_{i+1}) - {}^t(\delta_i\delta_{i+1})|{}^t(\beta_i\beta_{i+1})),$$

for $1 \leq i \leq (m-1)$. Let X, Y be two picture arrays, each with m rows, for some
$m \geq 1$, and given by

$$X = X_1 \circ^t(\alpha_1\alpha_2 \dots \alpha_m) \circ^t(\beta_1\beta_2 \dots \beta_m) \circ X_2,$$

$$Y = {}^t(\gamma_1\gamma_2 \dots \gamma_m) \circ Y' \circ^t(\delta_1\delta_2 \dots \delta_m),$$

where X_1, X_2, Y' are arrays over Σ with m rows, $\alpha_i, \beta_i, \in \Sigma \cup \{\lambda\} (1 \leq i \leq m)$, with $|\alpha_1| = |\alpha_2| = \dots = |\alpha_m|, |\beta_1| = |\beta_2| = \dots = |\beta_m|, \gamma_i, \delta_i, (1 \leq i \leq m), \in \Sigma \cup \{\lambda\}$ with $|\gamma_1| = |\gamma_2| = \dots = |\gamma_m|, |\delta_1| = |\delta_2| = \dots = |\delta_m|$. An application of the column flat splicing rules $r_1, r_2, \ldots, r_{m-1}$ to the pair of arrays (X, Y) yields the array Z

$$= X_1 \circ^t (\alpha_1 \alpha_2 \ldots \alpha_m) \circ^t (\gamma_1 \gamma_2 \ldots \gamma_m) \circ Y' \circ^t (\delta_1 \delta_2 \ldots \delta_m) \circ^t (\beta_1 \beta_2 \ldots \beta_m) \circ X_2.$$

The pair (X, Y) yielding Z is then denoted by $(X, Y) \vdash_c Z$.

(iv) Let $s_1, s_2, \ldots, s_{n-1}$ be a sequence of $(n-1)$ row flat splicing rules given by

$$s_j = (\eta_j \eta_{j+1} | (\mu_j \mu_{j+1}) - (v_j v_{j+1}) | \theta_j \theta_{j+1}),$$

for $1 \leq j \leq (n-1)$. Let U, V be two picture arrays, each with n columns, for some $n \geq 1$, and given by

$$U = U_1 \diamond (\eta_1 \eta_2 \ldots \eta_n) \diamond (\theta_1 \theta_2 \ldots \theta_n) \diamond U_2,$$

$$V = (\mu_1 \mu_2 \ldots \mu_n) \diamond V' \diamond (\delta_1 \delta_2 \ldots \delta_n)$$

where U_1, U_2, V' are arrays over Σ with n columns, $\eta_j, \theta_j, (1 \leq j \leq n), \in \Sigma \cup \{\lambda\}$ with $|\eta_1| = |\eta_2| = \cdots = |\eta_n|, |\theta_1| = |\theta_2| = \cdots = |\theta_n|, \mu_j, v_j, (1 \leq j \leq n), \in \Sigma \cup \{\lambda\}$ with $|\mu_1| = |\mu_2| = \cdots = |\mu_n|, |v_1| = |v_2| = \cdots = |v_n|$. An application of the row flat splicing rules $s_1, s_2, \ldots, s_{n-1}$ to the pair of arrays (U, V) yields the array W

$$= U_1 \diamond (\eta_1 \eta_2 \ldots \eta_n) \diamond (\mu_1 \mu_2 \ldots \mu_n) \diamond V' \diamond (\delta_1 \delta_2 \ldots \delta_n) \diamond (\theta_1 \theta_2 \ldots \theta_n) \diamond U_2.$$

The pair (U, V) yielding W is then denoted by $(U, V) \vdash_r W$.

(v) An array flat splicing rule is either a column flat splicing rule or a row flat splicing rule. The notation $\vdash$ denotes either $\vdash_c$ or $\vdash_r$.

(vi) For a picture language $L \subseteq \Sigma^{**}$ and a set R of array flat splicing rules, we define

$$f(L) = \{M \in \Sigma^{**} \mid (X, Y) \vdash M, \text{for} X, Y \in L, \text{and some rule in } R\}.$$

(vii) An array flat splicing system (AFS) is $\mathcal{A} = (\Sigma, M, R_c, R_r)$ where Σ is an alphabet, M is a finite set of arrays over Σ, called initial set, R_c is a finite set of column flat splicing rules and R_r is a finite set of row flat splicing rules. The picture language $L(\mathcal{A})$ generated by $\mathcal{A}$ is iteratively defined as follows:

$$f^0(M) = M; \text{For } i \geq 0, f^{i+1}(M) = f^i(M) \cup f(f^i(M));$$

$$L(\mathcal{A}) = f^*(M) = \cup_{i \geq 0} f^i(M).$$

The family of picture languages generated by array flat splicing systems is denoted by $L(AFS)$.

An illustration of the application of column and row flat splicing rules as well as the working of an array flat splicing system is given in the following example.

Example 2 Consider the array flat splicing system $\mathcal{A}_R$ with alphabet $\{a, b\}$ and the initial set consisting of the array $M = \{\begin{smallmatrix} a & b \\ a & b \end{smallmatrix}\}$. The column flat splicing rule is c where $c = (\begin{smallmatrix} a \\ a \end{smallmatrix} | \begin{smallmatrix} a \\ a \end{smallmatrix} - \begin{smallmatrix} b \\ b \end{smallmatrix} | \begin{smallmatrix} b \\ b \end{smallmatrix})$. The row flat splicing rules are r_1, r_2, r_3 where $r_1 = (aa|aa - aa|aa)$, $r_2 = (ab|ab - ab|ab)$, $r_3 = (bb|bb - bb|bb)$.

The column flat splicing rule c is applicable to the pair (M, M). Note that both the components in the pair being the same initial array M, the requirement of equal number of rows for the application of column flat splicing rule c is satisfied. Also the second array in the pair begins with the column $\begin{smallmatrix} a \\ a \end{smallmatrix}$ and ends with the column $\begin{smallmatrix} b \\ b \end{smallmatrix}$, as required in the rule c. The first array is "cut" between the columns $\begin{smallmatrix} a \\ a \end{smallmatrix}$ and $\begin{smallmatrix} b \\ b \end{smallmatrix}$ while the second array is "inserted" between them, yielding the array $\begin{smallmatrix} a & a & b & b \\ a & a & b & b \end{smallmatrix}$.

On the other hand, the row flat splicing rules r_1, r_2 can be used to expand the array rowwise. For example, the sequence of rules r_1, r_2, r_3 could be applied to the pair of arrays with both components having the same array $\begin{smallmatrix} a & a & b & b \\ a & a & b & b \end{smallmatrix}$. Again note that both the components in the pair being the same array, the requirement of equal number of columns for the application of a sequence of row flat splicing rules, is satisfied. In fact, the first array is "cut" between the first row $a\,a\,b\,b$ and the second row which is also $a\,a\,b\,b$. The second array satisfies the requirements of the sequence of rules r_1, r_2, r_3. The second array therefore can be "inserted" into the first array to yield the array $\begin{smallmatrix} a & a & b & b \\ a & a & b & b \\ a & a & b & b \\ a & a & b & b \end{smallmatrix}$. Proceeding like this, we compute the successive terms $f^0(M), f^1(M), \ldots$. In fact

$$f^0(M) = M = \{\begin{smallmatrix} a & b \\ a & b \end{smallmatrix}\},$$

$$f^1(M) = M \cup f(M) = \{\begin{smallmatrix} a & b \\ a & b \end{smallmatrix}, \begin{smallmatrix} a & a & b & b \\ a & a & b & b \end{smallmatrix}, \begin{smallmatrix} a & b \\ a & b \\ a & b \\ a & b \end{smallmatrix}\},$$

$\cdots$.

Fig. 2 A member of the
language of $\mathcal{A}_R$

a a a b b b

a a a b b b

a a a b b b

a a a b b b

Thus the picture language $L(\mathcal{A}_R) = f^*(M)$ consists of rectangular arrays of even side length over the symbols a, b. Also if the number of columns in such an array is $2n$, then the first n columns are over a while the next n columns are over b.

One such picture array is shown in Fig. 2.

3 Flat Splicing Regular Array Grammar System

We now introduce a new model of picture generation, called flat splicing array grammar system.

Definition 3 A flat splicing regular array grammar system (*FSAGS*) is a construct $G_{as} = (V_h, V_v, V_i, T, (S_1, R_1^h, R_1^v), \ldots, (S_n, R_n^h, R_n^v), F)$ where V_h, V_v, V_i are respectively, the finite sets of horizontal, vertical and intermediate nonterminals; $V_i \subseteq V_v$; T is the terminal alphabet; $S_i, 1 \leq i \leq n$ is the start symbol of the corresponding component; $R_i^h, 1 \leq i \leq n$ is a finite set of horizontal rules, which are regular of the forms $X \to AY, X \to A, X, Y \in V_h, A \in V_i$; $R_i^v, 1 \leq i \leq n$ is a finite set of right-linear vertical rules of the forms $A \to aB, A \to B, A \to a, A, B \in V_v, a \in T$; F is a finite set of array flat splicing rules The derivations take place in two phases as follows:

Each component grammar generates a word called intermediate word, over intermediates starting from its own start symbol and using its horizontal rules; the derivations in this phase are done with the component grammars working in parallel.

In the second phase any of the following steps can take place:

(i) each component grammar can rewrite as in a two dimensional matrix grammar using the vertical rules, starting from its own intermediate word generated in the first phase. (The component grammars rewrite in parallel and the rules are applied together). Note that the component grammars together terminate with all rules used in the form $A \to a$ or together continue rewriting in the vertical direction with all rules used either in the form $A \to aB$ or in the form $A \to B$.

(ii) At any instant the picture array X generated in the ith component for some $1 \leq i \leq n$ and the picture array Y generated in the jth component for some $1 \leq j \leq n$ can be flat spliced using array flat splicing rules, either column or row flat splicing rules as in Definition 2, thus yielding picture array Z in ith component. In any other component (other than ith component), the arrays generated at this instant will remain unchanged during this flat splicing process.

There is no priority between steps (i) and (ii).

The language $L_i(G_{as})$ generated by the *ith* component of G_{as} consists of all picture arrays, generated over T, by the derivations described above. This language will be called the individual picture array language of the system and we may choose this to be the language of the first component. The family of individual picture array languages generated by *FSRAGS* with at most n components is denoted by $L_n(FSRAGS)$.

We illustrate the working of *FSRAGS* with an example.

Example 3 Consider the *FSAGS* G_1 with two components given by

$$(\{S_1, S_2, X, Y\}, \{A, C, D\}, \{A, C, D\}, \{a, c, d\}, (S_1, R_1^h, R_1^v), (S_2, R_2^h, R_2^v), F)$$

where

$$R_1^h = \{S_1 \rightarrow AX, X \rightarrow CX, X \rightarrow A, \}, R_2^h = \{S_2 \rightarrow Z, Z \rightarrow DZ, Z \rightarrow D\},$$

$$R_1^v = \{A \rightarrow aA, C \rightarrow cC, A \rightarrow a, C \rightarrow c\},$$

$$R_2^v = \{A \rightarrow aA, D \rightarrow dD, A \rightarrow a, D \rightarrow d\}$$

and F consists of the column flat splicing rule

$$c_1 = (\begin{smallmatrix} a \\ a \end{smallmatrix} \begin{vmatrix} d \\ d \end{vmatrix} - \begin{smallmatrix} d \\ d \end{smallmatrix} \begin{vmatrix} c \\ c \end{vmatrix}).$$

Starting from S_1, the horizontal regular rules in the first component generate a word of the form $AC^{n-1}A$ on applying the rule $S_1 \rightarrow AX$ once followed by the application of the rule $X \rightarrow CX$ $(n - 1)$ times and finally the rule $X \rightarrow A$ once, terminating the derivation in the first component in the horizontal phase. At the same time in the second component derivations in the horizontal phase take place in parallel starting from S_2 and applying the rule $S_2 \rightarrow Z$ once followed by the application of the rule $Z \rightarrow DZ$ $(n - 1)$ times and finally the rule $Z \rightarrow D$ once, yielding the word D^n for the same n. In the second phase vertical derivations in both the components take place in parallel. In the first component an $m \times (n + 1)$ picture array as in Fig. 3a is generated while in the second component an $m \times n$ picture array as in Fig. 3b is generated. At this point, using the column flat splicing rule c_1 as many times as needed, the array in Fig. 3b is inserted in the array in Fig. 3a between the first column of $a's$ and the second column of $c's$ to yield a picture array of the form as in Fig. 4.

Fig. 3 a Picture array generated in the first component of G_1 **b** Picture array generated in the second component of G_1

(a)

a	c	c	c	a
a	c	c	c	a
a	c	c	c	a
a	c	c	c	a

(b)

d	d	d	d
d	d	d	d
d	d	d	d
d	d	d	d

Fig. 4 A picture array
generated by G_1

```
a d d d d c c c a
a d d d d c c c a
a d d d d c c c a
a d d d d c c c a
```

Theorem 1 $L_2(FSRAGS) \setminus L(AFS) \neq \emptyset$.

Proof The picture array language L generated by the *FSRAGS* G_1 in Example 3 consists of picture arrays M each of which has the following property P : M has *exactly two columns of $a's$ with one, the leftmost column and the other, the rightmost column, besides other columns made of either $c's$ or $d's$*. But L cannot be generated by any *AFS* since the flat splicing rules will "insert" one array of L into another, thereby yielding picture arrays which will violate the property mentioned above.   □

Theorem 2 $L(2RLG) = L_1(FSRAGS) \subset L_2(FSRAGS)$.

Proof The equality $L(2RLG) = L_1(FSRAGS)$ follows by noting that the *FSRAGS* with one component will have rewriting rules as in a *2RLG* and the set of array flat splicing rules can be taken to be an empty set as these rules can be applied only when there are at least two components. The picture array language L of Example 3, is generated by a *FSRAGS* with two components while an *FSRAGS* with one component is not enough since the property that each array of L having $d's$ and $c's$ has $(n + 1)$ consecutive columns of $d's$ (starting from the second column) followed by n columns of $c's$, cannot be maintained by regular horizontal rules of a *2RLG* and hence L cannot belong to $L_1(FSRAGS)$.   □

4 Application to Generation of Floor Designs

Generation of certain picture patterns, such as "floor designs" (see, for example, Fig. 5), using array generating grammar models has been done [8]. The idea is to consider the picture pattern as an array over certain terminal symbols and generate the array with an array grammar. Then substitute for each symbol some relevant primitive pattern of the picture to be generated, yielding the pattern.

Here we explain how to construct a *FSRAGS* with two components to generate the collection of floor designs, one member of which is shown in Fig. 5. The primitive patterns involved in this floor design are shgown in Fig. 6.

The picture array encoding the floor design is given in Fig. 7. The picture array language consisting of picture arrays such as the one given in Fig. 7, along with picture arrays as the one given in Fig. 8, can be generated by a *FSRGS* with two components. The idea is to allow the first component generate the left half of the picture array (with the rightmost column made of only e_h) while the second compo-

Fig. 5 A floor design

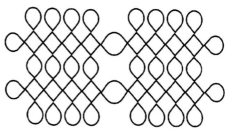

Fig. 6 Primitive Patterns

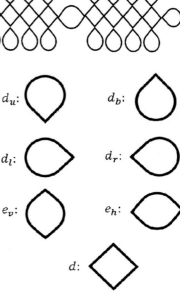

Fig. 7 A picture array
generated by G_1 using array
flat splicing

$$
\begin{array}{ccccccccccc}
b & d_u & d_u & d_u & d_u & b & d_u & d_u & d_u & d_u & b \\
d_l & d & d & d & d & e_h & d & d & d & d & d_r \\
b & e_u & e_u & e_u & e_u & b & e_u & e_u & e_u & e_u & b \\
d_l & d & d & d & d & e_h & d & d & d & d & d_r \\
b & d_b & d_b & d_b & d_b & b & d_b & d_b & d_b & d_b & b
\end{array}
$$

Fig. 8 A picture array
generated by G_1

$$
\begin{array}{cccccc}
b & d_u & d_u & d_u & d_u & b \\
d_l & d & d & d & d & e_h \\
b & e_u & e_u & e_u & e_u & b \\
d_l & d & d & d & d & e_h \\
b & d_b & d_b & d_b & d_b & b
\end{array}
$$

nent generates the second half and array flat splicing rules are created to concatenate
the "right half" to the right of the "left half", yielding the picture array of the form
in Fig. 7. When the relevant primitive patterns are substituted for the corresponding
symbols, this yields the floor design patterns as in Fig. 5.

5 Conclusion and Discussion

We have considered here a grammar system that involves regular rules as defined in the two-phase 2RLG and array flat splicing rules introduced in [6]. One could also consider CF rules as in the two-phase two-dimensional grammar called context-free matrix grammar [8], and examine the generative power of the resulting system, which might help describe more complex floor designs. In fact it will be of interest to explore the theoretical model proposed here for other possible applications in terms of experimental studies.

Acknowledgments The authors are grateful to the reviewers for their very useful comments.

References

1. Berstel, J., Boasson, L., Fagnot, I.: Splicing systems and the Chomsky hierarchy. Theoret. Comput. Sci. **436**, 2–22 (2012)
2. Csuhaj-Varjú, E., Dassow, J., Kelemen, J., Păun, Gh.: Grammar Systems: A Grammatical Approach to Distribution and Cooperation. Gordon and Breach Science Publishers, New York (1994)
3. Dassow, J., Mitrana, V.: Splicing grammar systems. Comput. Artif. Intell. **15**, 109–122 (1996)
4. Giammarresi, D., Restivo, A.: Two-dimensional languages. In: [7], Vol. 3, pp. 215–267 (1997)
5. Head, T.: Formal language theory and DNA: an analysis of the generative capacity of specific recombinant behaviours. Bull. Math. Biol. **49**, 735–759 (1987)
6. Isawasan, P., Venkat, I., Muniyandi, R.C., Subramanian, K.G.: A membrane computing model for generation of picture arrays. In: Proceedings of 4th International Visual Informatics Conference (2015)
7. Rozenberg, G., Salomaa, A. (eds.): Handbook of Formal Languages, vol. 1–3. Springer, Berlin (1997)
8. Siromoney, G., Siromoney, R., Krithivasan, K.: Abstract families of matrices and picture languages. Comput. Graph. Imag. Process. **1**, 284–307 (1972)
9. Subramanian, K.G., Roslin Sagaya Mary, A., Dersanambika, K.S.: Splicing array grammar systems. In: Proceedings of the International Colloquium Theoretical Aspects of Computing: Lecture Notes in Computer Science 3722, pp. 125–135(2005). Springer (2005)
10. Subramanian, K.G., Rangarajan, K., Mukund. M. (Eds.): Formal Models, Languages and Applications. Series in Machine Perception and Artificial Intell, vol. 66. World Scientific Publishing, Singapore (2006)

Identification of Multimodal Human-Robot Interaction Using Combined Kernels

Saith Rodriguez, Katherín Pérez, Carlos Quintero, Jorge López, Eyberth Rojas and Juan Calderón

Abstract In this paper we propose a methodology to build multiclass classifiers for the human-robot interaction problem. Our solution uses kernel-based classifiers and assumes that each data type is better represented by a different kernel. The kernels are then combined into one single kernel that uses all the dataset involved in the HRI process. The results on real data shows that our proposal is capable of obtaining lower generalization errors due to the use of specific kernels for each data type. Also, we show that our proposal is more robust when presented to noise in either or both data types.

1 Introduction

Interacting robots have been considered to play significant roles within the human society long before they actually existed. Applications range from education, service assistance, rescue and entertainment among many others [1]. Challenges in each specific application usually vary depending on the proximity of interaction between the robots and the humans (or other robots), the support of effective social interactions

S. Rodriguez (✉) · K. Pérez (✉) · C. Quintero (✉) · J. López (✉) · E. Rojas (✉) · J. Calderón (✉)
Universidad Santo Tomás, Bogotá, Colombia
e-mail: carlosquinterop@usantotomas.edu.co

S. Rodriguez
e-mail: saithrodriguez@usantotomas.edu.co

K. Pérez
e-mail: andrea.perez@usantotomas.edu.co

J. López
e-mail: jorgelopez@usantotomas.edu.co

E. Rojas
e-mail: eyberthrojas@usantotomas.edu.co

J. Calderón
University of South Florida, Tampa, FL, USA
e-mail: juancalderon@mail.usf.edu

© Springer International Publishing Switzerland 2016
V. Snášel et al. (eds.), *Innovations in Bio-Inspired Computing and Applications*,
Advances in Intelligent Systems and Computing 424,
DOI 10.1007/978-3-319-28031-8_23

263

and the robot's autonomy. The RoboCup initiative [2], for example, contains at least two competitions that require high levels of human-robot interaction (HRI), namely the RoboCup Rescue League and the RoboCup @ Home League [3]. In both cases, especially in the @ Home League it is important to provide the robots with the capabilities of understanding human interactions in different levels; from the most simple instructions to more complicated communications.

Human communications are complex and may be multimodal [1], i.e., when a human says something, she could also use gestures at the same time. These gestures may be meaningful, superfluous or redundant. As an example, a human may issue a spoken command of: "hand me those keys" and at the same time hand pointing to the key's physical location. Other example is a human issuing the order: "come here" and gesturing such order with her hands. In the former case, the instruction requires both data types; the spoken command and the visual gesture to execute the expected action. In the latter, only one of them is necessary, as the other is redundant. However, the usage of both sources of communication may help the robot to assert that the given instruction was clearly understood. Furthermore, having access to this redundant information, the interaction system may be more robust to noise in one data source. For instance, imagine that the robot is made to interact within a place with controlled light conditions but with high levels of auditive noise. In such scenario, the spoken issue may be polluted with noise, while the gesture may be easier to detect.

In this paper, we show the proposal and development of a methodology based on statistical learning that builds classifiers for the multimodal human-robot interaction problem. Our proposal uses kernel-based classifiers, specifying one kernel per data type that aims at exploiting the data type similarity and combining them into a single classifier. We use One-class classifiers in order to capture the probability distribution of each instruction avoiding the problem of retraining when a new instruction requires to be added. The main contributions of this paper are as follows:

- Modeling the detection of multimodal human-robot interaction as a multiclass statistical classification problem.
- Proposal of a suitable scalable architecture for a multiclass classifier based on One-class classifiers.
- Development of combined kernels by using data from different sources, each one with an appropriate kernel.
- Construction of a database of multimodal instructions (audio and video) with several test subjects and feature extraction.
- Validation of the proposed model using real data that shows improved classification when exposed to noise in any of the sources.

This paper starts by showing the work done by other researchers related to the HRI problem using statistical learning and combining input from several sources. Then, we describe the steps of our methodology, from the data acquisition step until the construction of a multiclass classifier using combined kernels. Finally, we show the implementation details of the methodology and its results when compared to two standard statistical learning classifiers and showing the behavior of our proposal in the presence of noise. Finally, we conclude and show future work.

2 Related Work

Many different strategies have been proposed in order to recognize instructions and gestures given by humans over the last years aiming at attaining a more natural communication between humans and robots. In this sense, [4] presents a proposal in which a telepresence robot is capable of heading its attention to places where a human is indirectly pointing by using gestures. They use a Kinect in order to recognize the gestures and head tracking as well as a microphone to achieve voice detection and localization of the audio source. Other proposals apply machine learning techniques to recognize only visual gestures, only voice commands or combined information sources, such as in [4]. In this work, the authors obtain data from a Kinect to recognize gestures using human body joints as movement markers.

On the other hand, in [5, 6] the authors show how to implement a gesture classifier capable of identifying signals and gestures with no meaning and others with well known meaning. This project also uses a Kinect for data acquisition. They consider four sets of gestures, each with 300 examples made by different subjects. In addition, they recorded a data set made of random movements, in order to be included in the validation process to assess the precision of the model. In [7], the authors present an audio classification system that uses SVM and RBFNN. In this work, the authors propose extraction of a number of audio features using Mel-frequency cepstral coefficients (MFCCs). The extraction of related coefficients is fundamental for classification systems as mentioned in [8].

In [9], the authors presented a classification system for five categories analyzing visual features. These features are the base of the works presented in [9], and is of high importance in the field of classification of directions given by gestures and audio. The authors propose a combination of data from gestures and audio to classify content in five categories (news, advertisement, sport, serial and movie). As the other mentioned works, they employ MFCC coefficients to extract audio features, however, for gestures they used color histograms to video segmentation. With the compiling all these features, they assemble a classifier using Support Vector Machines allowing them to build a classifier to every category for audio data and also for video data and through a combination method of weighted sum of audio and video, obtain a category for the processed information.

3 Proposed Methodology

We have designed a methodology that aims at solving the problem of multimodal human-robot interaction, initially for three instructions given by the human, namely: up, down and unknown. The human shall provide an instruction composed of a gesture (body movements) and a spoken word that the robot should be able to identify and execute. In the following sections we show a brief description of each step in the proposed methodology.

3.1 Data Acquisition

The first step in the proposed methodology consists on the construction of a database that contains examples of humans performing gestures and spoken instructions for the "Up" and "Down" classes. For the class "Up", the oral instruction is the verbal spanish word "arriba", while for the class "Down" the instruction is the word "abajo". The gestures for each class correspond to the act of moving both hands upward, for the class "Up" and downward for the class "Down".

Overall, we performed the experiment with 21 subjects who executed 20 repetitions for each instruction for a total of 420 observations for each class. From this dataset, the 90 % is used in the training phase and the remaining 10 % is used to validate the performance of each classifier. Additionally, the validation dataset also includes 80 samples of instructions that do not correspond to either class. Note that these 80 additional observations are not used in the training process of the classifiers, but only as a validation set to assess the performance of the various multiclass classifiers shown henceforth. On one hand, for the "audio" dataset that contains the oral instructions, we acquired the raw data using a microphone. On the other hand, the "video" dataset that corresponds to the gestures performed by the humans for both classes was taken using the Kinect.

3.2 Feature Extraction

3.2.1 Video

Using the Kinect, we have captured a set of frames taken at 30 fps per observation. At each frame, the Kinect provides spatial coordinates of the human skeleton joints for each subject, while the subject performed gestures. In this context, one observation corresponds to the execution of the complete gesture performed by the subject from the start of the data acquisition, following the skeleton tracking until no more movements are detected. For computational tractability and also to avoid the curse of dimensionality, we have only included the information corresponding to the tip of both hands. Figure 1 shows an example of a "Up" gesture performed by one subject and the skeleton detection executed by the Kinect.

The chosen set of features for each observation is composed of the following values:

- (x, y) coordinates of the final and initial positions of the gesture for the tip of the hands
- Mean velocity of the complete gesture for each hand
- Angle of the vector that points from the initial position to the final position for each hand

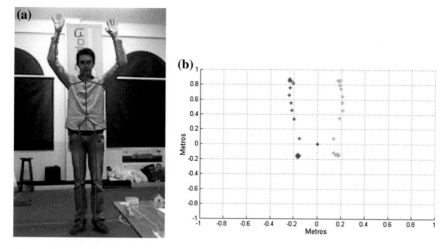

Fig. 1 a Example of Kinect skeleton tracking at the end of the "up" gesture for one subject.
b Example of coordinate trajectory for the tip of both hands when performing the "up" gesture

3.2.2 Audio

We have characterized the audio dataset using the Mel Frequency Cepstral Coefficients (MFCC). According to [8], the first 13 initial coefficients are sufficient to properly represent the original signal without losing important information. The procedure consists on dividing the entire audio signal in time windows of 15ms each where the Fourier coefficients are computed. Then, we compute an estimation of the power spectral density by using triangular overlapping windows, then, the we take the log of the power at each Mel frequency and finally take the Discrete Cosine Transform. The amplitude of such spectrum are the MFCCs. Each coefficient is then averaged over the time windows obtaining 13 coefficients that describe the complete signal.

4   SVM Classifier with Combined Kernels

This section contains the different stages followed to build an appropriate SVM-multiclass classifier using combined kernels. First, Sect. 4.1 shows the methodology we used to build multiclass classifiers based on multiple one-class classifiers. Afterwards, we evaluate the performance of using these classifiers with three different kernels for each type of data, i.e., audio and video. Subsequently, we show how to build a new classifier that uses a kernel that is the result of the combination of the two kernels that have shown improved performance for each data type. Finally, the result is compared and validated with multiclass classifiers that use a single kernel.

4.1 Multiclass Classifier Based on SVM One-Class Classifiers

The multiclass classifiers built for the task at hand were based on multiple SVM one-class classifiers. The reason to follow this approach, in contrast with building common multiclass classifiers, is mainly inspired on the specific problem that we are tackling, and is twofold:

- In the context of human-robot interaction, the potential amount of comands issued by a human that a robot would be required to identify may be high, which may cause the number of classes in a common multiclass problem to grow rapidly. This would require a potentially high amount of samples per class. Besides, every time a new class were added (i.e., a new instruction were included in the human-robot communication), the multiclass problem would require retraining, which is, in general, a hard process.
- In addition to the set of instructions that the robot should be able to identify and execute, we should also consider the fact that there is a need of an additional class where the robot identifies a given instruction as "invalid". This class encompasses all instructions for which the robot should perform no action. It is not feasible to collect data that represent the underlying distribution of such class since it contains all actions (potentially infinite) but the ones identifiable by the robot.

For these reasons, we proposed to build multiclass classifiers based on SVM one-class classifiers. Figure 2 shows our implementation of a multiclass classifier based solely on one-class classifiers. In this architecture, the complete dataset is used as input to each one-class classifier. Each individual classifier outputs its own hypothesis recognizing only the portion of the data that belongs to such class (i.e., each classifier is capable of identify data points that belong to one single class) which then needs to be combined into an output function that generates a single multiclass output. For the particular case, one observation belongs to the "invalid" class

Fig. 2 Multiclass classifier based on one class SVM

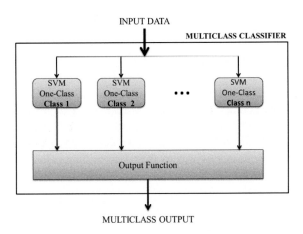

if any of the two following situations is true: (a) more than one classifier claims the observation belongs to its own class or (b) all classifiers label the observation as not belonging to their own class (i.e., the instructions is not recognized by any classifier). In situations where only one classifier recognize the observation as belonging to its class (and everyone else rejected it), the observation is assigned the label of the class that correspond to the one-class classifier that recognized it as part of its probability distribution. All the SVM multiclass classifiers built henceforth used this approach of combining several one-class classifiers.

4.2   Kernel Selection

The selection of the kernel is a highly relevant process and is at the center of the methodology proposed here. The idea is that each data source (i.e., audio and video) will be better represented using a kernel that exploit its similarities. This means that potentially the use of an appropriate kernel in a specific type of data will attained improved performance over other general kernels. In our methodology, we require an identification of the kernel for each type of data that captures the most representative similarity for such type of data. In other applications, special kernels may be constructed for specific types of data.

For this selection step we have split our dataset in two disjoint sets, one containing the audio data and other with the video data, and built a multiclass classifier for each using the polynomial, gaussian and sigmoidal kernels. The result of this process is one classifier for each data type for each kernel. Then, each classifier is evaluated using the error on the validation data and finally keeping the kernel for each data type that delivers the minimum validation error.

4.3   Construction of the Combined Kernel

The next step consists on finding a new kernel by combining the kernels that showed improved performance for each data type. According to [11], it is possible to combine several kernels by means of a linear combination of the individual kernels. The rationale behind this idea is that different kernels could be using inputs from representations of different sources. For this reason, they have different measures of similarity which correspond to different kernels. This procedure allows us to combine multiple information sources as required in HRI applications. Equation (1) shows the calculation of the combined kernel as a linear combination of the audio and video kernels. The coefficients of the linear combination requires a careful tuning process.

$$\kappa = \omega_a \kappa_a + \omega_v \kappa_v \tag{1}$$

5 Implementation and Results

We have used the LibSVM library [14] for the implementation of the SVM one-class classifiers using the precomputed kernel option. For the construction of the multiclass classifiers, we have performed cross-validation in which the entire dataset is split in two disjoint sets: one for the training process and one for validation. The performance of each classifier is measured as the accuracy attained in the validation set after the training process has been completed using only training data.

5.1 Kernel Selection

The performance of each SVM One-class classifiers (and hence that of the multiclass classifier) using each kernel heavily depends on a proper tuning of the kernel parameters. For this task, we have performed a grid search over the kernel parameters space for each classifier together with the cross validation procedure in order to asses the performance of each kernel combination in the classification process. Table 1 shows the generalization errors for each One-class classifier (Up and Down), for each data type (audio and video) and each kernel and the error obtained when building the multiclass classifier based on the One-class classifiers.

By using this experiment, we have shown that the kernel that provides the best representation for the audio type of data is the **Sigmoidal kernel** (with a generalization error of 28.05 %), while that for the video data is the **Gaussian kernel** (with a generalization error of 5.49 %). These kernels will be used for each data type in the kernel combination process below.

Table 1 Generalization errors of each One-class classifier for each kernel and for each type of data and generalization errors of the multiclass classifier using the three chosen kernels for each type of data

	Audio			Video		
	Polynomial (%)	Gaussian (%)	Sigmoidal (%)	Polynomial (%)	Gaussian (%)	Sigmoidal (%)
Up	21.34	16.46	19.51	3.05	2.44	2.44
Down	16.46	17.07	9.76	5.49	3.05	4.27
Multiclass	36.58	29.87	**28.05**	8.53	**5.49**	6.71

5.2 Construction of the Multiclass SVM Using Combined Kernel

Our proposed classifier is one multiclass SVM with the following characteristics:

- It is composed of several One-class SVM classifiers; one per each instruction. In our case study, we have two classifiers; one for the "up" class and other for the "down" class.
- Each One-class classifier is constructed using a combined kernel classifier using the sigmoidal kernel for the audio data and the gaussian kernel for the video data.

Overall, we solved two One-class combined kernel classifiers; one for each class which means that each SVM problem contained five parameters that needed to be tuned, namely: the regularization parameter v, the linear combination coefficients ω_a and ω_v and the kernel parameters for the sigmoidal and gaussian kernel γ_s and γ_g respectively. The procedure followed to tune such parameters was cross validation as described before.

After the tuning process, our multiclasss SVM classifier that uses combined kernels attained a generalization error of 3.66 % showing that it is possible to classify two human instructions using two different data sources with low generalization errors. In the following section we implement two standard multiclass classifiers and compared their results with ours.

5.3 Validation

In order to validate the results of our proposal, we have constructed two standard multiclass classifiers using Support Vector Machines and Neural Networks (NN) with the training dataset. On one hand, for the NN, we have chosen a multilayer perceptron (MLP) architecture with one hidden layer using sigmoidal activation functions. The only parameter that needs to be decided is the number of neurons in the hidden layer. In the case of the SVMs, we need to decide which kernel will be used and the values of the kernel parameters as well as the γ value.

The NN that presented the lowest error was one that uses 6 neurons in its hidden layer while the SVM with lowest generalization error was the one using sigmoidal kernel. These two classifiers will be used later to compare the performance of the classifier that combines kernels. The neural network with one hidden layer with 6 neurons attained 12.01 % of generalization error while a multiclass SVM using one sigmoidal kernel attained 10.37 %. These results show that the classifier built by using combined kernels showed improved performance when compared to two standard classifiers.

5.4 *Analysis of Noise in Generalization*

We have performed one additional experiment that aims at exposing our methodology to situations where the dataset is in the presence of noise. For this, we have included additive gaussian noise to the raw data, i.e., to the audio file and to the skeleton structure from the Kinect and performed three different tests, namely:

- Noise is added only to the audio dataset with different variance levels $\sigma \in [0\ 0.1]$
- Noise is added only to the video dataset with different variance levels $\sigma \in [0\ 0.05]$
- Noise is added to both, audio and video dataset with different variance levels.

Figure 3a, b, c show the results of such experiments. It is important to notice that the SVM classifier that uses combined kernels has lower generalization errors for the different noise variances in the audio dataset, showing that such classifier is more resilient to noise in the audio channel. For the noise in the video dataset we can see that the NN and the SVM with combined kernels achieve similar results. However, when the noise is presented in the complete dataset, the SVM that uses combined kernel, shows improved performance. This result shows that it is possible to build robust classifiers by exploiting each data type similarities by choosing appropriate kernels and then combining them into one single kernel.

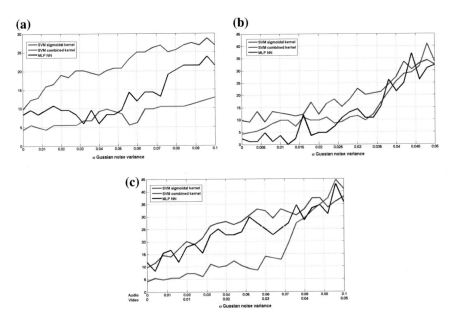

Fig. 3 a Generalization error of the three classifiers when exposed to noise in the audio dataset. **b** Generalization error of the three classifiers when exposed to noise in the video dataset. **c** Generalization error of the three classifiers when exposed to noise in the audio and the video dataset

6 Conclusions

In this work we have shown a methodology designed to build classifiers capable of identifying between several human instructions given using oral commands and gestures. The methodology is based on statistical learning and uses support vector machines with special kernels per data type that are combined into one single kernel by means of a linear combination. The classifier supports the addition of new instructions without the need to retraining. This is obtained by building classifiers based on several SVM One-class classifiers, one per each class. This methodology also solves the problem of acquiring data for unknown instructions for which it may be unfeasible to collect representable data. Finally, we have shown experimentally that our proposal attain improve performance over standard classifiers by using combined kernels which have also shown greater robustness when the data are corrupted with noise.

References

1. Goodrich, M., Schultz, A.: Human-robot Interaction: a survey. Found. Trends Hum.-Comput. Interact. **1**, 203–275 (2007)
2. Behnke, S., Veloso, M., Visser, A., Xiong, R. (eds.): RoboCup 2013: Robot World Cup XVII, LNCS. Springer, Berlin (2014)
3. Van Beek, L., Chen, K., Holz, D., Matamoros, M., Rascon, C., Rudinac, M., Ruiz del Solar, J., Sugiura, K., Wachsmuth, S.: RoboCupHome 2015: Rule and Regulations (2015)
4. Bhattacharya, S., Czejdo, B., Perez, N.: Gesture classification with machine learning using kinect sensor data. In: Third International Conference on Emerging Applications of Information Technology, pp. 348–351. IEEE Press, Kolkata (2012)
5. Kinect Gesture Recognition for Interactive System. http://cs229.stanford.edu/proj2012/ZhangDuLiKinectGestureRecognitionforInteractiveSystem.pdf
6. Huang, J., Lee, C., Ma, J.: Gesture Recognition and Classification using the Microsoft Kinect. Final Project CS229 Machine Learning. Stanford University, Stanford (2012)
7. Dhanalakshimi, P., Palanivel, S., Ramaligam, V.: Classification of audio signals using SVM and RBFNN. Expert Syst. Appl. **36**, 6069–6075 (2009)
8. Sarachaga, G., Sartori, V., Vignoli, L.: Identificacin Automtica de Resumen en Canciones. Proyecto de fin de carrera, Universidad de la Repblica, Uruguay (2006)
9. Suresh, V., Mohan, C., Kumaraswamy, R., Yegnanarayana, B.: Content-based video classification using support vector machines. In: Pal, N.R., et al. (eds.) ICONIP 2004. LNCS, vol. 3316, pp. 726–731. Springer, Heidelberg (2004)
10. Subashini, K., Palanivel, S., Ramalingam, V.: Audio-Video based classification using SVM and AANN. Int. J. Comput. Appl. **53**(18), 43–49 (2012)
11. Gönen, M.M., Alpaydin, E.: Multiple Kernel Learning Algorithms. J. Mach. Learn. Res. **12**, 2211–2268 (2011)
12. Manevitz, L., Yousef, M.: One-Class SVMs for document classication. J. Mach. Learn. Res. **2**, 139–154 (2001)
13. Scholkopf, B., Platt, J.C., Shawe-Taylor, J., Smola, A.J., Williamson, R.C.: Estimating the support of a high-dimensional distribution. Technical report, Microsoft Research MSR-TR-99-87 (1999)
14. Chang, C.C., Lin, C.J.: LIBSVM : a library for support vector machines. ACM Trans. Intell. Syst. Technol. **2**(3), 27:1–27:27 (2011)

Debris Detection and Tracking System in Water Bodies Using Motion Estimation Technique

T. Senthil Kumar, K.S. Gautam and H. Haritha

Abstract The paper proposes a Debris tracking system that associates Debris in water bodies in sequential video frames and gives the movement of debris between the frames. Hence we can identify and track Debris in water bodies in both front and aerial views. The system measures the displacement of each pixel when compared to the previous frame. Pixels show the motion and they are represented by displacement vectors. The system built automatically computes the displacement vector for every pixel. Experimental results prove that the system is capable to detect and track the Debris in water bodies under varying views.

Keywords Optical flow · Motion estimation · Motion vector optimization · Dynamic programming · Block matching algorithm

1 Introduction

This Human computer interaction is an ongoing research sector holding many applications including vision, management of information etc. It builds an interface between the user and the system, thus bridging the interaction gap between them. Object detection and tracking is one among the technologies that interface the interaction between human and computers. Detection and tracking of objects in motion is a simple task for human beings, but a challenging one for machines. The optical flow algorithm is employed to detect and estimate the movement of debris in water bodies. In a particular voxel position the algorithm estimates the motion

T.S. Kumar (✉) · K.S. Gautam · H. Haritha
Department of Computer Science and Engineering, Amrita Vishwa Vidyapeetham,
Coimbatore 641042, Tamil Nadu, India
e-mail: t_senthilkumar@cb.amrita.edu

K.S. Gautam
e-mail: lancer1589@gmail.com

© Springer International Publishing Switzerland 2016
V. Snášel et al. (eds.), *Innovations in Bio-Inspired Computing and Applications*,
Advances in Intelligent Systems and Computing 424,
DOI 10.1007/978-3-319-28031-8_24

between two successive image frames at time interval t_i and t_j. Motion smoothness constrain is involved in enhancing the detection accuracy and estimation stability.

It is an open secret that many issues remain unsolved when an electronic eye views a scenario such as noise, occlusion, spatio temporal change of background due to illumination change, wind driven motion. Another step backs to be considered are the speed of the camera and objects different appearance. Optical flow estimation is being used in multiple applications such as finding the direction of movement of vehicles, tracking objects, and reconstructing missing image regions [1]. The survey stresses on detecting debris in water bodies. Debris in water body are detected to identify the floating debris from the background. The algorithm gives high accuracy in detecting and tracking more number of moving objects in complex scenarios. Though the motion parameters of the moving objects are determined the phenomenon of occlusion and overlapping of objects still remains challenging.

The objective of the work is to design a debris detection system in water bodies. Detection of debris in the water is a challenging task in the current scenario, since we face difficulty in detecting them. The motivation for the work is the problem faced by the rescue team for Air Asia Flight crash in the Java Sea. The paper depicts how an optical flow algorithm can be used in the detecting debris in water bodies. After detecting debris and if the user needs further clear output, noise filtering can be done using Median filter [2] and if required segmentation can also be done based on the necessity.

2   Literature Survey

Zhiwen et al. [3] proposed an optical flow based object tracking method for moving objects using object contour. Here the velocity of moving object is estimated by Horn-Shunk algorithm which gets the moving pixels position between the frames. For tracking objects first segmentation is done which divides the digital image into multiple segments. Then histogram is computed for each image pixel. Based on the histogram value the objects are classified from the background. At the next step difference between the frames are calculated and their absolute difference is stored. At each pixel position the mean and variance are calculated that discriminates background from foreground. Histogram based segmentation is one of the segmentation techniques where histogram is computed for each image pixel and based on the histogram value and threshold the low intensity values are considered and the objects are classified. At every pixel position the difference between the frames are calculated and the absolute difference is stored. Using Single Gaussian Background method at each pixel position the mean and variance is calculated there by separates background and foreground. Features like colour, shape, edges any one of these is used to detect the object in a frame. Then the velocity and position is being calculated on an object in motion.

Koji et al. [4] proposed 1D optical flow for tracking the moving object with a rotating observer. They proposed a mapping method where a linear signal trajectory is created by converting the motion of a stationary environment object. From the mapped image, the object is detected by applying the block gradient method. This paper considers the case of the moving camera and stationary objects. The pinhole camera model is used for capturing the images. One dimensional optical flow axis spans both the sides including moving direction. The location of axis intersection is at the centre of object. If the viewer is permanent and the object is translated along the axis, then it takes positive values in one dimensional optical flow. By estimating the sum of the positive values the direction of the object is estimated. In case of the moving camera and stationary object, the apparent object motion has been eliminated by the estimation of one dimensional optical flow. The one dimensional optical flow window and direction is estimated by calculating the average brightness of window to get window data in each image sequence. The proposed method works well for a stable angular velocity of the revolving observer.

Kauleshwar et al. [5] proposed a competent methodology for tracking and extracting shape of the object in motion from a sequence of video. Diverse disturbances like noises object full and partial occlusions, scene illumination, object motion, complex object shape, requirements for real time processing has been eliminated using the proposed method. The steps involved in the generation of the system are taking the video input from an infra-red camera and generating frames. After frame generation the frames are grey sampled where the pace of every pixel is a sole sample. On a non-linear scale it allows 256 different intensities to be recorded. The image is then pre-processed thereby increasing the quality of the input image. Next background subtraction is being performed to filter out the objects from the frame that changes significantly from the background. This results in generating the object of interest. After this process image segmentation is being performed there by extracting the shape of the moving objects. Algorithm is easily to implement for tracking objects and estimating shapes.

3 Optical Flow

A series of consecutive frames allow motion estimation in any of the two ways as follows [6].

1. Instant image velocities
2. Discrete image displacements

The optical flow technique calculates the difference caused by the movement in between frame one and frame two taken at times t and $t + \Delta t$ respectively. Each pixel is assigned with a 2D vector, carrying direction and velocity of motion at a specific spot of the picture. To make the process simple, we transfer the 3D objects (coordinates of the real world) with time to 2D objects with time, so that the frame can be represented by 2D brightness function of time and location I (x, y, t). In case of change in the pixels nearby, the motion field doesn't influence the change in brightness intensity. The expression given below represents the above statement.

$$I(x, y, t) = I(x + \delta x, y + \delta y, t + \delta t) \tag{1}$$

Using Taylors series approximation for Eq. (1) we get

$$I(x + \delta x, y + \delta y, t + \delta t) = I(x, y, t) + \frac{\delta I}{\delta x} \delta x + \frac{\delta I}{\delta y} + \delta y + \frac{\delta I}{\delta t} + \delta t \tag{2}$$

After neglecting the higher order terms in Eqs. (1) and (2) we get,

$$I_x . v_x + I_y . v_y = -It \tag{3}$$

In vector representation

$$\nabla . I . \tilde{v} = -I_t \tag{4}$$

The above equation is called 2D Motion Constrain Equation or Gradient Constrain.

- ΔI—spatial gradient of brightness intensity,
- $\tilde{v}$—is the optical flow of the image pixel.
- I_t—time derivative of the brightness intensity.

The optical flow is estimated by calculating the partial derivatives of the image signal. The common methods are

- Lucas-Kanade
- Horn-Schunck

3.1 Lucas-Kanade

Lucas and Kanade used ρ_{LK}, the term of error for all the pixels. The following relation given below says that the addition of weighted least squares of gradient in pixels near by as shown in Eq. 4.

$$\rho_{LK} = \sum_{x, y \in \Omega}^{n} w * w(x, y)[\nabla I(x, y, t) . \tilde{v} + I(x, y, t)]^2 \tag{5}$$

Ω is the neighboring pixel, $W(x, y)$ is the weight of individual pixel in Ω.

In order to find the minimal error it is necessary to compute ρ_{LK}, the error term derived from the individual velocity components and making the result obtained equal to zero. At least the matrix representation of optical flow looks like

$$\tilde{v} = \left[A^T W^2 A\right]^{-1} A^T W^2 \dot{b} \tag{6}$$

For N number of pixels $N = n*n$ of Ω neighborhood and $(xi, yi) \in \Omega$ at time duration t

$$A = [\nabla I(xi, yi), \ldots, \Delta I(x_N, y_N)] \tag{7}$$

$$W = \mathrm{diag}[W(x_1, y_1), \ldots W(x_n, y_n)] \tag{8}$$

$$b = -[I_t(x_1, y_1), \ldots, I_t(x_n, y_n)] \tag{9}$$

We get the velocity for a single pixel as solution which is shown in Eq. (6). To reduce the computational complexity instead of calculating the sum, difference temporal gradient filter o Gaussian filter is used.

3.2 Horn-Schunk

In Horn-Schunk method in addition to gradient constrains, another error term named global smoothing term is added to deal with the limitations of major changes in the optical flow components (vx, vy) in Ω. The error minimization term is given by ρ_{HS}

$$\rho_{HS} = \int_D^0 \nabla I.\tilde{v} + It + \lambda*\lambda\left[\left(\frac{\delta vx}{\delta x}\right)*\left(\frac{\delta vx}{\delta x}\right) + \left(\frac{\delta VX}{\delta Y}\right)*\left(\frac{\delta VX}{\delta Y}\right) + \left(\frac{\delta VY}{\delta X}\right)*\left(\frac{\delta VY}{\delta X}\right) + \left(\frac{\delta vy}{\delta y}\right)*\left(\frac{\delta vy}{\delta y}\right)\right] dxdy \tag{10}$$

D is the area representing the entire image, λ. the relative error of the error term added after the first error term [7]. Using the relation we derive the system of equations which suits Gauss-Seidel or Jacobi methods [8] as shown in Eq. (10). Horn-Schunk method works comparatively good with better accuracy [9]. The demerit that Horn-Schunk hold is incase of a large number of iterations, it is slower.

4 Noise Filtering

The noise distribution is an annoying factor when they change the intensity level in a video frame. The corrupt pixels are divided into levels of intensity when compared to the neighbouring pixels [6], they are the pixels with relatively high and low intensity levels. To handle noises smoothing techniques have to be carried out for handling different noises. Hence, image filters are used for noise removal. Filters suppress noise in the homogenous regions, preserves the edges and removes random or constant values. Median filter runs through the image segment pixel by pixel and replaces each pixel with a median of pixels surrounding nearby. Hence, they produce considerably less blurring than linear filters [10].

5 Experimental Analysis

The dataset used for detecting debris in water bodies is a live footage of debris floating in the Java sea during the Air Asia flight crash and a manually footage of a ball floating in local water as shown below (Figs. 1, 2, 3 and 4).

Data set used:

Format	.mp4
Length	00:00:21
Frame Width	320
Frame Height	240
Frame Rate	29 frames/s
Data Rate	872 kbps
Total bit rate	1004 kbps

From the experimental results we say that the Exhaustive Search has highest PSNR, so it can be used for block matching. Whereas Three Step Search has minimum PSNR, hence it is not applicable for block matching.

Fig. 1 Ariel view of debris floating Java Sea during Air Asia flight crash (Video footage aerial view)

Fig. 2 Tracking debris in Java Sea

Fig. 3 Tracking the moving ball (Lucas-Kanade)

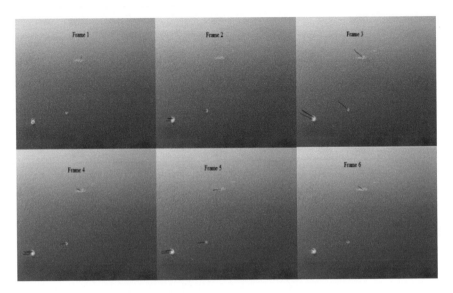

Fig. 4 Tracking object in motion in consecutive frames

5.1 Optimization of Motion Vectors Using Through Iterative Dynamic Programming and Block Matching Algorithm

Iterative optimization techniques are used to for better compensation performance and find optimal motion vectors. In case of using block matching algorithm, optimization task is done by obtaining an initial estimate and then each motion vectors are successively optimized. This result in over relaxed motion vectors, that leads to performance degradation. To handle the above problem we put forth an optimization scheme based on dynamic programming which solves dependent motion vector optimization problem. Our approach decomposes two dimensional dependency problems into a sequence of one-dimensional dependency problem. A reliable initial estimate of motion vectors can be obtained efficiently by only considering the dependency in the rate term. At the iterative optimization stage logarithmic search strategy can be used in combination with the dynamic programming to reduce the necessary complexity involved in distortion computation. In comparison with traditional iterative approaches, the experimental results demonstrate that our algorithm gives a superior rate and distortion performance and simultaneously maintains reasonable complexity. The performance of the block matching algorithms is tabulated in Table 1 and the best block matching algorithm suitable for optimization task is found to be Exhaustive search as per the experimental results shown in Fig. 5. Comparison of Peak Signal to Noise Ratio between Block Matching Algorithm and Iterative Dynamic Programming is shown below in Table 2.

Table 1 Performance metric in terms of mean square error for block matching algorithm

Search methods	Frame 1	Frame 2	Frame 3	Frame 4	Frame 5	Frame 6
Adaptive root pattern	35.51,612	37.41754	31.63679	23.87818	23.9247	26.85542
Exhastive search	36.87992	37.92657	32.0024	23.97436	24.00994	26.969
Theree step search	36.04878	36.7704	31.59786	23.83821	23.9533	26.8573
Simple and efficient search	35.68687	36.57812	31.31549	23.43987	23.4987	26.28658
New three step search	36.01888	36.48988	31.5957	23.835	23.95094	26.857
Four step search	35.97356	37.14228	31.10789	23.7251	23.85349	26.66327
Diamond search	35.58366	37.08438	31.25123	23.75781	23.86046	26.76388

Fig. 5 Graphical representation of mean square error

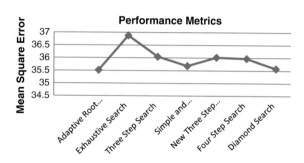

Table 2 Comparison of peak signal to noise ratio between block matching algorithm and iterative dynamic programming

Optimization technique	PSNR	Motion vector rate (bits)
Block matching algorithm (Exhaustive search)	36.87992	1904
Iterative dynamic programming	36.9235	1412

6 Conclusion

The paper presents a system for detecting and tracking motion in sequence of image from both aerial and stationary cameras. Debris detection in the huge water body is still a challenging issue. The proposed system employs Lucas Kanade and Horn-Schunk algorithms for optical flow estimation. The median filter plays a major role in noise reduction on frames compared to other filters. Experimentation is done using the real aerial video taken from the Java Sea during Air Asia flight crash. Experimental result says that both the algorithms work with better accuracy in detecting debris in water bodies from both front and aerial view.

References

1. Rakshit, S., Anderson, C.H.: Computation of optical flow using basis functions. IEEE Trans. Image Process. **6**(9), 1246–1254 (1997)
2. Tsekeridou, S., Constantine, K., Ioannis, P.: Morphological signal adaptive median filter for still image and image sequence filtering. In: Proceedings of the 1998 IEEE International Symposium on Circuits and Systems, 1998. ISCAS'98, vol. 4. IEEE (1998)
3. Chen, Z., Cao, J., Tang, Y., Tang , L.: Tracking of moving object based on optical flow detection. In: IEEE International Conference on Computer Science and Network Technology (ICCSNT) (2011)
4. Kinoshita, K., Enokidaniy, M., Izumida, M., Murakami, K.: Tracking of a moving object using one-dimensional optical flow with a rotating observer. In: IEEE 9th International Conference on Control, Automation, Robotics and Vision ICARCV'06
5. Prasad, K., Sharma, R., Design and simulation of a system for object tracking in video. Int. J. u- and e- Serv. Sci. Technol. **6**(4) (2013)
6. Brox, T., et al.: High accuracy optical flow estimation based on a theory for warping. In: Computer Vision-ECCV 2004, pp. 25–36. Springer, Berlin, Heidelberg, 2004
7. Horn, B.K., Schunck, B.G.: Determining optical flow. In: 1981 Technical Symposium East. International Society for Optics and Photonics (1981)
8. Hageman, L.A., Young, D.M.: Applied Iterative Methods. Courier Corporation (2012)
9. Silar, Z., Dobrovolny, M.: Comparison of two optical flow estimation methods using Matlab. In: 2011 International Conference on Applied Electronics, vol. 5 (2011)
10. Gonzalez, R.C., Woods, R.E., Eddins, S.L.: Digital Image Processing using MATLAB. Pearson Education India (2004)
11. Ke, Y., Sukthankar, R., Hebert, M.: Event detection in crowded videos. In: IEEE 11th International Conference on Computer Vision, 2007. ICCV 2007. IEEE, vol. 13 (2007)
12. Yamamoto, S., Mae, Y., Shirai, Y., Miura, J.: Realtime multiple object tracking based on optical flows. In: IEEE International Conference on Robotics and Automation, 0-7803-1965-6/95 (1995)
13. Müller, T., Rannacher, J., Rabe, C., Franke. U.: Feature- and depth-supported modified total variation optical flow for 3D motion field estimation in real scenes. In: IEEE Conference on Computer Vision and Pattern Recognition (CVPR) (2011)
14. Afonso, M.V., Marques, J.S., Nascimento, J.C.: Automatic estimation of multiple motion fields using object trajectories and optical flow. In: 1st International Conference on Pattern Recognition Applications and Methods, pp. 457–462 (2012)
15. Zoran, K.C., Slobodan, R.C., Vladimir, S.: Real-time object tracking based on optical flow and active rays. In: IEEE Electro Technical Conference (2000)

Understanding the Consequences of Social Isolation Using Fireworks Algorithm

Lourdes Margain, Alberto Ochoa, Teresa Padilla, Saúl González, Jorge Rodas, Odalid Tokudded and Julio Arreola

Abstract Social isolation, also known as "social withdrawal" occurs when a person is away from their environment completely involuntarily but might think otherwise. This condition occurs in people of all ages and can be a result of traumatic events in their history, such as being the victim of bullying or as part of any medical condition, such as depression. In this research we try to explain this social concept using a novel Bioinspired Algorithm named Firework Algorithm. Four minority groups in Chihuahua, have high levels of social isolation, thereby generating between different situations, principally school drop due to the lack of equal opportunities for the majority group. This research seeks to elucidate the reasons why it happens this by simulating social behavior. As future work, the research will be replicated in Aguascalientes.

Keywords Social isolation · Firework algorithm · School droops

L. Margain (✉) · T. Padilla
Polytechnic University of Aguascalientes, Aguascalientes, Mexico
e-mail: lourdes.margain@upa.edu.mx

T. Padilla
e-mail: mc140007@alumnos.upa.edu.mx

A. Ochoa (✉) · S. González · J. Rodas · J. Arreola
Juarez City University, Juarez, Mexico
e-mail: alberto.ochoa@uacj.mx

S. González
e-mail: saugonza@uacj.mx

J. Rodas
e-mail: jr.platon@gmail.com

J. Arreola
e-mail: julio.arreola@uacj.mx

O. Tokudded
National University of East Timor, Díli, Timor-Leste
e-mail: odalidtokudded@gmail.com

© Springer International Publishing Switzerland 2016
V. Snášel et al. (eds.), *Innovations in Bio-Inspired Computing and Applications*,
Advances in Intelligent Systems and Computing 424,
DOI 10.1007/978-3-319-28031-8_25

285

1 Introduction

At some point in our lives we have all felt the need to be alone and we leave a little of those around us, being totally normal. However, when this isolation is undefined and the person fails to maintain any kind of relationship with the rest of your environment, the situation should be seen as a problem that needs attention. Chihuahua, a state in northern Mexico has a social composition characterized by four minority: "Mormons, Mennonites, Rarámuris and Immigrants to the rest of Federation", in most elementary schools, these four minority groups are not considered for various activities due to differences: ethnic, religious and cultural factors that have, resulting have low rates of school performance due to various situations associated with social blockade and the lack of opportunities for their communities. INEGI records that settlement areas of the Mennonite population are Chihuahua, Zacatecas and Durango with almost 11,000 Mennonites, who are concentrated in the north of the country. In contrast, in the Central West region, the state of Aguascalientes recorded a minority of Mennonites by proximity to Zacatecas. Concerning the minority of Mormons in Aguascalientes, Aguascalientes there is a Mormon in every five in Chihuahua.

1.1 Causes of Social Isolation

Although we have conducted several studies to identify the specific causes of social isolation, the reality is that these are very diverse and depend on each particular situation. In some cases it may be because the person has lived abnormal conditions in childhood, such as being bullied or has been under extreme overprotection which prevented them to interact normally with other people their age, creating a lack of security and knowledge to establish new relationships as an adult. Another case is when the person has some kind of medical condition that difficult or impossible to get out. In this situation, people may find that after a medical accident completely away from other people. You can also find this condition in people who suffer from a serious medical condition, but facing away medical conditions that generate or stereotypes, as some mental disorders.

2 Proposal Methodology: Fireworks Algorithm

TechFerry is a company that has experience in Predictive Business Analytics, Machine Learning & AI, Analytics of Things (AoT), Massively Scalable Applications, Big Data, Rich UI & Usability, Mobility and has published an article about Swarm intelligence as an emerging field of biologically-inspired artificial intelligence based on the behavioral models of social insects such as ants, bees, wasps, termites etc. (Table 1).

Table 1 Swarm technologies

Algorithm example	Inspiration
Ant system	It is inspired by the pheromone communication of the blind ants regarding a good path between colony and the food source in an environment, the phenomenon known as stigmergy. The probability of the ant following a certain route is not only a function of pheromone intensity but also a function of distance to that city, the function known as visibility
Bees algorithm	It is inspired by the foraging behavior of the honey bees. The hive sends out the Scout bees which when locate nectar (a sugary fluid secreted within flowers), return to the hive and communicate the other bees the fitness, the quality, distance and direction of the food source via waggle dance
Gravitational search algorithm	Gravitational search algorithm (GSA) is a newly developed stochastic search algorithm based on the Newtonian gravity—"Every particle in the universe attracts every other particle with a force that is directly proportional to the product of their masses and inversely proportional to the square of the distance between them" and the mass interactions
Fireworks algorithm	It is inspired by observing the firework explosion

2.1 Fireworks Algorithm Framework

Strategy: In the FA, two explosion (search) processes are employed and mechanisms for keeping the diversity of sparks are also well designed. The explosion process of a firework can be viewed as a search in the local space around a specific point where the firework is set off through the sparks generated in the explosion [1] (Fig. 1).

We consider this behavior to simulate different bio-inspired algorithms evaluated in the literature, such as algorithms Cultural and Multi-Agent Systems Algorithm Fireworks, determining that the latter properly considered clustering between

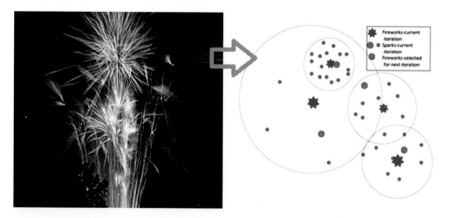

Fig. 1 Fireworks algorithm

Fig. 2 Framework of
fireworks algorithm

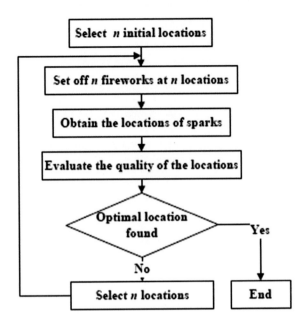

communities and how to visually show how social isolation increased with time in
the absence of a model of social integration, and public policies for this. Inclusive
we consider evaluate a model of predator-prey game to analyze the relationships
between these minorities and the rest of the majority group [2, 3].

When a firework is set off, a shower of sparks will fill the local space around the
firework [4, 5]. In our opinion, the explosion process of a firework can be viewed as
a search in the local space around a specific point where the firework is set off
through the sparks generated in the explosion. When we are asked to find a point xj
satisfying f(xj) = y, we can continually set off 'fireworks' in potential space until
one 'spark' targets or is fairly near the point xj. Mimicking the process of setting off
fireworks, a rough framework of the FA is depicted in Fig. 2. In the FA, for each
generation of explosion, we first select n locations, where n fireworks are set off.
Then after explosion, the locations of sparks are obtained and evaluated. When the
optimal location is found, the algorithm stops. Otherwise, n other locations are
selected from the current sparks and fireworks for the next generation of explosion.
From Fig. 2, it can be seen that the success of the FA lies in a good design of the
explosion process and a proper method for selecting locations, which are respectively elaborated in Sects. 2.2 and 2.3.

2.2 Design of Fireworks Explosion

Through observing fireworks display, we have found two specific behavior of
fireworks explosion. When fireworks are well manufactured, numerous sparks are

(a) **(b)**

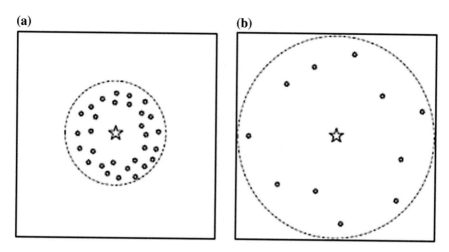

Fig. 3 Two types of fireworks explosion. **a** Good explosion. **b** Bad explosion

generated, and the sparks centralize the explosion center. In this case, we enjoy the spectacular display of the fireworks. However, for a bad firework explosion, quite few sparks are generated, and the sparks scatter in the space.

The two manners are depicted in Fig. 3. From the standpoint of a search algorithm, a good firework denotes that the firework locates in a promising area which may be close to the optimal location. Thus, it is proper to utilize more sparks to search the local area around the firework. In the contrast, a bad firework means the optimal location may be far from where the firework locates. Then, the search radius should be larger. In the FA, more sparks are generated and the explosion amplitude is smaller for a good firework, compared to a bad one [6].

Number of Sparks. Suppose the FA is designed for the general optimization problem:

$$\text{Minimize } f(x) \in \mathbb{R}, \; x_{\min} \leqslant x \leqslant x_{\max}, \tag{1}$$

where $x = x1, x2, ..., xd$ denotes a location in the potential space, $f(x)$ is an objective function, and $x$min and $x$max denote the bounds of the potential space.

Then the number of sparks generated by each firework xi is defined as follows.

$$s_i = m \cdot \frac{y_{\max} - f(x_i) + \xi}{\sum_{i=1}^{n} (y_{\max} - f(x_i)) + \xi}, \tag{2}$$

where m is a parameter controlling the total number of sparks generated by the n fireworks, ymax = max(f(xi)) (i = 1, 2, ..., n) is the maximum (worst) value of the objective function among the n fireworks, and ξ, which denotes the smallest constant in the computer, is utilized to avoid zero-division-error.

To avoid overwhelming effects of splendid fireworks, bounds are defined for si, which is shown in Eq. 3.

$$\hat{s}_i = \begin{cases} round(a \cdot m) & \text{if } s_i < am \\ round(b \cdot m) & \text{if } s_i > bm \, , \quad a < b < 1, \\ round(s_i) & \text{otherwise} \end{cases} \tag{3}$$

where a and b are const parameters.

Amplitude of Explosion. In contrast to the design of sparks number, the amplitude of a good firework explosion is smaller than that of a bad one. Amplitude of explosion for each firework is defined as follows.

$$A_i = \hat{A} \cdot \frac{f(x_i) - y_{\min} + \xi}{\sum_{i=1}^{n}(f(x_i) - y_{\min}) + \xi}, \tag{4}$$

where $\hat{A}$ denotes the maximum explosion amplitude, and $Y_{\min} = \min(f(x_i))$ $(i = 1, 2, \ldots, n)$ is the minimum (best) value of the objective function among the n fireworks.

Generating Sparks. In explosion, sparks may undergo the effects of explosion from random z directions (dimensions). In the FA, we obtain the number of the affected directions randomly as follows.

$$z = round(d \cdot rand(0, 1)), \tag{5}$$

where d is the dimensionality of the location x, and rand(0, 1) is an uniform distribution over [0, 1].

The location of a spark of the firework xi is obtained using Algorithm 1.

Mimicking the explosion process, a spark's location $\tilde{x}j$ is first generated. Then if the obtained location is found to fall out of the potential space, it is mapped to the potential space according to the algorithm.

Algorithm 1. Obtain the location of a spark

Initialize the location of the spark: $\tilde{x}_j = x_i$;
$z = round(d \cdot rand(0,1))$;
Randomly select z dimensions of $\tilde{x}_j$;
Calculate the displacement: $h = A_i \cdot rand(-1,1)$;
for each dimension $\tilde{x}_k^j \in \{$ pre-selected z dimensions of $\tilde{x}_j\}$ **do**
    $\tilde{x}_k^j = \tilde{x}_k^j + h$;
    if $\tilde{x}_k^j < x_k^{\min}$ or $\tilde{x}_k^j > x_k^{\max}$ **then**
        map $\tilde{x}_k^j$ to the potential space: $\tilde{x}_k^j = x_k^{\min} + | \tilde{x}_k^j | \% (x_k^{\max} - x_k^{\min})$;
    end if
end for

To keep the diversity of sparks, we design another way of generating sparks — Gaussian explosion, which is show in Algorithm 2. A function Gaussian(1, 1), which denotes a Gaussian distribution with mean 1 and standard deviation 1, is utilized to define the coefficient of the explosion. In our experiments, ^m sparks of this type are generated in each explosion generation.

Algorithm 2. Obtain the location of a specific spark

Initialize the location of the spark: $\hat{x}_j = x_i$;
$z = round(d \cdot rand(0,1))$;
Randomly select z dimensions of $\hat{x}_j$;
Calculate the coefficient of Gaussian explosion: $g = Gaussian(1,1)$;
for each dimension $\hat{x}_k^j \in \{$pre-selected z dimensions of $\hat{x}_j\}$ do
    $\hat{x}_k^j = \hat{x}_k^j \cdot g$;
    if $\hat{x}_k^j < x_k^{min}$ or $\hat{x}_k^j > x_k^{max}$ then
        map $\hat{x}_k^j$ to the potential space: $\hat{x}_k^j = x_k^{min} + |\hat{x}_k^j| \%(x_k^{max} - x_k^{min})$;
    end if
end for

2.3 Selection of Locations

At the beginning of each explosion generation, n locations should be selected for the fireworks explosion. In the FA, the current best location x∗, upon which the objective function f(x∗) is optimal among current locations, is always kept for the next explosion generation. After that, n − 1 locations are selected based on their distance to other locations so as to keep diversity of sparks. The general distance between a location xi and other locations is defined as follows.

$$R(x_i) = \sum_{j \in k} d(x_i, x_j) = \sum_{j \in k} \left\| x_i - x_j \right\|, \tag{6}$$

where K is the set of all current locations of both fireworks and sparks.

Then the selection probability of a location xi is defined as follows.

$$p(x_i) = \frac{R(x_i)}{\sum_{j \in K} R(x_j)}. \tag{7}$$

When calculating the distance, any distance measure can be utilized including Manhattan distance, Euclidean distance, Angle-based distance, and so on [7]. When d(xi, xj) is defined as If(xi) − f(xj)l, the probability is equivalent to the definition of the immune density based probability in [8].

3 Results of Our Research

Consequences of social isolation to our study describe that this Bioinspired Algorithm can describe better whom living under this condition often face different situations and problems, the most common and severe depression. However, several studies have been conducted indicate that people living social isolation often have learning disabilities, attention and decision making [9, 10].

This is because when we interact with our environment, our brain does not receive the appropriate stimulus and does not work properly. For this reason, people who live in social isolation may seem a bit clumsy or slow when making decisions [7, 8].

For proper testing our algorithm Fireworks analyze a database related to dropout in Chihuahua, a state in northern Mexico, with four minorities, two ethnic, religious and one for ubiquity. To do this we detail each of the situations that occur in schools first level and thus try to properly determine how you could view each student to discover whether there were clusters among minorities or each decided survive in a lonely way the day by day according at its members [11].

There are several situations that cause dropouts is given, this due to social cohesion and the time, most of the communities that make Chihuahua cannot be justified as a cause of dropout only the relationship of coexistence with the largest group among the various causes unemployment affecting 9.87 % of the population economically and even Chihuahua Ciudad Juarez are reaching a figure of 14.17 % in January 2012, the other is primarily involved with food because indigenous groups have lower (21 %) than the average caloric intake of the majority group.

Once interviewed in written form, the selected sample a model of social distance Bogardo (See Fig. 4), which reflected us variant forms of ascertaining the situation in each of the four minorities analyzed in Chihuahua was developed and how it affects the daily lives of people who are not considered part of the majority group.

4 Recommendations of Our Research

There are many considerations that must be taken into account with respect to this research, the first is that while there are no adequate social programs that affect the minority population, may not achieve an integration model that allows to give the children and young people, equal opportunities, one of the crucial aspects in this proposed model using bio-inspired algorithm was to determine the number of individuals in these communities that were continuing at the higher level or graduate school was reduced to values less than 1 %.

A public policy proposal for its implementation, would be a self-evaluation as follows: If you've noticed that you live under this condition and you really want out of it, the first thing to do is find the social circles in which you feel comfortable and you start to visit. For example, if you like dogs and you have one; you can take a

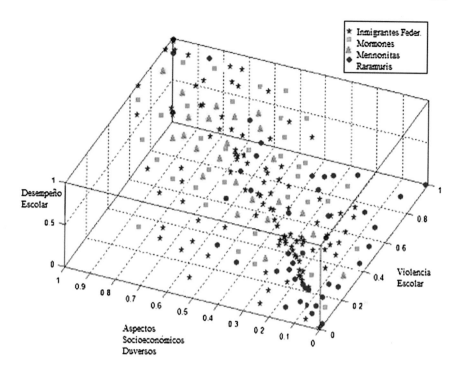

Fig. 4 Visual representation of the sample, featuring diverse aspects as: socioeconomic, school violence including social locking and reflected in school

walk in the park. This way, you will begin to socialize with other dog owners. To exit the social isolation is essential that you open to new experiences, to learn to get out of your comfort zone and put yourself in new situations.

One of the important aspects that are blurred in this study, is that individuals in these communities in urban areas considered not owning a pet to feel they could not display it, while people in rural areas had on average up to three pets, even squirrels, raccoons and spotted salamanders, unlike dogs, hamsters and fish orna-ment popular among children related with the majority social group.

There are few studies that adequately detail the use of a bioinspired algorithm to properly classify social aspects, and we believe it will be useful to predict properly serious problems such as bullying, social discrimination, or in our case the social isolation that many of cases carries with him as the final result suicide.

In some social modelling groups in videogames inclusive in serious games as "The Tribe", the people more different has less social skills or is considerate more vulnerable to die, simulations analyzed with another bioinspired algorithms as in [12].

In future research we try to: (a) testing our algorithm Fireworks analyze a database related to dropout in Aguascalientes and (b) consider another society with

Fig. 5 A Coahuilan
Seminole Lass

(a)   **(b)**

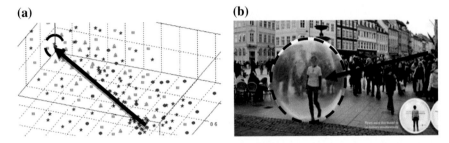

Fig. 6 Comparison between visual representation and solution search insolation problem.
a Visual representation. **b** Isolation problem

much multidiversity as: China, India or Federation Russia with almost over 47
minorities and analyze the social politic publics applied in them. We try to simulate
more minorities groups in Mexico, for example exists 57 dialects spoken, and
people that descending from African roots as Seminoles in Coahuila (see Fig. 5)
and Scottish and Welsh pirates as in Tabasco, but these minorities groups try to
hidden these situations in the school, the reason the social isolation, a pendent
signature to the next generation.

In another way to improve our future research, we will be use a model based on a
colored graph to understand the similarity between specific communities without
descriptive values associated with locations, as the propose in [13], with this model
we featuring different attributes in a society as is shown in Fig. 6a. With this
descriptive model we try to explain diverse situations associated with a gender roll
in a community and prevent complex social behaviors as cyberbullying, in our case
prevent discrimination associated with not have a written language.

References

1. Techferry Company, consulted September 10, 2015. http://www.techferry.com/articles/swarm-intelligence.html
2. Krivan, V.: Dynamic ideal free distribution: effects of optimal patch choice on predator-prey dynamics A. Nature **149**, 164–178 (1997)
3. Lina, S.L.: Putting predators back into behavioral predator-prey interactions. Trends Ecol. Evol. **17**, 70–75 (2002)
4. Liu, J., Zheng, S., Tan, Y.: Analysis on global convergence and time complexity of fireworks algorithm. IEEE Congress on Evolutionary Computation 2014, pp. 3207–3213
5. Liu, J., Zheng, S., Tan, Y.: Adaptive fireworks algorithm. IEEE Congress on Evolutionary Computation 2014, pp. 3214–3221
6. Zheng, S., Janecek, A., Li, J., Tan, Y.: Dynamic search in fireworks algorithm. IEEE Congress on Evolutionary Computation 2014, pp. 3222–3229
7. Thomas, C.D.: Predator-herbivore interactions and the escape of isolated plants from phytophagous insects. Oikos **55**, 291–298 (1989)
8. Van Baalen, M., Sabelis, M.W.: Coevolution of patch selection strategies of predator and prey and the consequences for ecological stability. Am. Nat. **142**, 646–670 (1993)
9. Parker, G.A.: Sexual selection and sexual conflict. In: Blum, M.A., Blum, N.A. (eds.) Reproductive Competition and Sexual selection. Academic Press, New York, pp. 124–166 (1979)
10. Sih, A., McCarthy, T.M.: Prey responses to pulses of risk versus pulses of safety: testing the risk allocation hypothesis. Anim. Behav. **63**, 437–443 (2002)
11. Stiling, P.D.: The frequency of density-dependence in social conflicts. Soc. Modell. J. **21**, 844–856 (1987)
12. Ochoa, A., Quezada, S., Ponce, J., Ornelas, F., Torres, D., Correa, C., de la Torre, D., Meza, M.: From Russia with Disdain: simulating a civil war by means of predator/prey game and cultural algorithms. In: Sidorov, G. (ed.) Artificial Intelligence for Humans: Service Robots and Social Modeling, Oct 2008, Mexico (2008). 978-6-07000-478-0 137–145
13. Cruz-Álvarez, V.R., Montes-Gonzalez, F., Ochoa, A., Palacios-Leyva, R.E.: Distribution and selection of colors on a diorama to represent social issues using cultural algorithms and graph coloring. DCAI 2012:57–64

Ant Pheromone Evaluation Models Based Gateway Selection in MANET

Naveen Kumar Gupta, Rakesh Kumar, Amit Kumar Gupta
and Prakash Srivastava

Abstract A mobile ad hoc network (MANET) is an infrastructure less, low cost, small range autonomous network where mobile devices can share the data as well as resources. If any mobile node in MANET wants to communicate with fixed host, it requires discovering an appropriate Internet gateway. A novel adaptive gateway discovery scheme using ant like mobile agent has been proposed in this paper. Ant releases a chemical substance called Pheromone. In this paper, different pheromone evaluation models are used to calculate this pheromone value. On the basis of pheromone value, traffic and route stability has been analyzed. These models provide an adaptive route to a gateway in different conditions. A discovery value is computed and used in selecting an optimal gateway in case of multiple gateways. The analytical study carried out in this paper validate that ant scheme with different model provides a better route discovery than existing ones. The proposed scheme also provides a stable and an optimal route between a mobile node and particular gateway.

Keywords Ant-like mobile agent · Pheromone value · Pheromone · Evaluation model · Discovery value

N.K. Gupta (✉)
Department of Computer Science and Engineering,
Motilal Nehru National Institute of Technology, Allahabad, India
e-mail: naveenkumar.gpt@gmail.com

R. Kumar · P. Srivastava
Department of Computer Science and Engineering,
Madan Mohan Malaviya University of Technology, Gorakhpur, India
e-mail: rkiitr@gmail.com

P. Srivastava
e-mail: prakash2418@gmail.com

A.K. Gupta
Image Enterprises, Gurgaon, Haryana, India
e-mail: amitguptaedu053@gmail.com

© Springer International Publishing Switzerland 2016
V. Snášel et al. (eds.), *Innovations in Bio-Inspired Computing and Applications*,
Advances in Intelligent Systems and Computing 424,
DOI 10.1007/978-3-319-28031-8_26

297

1 Introduction

A MANET is an autonomous, self configured, infrastructure less in nature where each mobile node can move arbitrary within a particular range. A node in MANET can use the resources of another node but a node in MANET cannot communicate outside the network or Internet without an Internet gateway. So gateway discovery techniques are needed as discussed in this paper. If more than one gateways are discovered then need to select the best one on the basis of different parameters e.g. hop counts, traffic, congestion, load etc. so a new parameter "Discovery Value (DV)" is also proposed in this paper. If more than one gateway is discovered then selection of gateway will depend on DV. A gateway with minimum discovery value will be selected. This discovery value is a well defined calculation of hop counts, stability, load, traffic and security parameters, etc.

Ant like mobile agents (ALMA) provides an optimal route to Internet gateways. Ants use routing algorithms [5, 6, 10] and provide a route to the destination. When the ant moves towards the food source then they release a special chemical substance called pheromone. Other ants follow the same path based on pheromone concentration i.e. decreases with time. If any ant found a shorter path then this path will have higher pheromone concentration. So other ants will follow a path with maximum pheromone concentration. So in this way, an optimal path to the destination is discovered. The pheromone values are network state indicator. If any variation occurs because of link breakage or congestion then pheromone values will change which results traffic flow variation.

The remainder paper is organized as follows: Related work and literature reviews are discussed in Sect. 2. In Sect. 3 different ant pheromone evaluation models are discussed. The Proposed gateway selection scheme is described in Sect. 4. Section 5 describes the Analytical model. Analytical proof of proposed scheme is given in Sect. 6. Finally, Sect. 7 gives the conclusion about the paper.

2 Related Work

Yi et al. [10] showed that the ant system can be utilized for multicast routing in the MANET and proposed an improved ant based routing algorithm for MANET and compared with other routing protocols. After simulation he analyzed that ant based routing algorithm works better than proactive and reactive routing algorithms in some situations. Fernando Correia et al. [4] studied ants behavior based on pheromone concentration and simulate with MANET routing problems. After studying, he proposed some models for pheromone value evaluation in MANET. In this paper these models are used for gateway discovery. We have studied the evaluation of ant system and analyzed that ant related algorithms can easily solve any routing related problems.

Ayyadurai et al. [1] proposed a hybrid routing scheme which combines Ant and AODV, where AODV finds the path in local MANET and ALMA finds the path towards gateway. In this scheme, they used AODV which suffers from initial delay and they did not consider network load and stability factor to select appropriate gateway. Bin et al. [2] proposed an adaptive Gateway discovery scheme. In this approach, TTL values of gateway advertisement messages are adjusted dynamically according to the Internet traffic generated by the mobile nodes and their relative location from Internet Gateways with to which they are registered but they did not consider the stability factor.

An efficient load aware gateway discovery approach was proposed by Srivastava et al. [9] where load is considered but path may be unstable and continuously changing intermediate routes. Pandey et al. [8] proposed congestion avoiding gateway selection scheme and used proxies to reduce the network overhead but path with less congestion may not be optimal path. Bo et al. [3] uses mobile IP to discover optimal gateway but mobile IP suffers from frequent handovers which is undesirable process. Yuste et al. [11] proposed an adaptive gateway discovery scheme where interval of emission of gateway advertisement is dynamically adjusted. This technique is shown to give better performance than conventional proactive gateway discovery scheme but it is not efficient.

Zaman et al. [12] proposed an adaptive gateway discovery approach based on path load balanced. This approach primarily focuses on maximal source coverage. To cover maximum geographical area, this technique can provide better end-to-end delay and packet delivery ratio and in some specific cases only. Zhanyang et al. [13] proposed a simplified scheme for Internet connectivity where they used the concept of virtual MANET and consider that all Internet gateways are at fixed position but moving (dynamic) gateway node may exist.

3 Pheromone Evaluation Models

When the ants go to search food, they always want to search for a better route. So initially they go in different (arbitrary) direction and release pheromones. Pheromones concentration decrease with time and long route takes long time. So a route with the higher pheromone concentration is a better path. Finally best route is selected and other ants follow the same route for food collection. Similarly in a networking environment food source is an Internet gateway and ants are data packets. In the networking environment virtual ants are considered so ants will release pheromone trail in the networks. Pheromone evaporates with time and reaches to 0 and if any other ant passes through the same route then it will reinforces (increases). These pheromone values are calculated by formula:

$$\emptyset_{j,k,t} = \varepsilon . \emptyset_{j,k,t-1} + \sum_{i=1}^{n} \Delta\emptyset_{j,k,t}^{i} \tag{1}$$

where

$$\Delta\emptyset_{j,k,t}^{i} = \begin{cases} \frac{1}{d_{j,k}} & if\,(j,k) \in path\,use\,by\,ant\,i \\ 0 & otherwise \end{cases}$$

where $\emptyset_{j,k,t}$ is pheromone value of link between node j to node k at time t. $\Delta\emptyset_{j,k,t}^{i}$ represents the value of pheromone evaporation reinforcement and ε $(0 < \varepsilon < 1)$ is evaporation rate. The total numbers of ants are 'n'. This paper intended to identify the problems related to MANET, like frequent change in topology, broken link, congestion etc. To achieve this objective, four pheromone evaluation models [4] are discussed as follows.

3.1 Ant System Pheromone Model

In Eq. (1) pheromone value is not depending on the previous network state, but in network environment routing always depends on the previous state. So it is modified to Eq. (2) as follows:

$$\emptyset_{j,k,t} = \varepsilon . \emptyset_{j,k,t-1} + \sum_{i=1}^{n} \Delta\emptyset_{j,k,t}^{i} \wedge \Delta\emptyset_{j,k,t-1}^{i} \tag{2}$$

This model identifies pheromone values on three following different phases:

- Learning phase: In this phase, the pheromone value increases in a logarithmic manner until reaching a stationary value.
- Maintenance phase: In this phase, the pheromone value remains constant with small variations. Here, the reinforcement rate is equal to evaporation rate.
- Evaporation phase: When data transfer through the link is over then pheromone value decreases rapidly until reaches '0'.

The Ant System pheromone model increases its pheromone intensity until reaches a stationary value when the link capacity has enough resources to transport the packets. It presents a fast growth rate to reach a stable value. When the data transfer session ends, or when in a presence of a bottleneck due link problem, the AS model will react to this and present a fast decrease of the pheromones. This model can be useful when fast response is required. However, on a network, changes on traffic transfer rate like packet jitter or a burst of packets could create a response similar to link bottleneck and give wrong information about the network state.

3.2 Temporal Active Pheromone (TAP) Model

The ant system pheromone model suffers from incorrect result in case bottleneck. In TAP model, even if the path is idle, pheromone value will active for a certain time.

$$\emptyset_t = \begin{cases} 1\,(active) & t\_set < t < t\_set + \delta \\ 0\,(evaporation) & t > t\_set + \delta \end{cases}$$

$$T_t = \sum \emptyset_t \tag{3}$$

where $\emptyset_t$ and δ represents the single pheromone state at time t and the duration of activity respectively. The pheromone intensity of a link (T_t) is the addition of all deposited pheromones on the link. On completion of data transfer, the reinforcement rate decreases to 0. So network state can be analysed during the maintenance phase. This model needs to simulate at each node which requires more memory and computational capacity at each node. If the network has sufficient resources to transfer the packet then pheromone variation will be about to zero. In case of broken links or congestion, the pheromone refreshes value will change. In this model, decrease rate is linear, so these effects can be neglected and the link can get its pheromone value when lost. Due to these reasons, the temporal active pheromone model is more stable to the network state changes than the Ant System pheromone model.

3.3 Progressive Pheromone Reduction (PPR) Model

This model uses relatively less resources in computation of intensity of pheromone. Traffic variation can be analysed by pheromone increase rate. The formulas of PPR model are described as follows:

$$\emptyset_t = \emptyset_{t-1} + \rho$$

$$\emptyset_t = \emptyset_{t-1} - \varepsilon \tag{4}$$

where

$$\varepsilon = \begin{cases} 1 & \text{with link activity} \\ \varepsilon*2 & \text{without link activity} \end{cases}$$

where ρ is the increase rate (initially set to 1) and $\emptyset_t$ is pheromone value. When this increase rate changes, it means that the network is suffering from some problem. If the packets are passing through the link then it represents the activeness of the link and ε is set to 1, however, in case of traffic jam, ε updates its value every time the evaporation procedure is called and it represents the idle state of that link. Thus, minor traffic problems like packet burst, jitter or route repair procedure cannot affects the pheromone value, but if any link breaks during data transfer session then it will cause progressive pheromone evaporation and without reinforcement, the pheromone value decreases quickly to 0. On stopping the pheromone activity, the route can be released.

3.4 Progressive Pheromone Reduction with Maximum Value (PPR-MV) Model

This model is about equivalent to the PPR model, but here maximum pheromone value can't be greater than MAX_VAL. The formula of PPR-MV model is defined as follows:

$$\emptyset_t = \begin{cases} \emptyset_{t-1} + \rho & \text{if} \quad \emptyset_{t-1} < MAX\_VAL \\ MAX\_VAL & \text{otherwise} \end{cases} \tag{5}$$
$$\emptyset_t = \emptyset_{t-1} - \varepsilon$$

where

$$\varepsilon = \begin{cases} 1 & \text{with link activity} \\ \varepsilon * 2 & \text{without link activity} \end{cases}$$

PPR-MV model also considers learning phase, maintenance phase and evaporation phase. The learning phase is related to the session packet rate in the path and it ends when the pheromone value reach the maximum that has been selected. Maintenance phase represents that data transfer session is active and the maximum pheromone value is maintained. In this model, network state identification is difficult because of slow reaction to the network state changes. This model has ability to differentiate between active routes and idle routes.

4 Proposed Gateway Discovery Approach with Ant

In our proposed scheme, each mobile node is considered as home of ants, where each data packet is considered as a single ant. The gateway node is equivalent to the food source of ants. In the proposed algorithm we have used proper balance of hop count, load and stability factor. So we can get an optimal gateway with suitable route.

4.1 ANT Based Routing Protocol

Ant-based routing algorithms have been explored by Marwaha et al. [5–7]. The ants move node to node randomly and update the routing tables of visiting nodes. In ant-based routing algorithm, each ant works independently. In the conventional ant

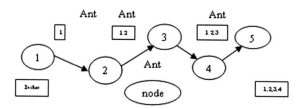

Fig. 1 Ant traversing the network and providing routing information to nodes

Table 1 Entries in routing table of mobile node

Destination address	Next hop address	Number of hops
Pheromone value	Pheromone decay time	Balance index

algorithms the next hop is selected randomly, but this paper implements "no return rule" [7] while selecting the next hop. Routing ants contain a history of visited nodes. The General behavior of ant is shown in Fig. 1.

In our gateway discovery scheme, discovery time is reduced because ALMA takes the responsibility to discover the gateway and provides a highly dynamic approach to discover an optimal gateway in different conditions. It also provides automatic handover to another gateway while the running gateway becomes unavailable. Each node frequently broadcasts ADV messages to its neighbour nodes. The ADV message helps in maintaining the neighbor list. This used for selecting the next neighbour node by the ants. The hybrid mechanism operates simultaneously when a route to the destination is needed by the source mobile node. In proposed scheme, the routing table of the mobile node is updated dynamically as shown in Table 1.

4.2 Gateway Discovery Scheme

The source mobile node initiates a gateway discovery process by sending forward ant (FANT) message to its neighbours. FANT maintains a history field of visited intermediate nodes. Each neighbor multicasts FANT until it reaches gateway node. The intermediate node receives FANT, update ant's history and resend to a neighbour with maximum probability. On reaching FANT to Internet gateway, it checks whether it is a first ant with the same source. If any other ant reaches to the same gateway previously then drop this ant otherwise it will update the routing table of gateway with source node, next hop, hop count and pheromone value and then gets converted into backward ant (BANT) message.

The BANT packet contains information about gateway (address, pheromone value, balance index etc.), intermediate nodes and source node. The balance index is a ratio of current gateway load and gateway capacity. The BANT packet is unicast from the gateway to the source mobile node will pass through the list of intermediate nodes visited by the FANT packet. When BANT reaches to its source then update the source routing table with information generated. The source mobile node receiving more than one BANT packet selects the adaptive gateway and a route with minimum DV.

In the proposed algorithm, the PPR-MV model is used for calculating the pheromone values because if nodes have relatively less resources and gateway node is much far from the mobile node then PPR-MV model works efficiently. This model gives the best result in normal conditions. The proposed algorithms are as follows:

Algorithm: Gateway Discovery and Route Selection

'no return rule' is considered in the algorithm. Main function for gateway discovery is given as follows:

```
1. begin
2. Initially set pheromone value for each link Øi,j =0
3. if (S wants to send/receive data to/from internet host)
4.      set S as a 'source'
5.      GW_Discovery(source)
6. Select a gateway with minimum DV value
7. end
```

The above algorithm is a main algorithm for gateway discovery. In this algorithm pheromone values of all links are initialized to zero. If any mobile node wants to use Internet then it calls GW_Discovery (source) algorithm. The main algorithm of the gateway discovery uses the following algorithms:

Algorithm I. (Source mobile node)

```
GW_Discovery (source)
1. initiate FANT[], fant_id, index ← 0
2. FANT[index++] ← fant_id
3. FANT[index++] ← source
4. for (each neighbour Ns ∈ neighbour of source)
5.      if (Ns == GW) then  GW_Discovered (Ns, index, FANT)
6.      else    GW_FRouting (Ns, index, FANT)
```

The source node initiates ant packet. This ant packet is in form of array. The first index of this ant contains ant ID and second index contains source ID of ant. After initialization it is broadcasted to all of its neighbours. When it reaches to any neighbour then it calls IGW_Discovered (Ns, index, FANT) or IGW_FRouting (Ns, index, FANT) algorithm based on given condition. Here 'Ns' is neighbour of source node.

Algorithm II. (Intermediate mobile node)

```
GW_FRouting (MN, index, FANT[])
1. if(MN ∈ FANT[]) then drop packet and exit
2. FANT[index++] ← MN
3. for (each neighbourNmn ∈ neighbour of MN)
4.      if(Nmn == GW) then  GW_Discovered (Nmn, index, FANT)
5.      else if(∅MN,Nmn,t−1 < MAX_VAL)
6.              ∅MN,Nmn,t  =∅MN,Nmn,t−1 + ρ
7.      else    ∅MN,Nmn,t = MAX_VAL
8.      if (t % max_time ==0)
9.              ∅MN,Nmn,t  =∅MN,Nmn,t−1 − ε
10.             if (link is active) then    ε = 1
11.             else    ε = ε * 2
12.         update entries of Table of node
13.         GW_FRouting (Nmn, index, FANT)
14. end for
```

The above algorithm checks the ant history. If it already contains the current node ID then ant packet will be dropped. Otherwise ant packet will update node ID in next empty slot. The current intermediate node will update its entries and again broadcast the FANT to its neighbors until gateway node is discovered.

Algorithm III. (Intermediate mobile node)

```
GW_BRouting (MN, N, index, BANT[])
1. if (index==1)
2.       print(path information is present in BANT)
3.       return BANT
4. else if (∅BANT [index ],BANT [index +1],t−1 < MAX_VAL)
5.       ∅BANT [index ],BANT [index +1],t  =∅BANT [index ],BANT [index +1],t−1 + ρ
6. else  ∅BANT [index ],BANT [index +1],t = MAX_VAL
7. if (t % max_time ==0)
8.       ∅BANT [index ],BANT [index +1],t =∅BANT [index ],BANT [index +1],t−1 − ε
9.       if (link is active) then ε = 1
10.      else  ε = ε * 2
11. update entries of Table of node
12. index=index-1
13. GW_BRouting(BANT[index], N, index, BANT)
```

The above algorithm is also applied on intermediate mobile node but this algorithm is applies when gateway is discovered. This procedure applies in reverse direction i.e. from gateway node to source node so this ant packet is also known as BANT. The BANT packet contains reverse path. The ant packet follows the route according to intermediate node entry present in BANT packet. The selected route is present in BANT packet.

Algorithm IV. (Gateway node)

```
GW_Discovered (GW, index, FANT[])
1.  if(GW ∈ FANT[]) then drop packet and exit
2.  initialize BANT[]
3.  balance_index ← (current_gateway_load / gateway_capacity)
4.  DV ← Hop*F_H +balance_index*F_L + F_S/RS
5.  FANT[index] ← DV
6.  for(p=0 to index)
7.         BANT[p] ← FANT[p]
8.  N ← index
9.  if(∅_GW,FANT [index 1],t 1 < MAX_VAL)
10.        ∅_GW,FANT [index −1],t =∅_GW,FANT [index −1],t−1 + ρ
11. else  ∅_GW,FANT [index −1],t = MAX_VAL
12. if (t % max_time ==0)
13.        ∅_GW,FANT [index −1],t =∅_GW,FANT [index −1],t−1 − ε
14.        if (link is active)  then  ε = 1
15.        else      ε = ε * 2
16. update entries of Table of node
17. index = index - 1
18. GW_BRouting (BANT[index], N, index, BANT)
```

The above algorithm IGW_Discovered (GW, index, FANT []) used when gateway is discovered by FANT packet. If FANT is visited previously then simply drop this packet otherwise gateway node calculates balance index and Discovery Value. We decided fraction of each parameter in simulation part. So, gateway node will calculate discovery value and put it into last slot of FANT packet. After calculating DV at gateway node, FANT will convert to BANT and all values with index copied to BANT packet. Gateway node also contains a routing table and stores all information same as intermediate mobile node with one extra field i.e. discovery value. At the end, it calls IGW_BRouting algorithm until BANT reached at source mobile node.

5   Analytical Model

Through our analytical model, the proposed approach is proved analytically. Each node has a communication range (R), where it can communicate with other node i.e. neighbour node. In the proposed scheme, a pheromone value $\emptyset_{i,j}$ is assigned to every link between node 'i' and node 'j'. The Eq. (6) computes the total probability of the link towards the gateway 'g'. The mobile agent located at particular node uses pheromone value $\emptyset_{i,j,g}$ to calculate the link probability $P_{i,j,g}$ towards the gateway g. The specific next hop neighbour is chosen according to the probability distribution in each link.

Table 2 Description of symbols used

Symbol	Description
$\emptyset_{i,j}$	Pheromone value of edge $e(i,j) \in E$
$P_{i,j}$	Probability value of edge $e(i,j) \in E$
Vs	Source node (vertex)
R	Mobile node's communication range
T	Pheromone threshold value (constant)
S	Pheromone sensitivity value (constant)
E	Set of all edges (links) in wireless network
N_i	Set of neighbors of node i
N	n = |V| number of nodes in the network
V_g	Gateway node (vertex)

$$P_{i,j,g} = \begin{cases} \dfrac{(\emptyset_{i,j,g}+T)^S}{\sum_{j=1}^{N_i}(\emptyset_{i,j,g}+T)^S} & \text{if } j \in N_i \\ 0 & \text{otherwise} \end{cases} \tag{6}$$

In the above equation, S and T are pheromone sensitivity value and threshold value respectively. $S \geq 0$, used to modulate the differences between pheromone amounts present in link probability. The value $S < 1$ evaporate the link, while $S > 1$ will reinforce the differences between links. The value S equal to 1 gives the normal form. If $T \geq 0$ larger then large amounts of pheromone will be present before an appreciable effect will be seen in the link probability. The link probability $P_{i,j,g}$ of the node i fulfills the constraint in giving transition probabilities as in Eq. 7.

$$\sum_{j \in N_i} P_{i,j,g} = 1, \ i \in [1 \ldots N] \tag{7}$$

During the route finding process to the gateway, mobile agents deposit pheromone on the edges. The pheromone concentration is varying from one edge to another between the nodes connecting them by an amount $\Delta\emptyset$. The pheromone value changes at the edge e(i, j) when mobile agents moving from node i to node j are given in Eq. 8.

$$\emptyset_{i,j,g} = \emptyset_{i,j,g} + \Delta\emptyset \tag{8}$$

The selection decision gives the fact that the path length between the nodes connecting the gateway is less than R and the route selected is stable. The various symbols used in this analytical model are briefly described in Table 2.

5.1 Route Maintenance

The route maintenance algorithm is responsible for maintaining the same established route to the gateway or establishing any other better route. Ants continuously move from one node to another node and pheromone value evaporate after time max_time. When a data packet passes through node i toward the gateway g to a neighbour node j, it increases the pheromone value if link e(i, j) by $\Delta\emptyset$ i.e. represented in Eq. 8. The evaporation process is represented in Eq. 9 given below.

$$\emptyset_{i,j} = (1-q)\emptyset_{i,j}, \ q \in [0\ldots1] \tag{9}$$

In both the cases pheromone value of the route keeps on changing. In this case new route with minimum DV may discover but continuously route cannot be changed because it will represent the instability of the route. The proposed algorithm for route maintenance is given as follows:

Algorithm V. (Overall network)

```
Gateway and Route maintenance
Pre-assumption: Time t will increase at constant rate
1.  set t=0
2.  for (each link L_{i,j,g}∈ network)
3.      if (ANT passes through L_{i,j,g})
4.          Ø_{i,j,g} ← Ø_{i,j,g} + ΔØ    //ΔØ is reinforcement constant
5.      if (max_time % t == 0)
6.          Ø_{i,j,g} ← (1-q)Ø_{i,j,g}  // q is evaporation constant
7.          set t=0
8.  calculate DV periodically based on information given by ANT
9.  if (DV_new< (DV_old - 0.2×DV_old) )
10.     update route and Internet gateway according DV_new and ANT
```

The algorithm represents the route maintenance procedure from source node to gateway node. During data transfer if ant finds new route with minimum discovery value then we should change the route but in this way discovery process will be unstable and we need to handover the connection again and again. On other hand, if we do not change the route then end to end delay will be increased. To manage the stability of discovery process we considered a new logic, if newly discovered route has 20 % less DV then only route will change otherwise route will remain same.

5.2 Connection Recovery

If node moves from the network then connection to the gateway may be lost. This may happen due to link failure, bandwidth limitation, node movement or power failure. In this case connection recovery mechanism is required. When connection is lost, link probability of related nodes goes to zero, i.e. identified by neighbor nodes. Then neighbor node again establishes the connection with maximum link probability and connection to the gateway is recovered.

6 Results and Discussion

On the basis of result of analytical study, we observed that ant scheme works better than proactive and reactive gateway discovery schemes. Figures 2 and 3 represent the comparative analysis of the packet delivery ratio and end to end delay respectively.

Packet Delivery Ratio (PDR): It is the ratio of total data send from source and received at the destination. Figure 2 represents the result of the comparison of ant scheme with proactive and reactive scheme and proved that ant scheme works better than proactive and reactive scheme, in terms of packet delivery ratio.

End-to-end packet delivery latency: End to end delay is the time difference between data send from source and received at the gateway. In initial stage, proper path is not established. So, ants pass through random path. In this case

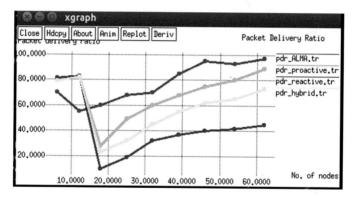

Fig. 2 Packet delivery ratio

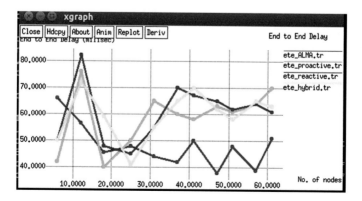

Fig. 3 End-to-end delay

end to end delay increased, but after finding the optimal path, packet directly received as destination without much delay. Figure 3 describes the comparative analysis of end to end delay with proactive and reactive gateway discovery schemes and shows that on increasing time, end to end delay get decreased.

7 Conclusion

In this paper, four models for pheromone value computation are discussed. Each model has different characteristics, so each model can be used in some specific condition. The Pheromone variation, used in this paper is an indicator of network state. The proposed approach overcomes the shortcomings of other existing gateway discovery protocols and provides the optimal route to the Internet gateway with less discovery time, high packet delivery ratio and also minimizes end to end delay. These improvements are obtained at the cost of slightly higher average overhead. This is possible with the help of a new metric i.e. discovery value. This new metric combines hop count, stability factor, congestion and load on the gateway. A route with minimum discovery value has been selected.

References

1. Ayyadurai, V., Ramasamy, R.: Internet connectivity for MANET using hybrid adaptive mobile agent protocol. Int. Arab J. Inf. Technol. 5(1) (2008)
2. Bin, S., Bingxin, S., Bo, L., Zhonggong, H., Li, Z.: Adaptive gateway discovery scheme for connecting mobile ad hoc networks to the internet. In: Proceedings of the IEEE International Conference on Wireless Communications, Networking and Mobile Computing, pp. 795–799 (2005)
3. Bo, L., Bin, Y., Bin, S.: Adaptive discovery of internet gateways in mobile ad hoc networks with mobile IP-based internet connectivity. In: Proceedings of the 5th IEEE International Conference on Wireless Communications, Networking and Mobile Computing. pp. 1–5 (2009)
4. Correia, F., Lobo, V.J., Vazao, T.: Models for pheromone evaluation in ant system for mobile ad hoc networks. In: Proceedings of the First IEEE International Conference on Emerging Network Intelligence. pp. 85–90 (2009)
5. Huang, L., Han, H., Hou, J.: Multicast routing based on ant system. Appl. Math. Sci. 1, 2827–2838 (2007)
6. Karthikeyan, D., Dharmalingam, M.: Ant based intelligent routing protocol for MANET. In: Proceedings of the IEEE International Conference on Pattern Recognition, Informatics and Mobile Engineering (PRIME), Salem, India, pp. 11–16 (2013)
7. Marwaha, S., Tham, C.K., Srinivasan, D.: A novel routing protocol using mobile agents and reactive route discovery for ad hoc wireless networks. In: Proceedings of the 10th IEEE International Conference on Networks, pp. 311–316 (2002)
8. Pandey, A., Gupta, N., Agrawal, A., Prakash, J., Kumar, R.: Congestion avoiding gateway discovery and selection scheme for MANET. In: Proceedings of the Eighth International Conference on Computer Communication Network (ICCN 2014), Bangalore, India, pp. 33–38 (2014)

9. Srivastava, P., Kumar, R.: An efficient load-aware gateway discovery for internet connectivity in MANET. In: Proceedings of the Eighth International Conference on Computer Communication Network (ICCN 2014), Bangalore (2014)

10. Yi, Z., Chun, L.Y.: An improved ant colony optimisation and its application on multicast routing problem. Int. J. Wirel. Mobile Comput. 5(1), 18–23 (2011)

11. Yuste, J., Triviño, A., Trujillo, F.D., Casilari, E.: Improved scheme for adaptive gateway discovery in hybrid MANET. In: Proceedings of the 30th IEEE International Conference on Distributed Computing System Workshop (ICDCSW). pp. 270–275, (2010)

12. Zaman, R.U., Khan, K.U.R., Reddy, A.V.: Path load balanced Adaptive Gateway Discovery in integrated Internet-MANET. In: Proceedings of the Fourth IEEE International Conference on Communication Systems and Network Technologies (CSNT) (2014)

13. Zhanyang, X., Xiaoxuan, H. and Shunyi, Z.: A simplified scheme of internet gateway discovery and selection for MANET. In: Proceedings of the 5th IEEE International Conference on Wireless Communication, Networking and Mobile Computing, pp. 1–4 (2009)

A Review on How Human Aging Influences Facial Expression Recognition (FER)

Robson Mary and T.V. Jayakumar

Abstract An active research topic in computer vision for a while has been Facial Expression Recognition (FER), but study on how human aging affects computational FER, is only slowly budding. A review has been conducted on this study based on two collected databases from the psychology society—the Lifespan and FACES, and it was found that human aging has significant impact on computational FER. The reasons were also analyzed, based on which two schemes have been proposed—the age group constrained and aging details removal. The success of these schemes for Lifespan FER has been evaluated, and high accuracies were obtained on both databases. A solution has been proposed for facial aging during growing years, which would account for the gradual change in feature dimensions of the face. This could be implemented in the future perhaps.

Keywords Computational facial expression recognition · Human aging · Cross-age FER · Lifespan FER

1 Introduction

A facial expression is one or more motions or positions of the muscles beneath the skin of the face. These movements express the individual's emotional state to the observer. Facial expressions form a part of our nonverbal communication. Expression reveals the characteristics of a person and usually displays a change in visual pattern over time. The concept of facial expression, thus, includes: (1) a trait of a person that is signified; (2) a visual pattern that shows this characteristic, i.e., the signifier; (3) the physical foundation of this appearance like the skin, muscle movements, wrinkles etc.; (4) an observer that perceives and interprets these visual facial signs. The presence and relationships among these components has been a

R. Mary (✉) · T.V. Jayakumar
8F, Cheloor Towers, Poothole 680004, Thrissur, Kerala
e-mail: mary4robson@gmail.com

© Springer International Publishing Switzerland 2016
V. Snášel et al. (eds.), *Innovations in Bio-Inspired Computing and Applications*,
Advances in Intelligent Systems and Computing 424,
DOI 10.1007/978-3-319-28031-8_27

313

wide scope for study in the psychological and behavioral sciences. Facial expressions can play an important role wherever humans interact with machines due to the information that they carry [1]. Automatic recognition of facial expressions will soon be a necessary part of natural human machine interfaces. Although humans can recognize facial expressions without much effort, consistent expression recognition by computers is still a hurdle. These interfaces provide automation of services that need a good approval of the emotional state of the service user, as in cases of transactions that need some negotiation [2].

Facial expression recognition (FER) is thus a process which includes: (1) Face detection: Locating faces in the scene (2) Facial feature extraction: Extracting facial features from the detected face region (3) Facial expression interpretation: Analyzing the changes in the appearance of facial features and classifying this information into six global emotion categories: Neutrality, Sadness, Disgust, Fear, Anger, Happiness [3].

2   Facial Image Databases

Face images representing only a small span of human lifetime are generally used in FER calculations. This is seen in some FER databases clearly: the JAFFE database [4] contains only the expressions of ten Japanese young women; the CMU expression database [5] has only young adults and some middle-aged. Recently, researchers in psychology found that the lack of larger age ranges has confined the simplification of psychology research studies [6]. Hence the need for facial databases with large span of ages arose. The *lifespan* database is a pioneer work in this field, collected by Minear and Park [7] which includes ages of 18–94 years, for psychology studies. The FACES database [8], by Ebner et al., includes six expressions and age range from 19 to 80 years. Ebner et al. showed that the age group differences do have an impact on the expressions of human emotions. Unfortunately, no study in computer vision has studied the impact of human aging on FER yet.

2.1   Databases Used in Research

When an algorithm is standardized, it is recommended to use a customary test data set for researchers so that results can be compared directly. The choice of an apt database to be used should be based on the job given. In the psychology society, there has been two databases of facial expressions ranging over a wide span of ages, which is used here in this review. (1) The Lifespan: The Lifespan database contains 844 images of frontal faces. Compared to some well-liked databases used in previous works, like the JAFFE database [4] and the CMU expression database [5], a major difference is that the lifespan expression database has a greater range of age

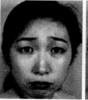

Fig. 1 Example images in JAFFE database showing 5 expressions

Table 1 Two Databases (I. Lifespan and II. FACES) of facial expressions with age group divisions [6]

DB	Expression	Age group			
		18–29	30–49	50–69	70–94
I	Neutral	223	76	133	158
	Happy	192	19	48	45
II	Six expressions	51	35	31	54

variations (from 18 to 94 years old). The database contains 590 individuals with neutral faces, with a special stress on employing older adults. Thus there are four age groups with each individual's one neutral face image taken. Some happy expressions have also been taken of a few people [6]. (2) FACES: The FACES database [8] contains 171 individuals with 6 expressions (neutrality, sadness, disgust, fear, anger, and happiness) for each person in frontal views. The total number of face images is 1026. The Lifespan database contains more diverse number of individuals than FACES, but only neutral and happy images while the latter one has six expressions (Fig. 1 and Table 1).

3 FER Within and Across Age Groups

To study the aging effect on facial expression recognition (FER), FER is performed in two cases: (1) Within each age group and (2) across age groups. By comparing the expression recognition accuracies in the two cases, it is found whether there is any aging influence on computational FER. If the results are very different, the aging effect will be noteworthy on FER; otherwise, the aging influence can be ignored [6].

3.1 Facial Features and Classifiers

FER is a class of pattern recognition problem. FER is applied within and across age groups in two sessions: (1) Facial feature extraction (2) Classification Units

(a) *Gabor Filters*: Gabor filters [9] are used to take out facial features in the traditional methods. A Gabor filter, named after Dennis Gabor, is a linear filter used for edge detection. Frequency and orientation representations of Gabor filters are parallel to the human visual system, and they are good for texture representation and classification. In the spatial domain, a 2D Gabor filter is a Gaussian kernel function modulated by a sinusoidal plane wave [6].

(b) *Fiducial Points*: A fiducial marker/point is an object kept in the view of an imaging system for use as a point of reference. To avoid possible wrong automatic fiducial point localization, they have been manually labeled in each face image in both databases. 31 fiducial points have been marked and used here. 3 scales and 6 orientations of the Gabor filter bank have been used at each fiducial point. The features are then normalized and fed into the SVM [10] for classifier learning. Support vector machines (SVMs) are supervised learning models with associated learning algorithms that scrutinize data and find patterns.

(c) *Performance Test*: The tenfold cross validation scheme is used to test the performance of FER in this review. Face images in each age group are randomly separated into ten parts. (1) For FER within each age group, the test is a normal ten-fold cross checking scheme. (2) For cross-age-group FER, the test is different from the standard scheme. For example, if the crossing is between age group A and age group B, the test images will come from age group B, and the training data are from age group A. Each of the ten parts in age group B will be used for testing in each round, while each of the equivalent nine parts in age group A is used for training. The average over the ten runs is then calculated as the recognition accuracy (Fig. 2).

Fig. 2 The 31 fiducial points labeled on a face image [6]

3.2 Analysis of Aging Effects on FER

A stranger who wants to know another person's age can either ask for identification or look at the person's face. People are able to determine age from the face with surprising accuracy, to within a few years when the person judged is between 20 and 60, a skill which even babies have so that they are capable of distinguishing children from adults.

(a) *How the Face ages*: When immature facial features mature, we say the face ages: wrinkles, blemishes, changed positions. The signs of age in the face are clear: expansion of the eye, nose and mouth features; less forehead and more chin; changes in hair, skin color and texture, pores open and wrinkles increase. These changes are the reasons for the aging of the face. The greatest changes span from infancy to puberty as the face and head take on the adult form. The eyes, nose, and mouth features of the face expand to fill a relatively greater area of the surface of the cranium. The relative area occupied by the forehead reduces as the eyes move up into this area. The eyes become proportionately smaller and the forehead slopes back more; the face becomes smaller in respect to the rest of the body; and the chin tends to become larger and more protrusive. Some of these changes may continue into adulthood, but other changes begin to occur to mark old age.

(b) *Continuity of Identity through the Ages*: Despite the often remarkable changes in the face as it ages, the identity of the face is conserved throughout a person's lifetime. The facial identity of each person is seen from early days till old age, though it is difficult to match a face using only the extremes of age, versus seeing the steps as the face develops and ages. There is a likeness from age to age that can be seen even between the youngest and oldest image. Humans interpret this consistency in the appearance of the face as part of the unique identity of that person.

(c) *Age variation or Time delay*: The variation in face over age is not minor. Many face recognition methods fail if the time lapse between the training and test image is not small. Facial aging is a complex process that affects both the shape and texture of a face. These shape and texture changes lower the performance of automatic face recognition systems. Facial aging has not been looked into much as compared to other facial variations due to pose, lighting, and expression. One way to prevail over the problem of aging, is to upgrade the database or retrain the system as required. Better solution is to simulate the age or age modeling can be done. This aging issue can serve many applications such as in surveillance for identifying missing children, screening, and detection of multiple enrollments. The three database containing images of persons with age variations available today are FG-NET, MORPH and BROWNS (Table 2).

(d) *Analysis*: Thorough examination of face images depict the fact that older adults perform facial expressions with lesser clarity than young adults. As shown in Fig. 3, young people perform the happy expression with obvious

Table 2 Average accuracies of FER in two cases (cross and within age group) on the two databases [6]

DB	Case	FER accuracies (avg.) (%)
I	Cross age group	69.32
	Within age group	82.63
II	Cross age group	64.04
	Within age group	97.85

Fig. 3 Difference seen between young and old in facial expressions. Each column contains the same person with different expressions (neutral in *top* and happy at *bottom*) [6]

facial muscle movement, while old adults perform the expression very delicately. This observation is consistent with the psychology study results [11, 12], and it unravels one aspect of the aging effects on performing facial expressions. The second observation is that wrinkles and the facial muscle elasticity reduction can change the expression appearance. The nasolabial fold that separates the upper lip from the cheek can be noticed in happy expressions for youngsters, but it is also noticed in older adults performing a neutral expression due to the drooping of the cheek fat. Based on this understanding, several schemes to deal with *lifespan FER* are proposed. One scheme is to do FER for a given face within that age group itself to avoid the cross-age-group influence, and another scheme is to reduce the aging effect on faces within each age group by removing/reducing facial wrinkles and/or nasolabial folds [6].

3.3 Aging Details Removal

The appearance of wrinkles, nasolabial folds and other aging details has to be removed to reduce the aging effect in the second approach. The facial features has to be preserved most importantly i.e. eyes, nose, mouth, etc. and also the facial expression structures like muscle movements, while removing wrinkles and aging details. A facial smoothing mechanism is created to remove aging details but without changing the important structural information, which is a challenging problem. The current method is built on an edge-preserving image smoothing technique, called weighted least squares (WLS) [13]. Let l represent the gray level face image. The WLS method decomposes l into two layers, the base layer b and the detail layer d, respectively. The main idea of the WLS method is to seek b as close as possible to l while at the same time as smooth as possible everywhere except across significant gradients in l. This problem can be formally expressed as minimization of an energy function, where the subscript p denotes the spatial location of a pixel

$$\sum_{p} \left((b_p - l_p)^2 + \lambda \left(a_{x,p}(l)(\partial b/\partial x)^2_p + a_{y,p}(l)(\partial b/\partial x)^2_p \right) \right). \tag{1}$$

The smoothness requirement is enforced in a spatially varying manner via the smoothness weights a_x and a_y, which depend on the gradients. The parameter λ is used to balance the two terms; increased value of λ results in a smoother base layer b. Using only the edge-preserving smoothing technique [13] cannot keep intact the facial structure. A mask called β was proposed to keep the major facial structures, e.g., eyes, nose, and mouth, in smoothing [14]. But the facial makeup transfer [14] has nothing to do with facial aging. Recently a mask called γ was proposed to remove aging details in faces as needed [15], as facial wrinkles etc. may have different aging "degrees" in various areas. Here the aging details in face images are remove by eliminating the detail layer d for face re-synthesis. This validates our anti-aging approach in a new scenario of expression recognition [15]. The designed masks are shown in Fig. 3 for the corresponding face image [6].

(a) *Aging Details Removal Done In Face Images*: It is observed that the smoothing technique with masks remove the facial aging details to some extent. Next, this FER performance is checked and compared with the original face images [6].

(b) *Performance Test of FER with Aging Details Removal*: Facial expression recognition is done on the new face images with aging details removal. The ten-fold cross validation procedure is used. The same training and test examples, with the only change of aging details removal for all face images, are used. The new FER results are shown in Table 3. Since the age group

Table 3 Age group constrained FER on two databases (I. Lifespan and II. FACES) [6]

DB	Case	Joint classification accuracies	Age group classification accuracies (%)	FER accuracies (%)
I	Original	94.36	97.18	96.15
	Aging detail removal	95.51	97.95	96.79
II	Original	96.17	97.39	97.00
	Aging detail removal	97.06	98.11	97.89

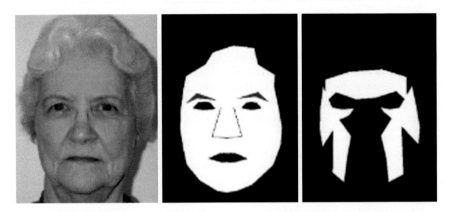

Fig. 4 A face image and its corresponding masks β and γ for aging details removal [6]

constrained (AGC) FER is effective for Lifespan FER, the new scheme's evaluation is built on it. From Table 3, one can see the improvement over the AGC-FER. The scheme of aging details removal can be combined with the age group constrained approach to improve the accuracies of lifespan FER [6] (Fig. 4).

3.4 Performance of Human Perception

The FACES database has also been tested by human raters for facial expression perception [8]. From the results of the research done in [6], it was reported that the perception accuracies are 81 % for the angry faces, 68 % for disgusted faces, 81 % for fearful faces, 96 % for happy faces, 87 % for neutral faces, and 73 % for sad faces, respectively, averaged over 154 raters [8]. Human perception of facial expressions has an average accuracy of 81 %, which is much lower than the accuracy of 97.89 % based on the computational FER done. It was also seen that the human perception performance is related to the ages of human evaluators [8], while the computer recognizes expressions without the aging bias [6].

4 Proposed System

A problem that is often met, but ignored, while handling face images of children, is the progressive change that is noticed on the size of their faces across years. A proposed solution for facial aging during growing years should account for the gradual change in feature dimensions of the face. The face images can be aligned using a set of fiducial features such as eyes, nose, mouth etc. and construct an algorithm to space out these fiducial points as one ages according to the current research findings about the human face growth patterns and speed of growth. Alterations to the ASM model may also pose as a solution for the spacing out of fiducial points. If this could be implemented, then the frequent change of faces captured in a database of individuals is not required, thus providing economic efficiency and time competence.

5 Future Enhancements

Faces are real-time projectors of our emotional and social behaviours. The automation of the entire process of FER is a highly captivating problem, the solution to which would be enormously beneficial for fields as diverse as medicine, law, communication, education, and computing. In future, avoiding the manual labeling of facial fiducial points, a method of labeling could be created. Similarly, faces being 3D objects, 3D facial aging methods need to be explored. Again, sufficient data in the form of laser scans obtained from many individuals across different ages would significantly help model the shape variations in 3D that are observed with increase in age [3].

Although two databases are introduced from the psychology society to the computer vision community, and we have used them for studying the aging effect on FER, both databases are not big. Further, there are no ages below 18 or 19 years in these databases. Larger databases could be collected by the computer vision researchers in the near future [6].

6 Conclusion

The aging effect on computational FER has been reviewed. Based on two databases collected in the psychology society, the Lifespan and FACES, it was found that human aging has great influence in computational FER. The reasons of aging influence on FER from a computational perspective were also analyzed based on which two schemes were proposed: the age group constrained (AGC-FER), and aging details removal. The effectiveness of these schemes for lifespan FER was shown, and obtained high accuracies on both databases. A solution that could be implemented in the future perhaps was proposed, for facial aging during growing years, which would account for the gradual change in feature dimensions of the face. Finally, lifespan FER was shown to deliver accuracy higher than human perception.

References

1. Chibelushi, C.C., Bourel, F.: Facial expression recognition: a brief tutorial overview. IEEE (2002)
2. Bruce: What the human face tells the human mind: some challenges for the robot-human interface. In: Proceedings of IEEE International Workshop Robot and Human Communication, pp. 44–51 (1992)
3. Pantic, M.: Facial Expression Recognition. Springer, Berlin (2009)
4. Lyons, M.J., Akamatsu, S., Kamachi, M., Gyoba, J.: Coding facial expressions with gabor wavelets. In: IEEE International Conference on Face and Gesture Recognition, pp. 200–205 (1998)
5. Kanade, T., Cohn, J., Tian, Y.: Comprehensive database for facial expression analysis. In: IEEE International Conference on Face and Gesture Recognition, pp. 46–53 (2000)
6. Guo, g., Guo, R., Li, X.: Facial expression recognition influenced by human aging. IEEE Trans. Affect. Comput. (2013)
7. Minear, M., Park, D.C.: A lifespan database of adult facial stimuli. Behav. Res. Methods Instrum. Comput. 36, 630–633 (2004)
8. Ebner, N., Riediger, M., Lindenberger, U.: Faces—A database of facial expressions in young, middle-aged, and older women and men: development and validation. Behav. Res. Methods 42(1), 351–362 (2010)
9. Daugman, J.: Uncertainty relation for resolution in space, spatial frequency and orientation optimized by tow-dimensional visual cortical filters. J. Opt. Soc. Am. A 1160–1169 (1985)
10. Vapnik, V.N.: Statistical Learning Theory. Wiley, New York (1998)
11. Ebner, N., Johnson, M.: Young and older emotional faces: are there age-group differences in expression identification and memory? Emotion 9(3), 329–339 (2009)
12. Ebner, N., Johnson, M.: Age-group differences in interference from young and older emotional faces. Cogn. Emot. 24(7), 1095–1116 (2010)
13. Farbman, Z., Fattal, R., Lischinski, D., Szeliski, R.: Edge preserving decompositions for multi-scale tone and detail manipulation. ACM Trans. Graph. 27(3), 1–10 (2008)
14. Guo, D., Sim, T.: Digital face makeup by example. Proc. CVPR (2009)
15. Guo, G.-D.: Digital anti-aging in face images. In: IEEE International Conference on Computer Vision (2011)

An Intelligent Approach for Diabetes Classification, Prediction and Description

Tarik A. Rashid, Saman M. Abdullah and Rezhna Mirza Abdullah

Abstract A number of machine learning models have been applied to a prediction or classification task of diabetes. These models either tried to categorise patients into insulin and non-insulin, or anticipate the patients' blood surge rate. Most medical experts have realised that there is a great relationship between patient's symptoms with some chronic diseases and the blood sugar rate. This paper proposes a diabetes-chronic disease prediction-description model in the form of two sub-modules to verify this relationship. The first sub-module uses Artificial Neural Network (ANN) to classify the types of case and to predict the rate of fasting blood sugar (FBS) of patients. The post-process module is used to figure out the relations between the FBS and symptoms with prediction rate. The second sub-module describes the impact of the rate of FBS and the symptoms on the patient's health. Decision Trees (DT) is used to achieve the description part of diabetes symptoms.

Keywords Diabetes disease · Blood sugar rate and symptoms · ANN · Prediction and classification models

1 Introduction

While approaches are diverse to solve for Diabetes, Machine Learning techniques become important tools in medical field. Several arguments have been discoursed to demonstrate the efficacy of these techniques in health segment, such as analysis symptoms, early detection of diseases, prevention of medical errors and medical description [1]. Chronic diseases prediction has been extended to a large space of

T.A. Rashid (✉) · R.M. Abdullah
Software Engineering Department, Salahaddin University-Erbil, Kurdistan, Iraq
e-mail: tarik.rashid@su.edu.krd

S.M. Abdullah
Software Engineering Department, Koya University, Kurdistan, Iraq
e-mail: saman.mirza@koyauniversity.org

© Springer International Publishing Switzerland 2016
V. Snášel et al. (eds.), *Innovations in Bio-Inspired Computing and Applications*,
Advances in Intelligent Systems and Computing 424,
DOI 10.1007/978-3-319-28031-8_28

using machine learning methods such as prediction models or classifiers in the medical sector. Diabetes Mellitus (DM) is one of the widespread chronic diseases. The International Diabetes Federation stated that within the next 20 years, the figure of diabetes persons will stretch to 285 million in the world [2]. Consequently, numerous research works in preceding and existing have been conducted to analyse and categorise the DM patient types [3, 4]. In the field of computer science or engineering fields, particularly in software engineering, numerous machine learning based categorising models have been established to separate types of diabetes disease. Most of the researchers have depended further on Artificial Intelligent (AI) techniques and data mining tools for constructing their classifier or forecaster models. Most researchers are aiming to target two important objectives to establish any AI based classifier models, these are; the first is to point out the most related features and predictors or statically so called independent variables that should have no correlation among each other and have strong correlation with the desired target. The second is to select a suitable AI technique as a classifier or predictor tool which would possibly produce highest accuracy rate as in [5, 6]. In other words, at this stage, most of AI models would not provide or improve something to the knowledge of the physicians and medical staffs who are observing DM cases. The only support that they can provide is to categorize the type of DM cases or predict glucose rate in the blood. In addition, the most important benefit to the physician staffs and even to the patients themselves who are in need is to describe the future of DM patients. It is so crucial to study the symptoms of DM patients not only to categorise their types, but also to envisage what side-effects or more chronic diseases a patient should anticipate.

For that reason, this work goes more beyond just classifying DM cases. It utilises some independent variables to diagnose or predict the rate of blood surge for patients through ANN model. After diagnosis or prediction, the work utilises more variables (symptoms of the patients) and the predicted blood sugar rate for the same patients to find out the relation between the symptoms and five major chronic diseases that diabetes patients have high probability to get them. The next sub section describes the background of DM and AI techniques.

1.1 Literature Work on DM and AI Techniques

DM is considered as a chronic disease in which the body of a patient is incapable to produce, use and store glucose, which is a form of sugar. This is a lifelong disorder that affects the ability of human's body to utilise the energy found in food. The disease would have several types and forms, yet, the most prevalent types are Type1 which is branded as insulin based DM and Type2 which is recognised as noninsulin based DM [2]. Even though DM would have diverse types, symptoms and side-effects of both types which are not much different, and there are hideous statistical figures for all types of the DM patients worldwide. In 2014, it was projected that 387 million people were with diabetes worldwide, it was also stated that from 2012 to 2014, estimated that DM caused in 1.5–4.9 million deaths each year. In addition, the number of people with diabetes is expected to rise to 592 million by the year 2035 and the diabetes' global economic cost was expected to be 612 billion US$ in the year 2014 [7].

Clinically speaking, the physicians and doctors can only describe the type of DM patients so that appropriate medications could be given to them, additionally, they would be able to provide medical advices to patients so that to reduce their side effects of these chronic diseases. Of course these medical advices which are provided by physicians are more vigorous than medications as they plan the patient's future line. The symptoms and the patient's age will help physicians to provide appropriate recommendations so that the bad circumstances of a DM patient will not advance to worst [8]. Nevertheless, previous constructing models for diagnosing and classifying DM types were dealt only with the process of distinguishing type-1 DM from type-2. Furthermore, models might forecast the rate of glucose in the blood for DM patients depending on some predictors. Though, describing the situation of DM patients through symptoms is still inactive all the way through the history of AI based medical models.

In General, researchers must form some intentions before starting to design the model and ought to examine other objectives during the implementation process to construct any AI based medical models. The intentions must be around the type of data set that ought to be collected, indicating the target output (which is either class definer or rate predictor) that the proposed model should find and finally the feature selection methods. The other intentions must cover the types of AI techniques that might be suggested, learning algorithms, and finally the performance measure via which the accuracy of the model can be evaluated [3, 9–12].

Basically, there are two categories of records that could be provided for any AI based medical model. Some research works rely on records called primary data, where data are collected by investigator, whereas others rely on secondary data, where data are collected by someone different than investigator. As a matter of fact, both types of data would work for AI based medical diagnosing and classification [13, 14]. Though several AI techniques used by researchers, the most important techniques are Artificial Neural Network (ANN), Support Vector Machine, Fuzzy Logic systems, K-mean classifier, and many others [3, 12, 15–17]. ANN is considered to be the most popular one among all.

A review work on using ANN in medical diagnoses is accomplished by [10], and it has been displayed that ANN can have several practices and can have different algorithms for training in [18]. Most of research works utilised the multilayer ANN with feed-forwarded back propagation (FFBP) algorithms to achieve DM classification. A research work has been done by [5] to categorise diabetes patients into insulin and non-insulin. The work depended on datasets collected in India called Pima Indian diabetes dataset [12]. Another research work used the same FFBP neural algorithm to diagnose the DM cases [8]. They collected the database from Sikkim Manipal Institute of Medical Sciences Hospital, Gangtok, Sikkim for the diabetic patients.

Fuzzy logic classifier model is another type of AI tools that have been utilised by researchers [17] to categorise cases into type-1 and type-2. Their work relied on secondary type of database called Pima dataset. The accuracy of their work was evaluated against the rate of misclassified cases. A particular work utilised Decision tree which is also considered as an AI tool for diagnosing diabetes to achieve classification and compared to ANN, their results indicated that that DT demonstrated better accuracy [19].

Several approaches and algorithms were used to extract hidden information from biomedical datasets. A research work conducted to classify diabetes cases using Principal Component Analysis and Neuro-Fuzzy Inference. The diabetes disease dataset that used in this study was taken from the UCI (from Department of Information and Computer Science, University of California) Machine Learning Database and the obtained classification accuracy was 89.47 % [20]. Another work conducted to classify diabetes cases and they obtained 78.4 % classification accuracy with 10-fold cross-validation (FCV) using Evolving Self-Organizing Map [21]. A combination approach was followed to combine Quantum Particle Swarm Optimization (QPSO), Weighted Least Square (WLS) and Support Vector Machine to diagnose Type-II diabetes. More Research works recorded in this area as the one that applied c4.5 algorithm for classification and it obtained 71.1 % accuracy rate [22].

As mentioned in all above works, researchers continuously were busy to classify or diagnose diabetes cases into some defined categories. Nevertheless, lately, physicians and doctors are concerned about the likelihood of more chronic diseases or problems that might attack the diabetes patients. Thus, this work is given more descriptive information about diabetes patients through predicting which chronic diseases or problems more probably can attack a diabetes case based on symptoms. Therefore, it is important to define the role of physicians in the next sub section as a baseline for the proposed model before introducing it in Sect. 2.

1.2   The Role of Physicians

In general, most of physicians are concerned about three points while they screen any DM case. The points are; first getting the rate of FBS of the patient, then describing the treatment type of DM (the type and dosage of the treatments) based on the FBS rate, and examining their symptoms so that to be able to provide good health care recommendations. It can be stated that the first two points can be modeled as case classification or prediction and there are several AI models via which these classification or prediction activities can be conducted. Current technology models can help DM cases for this classification. These models can give several figures about a DM case, such as FBS rate. Besides, these models can provide these supports at home. Most of DM patients are more interested in knowing the analysis of their symptoms which still a consultant physician needs to explain to them. Thus, the main objective of this work is to build an AI model that can involve all three points which physicians can provide them to DM patients.

The rest of this paper is structured as follows: Sect. 2 describes the proposed method, Sect. 3, presents the results and discussions, and finally the conclusion of the paper is outlined.

2 The Proposed Method

Although the new proposed approach in this work has a combination form, the flow operation is shown in Fig. 1. Future details on the proposed system are elaborated in coming subsections.

2.1 Data Collection and Analysis

Collecting primary records on diabetes cases is conducted in this work which involved over 500 patients and their cases were examined. For each patient, 26 records are collected. Each record can be represented as an independent variable (predictor or feature). The collected features are: Gender, Age, Weight (kg), Height (cm), Blood Pressure (systolic) (mm Hg), Blood Pressure (diastolic) (mm Hg), Inheritance, Losing weight (symptoms), Polyuria (symptoms), Nocturia (symptoms), Polydipsia (symptoms), thirsty, inactive, Paraesthesia (symptoms), Numbness (symptoms), Frequent urination (symptoms), losing weight (symptoms), Coma (symptoms), since when, Heart (weakness), Teeth (weakness), Kidney (weakness), Skin injury (weakness), Eyes (weakness), Blood glucose concentration mg/dl. The last record is the type of DM (insulin/non-insulin).

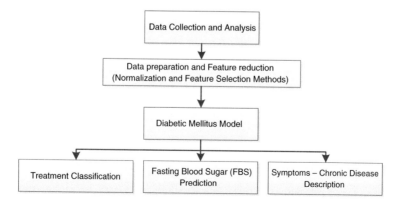

Fig. 1 Steps of building an intelligent system for diabetic mellitus model

The process of the collecting data took 60 days, and through this process 154 males and 346 females of patients were examined. The range of the age for the patients is started from four years up to 90 years old. Out of 501 cases, 222 patients are type-1 DM and 279 patients are type-2. About 94 % of type-1 cases are located between four years old to 35 years old, while only 6 % of type-2 is found in that range. For the range of 35–90 years old, dataset has 395 cases; 68 % was type-1 and 32 % was type-2. As a gender, about 56 % of the 346 females are type-1 cases and 44 % are type-2, while about 54 % of the 155 males are type-1 and 46 % are type-2. The analysis showed that cases of type-1 and type-2 are almost same in viewpoint of gender.

2.2 Data Preparation and Feature Selection

At this stage the collected data will be normalised, first, and later features will be selected. In this work two well-known normalisation methods are tested, these are; min-max and z-score [23]. Choosing a method depends on getting best standardisation and less zero values in features of the dataset. Z-score generated sequence of zero values in a few features for the data set that utilised. Moreover, both methods generated standardisation within a range of [0, 1]. Therefore, the one that has less zero values was a min-max method [23].

Feature selection is another method applied on collected dataset to minimize the dimensionality size [24]. In this work two feature selection methods are applied; the first is called Sequential Feature Selection (SFS), which utilized for the classification sub-model. The second is finding the P-value of correlation, which used for prediction sub-model. Applying these two feature selection methods, extracted features will be as below:

1. For classification issue, 11 variables have been selected as significant features which are chosen through using Sequential Forward Selection (SFS) methods. Age, Weight (kg), Height (cm), Blood Pressure (systolic) (mm Hg), Blood Pressure (diastolic) (mm Hg), Losing weight (symptoms), Nocturia (symptoms), Numbness (symptoms), Frequent urination (symptoms), Thirty (symptoms), and the targeted output which is the type of DM (insulin/non-insulin). The target variable (dependent variable) for this sub-model is treatment type (insulin or noninsulin).
2. For the prediction sub model, the variables that have been selected through SFS method are 10; Height, D.B.P, Polyuria, Nocturia, Polydpsia, thirsty, inactive, Paraesthesia, Urinal Frequency, Losing Weight, and Numbness. The target for this sub model is FBS rate.
3. For the description sub model, this work involves 10 symptoms and six chronic diseases cases plus normal case. The symptoms become inputs for the sub model. The involved symptoms are S. Blood Pressure, D. Blood Pressure, Polyuria, Nocturia, Polydpsia, Weakness, Paraesthesia, Urinal frequency,

Losing weight, and Numbness. The six chronic diseases are the targets of this sub model. The involved chronic diseases are Eyes Problem, Heart Problem, Teeth Problem, Kidney's Problem, Coma Problem, and Diabetic Foot (injury or damages). Another target this sub model has is normal, which means no problem is expected for a DM patient.

2.3 DM Medical Model

The main model has three sub models. The first part receives five independent variables for each patient, and it does the prediction and/or classification. The second sub-part is more important than the first one as it provides more important information to physician staffs. Details of each sub models are as follows.

2.3.1 The Classification Sub Models

An ANN is proposed to build a classifier model that can distinguish the type of treatment between insulin and noninsulin. ANN is a computation system that simulates the human brain for solving nonlinear problems. It involves strong interconnected elements called nodes. As shown in Fig. 2, a typical structure of any ANN has three different layers; input layer, hidden layer, and output layer. Nodes are distributed over these layers based on the type and the complexity of the simulated problem. Nodes at input layer depend on the number of the significant

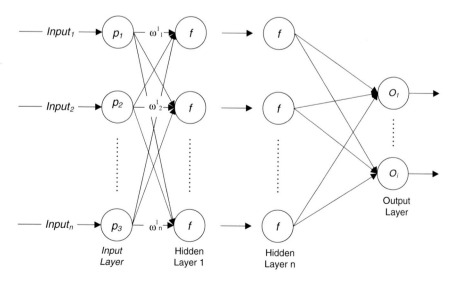

Fig. 2 The typical structure of ANN

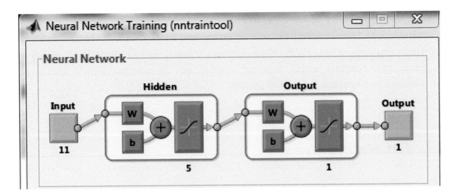

Fig. 3 The implemented structure of the DM treatment based classifier sub-model. The model took only 42 iterations to learn until 0.0974 of errors. This performance has been checked using *mse* method of checking errors between actual and predicted output

features that selected, while the hidden number and the nodes number inside each hidden layer is changing based on the complexity of the simulated problem. Finally, the output layer has nodes based on the type and number of the dependent variables (targets).

Matlab program is used to build the DM treatment based ANN classifier sub model. The network has been trained based on back propagation algorithm. The network receives the 11 features that selected in the Sect. 2.2. Figure 3 shows the designed structure of this sub model.

2.3.2 FBS Prediction for DM Patients

ANN is not only used for object classification, it can be adopted also as prediction models too. In this work, the same algorithm is used for training the ANN for predicting the FBS rate for DM patients. The only thing that changed with the ANN presented in the Sect. 2.3.1 is the number of the selected significant features. Figure 4 shows the all necessary details about the structure and the training parameters of this prediction model.

2.3.3 DM Description Sub-model

The last sub-part of the main medical model is giving more details and information on a DM case with reference to symptoms with common chronic diseases relationships.

The input part of this sub-model is involved (10) features (Symptoms) and the output is (6) chronic diseases and problems, which are more common among DM cases. Through this prediction part, it will be easy for any physician to provide

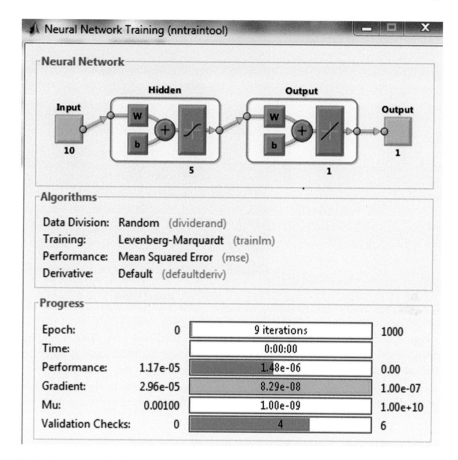

Fig. 4 As shown in the figure, the number of input features is 10, and the reached performance is 1.17 e–05. Only nine iterations are needed to learn the ANN

more details and information about the problems and chronic diseases that a DM might get them based on his/her symptoms. Two AI models were used for evaluation, which are namely; c45 and Self-Organizing Map (SOM) [7].

3 Results and Discussions

After preparing the data set and using the feature selection technique then the sub models were fed with input data and finally several experiments were conducted on the sub models. Figure 5 shows the confusion matrix of the classification part.

Fig. 5 The confusion matrix
for the classifier sub model

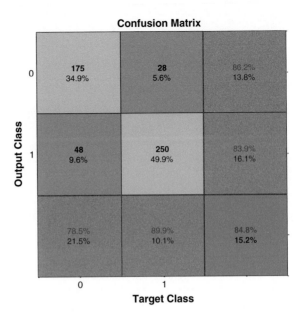

Figure 6 Shows the error histogram for the predictor part. This work has depended on decision tree for predicting the response (problems and chronic disease) as a function of the predictors (symptoms) to achieve this prediction.

The accuracy of the tree prediction sub-model is shown in the Fig. 7. The Figure shows the relation between a specific cost that required visiting a node and

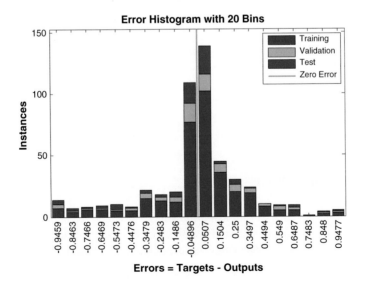

Fig. 6 Error histogram for the predictor sub model

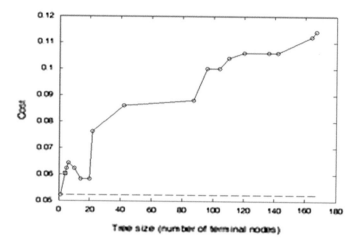

Fig. 7 The cost accuracy of the proposed decision tree sub model

the number of nodes that have the same cost. Because the proposed decision tree is working as regression mode, this cost is the average squared error over the observations in that node. The dash line in the cost is representing the minimum cost among the set of costs. In this work the proposed model evaluated against supervised and unsupervised AI models. Table 1 shows the performance evaluation for classification AI models with/without feature selection process and prediction FBS with/without feature selection process.

In this work the proposed model evaluated against supervised and unsupervised AI models. Table 1 shows the performance evaluation for classification AI models with/without feature selection process and prediction FBS with/without feature selection process.

Table 2 shows the performance evaluation two AI models for Description of Chronic Disease.

It is evident that FFBP-NNs with feature selection for classification type of DM and prediction FBS are better than c4.5 tree, however, for description of chronic disease the c4.5 is better than SOM.

Table 1 Evaluation for classification and prediction FBS

Classification (Accuracy rate)			Prediction FBS (RMSE)	
Models	FFBP-NN (%)	Decision tree (c4.5) (%)	FFBP-NN	Regression tree (c4.5)
Without feature selection	76.2	57	0.1337	0.1600
With feature selection	84.8	80	0.0013	0.1547

Table 2 Description of chronic disease

Models	C4.5	SOM
Description of chronic disease	0.0122	0.1623

4 Conclusions

This work intended at designing an intelligent based classification, prediction, and description model to provide comprehensive knowledge that required by DM patients. As a matter of fact, classes of the DM type and FBS rate are important for any DM patient. Physicians always depend on these two figures to adopt the type and dosage of treatments. Based on some symptoms extra health care recommendations are also specified by physicians to put DM patients away for some potential side effects. Intelligent models can play the roles of physicians for all these requirements.

Acknowledgments The authors would like to thank Salahaddin University-Erbil.

References

1. Srimani, P., Koti, M.S.: Medical diagnosis using ensemble classifiers-a novel machine-learning approach. J. Adv. Comput. **1**, 9–27 (2013)
2. Wild, S., et al.: Global prevalence of diabetes estimates for the year 2000 and projections for 2030. Diabetes Care **27**(5), 1047–1053 (2004)
3. Kumari, V.A., Chitra, R.: Classification of diabetes disease using support vector machine. Int. J. Eng. Res. Appl. **3**(2), 1797–1801 (2013)
4. Deekshatulu, B., Chandra, P.: Classification of heart disease using artificial neural network and feature subset selection. Global J. Comput. Sci. Technol. **13**(3) (2013)
5. Zainuddin, Z., Pauline, O., Ardil, C.: A neural network approach in predicting the blood glucose level for diabetic patients. Int. J. Comput. Intell. **5**(1), 72–79 (2009)
6. Adeyemo, A.B., Akinwonmi, A.E.: On the diagnosis of diabetes mellitus using artificial neural network models artificial neural network models. Afr. J. Comput. ICT Ref. Format **4**(1), 1–8 (2011)
7. Federation, I.D.: IDF. International Diabetes Federation. IDF Diabetes Atlas. Accessed 29 November 2014
8. Association, A.D.: Diagnosis and classification of diabetes mellitus. Diabetes Care **37** (Supplement 1), S81–S90 (2014)
9. Luger, G.F.: Artificial Intelligence: Structures and Strategies for Complex Problem Solving. Pearson education, Boston (2005)
10. Amato, F., et al.: Artificial neural networks in medical diagnosis. J. Appl. Biomed. **11**(2), 47–58 (2013)
11. Lee, S.-C.L., Embrechts, M.: I-N, Data mining techniques applied to medical information. Inform. Health Soc. Care **25**(2), 81–102 (2000)
12. Karegowda, A.G., et al.: Rule based classification for diabetic patients using cascaded k-means and decision tree C4. 5. Int. J. Comput. Appl. vol. 45 (2012)
13. Quinlan, J.R.: C4. 5: Programs for Machine Learning, vol. 1. Morgan kaufmann, San Mateo (1993)

14. Liberti, L., et al.: Euclidean distance geometry and applications. SIAM Rev. **56**(1), 3–69 (2014)
15. Dey, R., et al.: Application of artificial neural network (ANN) technique for diagnosing diabetes mellitus. In: IEEE Region 10 and the Third international Conference on Industrial and Information Systems, ICIIS 2008. IEEE, Kharagpur (2008)
16. Deng, D., Kasabov, N.: On-line Pattern Analysis by Evolving Self-Organizing Maps. Neurocomputing. **51**(1), 87–103. Elsevier, April 2003
17. Baldwin, J.F., Xie, D.W.: Simple fuzzy logic rules based on fuzzy decision tree for classification and prediction problem. In: Intelligent Information Processing II, pp. 175–184. Springer (2005)
18. Yegnanarayana, B., Artificial Neural Networks. PHI Learning Pvt. Ltd, New Delhi (2009)
19. Caballero-Ruiz, E., et al.: Automatic blood glucose classification for gestational diabetes with feature selection: decision trees vs. neural networks. In: XIII Mediterranean Conference on Medical and Biological Engineering and Computing 2013. Springer (2014)
20. Yegnanarayana1, B.: Artificial Neural Networks for Pattern Recognition Sadhana, **19**(2), 189–238 (1994)
21. Feizollah, A., et al.: A review on feature selection in mobile malware detection. Digital Investigation **13**, 22–37 (2015)
22. Berglund, E., Sitte, J.: The parameterless self-organizing map algorithm. IEEE Trans. Neural Netw. **17**(2), 305–316 (2006)
23. Lu, J.: A novel feature selection method based on data normalization. In: International Conference on Computer Application and System Modeling (ICCASM). IEEE (2010)
24. Balsamo, S., et al.: Model-based performance prediction in software development: a survey. IEEE Trans. Softw. Eng. **30**(5), 295–310 (2004)

A Novel Algorithm for Utility-Frequent Itemset Mining in Market Basket Analysis

M.A. Jabbar, B.L. Deekshatulu and Priti Chandra

Abstract Data mining has made a significant impact on business and knowledge management in recent years. Data mining also known as KDD is a powerful new technology having great potential to analyze useful information stored in large data bases. Association rule mining is an active research area, used to find association and/or correlation among frequent item sets. Association rule mining discover frequent item sets with out considering total cost, quantity and number of items in a transaction. Utility based frequent item set mining is a new research area, which provide importance to sale quantity and price among items in a transaction. In this paper, we have proposed a novel approach for utility frequent item set mining. Our method mines novel frequent item sets by giving importance to items quantity, significance weightage, utility and user defined support. Our approach can be used to provide valuable recommendation to the enterprise to improve business utility.

Keywords Data mining · Utility mining · Significance weight age · Frequent item sets

1 Introduction

Data mining, one of the steps in knowledge discovery in data a base is the extraction of the hidden information from huge repositories. There are several data mining techniques, among association rule mining is important data mining

M.A. Jabbar (✉)
Muffakham Jah College of Engineering and Technology, Hyderabad, India
e-mail: jabbar.meerja@gmail.com

B.L. Deekshatulu
IDRBT, RBI, Hyderabad, India
e-mail: deekshatulu@hotmail.com

P. Chandra
ASL, DRDO, Hyderabad, India
e-mail: priti_murali@gmail.com

© Springer International Publishing Switzerland 2016
V. Snášel et al. (eds.), *Innovations in Bio-Inspired Computing and Applications*,
Advances in Intelligent Systems and Computing 424,
DOI 10.1007/978-3-319-28031-8_29

337

strategy. Association rule mining first introduced by Agrawal et al. [1]. Traditional association rule mining algorithm (ARM) is to find association rules given support and confidence. ARM focuses on deriving associations and/or correlations among frequent item sets. Frequent item sets discovered by ARM only contribute to a small portion of the item utility. Utility item set mining is to discover high utility item sets [2]. Utility item set mining has gained popularity in data mining research [3]. Utility item set mining focused on users objective and it's utility, which is not reported in traditional association rule mining algorithms. Utility based mining encompasses descriptive method to detect rare events of high utility [4]. Utility item set mining focus to identify the item sets with high utility by considering the components such as quantity, profit and cost [5]. For example from business perpective, it is important to identify product combinations which are having highest revenue generating power [6]. High utility item sets consist of rare items. Rare item sets provide important information in many decision making domains such as business transactions, retails communities, medical, fraudulent transactions etc. [7]. In a supermarket, customers purchase rice maker or idly cooker rarely as compared to soaps, washing powder, Tea Powder. Super market manager may be interested in identifying customers who contribute major role of company profit.

The main objective of our approach is to identify novel and useful utility item sets which are statistically and semantically important to improve the business.

The remainder of this paper is organized as follows. Section 2 presents brief review on utility mining review. In Sect. 3, we present our proposed approach. In Sect. 4 experimental results are presented. Finally conclusions are drawn in Sect. 5.

2 Related Work

In this section, we will review some articles related to utility mining and association rule mining.

Association rule mining was first introduced by Agrawal et al. [1]. They proposed Apriori algorithm which generates significant association rules from repositories. Apriori is a famous algorithm to prune frequent item sets. It is a mulipass algorithm, which scans data base number of times. Later many efficient methods have been proposed to extract useful rules from the data bases.

In real life applications, an item can be valued because of its importance or utility. Utility frequent item mining (UFM) was proposed by Yao et al. [2]. Their proposed approach overcomes the shortcomings of traditional rule mining approaches, which ignores price and sale quality.

Two phase algorithm for a novel utility mining was proposed by Jich-Dhan et al. [8]. Authors proposed a bottom up two phase algorithm, for mining utility frequent item sets.

Fast utility frequent item set mining was proposed by Vid Pod pecan et al. [4]. Authors proposed a approach which finds utility frequent item sets within support and utility threshold.

Preetham Kumar et al. proposed a novel approach to discover frequent item sets based on user defined support and items quantity [9]. Authors compared their approach with Apriori and FP-Tree.

Novel utility sentient approach for mining association rules was proposed by Shankar et al. [3]. Their approach mines novel patterns by giving importance to utility, interesting ness of the users.

Chu et al. proposed Temporal high utility item sets mining from data streams. Their approach effectively identify high utility item sets. The method generates few candidates and thus execution time is reduced [10].

G.C. lan et al. proposed rare utility item sets which considers profits and quantities. Their method also considers existing period of the transaction and branches of the items [11].

Association rule mining for heavy item sets was proposed by Girish et al. [12]. Authors proposed efficient greedy algorithm to generate heavy item sets. Their approach reduced the generation of association rules, which helps the user in understanding business.

Above literature on association rule mining and utility mining motivates us to develop new algorithm, which efficiently mine utility item sets from the data bases. Our approach focuses on interestingness, significance and utility of an item.

3 Proposed Approach

In this section, we propose a novel utility frequent item set mining, which overcomes the shortcomings of traditional rule mining algorithms, which ignores the price and quantity of items in data base.

We first define some associated terms with utility item set mining.

(i) **Item count**: The item count of an item $i_C \in I$ in transaction T_D is the number of item i_C purchased in an transaction T_D.

(ii) **External utility of an item**: External utility of an item i_C is the value associated with an item defined by user. External utility is transaction independent.

(iii) **Utility of an item**: This is the quantitative measure of utility for item i_C in a transaction T_D. Utility of item = External utility of an item*item count.

(iv) **Utility of item set**: Utility of item set is defined as
Utility of item set $U(X, T_D) = \sum ic \in U(i_{C,} T_D)$.

(v) **Transaction utility**: Transaction utility is defined as the value of an item in a transaction.

Our proposed approach uses item-tree, which consists of information about items, quantity of items and transactions in which these items are not purchased.

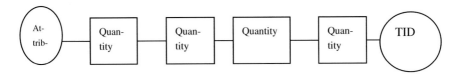

Fig. 1 Item-tree [9]

(vi) **Structure of Item Tree**: Item-Tree has three different nodes.

    (1) First node (circle) represents Attributes

    (2) Second node (square node) represents which particular item is purchased

    (3) Third node (oval shaped) hold information regarding transactions which do not contain the particular item. This node is the last node in the tree. Information regarding item, which is not present in transaction is a whole number which is obtained by multiplying prime numbers. If we factorize this product, we will get original transactions that do not contain the particular item. Figure 1 shows item-tree of a transaction. Last node represents item which is not appear in transactions.

(vii) **Significance of prime number**:

    The prime numbers are assigned to each transaction in a transactional data base.

    Assume a and b are two prime numbers their product $a^m\, b^m$. We can obtain original numbers from this product (Table 1).

    Example: If the value of last node of an item tree is 33. Factors of 33 are 3 and 11. Transaction numbers assigned to 3 and 11 are 2 and 5. Hence item is not present in 2 and 5th transaction. Here prime numbers are used as complimentary operator.

(viii) **Significance weightage (S.W)**: significance weightage of each item in transaction table is calculated as

$$S.W = P/\sum Pi \quad i = 1 \text{ to } n$$

where P is the profit and n is the number of items.

(ix) **Utility weight age (U.W)**: Utility weight age of item set is calculated by the following formulae

$$U.W = S.W \times \text{Frequency count}$$

Table 1 Prime numbers assigned to transactions

TID	1	2	3	4	5	6	7	–	n
Prime number	2	3	5	7	11	13	17	–	Nth prime number

Pseudo code of our approach is as follows.
Algorithm: Novel utility frequent item set mining (NUFM)
Input: Transactional data base
Constraints: Minimum support, Utility weightage, Transaction utility
Output: All frequent utility item sets
Method:

(1) L = 1
(2) Find the set of one item sets with support ≥ minimum support
(3) Construct the item-tree
(4) Use join approach borrowed from Apriori algorithm to obtain candidate item
    sets of length L from the set of old frequent item sets.
(5) Check the validity of item sets. Item set should satisfy three constraints.

  (i)   Minimum support (frequency count of item)
  (ii) Utility weight age
  (iii) Transaction utility
  In high utility mining, antimonotone property has to be satisfied. If transaction
  Utility of an item set < minimum utility, any super set of X can't be high
  utility Item set.
(6) Stop the algorithm if the new set is empty, otherwise go to step 5.

4   Experimental Results

In this section, we have presented experimental results of our proposed approach.
The proposed algorithm is implemented in Java.

Table 2 gives customer transactions and Table 3 gives weight (Profit) of an item.

Table 2 Customer transactions

TID	A	B	C	D	E	Transaction utility
1	10	1	4	1	0	820
2	0	1	0	3	0	280
3	2	0	0	1	0	210
4	0	0	1	0	0	30
5	1	2	0	1	3	190
6	1	1	1	1	1	210
7	0	2	3	0	1	130
8	0	0	0	1	2	110
9	7	0	1	1	0	540
10	0	1	1	1	1	150
Significance weightage	0.28	0.047	0.142	0.428	0.095	Total transaction utility = 2700

Table 3 Weights assigned to each item

Item	Profit
A	60
B	10
C	30
D	90
E	20

Explanation:

Significance weight age of item A = 60/(60 + 10 + 30 + 90 + 20) => 0.28.

Similar approach will be adopted to calculate significance weight age for other items.

Cost of transaction will be calculated as

T1 = 10 * cost of item A + 1 * cost of item B + 4 * cost of item C + 1 * cost of item D
= 10 * 60 + 10 + 120 + 90 = 820

Constraints: Minimum support 5, utility weight age 3, and Minimum Transaction Utility threshold as 1350.

Reason: To find high utility item sets, we have to specify user defined utility threshold i.e. Minimum Utility Threshold.

Minimum transaction utility threshold (∂) is given by percentage of total transaction utility values of the database. Assume Minimum transaction utility taken (∂) as 50 %, so min utility = 1350 (50 % of total transaction utility).

Step 1:

Calculate minimum support of all items. All items have minimum support above threshold 5. As per minimum support all items are frequent 1 item sets, because all have support above 5. These are treated as frequent 1 item sets.

Step 2:

Assign prime number to each transaction i.e. T1 = 2, T2 = 3, … T10 = 29 as shown in Table 1.

Last nodes indicate the absence of an Item. Attribute A is shown in Fig. 2.

Item A is absent in transactions T2, T4, T7, T8, and T10. Prime numbers assigned for these transactions are 3, 7, 17, 19, and 29. Multiply the prime numbers we get **196707**.

Item B is absent in transactions T3, T4, T8, T9. Prime numbers assigned for these transactions are 5, 7, 19, and 23. Item-tree for attribute B is shown in Fig. 3.

Similar procedure is applied for C, D, and E items also.

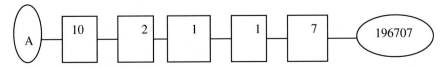

Fig. 2 Item-tree for attribute A

Fig. 3 Item-tree for attribute A

The values at the end of Item-Tree for C are 3135 and for D is 119 and for is E 4830.

Step 3:

Item set is said to be utility frequent if it satisfies the following three conditions

(1) Minimum support
(2) Utility weight age
(3) Transaction weight utilization (this should be >= min utility threshold i.e. ∂)

In high utility pattern mining, we have to maintain the antimonotone property. If Transaction weight utilization of an item set X < min utility, any super pattern of X Can't be high utility item set.

Step 4:

Now from the frequent 1 item set, we make the combinations of 2 item sets.

First we make the combination of A, D. A value is 196707 and its factors are 3, 7, 17, 19, 29 and D value in the tree is 119 and its factors are 7, 17. Union of A, D factors are 3, 7, 17, 19, 29 its equivalent transaction numbers are 2, 4, 7, 8, 10. So A, D will not appear in 2, 4, 7, 8, 10, and appear in 1, 3, 5, 6, 9 i.e. in 5 transactions.

Satisfying 1st condition. Utility weight age of A, D = (0.28 + 0.428) * 5 = 3.54. Satisfying 2nd condition i.e. Transaction weight age of A, D. The combination appears in 1, 3, 5, 6, 9 so add the cost of the 5 transactions. Value of transaction utility is 2040.

Satisfying the 3rd condition. Hence A, D is a frequent item set. Similar procedure is applied to A, B. A value is 196707 and its factors are 3, 7, 17, 19, 29 and for B is 15295 and factors are 5, 7, 19, 29. Union of A, B is 3, 5, 7, 17, 19, 23, 29. Equivalent transactions are 2, 3, 4, 7, 8, 9, 10. So A, B appears in 1, 5, 6 only. As min support is 5, it is not satisfying the 1st condition. Utility weightage of A, B is (0.28 + 0.047) * 3 = 0.98 < min utility weightage.

So condition 2 also fails. A, B appears in 1, 5, 6, so the cost of the transactions are 820 + 190 + 210 = 1220 < Min transaction weight. All the conditions fail so A, B is not a frequent item set. Calculating for B, C. B, and C appears in 1, 6, 7, 10. So min support fails. Utility weight age of B, C is 0.756, which fails the second condition. Transaction weight is 1310, which also fails to meet 3rd condition. Similarly for D, E. D, E appears in 5, 6, 8, 10. So min support fails. Utility weight age = 2.092 also fails. Transaction weight is 660 so 3rd condition also fails. So it is not a frequent item set. Frequent item set is only A, D. Table 4 shows calculations

Table 4 Calculations of utility weightage for 2 item sets

Sl. no	Item set	Support count	Transaction weightage	Utility weightage	Frequent item (Yes/No)
1	A, D	5	2040	3.54	Yes
2	A, B	3	1220	0.98	No
3	B, C	4	1310	0.756	No
4	D, E	4	660	2.092	No
5	B, D	4	1650	1.5	No
6	C, D	4	1820	2.28	No

Table 5 Support and confidence calculation

Sl.no	Rule	Support count	Satisfying user threshold (Yes/No)	Confidence (%)	Satisfying user threshold (Yes/No	Strong association rule (Yes/No)
1	A => D	5	Yes	100	Yes	Yes
2	A => B	3	No	60	No	No
3	B => C	4	No	66	No	No
4	D => E	4	No	50	No	No
5	B => D	4	No	66	No	No
6	C => D	4	No	66	No	No

for 2 item sets. Table 5 shows the support and confidence of item sets (minimum support 4 and confidence 100 %). This procedure will recursively applied to get n-frequent weighted utility item sets.

5   Conclusion

In this research paper, we have presented a novel and efficient approach for generation of utility frequent item sets. We analyzed problems of market basket data and made important contributions. Our method mines novel frequent item sets by giving importance to items quantity, significance weightage, transaction utility and user defined support. The outcome of utility mining will also enable the business people in taking important business decisions, such as applying discount, shelf space in super market based on utility frequent item set mining. Our approach can be used to provide valuable recommendation to the business enterprise to improve business utility.

References

1. Agrawal, R., Imielinski, T., Swami, A.: Mining association rules between sets of items in large databases. In: Proceedings of the ACM SIGMOD, International Conference on Mobile Data Management, Washington, D.C., pp. 207–216, May 1993
2. Yao, H., Hamilton, et al.: A foundational approach to mining item set utilities from data bases. In. SIAM International Conference, pp. 428–486 (2004)
3. Shankar, Purusotham.: A novel utility sentient approach for mining interesting association rules. IJET, 454–460 (2009)
4. Podpecan, V., et al.: A fast algorithm for mining utility frequent item sets. http://citeseerx.ist.psu.edu
5. Pillai, J., Vyas, O.P.: Overview of item set utility mining and its applications. Int. J. Comput. Appl. (0975–8887), **5**(11), 9–13 (2010)
6. Hu, J., Mojsilovic, A.: High-utility pattern mining: a method for discovery of high-utility item sets. Pattern Recogn. **40**, 3317–3324 (2007)
7. Pillai, J.: User centric approach to item set utility mining in market basket analysis. IJCSE, 393–399 (2011)
8. Yeh, J.S., et al.: Two phase algorithms for a novel utility frequent mining model. In: PAKDD 2007, LNAI, pp. 433–444 (2007)
9. Kumar, P., et al.: Discovery of frequent item sets based on minimum quantity and support. IJCSS **3**(3), 216–225(2011)
10. Chu, et al.: An efficient algorithm for mining temporal high utility item sets from data streams. JSS, 1105–1117 (2008)
11. Lan, G.C., et al.: A Novel Algorithm for Mining Rare—Utility Item Sets in Multi Data Base Environment (2013)
12. Girish, et al.: Association Rules Mining Using Heavy Item Sets, pp. 148–155. CSI (2005)

Exact Computation of 3D Geometric Moment Invariants for ATS Drugs Identification

Satrya Fajri Pratama, Azah Kamilah Muda, Yun-Huoy Choo
and Ajith Abraham

Abstract The war on drug abuse involves all nations worldwide. Normally, molecular components are unique, and thus the drugs can be identified based on it. However, this procedure started to be more unreliable with the introduction of new ATS molecular structures which are increasingly complex and sophisticated. Hence, unique characteristics of molecular structure of ATS drug must be accurately identified. Therefore, this paper is meant for formulating an exact 3D geometric moment invariants to represent the drug molecular structure. The performance of the proposed technique was analyzed using drug chemical structures obtained from United Nations Office of Drugs and Crime (UNODC) and also from various sources. The evaluation shows the technique is qualified to be further explored and adapted in the future works to be fully compatible with ATS drug identification domain.

Keywords Exact geometric moments · 3D moments function · Moment invariants function · ATS drugs · Drugs identification · Molecular structure

S.F. Pratama (✉) · A.K. Muda (✉) · Y.-H. Choo · A. Abraham
Computational Intelligence and Technologies (CIT) Research Group, Center of Advanced
Computing and Technologies, Faculty of Information and Communication Technology,
Universiti Teknikal Malaysia Melaka, Hang Tuah Jaya, 76100 Durian Tunggal,
Malacca, Malaysia
e-mail: satrya@student.utem.edu.my; rascove@yahoo.com

A.K. Muda
e-mail: azah@utem.edu.my; azahkm@gmail.com

Y.-H. Choo
e-mail: huoy@utem.edu.my

A. Abraham
e-mail: ajith.abraham@ieee.org

A. Abraham
Machine Intelligence Research Labs (MIR Labs), Scientific Network for Innovation
and Research Excellence, Auburn, WA, USA

© Springer International Publishing Switzerland 2016
V. Snášel et al. (eds.), *Innovations in Bio-Inspired Computing and Applications*,
Advances in Intelligent Systems and Computing 424,
DOI 10.1007/978-3-319-28031-8_30

347

1 Introduction

Abuse of Amphetamine-type Stimulants (ATS) drugs, involving amphetamine, methamphetamine, and substances of the "ecstasy"-group (MDMA, MDA, MDEA, etc.) has become a global, harrowing social problem. National law enforcement authorities are still struggling to find a concrete solution to prevent drugs abuse in society due to the existence of new brand or unfamiliar ATS drug substances. However, the cheminformatics research community focuses more on the development of chemical drugs that causes desired biological effect. Furthermore, less attention is given to the shape similarity search that can lead to identification of unknown substances.

Generally, drugs can be identified based on the structure of its molecular components. This procedure is becoming more unreliable with the introduction of new ATS molecular structures which are increasingly complex and sophisticated. However, due to the limitations of the current test kit to detect new brand or unfamiliar ATS drug, it presents a challenge to both national law enforcement authorities and scientific staff of forensic laboratories. In addition, existing drug test kits are sometimes prone to false positive detection.

Both of 2-dimensional (2D) and 3-dimensional (3D) chemical structures are basically represented as a shape. 2D shape descriptors have been developed which can be generally divided into boundary and area based. Meanwhile, 3D shape descriptor focus on volume based and surface based descriptors. 3D shape descriptor also has been described as more powerful and accurately represent a component structure's shape. Thus, 3D descriptor is believed can be used to identify molecular structure of ATS drug's chemical components, even for a new brand of ATS drug due to their similar ring substitutes.

This paper aims to propose a novel 3D Exact Geometric Moment Invariants to represent the ATS drug molecular structure. The remainder of the paper is structured as follows. In next section, an overview of ATS drug identification is given. Section 3 provides an overview of 3D Geometric Moment Invariants. In Sects. 4 and 5, the proposed technique is introduced and the experimental setup describing the data source collection and experimental design are presented respectively, while the results are discussed in Sect. 6. Finally, conclusion and future work is drawn in Sect. 7.

2 ATS Drug Identification

Manual identification of ATS follows a set of standard methods outlined by United Nations Office of Drugs and Crime (UNODC). These standards however, are not closely followed by chemists, causing the results obtained possibly ranging from one testing laboratory from another. The most common method to identify a chemical substance is gas chromatography/mass spectrometry (GC/MS) [1–3].

A recent study shows that GC/MS is flawed while trying to identify several ATS drugs, most notably is methamphetamine [4]. Methamphetamine itself has two stereo-isomers, which is *l*-methamphetamine and *d*-methamphetamine. Reference [5] defines isomers as one of several species (or molecular entities) that have the same atomic composition (molecular formula) but different line formulae or different stereo-chemical formulae and hence different physical and/or chemical properties. GC/MS is also increasingly incapable to determine that several chemical structures are actually ATS drugs. While *l*-methamphetamine has very little pharmacodynamics effect, *d*-methamphetamine on the other hand is a controlled substance that has high potential for abuse and addiction [6].

The traditional identification of ATS drugs is fully dependent of the chemical composition of the drugs. Hence, existing drug test kits are sometimes prone to false positive detection. Therefore, the introduction of new chemical composition to ATS drugs will pose greater problems in the identification task. This problem can be resolved by relying on the shape of the chemical structure of the drug itself. Both 2D and 3D model are often used to depict the molecular structure. However 2D model hides the properties of the ring substitutes in a molecule. Therefore, a 3D model is essential to show and differentiate the unique features at a ring substitute.

Both 2D and 3D chemical structures are basically represented as shape. These representations are commonly referred as molecular descriptors. Molecular descriptors are obtained when molecules are transformed into a molecular representation enabling mathematical treatment [7]. Searching for an image by using the shape features gives challenges for many researches, since extracting the features that represent and describe the shape is a difficult task [8]. There are simple molecular descriptors, usually called topological or 2D-descriptors, and there are molecular descriptors derived from a geometrical representation that are called geometrical or 3D-descriptors. Because a geometrical representation involves knowledge of the relative positions of the atoms in 3D space, geometrical descriptors usually provide more information and discrimination power than topological descriptors for similar molecular structures and molecule conformations.

3 Existing 3D Geometric Moment Invariants

Moments Function (MF) can be used to generate a set of moments that uniquely represent the global characteristic of an image. Moments are scalar quantities used to characterize a function and to capture its significant features. Moment Invariants (MI) are very useful tools for pattern recognition [9].

The first introduction of MI to pattern recognition and image processing was the employment of algebraic invariants theory by [10], which derived his renowned seven invariants to the rotation of 2D objects. And thus ever since, it has been chosen as one of the most important and frequently used shape descriptors options, and has been extended to 3D images as well [11]. Even though they suffer from certain intrinsic limitations, they frequently serve as first-choice descriptors and a reference method for evaluating the performance of other shape descriptors [12]. 3D Geometric Moments (GM) of image intensity function $f(x, y, z)$ is generally defined as [11]

$$m_{pqr} = \int_{-\infty}^{+\infty} \int_{-\infty}^{+\infty} \int_{-\infty}^{+\infty} x^p y^q z^r f(x, y, z) dx dy dz \qquad (1)$$

where $p, q, r = 0, 1, 2, \dots$. However, some studies also define 3D GM as [13, 14]

$$m_{pqr} = \int_{-1}^{+1} \int_{-1}^{+1} \int_{-1}^{+1} x^p y^q z^r f(x, y, z) \, dx dy dz \qquad (2)$$

In case the intensity function $f(x, y, z)$ corresponds to an $N \times N \times N$ 3D image, (1) takes the following discrete form [12]

$$m_{pqr} = \sum_{i=1}^{N} \sum_{j=1}^{N} \sum_{k=1}^{N} i^p j^q k^r f_{ijk} \qquad (3)$$

where i, j, k are coordinates of the voxels (volume pixels) and f_{ijk} is the gray-level of the voxel i, j, k. The sum of the indices is called the order of the moment. The translation invariance can easily be provided by the 3D Central Geometric Moments [12]

$$\mu_{pqr} = \int_{-\infty}^{+\infty} \int_{-\infty}^{+\infty} \int_{-\infty}^{+\infty} (x - x_c)^p (y - y_c)^q (z - z_c)^r f(x, y, z) \, dx dy dz \qquad (4)$$

where the centroid of the image intensity function $f(x, y, z)$ is calculated by $x_c = \frac{m_{100}}{m_{000}}$, $y_c = \frac{m_{010}}{m_{000}}$, and $z_c = \frac{m_{001}}{m_{000}}$. The central moments are often normalized to produce scale invariants [12]

$$\eta_{pqr} = \frac{\mu_{pqr}}{\mu_{000}^{(p+q+r+3)/3}} \qquad (5)$$

Recently, [15] proposed a novel 3D rotation invariants from GM using moment tensor method, and also implemented an automatic method for generating those 3D rotation invariants. These invariants are built up from the moments of order 2 up to order 16, and consist of 1185 invariants. They define a moment tensor by using tensor calculus for derivation of MI

$$M^{i_1 i_2 i_3} = \int\limits_{-\infty}^{+\infty} \int\limits_{-\infty}^{+\infty} \int\limits_{-\infty}^{+\infty} x^{i_1} y^{i_2} z^{i_3} f(x^1, x^2, x^3) \, dx^1 dx^2 dx^3 \qquad (6)$$

where $x^1 = x$, $x^2 = y$, and $x^3 = z$. If p indices equal 1, q indices equal 2 and r indices equal 3, then $M^{i_1 i_2 i_3} = m_{pqr}$. In order to create rotation invariants, [15] work with the moment tensors as with Cartesian tensors. Each tensor product of the moment tensors, where each index is used just twice, is then the invariant to the 3D rotation. This technique is hereby dubbed as 3D Suk-Flusser Rotation Invariants Central Geometric Moments (3D SFRI-CGM).

4 Proposed 3D Exact Geometric Moment Invariants for Molecular Structure Representation

A digital 3D image of size $N \times N \times N$ is an array of voxels. Centers of these voxels are the points (x_i, y_j, z_k), where the image intensity function is defined only for this discrete set of points $(x_i, y_j, z_k) \in [-1, 1] \times [-1, 1] \times [-1, 1]$. $\Delta x_i = x_{i+1} - x_i$, $\Delta y_j = y_{j+1} - y_j$, and $\Delta z_k = z_{k+1} - z_k$ are sampling intervals in the x-, y-, and z-directions, respectively. In the literature of digital image processing, the intervals Δx_i, Δy_j, and Δz_k are fixed at constant values $\Delta x_i = \Delta x = 2/N$; $\forall i$, $\Delta y_j = \Delta y = 2/N$; $\forall j$, $\Delta z_k = \Delta z = 2/N$; $\forall k$. Therefore, the set of points (x_i, y_j, z_k) will be defined as follows:

$$n_a = -1 + (a - 0.5) \, \Delta n \qquad (7)$$

with $a = 1, 2, 3, ..., N$. For the discrete-space version of the image, (2) is usually approximated as:

$$m_{pqr} = \sum_{i=1}^{N} \sum_{j=1}^{N} \sum_{k=1}^{N} x_i^p y_j^q z_k^r f(x_i, y_j, z_k) \, \Delta x \Delta y \Delta z \qquad (8)$$

Equation (8) is so-called direct method for computing GM, which is the approximated version using zeroth-order approximation. As indicated by [16], (8) is not a very accurate approximation of (2). To improve the accuracy, they proposed to use the approximated form, which is then refined by [13]

$$m_{pqr} = \sum_{i=1}^{N} \sum_{j=1}^{N} \sum_{k=1}^{N} I_p(i) I_q(j) I_r(k) f(x_i, y_j, z_k) \qquad (9)$$

where

$$I_s(a) = \int\limits_{n_a - \frac{\Delta n}{2}}^{n_a + \frac{\Delta n}{2}} n^s dn = \frac{1}{s+1}\left[U_{a+1}^{s+1} - U_a^{s+1}\right] \tag{10}$$

and

$$U_{a+1} = n_a + \frac{\Delta n}{2} = -1 + a\Delta n; \ U_a = n_a - \frac{\Delta n}{2} = -1 + (a-1)\Delta n \tag{11}$$

Reference [16] proposed an alternative extended Simpson's rule to evaluate the triple integral defined by (2), and then use it to calculate exactly the GM, hence it is known as Exact Geometric Moments (EGM). Although it was meant for 2D GM, the formula can be easily extended into 3D GM. The idea of this paper is to incorporate EGM instead of regular GM prior to the formulation of 3D Suk-Flusser Rotation Invariants.

It is also necessary to produce translation invariants EGM by calculating its central moments. Reference [17] provides a relationship between central and non-central moments for regular GM, and defines a convenient formula to calculate central moments from non-central moments, and vice versa

$$\mu_{pqr} = \sum_{a=0}^{p}\sum_{b=0}^{q}\sum_{c=0}^{r} \binom{p}{a}\binom{q}{b}\binom{r}{c} m_{100}^{p-a} m_{010}^{q-b} m_{001}^{r-c} (-m_{000})^{-(p+q+r-a-b-c)} m_{abc} \tag{12}$$

$$m_{pqr} = \sum_{a=0}^{p}\sum_{b=0}^{q}\sum_{c=0}^{r} \binom{p}{a}\binom{q}{b}\binom{r}{c} m_{100}^{p-a} m_{010}^{q-b} m_{001}^{r-c} m_{000}^{-(p+q+r-a-b-c)} \mu_{abc} \tag{13}$$

Even though (12) and (13) were meant for regular GM, this paper proposes to employ (12) as a mean to calculate Central Exact Geometric Moments (CEGM). And thus, it is an efficient and elegant way of deriving rotation invariants from EGM in 3D based on 3D CEGM, namely 3D Suk-Flusser Rotation Invariants Central Exact Geometric Moments (3D SFRI-CEGM). This forms the major theoretical contribution of the paper. In the next section, the invariance property on ATS and non-ATS dataset is demonstrated, and shows the numerical stability of both 3D SFRI-CGM and 3D SFRI-CEGM.

5 Experimental Setup

With the goal stated in the section above, an extensive and rigorous empirical comparative study is designed and conducted. In this section, a detailed description of the experimental method is provided.

5.1 Data Source Collection

This section describes the process of transforming molecular structure of ATS drug into 2D and 3D computational data representation. ATS dataset used in this research comes from [2], which contains 60 molecular structures which are commonly distributed for illegal use. On the other hand, 60 non-ATS (n-ATS) drug chemical structures are also collected from various sources, ranging from everyday over-the-counter drugs to anti-malarial drugs, which will be used as benchmarking dataset.

These structures are drawn in 2D chemical structure format using MarvinSketch 15.8.0 [18]. After the 2D chemical structure is created, the structure will be cleaned and transformed to 3D chemical structure, also by using MarvinSketch 15.8.0 [18]. The structure will be then saved as Structure Data Format (SDF) file. The SDF file must be then converted to Virtual Reality Markup Language (VRML) format, because VRML format is the input type required for generating voxel data of 3D chemical structure. In order to convert SDF file to VRML file, Jmol 13.0 [19] is required.

VRML file will be then voxelized to voxel grid data with 512 voxel resolution using binvox 1.21 program [20] for the training dataset, and will be uniquely and randomly translated and rotated 50 different times for the testing dataset. After the voxel data has been generated, 3D SFRI-CGM and 3D SFRI-CEGM for training and testing dataset will be calculated from 16th order using existing and proposed technique respectively. The sample of the molecular structure is shown in Fig. 1 and the output of the existing and proposed technique is shown in Table 1, respectively.

5.2 Experimental Design

The traditional framework of pattern recognition tasks, which are preprocessing, feature extraction, and classification, will be employed in this paper. This paper will

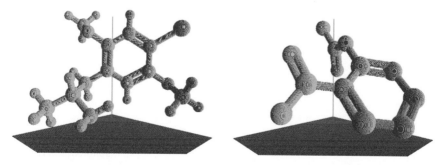

Fig. 1 Sample molecular structure of an ATS drug, 4-Iodo-2,5-dimethoxyamphetamine (*left*) and n-ATS drug, aspirin (*right*)

Table 1 Sample output of MI techniques

Structure name	MI	F1	F2	...	F1184	F1185
4-Iodo-2,5-dimethoxyamphetamine	3D SFRI-CGM	22.21497	15.49400	...	65536.87	6.361E09
	3D SFRI-CEGM	22.21510	15.49406	...	65539.87	6.362E09
Aspirin	3D SFRI-CGM	21.11886	14.64512	...	72363.25	9.107E09
	3D SFRI-CEGM	21.11898	14.64518	...	72366.03	9.108E09

compare the performance of 3D SFRI-CGM and 3D SFRI-CEGM. It is also worth mentioning that all MIs are magnitude normalized to their respective degree.

All extracted instances are tested using training and testing dataset aforementioned for its invariance on the same ATS molecular structure, intra-class and inter-class of ATS drugs, and also classification of unknown ATS molecular structure, all of which are executed for 50 times. In order to justify the quality of features produced by each MF, the features are tested against well-known classifier, Random Forest [21].

6   Experimental Results and Discussion

In this section, the proposed invariants will be tested numerically by constructing 3D SFRI-CGM and 3D SFRI-CEGM from invariant 1 to 1185 for all 60 ATS drugs molecular structure. The experiments are designed to verify rotation invariance and evaluate the numerical stability of the proposed invariants.

6.1   Invariance on the Same ATS Molecular Structure Analysis

To evaluate quantitatively the invariance, mean relative error (MRE) is used to measure the computational error of the ith invariant. The MRE of the ith invariant is defined as

$$MRE_i = \frac{1}{N} \sum_{j=i}^{N} \left| \frac{I_i^j - I_i}{I_i} \right| \times 100 \ \% \qquad (14)$$

where I_i and I_i^j are the ith invariants of the original molecular structure and the jth rotated version, respectively. N is the number of rotated versions. Consequently, it is found the average MRE (AMRE) and maximum MRE (MMRE) for each molecular structure. AMREs and MMREs from both 3D SFRI-CGM and 3D SFRI-CEGM exhibit the same result. AMREs for all 60 ATS molecular structures are below 1 %, while the value of MMREs is shown in Fig. 2. But nevertheless, this experiment exhaustively demonstrated desirable rotation invariance.

6.2   Intra-class and Inter-class Analysis

To validate the invariance and the effectiveness of the 3D SFRI-CGM and 3D SFRI-CEGM, the concept of intra-class and inter-class is used. Intra-class is

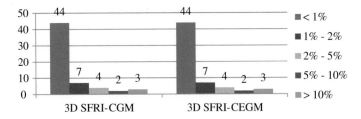

Fig. 2 Invariance frequency of MMREs for MI techniques

referred to a group of invariant features from ATS drug, while inter-class is referred
to a group of invariant features from n-ATS drug. For intra-class and inter-class
validation, the previously defined MRE can also be employed as distance measure,
although it requires several modifications. Instead of using the original molecular
structure, the MREs of *j*th rotated version of a molecular structure version will be
referencing all molecular structure in the training dataset, both ATS and n-ATS,
except for its original molecular structure.

MRE value for intra-class should be smaller compared to inter-class, regardless
of any shape of molecular structure in order to prove the presence of unique features
at a ring substitute of ATS drugs. The result of intra-class and inter-class analysis is
depicted in Fig. 3.

Based on the results shown in Fig. 3, 51 molecular stucture has lower intra-class
MREs compared to inter-class for both 3D SFRI-CGM and 3D SFRI-CEGM,
which is then prove the presence of unique features at a ring substitute of ATS
drugs.

6.3 Classification of Unknown ATS Drug Molecular Structure

The comparison of classification accuracy of 3D SFRI-CGM and 3D SFRI-CEGM
is also one of the primary consideration of this paper. Table 2 shows the results of
mean classification accuracy from 50 executions using Random Forest classifier.

Based on the results shown in Table 2, it is evident that 3D SFRI-CEGM
produces slightly better result of mean classification accuracy, although it is not
statistically significant. This is because, even though the calculation of the EGM is

Fig. 3 Intra-class and inter-class analysis for MI techniques

Table 2 Mean classification accuracy for MI techniques

MI	Mean accuracy (%)
3D SFRI-CGM	88.87
3D SFRI-CEGM	88.90

accurate, the moments itself will be approximated when converted into CEGM, and may lose some of its discriminative power. But nevertheless, this result inspire for further application into Orthogonal (OG) MF, because EGM can be used as a basis to indirectly calculate OG Moments, especially Continuous OG Moments on a cube or sphere, such as Legendre and Zernike Moments.

7 Conclusion and Future Works

An extensive comparative study on MI techniques for representing drug molecular structure has been presented. This paper proposed an incorporation of 3D CEGM into 3D Suk-Flusser Rotation Invariants and compared the merits of the proposed technique with 3D SFRI-CGM. Although the experiments have shown that the proposed technique produces the slightly better results compared to the existing one, but this study also serves as a stepping stone towards better molecular representation, especially for Continuous OG Moments.

Hence, future works to incorporate the proposed technique into OG MF and to better represent the molecular structure based on this experimental study is required. The proposed feature extraction technique will be using specifically-tailored classifiers for shape representation, and ATS drug molecular structure data from National Poison Centre, Malaysia, will also be used as additional dataset in the future works.

Acknowledgments This work was supported by Collaborative Research Programme (CRP) – ICGEB Research Grant (CRP/MYS13-03) from International Centre for Genetic Engineering and Biotechnology (ICGEB), Italy and Universiti Teknikal Malaysia Melaka, UTeM.

References

1. Langman, L.J., Bowers, L.D., Collins, J.A., Hammett-Stabler, C.A., LeBeau, M.A.: Gas Chromatography/Mass Spectrometry Confirmation of Drugs; Approved Guidelines, 2nd edn. Clinical and Laboratory Standards Institute, Pennsylvania (2010)
2. Langman, L.J., Bowers, L.D., Collins, J.A., Hammett-Stabler, C.A., LeBeau, M.A.: Gas Chromatography/Mass Spectrometry Confirmation of Drugs; Approved Guidelines, 2nd edn. In: Clinical and Laboratory Standards Institute, Pennsylvania (2010)
3. Lin, D.-L., Yin, R.-M., Ray, L.H.: Gas Chromatography-mass spectrometry (gc-ms) analysis of amphetamine, methamphetamine, 3,4-Methylenedioxyamphetamine and 3,4-methylenedioxymethamphetamine in human hair and hair sections. J. Food Drug Anal. **13**(3), 193–200 (2005)

4. McShane, J.J.: GC-MS is Not Perfect: The Case Study of Methamphetamine (2011)
5. International Union of Pure and Applied Chemistry: Compendium of Chemical Terminology. Gold Book, 2nd edn. Blackwell Scientific Publications, Oxford (2006)
6. Mendelson, J., Uemura, N., Harris, D., Nath, R.P., Fernandez, E., Jacob, P., Everhart, E.T., Jones, R.T.: Human Pharmacology of the methamphetamine stereoisomers. Clin. Pharmacol. Ther. **80**(4), 403–420 (2006)
7. Todeschini, R., Consonni, V.: Descriptors from molecular geometry. In: Handbook of Chemoinformatics. pp. 1004–1033. Wiley-VCH Verlag GmbH (2008)
8. Muda, A.K.: Authorship Invarianceness for Writer Identification Using Invariant Discretization and Modified Immune Classifier. Universiti Teknologi Malaysia (2009)
9. Sun, Y., Liu, W., Wang, Y.: United moment invariants for shape discrimination. In: International Conference on Robotics, Intelligent Systems and Signal Processing, Changsha, pp. 88–93. IEEE (2003)
10. Hu, M.K.: Visual pattern recognition by moment invariants. In: IRE Transactions on Information Theory 1962, pp. 179–187
11. Sadjadi, F.A., Hall, E.L.: Three-dimensional moment invariants. pattern analysis and machine intelligence. IEEE Trans. PAMI-2 **2**, 127–136 (1980). doi:10.1109/TPAMI.1980.4766990
12. Flusser, J., Suk, T., Zitová, B.: Moments and Moment Invariants in Pattern Recognition, vol. 1. John Wiley and Sons Ltd, West Sussex (2009)
13. Hosny, K.M.: Exact and fast computation of geometric moments for gray level images. Appl. Math. Comput. **189**(2), 1214–1222 (2007). doi:10.1016/j.amc.2006.12.025
14. Yap, P.-T., Paramesran, R.: An efficient method for the computation of legendre moments. IEEE Trans. Pattern Anal. Mach. Intell. **27**(12), 1996–2002 (2005). doi:10.1109/tpami.2005. 232
15. Suk, T., Flusser, J.: Tensor Method for constructing 3D moment invariants. In: Berciano, A., Díaz-Pernil, D., Kropatsch, W.G., Molina-Abril, H., Real, P. (eds.) Computer Analysis of Images and Patterns, Sevilla, Spain, pp. 213–219. Springer, Berlin (2011)
16. Liao, S.X., Pawlak, M.: On image analysis by moments. IEEE Trans. Pattern Anal. Mach. Intell. **18**(3), 254–266 (1996). doi:10.1109/34.485554
17. Xu, D., Li, H.: Geometric moment invariants. Pattern Recogn. **41**(1), 240–249 (2008)
18. ChemAxon.: Marvin. http://www.chemaxon.com (2014)
19. Jmol.: Jmol: an open-source Java viewer for chemical structures in 3D. http://www.jmol.org/ (2014)
20. Min, P.: http://www.cs.princeton.edu/~min/binvox (2014)
21. Breiman, L.: Random Forests. Machine Learning, 5–32 (2001)
22. Razali, N.M., Yap, B.W.: Power Comparisons of shapiro-wilk, kolmogorov-smirnov, lilliefors and anderson-darling tests. J. Stat. Model. Analytics **2**(1), 21–33 (2011)

D-MBPSO: An Unsupervised Feature Selection Algorithm Based on PSO

K. Umamaheswari and M. Dhivya

Abstract High dimensional data with limited number of samples is a challenging task in microarray data classification. Unsupervised gene selection methods handle such data of existing methods. Various methods are available to handle the data with class labels whereas some data are mislabeled and unreliable. We propose an unsupervised filter based method known as dynamic MBPSO (D-MBPSO) which integrates MBPSO into filter approach by defining new fitness function and it is independent of any learning model. The main aim of the filter approach is to quantify the relevance based on the intrinsic properties of the data. The proposed method is applied on benchmark microarray datasets and the results are compared with well known unsupervised gene selection methods using different classifiers. The proposed method has a remarkable ability to obtain reduced feature subset with good classification accuracy.

Keywords Microarray data · Feature reduction · Particle swarm optimization

1 Introduction

Microarrays play a significant role in genetic diagnosis, which enables researchers to identify the genetic disorders based on differences in gene expression. The main goal is to examine the differentially expressed genes as normal or abnormal. Doctors can make use of this information for disease diagnosis and treatment of patients. The most important challenge is "curse of dimensionality" [1] needs to be addressed where the high dimensional nature of microarray data with less number

K. Umamaheswari (✉)
Department of Information Technology, PSG College of Technology, Coimbatore, India
e-mail: uma@ity.psgtech.ac.in

M. Dhivya (✉)
Department of Information Technology, SRM University, Chennai, India
e-mail: dhivyapsg12@gmail.com

© Springer International Publishing Switzerland 2016
V. Snášel et al. (eds.), *Innovations in Bio-Inspired Computing and Applications*,
Advances in Intelligent Systems and Computing 424,
DOI 10.1007/978-3-319-28031-8_31

359

of samples need to be analyzed. Major demerits of this nature are irrelevant and redundant genes. Data preprocessing techniques are used to obtain accurate and relevant information of genes [2, 3]. One of the most common data preprocessing techniques is feature selection which selects subset of genes from the original dataset which increases the performance.

Feature selection methods are classified into four categories including filter, wrapper, embedded and hybrid approaches [3, 4]. The filter approaches are independent of learning model and estimate the relevance of genes based on the statistical properties of the data. There are various strategies available to assess the relevance of genes including univariate and multivariate strategies [2, 4, 5]. Some of the univariate methods are term variance [1], Laplacian score [6], Signal-to-Noise ratio [7], mutual information [8] and information gain [9]. There are various multivariate strategies such as mRMR [10], FCBF [11], RSM [12], and Mutual correlation [13]. Swarm intelligence-based methods such as ACO and PSO are considered to be very effective in selecting feature subset and have been successfully employed for the applications like face recognition [14], text classification [15], and financial domains [16]. In microarray datasets, the wrapper approach is not mostly used due to the time consumption. Therefore, the filter approach is suggested for the microarray data classification problem. If the class labels of the microarray data are available, then it is termed as supervised gene selection methods [7–10]. Neverthless, some of the microarray data samples incorrectly labeled or may have unreliable class labels [3, 17]. On that account, the significance of the unsupervised gene selection methods have been employed in the DNA microarray field.

The proposed method is to build a framework to combine the computational efficiency of the filter approach and the good performance of the PSO algorithm, in which the learning model and the class labels of the sample are not needed in the gene selection process. In this paper, we propose a novel unsupervised filter based gene selection method for microarray data classification called microarray gene selection based on PSO with dynamic genetic operators.

The rest of the paper is composed as follows. Section 2 briefly reviews the particle swarm optimization algorithm. Section 3 describes the system design and Sect. 4 presents the proposed gene selection method using the PSO algorithm. Sect. 5 provides the experimental results on five microarray datasets.

2   Related Work

Eberhart and Kennedy [18] proposed an optimization algorithm known as Particle Swarm Optimization which simulates the behaviors of bird flocking involving the scenario of a group of birds randomly looking for food in an area. PSO is initialized with a group of random particles and velocity. Each particle is treated as a point in a D-dimensional space. In every iteration, each particle is updated by following the two best values. After finding the two best values, the particle updates its velocity and positions according to the following equation:

$$V_{id} = V_{id} + c_1 r_1 \left(p_{id} - X_{id} \right) + c_2 r_2 \left(p_{gd} - X_{id} \right) \tag{1}$$

$$X_{id} = X_{id} + V_{id} \tag{2}$$

where c_1 and c_2 are two positive constant named as learning factors, r_1 and r_2 are random numbers in the range of (0, 1). The Eq. (1) is used to calculate the particle's new velocity according to it's previous velocity and the distances of its current position from its own best position and the group's best position. Then the particle flies toward a new position according to Eq. (2). Such an adjustment of the particle's movement through the space causes it to search around the two best positions. If the minimum error criterion is attained or the number of cycles reaches a user-defined limit, the algorithm is terminated.

The PSO is simple for implementation and there are few parameters to adjust. The PSO has been found to be robust and fast in solving some optimization problem [19]. It has been applied to different problems such as flow shop [20], antenna design [21] or healthcare [22]. Initially, particle swarm optimization have been developed for continuous problems. Later, it was extended for discrete problems, which resulted in the commonly known binary PSO [23]. The binary version of particle swarm optimization has the tendency to prematurely converge as noted in [20, 22, 24], especially in more challenging optimization tasks. There are many approaches to solve this problem: using the mutation operator from EA [22], reset the swarm best if the fitness stagnates [24] or using perturbation mechanisms [20]. The modified BPSO algorithm is used in this paper, associates the benefits of local search (mutations) [25, 26] with resetting the swarm best mechanism [24]. This technique is applied for variable selection in MLR and PLS modeling [27] and also for predicting mortality for septic patients using SVM [28].

3 Proposed Method

The proposed method includes two parts: feature selection and classification. The following steps clearly elaborate the proposed method:

1. Encoding and Initialization: The features are encoded as binary values (0 or 1) where '1' represents the feature to be selected and '0' represents the feature not selected. Initially, the velocity factor is set as 0. pbest defines the particles as size of the initial population. Initial population will be taken as the initial pbest and then it will be updated based on the velocity factor. gbest defines one particle that is to be selected based on high fitness value.
2. Fitness Function: The filter based method is to quantify the relevance between the intrinsic properties of the genes. Thereafter, subsets with maximum

relevance should get a greater fitness value. The fitness value of solution k is computed as follows:

$$fitness(k) = \frac{1}{|subset(k)|} \sum_{i=1}^{|subset(k)|} relevance\left(g_i^k\right) \tag{3}$$

where $subset(k)$ is the subset of genes selected by particle k, $|subset(k)|$ is the size of $subset(k)$, g_i^k is the ith gene in the $subset(k)$ and relavance is the function that evaluates the relevance of the each gene. In this paper the term variance [1] is used as a relevance function, which is defined as follows:

$$TV(g_i) = \frac{1}{p} \sum_{s=1}^{p} (g_{is} - \bar{g}_i)^2 \tag{4}$$

where p is the number of samples, g_{is} denotes the value of gene i for sample s, and $\bar{g}_i$ is the average value of all the samples corresponding to gene g_i. Also, the relevance value of each gene is normalized in the interval [0, 0.1] using the softmax scaling function [1]. Note that the number of selected genes by particles in each iteration is equal to a constant value NG. It can be seen from Eq. (3) that this specific kind of fitness function is independent of any learning model.

3. Updation: For each particle, its current fitness value is compared with the fitness of its pbest, if the current value is better, then update pbest and its fitness value. The best particle of group with the best fitness value is determined, if the current fitness value is better than the fitness value of gbest, then update the gbest and its fitness value with the position. The resulting change in position is defined by the following rule:

$$If\,(0 < V_{id} \le a), then = x_{id}(old) \tag{5}$$

$$If\left(a < V_{id} \le \left(\frac{1}{2}(1+a)\right)\right), then\; x_{id}(new) = p_{id} \tag{6}$$

$$If\left(\frac{1}{2}(1+a) < V_{id} \le 1\right), then\; x_{id}(new) = p_{gd} \tag{7}$$

If global best remains constant for specific number of iterations, set the swarm particle and update using the displacement rate by

$$\begin{cases} x_{ij}^{pb} = \neg x_{ij}^{pb} \; if\; r \le dr \\ x_{ij}^{pb} = x_{ij}^{pb} \; otherwise \end{cases}, \quad i = 1, \ldots, N, j = 1, \ldots, N_p, \tag{8}$$

where dr is the displacement rate, i.e. the probability of each bit in the x_{ij}^{pb} being flipped, and r is a random number $\in[0, 1]$. Here S(.) represented the logistic function, and it served as the probability distribution for the position x_{ij} used to define the velocity and position of particle x_{ij}

$$S(v_{ij}) = \frac{1}{1 + \exp(-v_{ij})} \tag{9}$$

4. Dynamic crossover and mutation rate. The values of the crossover and mutation rates are set dynamically. The crossover rate for two chromosomes is determined by the fitness values of the two chromosomes. The mutation rate of a chromosome is calculated only by the fitness value of the chromosome. The formulae for the crossover and mutation rates are shown as follows:

$$D_c = \begin{cases} \alpha\left[1 - \frac{f - f_{med}}{f_{max} - f_{med}}\right] & for\ f > f_{med}, \\ \alpha & for\ f \leq f_{med}; \end{cases} \tag{10}$$

$$D_m = \begin{cases} \beta\left[1 - \frac{f_{mut} - f_{med}}{f_{max} - f_{med}}\right] & for\ f_{mut} > f_{med}, \\ \beta & for\ f_{mut} \leq ff_{med}; \end{cases} \tag{11}$$

where D_c denotes the crossover rate; D_m denotes the mutation rate; f denotes the largest fitness value of the two chromosomes in a crossover operation; f_{mut} denotes the fitness value of the chromosome in a mutation operation; f_{med} and f_{max} are the median and maximum fitness values, respectively, of all the chromosomes in the current population. The values of both α and β are set to 1. According to Eq. (10), for a pair of chromosomes with small fitness value, high crossover rate is assigned to increase their chance of evolution. When the highest fitness values of a pair of chromosomes is less than or equal to the median fitness value of the current population, crossover rate of 1 is assigned to make them evolve. Similarly, from Eq. (11), higher mutation rate is assigned for a chromosome with lower fitness.

5. Termination Criteria: The steps 2–4 are iterated when any one of these two termination conditions satisfied: (1) when the fitness value of one solution in the current generation achieves 100 % classification accuracy, and (2) when the number of generations is larger than 100 and the best fitness value of the last 15 generations remains the same.

6. Classification: The final gene subset is classified using different types of classifiers including SVM, Naïve Bayes and Decision Tree.

4   Experimental Results

The datasets collected for our experiments are Leukemia [29], Lung [30], Colon
[29], Prostrate [30] and SRBCT [30]. The experiment is done using Java with
NetBeans IDE 7.1 in Windows 7. The results of various stages are presented below:

The initial value of velocity factor is set to be zero. The cognitive measures c_1.
and c_2 are set as 2.05 where $c_1 + c_2 > 4$. The total number of particles are 100 and
maximum number of iterations are 50. The WEKA machine learning software
library [31] is used for the implementation of the classifiers. SMO with the
polykernel is chosen as the SVM classifier which used the one-against-rest strategy
for the multiclass problems. The complexity parameter c is set to 1 and the tolerance
parameter is set to 0.001. Moreover, naïve bayes is used as the NB classifier.
Furthermore, J48 is adopted as the DT classifier, in which the post-pruning technique is used in the pruning phase where its confidence factor is set to 0.25, and the
minimum number of samples per leaf is set to 2.

Tables 1, 2 and 3 represents the comparisons of proposed method over various
supervised filter based method like UFSACO [32], RSM, MC, RRFS, TV and LS,
and also average classification accuracy of the datasets over 5 independent runs
using SVM, naïve bayes and decision tree algorithms.

Table 1   Average classification accuracy of datasets over 5 independent runs using SVM

Datasets	Avg no. of selected genes	Classification accuracy (%)						
		DMBPSO	UFSACO	RSM	MC	RRFS	TV	LS
Colon	11	77.09	78.19	75.46	61.82	75.46	78.19	66.37
SRBCT	14	79.32	71.73	62.07	54.49	68.28	60.69	63.45
Leukemia	18	60.04	58.98	62.36	61.77	76.48	79.41	64.71
Prostrate	21	61.78	59.43	77.15	65.72	69.15	72	52
Lung	13	85.34	82.86	64.29	71.43	80.86	72.29	82
Average		72.71	70.24	68.27	63.05	74.05	72.52	65.71

Table 2   Average classification accuracy of datasets over 5 independent runs using naive bayes

Datasets	Avg no. of selected genes	Classification accuracy (%)						
		DMBPSO	UFSACO	RSM	MC	RRFS	TV	LS
Colon	11	87.29	83.64	73.64	68.19	67.28	58.19	52.73
SRBCT	14	89.12	88.28	62.08	62.07	71.73	61.38	67.59
Leukemia	18	60.04	58.98	57.65	70.59	64.71	67.65	91.18
Prostrate	21	75.07	70.09	69.72	66.29	68.58	66.86	67.43
Lung	13	66.83	64.29	76.43	40.96	78.29	68.01	70.01
Average		75.67	73.05	67.90	61.62	70.12	64.42	69.79

Table 3 Average classification accuracy of datasets over 5 independent runs using decision tree

Datasets	Avg no. of selected genes	Classification accuracy (%)						
		DMBPSO	UFSACO	RSM	MC	RRFS	TV	LS
Colon	11	78.65	75.46	71.82	66.37	65.46	68.19	60.91
SRBCT	14	73.43	72.42	41.38	55.87	71.04	77.25	54.49
Leukemia	18	72.31	69.24	61.18	67.65	79.42	79.42	70.59
Prostrate	21	68.96	66.29	66.29	64	62.29	61.15	56.01
Lung	13	84.35	69.29	69.29	68.58	79.72	75.72	78.57
Average		75.54	70.97	61.99	64.49	71.59	72.35	64.11

It is inferred from Table 1 that the proposed method obtains the highest classification accuracy of 79.32 % for SRBCT and 85.34 % for Lung whereas for Colon, Leukemia and Prostrate results are obtained equivalently with other methods. The average classification accuracy over all of the datasets show that the proposed method with an accuracy of 72.71 % outperforms UFSACO, RSM, MC, TV and LS, where RRFS shows an increase of 1.34 % of accuracy than the proposed method.

The results of Table 2 illustrate that proposed method outperforms the other methods in terms of classification accuracy for the NB classifier on the Colon, SRBCT, Leukemia, and Lung Cancer datasets. The average values on all of the datasets, in the last row of Table 3, show that the MGSACO is superior to all the other methods. It outperforms UFSACO by 2.62 %, RSM by 7.77 %, MC by 14.05 %, RRFS by 5.55 %, TV by 11.25 %, and LS by 5.88 %.

From Table 3 it is observed that the classification accuracy of the DT classifier based on the proposed method is superior to that of the unsupervised filter-based methods as 78.65 % for Colon, 68.96 % for Prostate Tumor, and 84.35 % for Lung Cancer datasets where it gives statisfied results for SRBCT and Leukemia. The average values in the last row of Table 3 show that the proposed method performs the best on all the datasets in terms of classification accuracy. In other words, the proposed method outperformsUFSACO by 4.57 %, RSM by 13.55 %, MC by 11.05 %, RRFS by 3.95 %, TV by 3.19 %, and LS by 11.43 %.

It can be concluded from Tables 1, 2 and 3 that proposed method shows an improvement of 5–6 % over the existing unsupervised filter-based methods (i.e., UFSACO, RSM, MC, RRFS, TV, and LS) in terms of classification accuracy for each of the three classifiers over different datasets. The performance of the proposed method has been evaluated over different numbers of selected genes using various types of classifiers. Figure 1a, b report the graphical results of the different datasets using the SVM, NB, and DT classifiers correspondingly.

The x-axis denotes the type of datasets, while the y-axis shows the average classification accuracy (in %). Figure 1a shows the results of proposed method with UFSACO using SVM classifier which gives statisfied results for SRBCT, Colon and Prostrate datasets. It can be concluded that the classification accuracy rate of the

Fig. 1 Performance
comparison

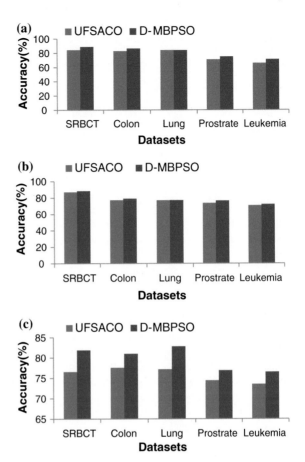

proposed method is significantly superior to that of the UFSACO method for
leukemia and similar results for lung dataset. As seen in Fig. 1b illustrates the
respective comparison results for different datasets with proposed method and
UFSACO using naïve bayes classifier. The different classification accuracy rates of
proposed method and UFSACO can be seen more prominently for prostrate dataset
where the proposed method acquires significantly higher classification accuracy for
SRBCT, Colon, Lung and Leukemia as 88.43, 79, 77.15, and 71.98 %, correspondingly, than UFSACO.

Figure 1c demonstrates that the overall performance of proposed method is
superior to that of the UFSACO method when the decision tree classifier is applied.
Especially, for SRBCT and Lung datasets, the classification accuracy of the proposed method was 81.88 and 82.81 %, while for the UFSACO this value was
reported as 76.55 and 77.15 % respectively. Moreover, it can be seen that the
performance of the proposed method is better than that of the UFSACO for colon,
prostrate and leukemia datasets, proposed method got classification accuracy of

80.99, 76.89, and 76.49 %, respectively, while in these cases the classification accuracy of UFSACO was reported 77.55, 74.41, and 73.5 %, respectively.

It can be concluded from Fig. 1a, c that although the proposed method is an unsupervised method and does not need class labels of the samples, it can be much better than the UFSACO method. The proposed method is a population based method which simultaneously explores the search space from different points. Moreover, it uses an iterative improvement process to select the subsets of genes. The computational complexity of the proposed method is calculated as $O(In^2p)$. It is inferred that the proposed method has significant improvement in the computational cost than the filter based methods and it performs faster than the wrapper based methods.

5 Conclusion

In this paper, unsupervised filter based method is proposed known as dynamic MBPSO based on PSO mechanism for gene selection process. The computational efficiency of the filter approach and good performance of the PSO were combined to improve the performance of the proposed method. Moreover, a new fitness function is used to evaluate the subsets of selected genes without using any learning model to enhance the efficiency of the proposed method. The proposed method outperforms the well known unsupervised filter-based gene selection methods including unsupervised feature selection based on ACO (UFSACO), relevance-redundancy feature selection (RRFS), random subspace method (RSM), mutual correlation (MC), term variance (TV), and Laplacian score (LS). The experimental results show that the proposed method was able to select a subset of genes with minimum redundancy and maximum relevance. Furthermore, the results show that the classification accuracy of the proposed method outperforms other unsupervised methods for various subsets of genes over all the three classifiers.

References

1. Theodoridis, S., Koutroumbas, K.: Pattern Recognition, 4th ed. Elsevier Science, Amsterdam (2008)
2. Lazar, C., Taminau, J., Meganck, S., Steenhoff, D., Coletta, A., Molter, C., Schaetzen, Vd, Duque, R., Bersini, H., Nowe, A.: A survey on filter techniques for feature selection in gene expression microarray analysis. IEEE/ACM Trans. Comput. Biol. Bioinf. **9**, 1106–1119 (2012)
3. Bolón-Canedo, V., Sánchez-Maroño, N., Alonso-Betanzos, A., Benítez, J.M., Herrera, F.: A review of microarray datasets and applied feature selection methods. Inf. Sci. **282** (2014). http://dx.doi.org/10.1016/j.ins.2014.05.042
4. Saeys, Y., Inza, I., Larrañaga, P.: A review of feature selection techniques in bioinformatics. Bioinformatics **23**, 2507–2517 (2007)

5. Lai, C., Reinders, M., van't Veer, L., Wessels, L.: A comparison of univariate and multivariate gene selection techniques for classification of cancer datasets. BMC Bioinf. **7**, 235 (2006)
6. He, X., Cai, D., Niyogi, P.: Laplacian score for feature selection. Adv. Neural Inf. Process. Syst. **18** (2005)
7. Golub, T.R., Slonim, D.K., Tamayo, P., Huard, C., Gaasenbeek, M., Mesirov, J.P., Coller, H., Loh, M.L., Downing, J.R., Caligiuri, M.A., Bloomfield, C.D., Lander, E.S.: Molecular classification of cancer: class discovery and class prediction by gene expression monitoring. Science **286**, 531–537 (1999)
8. Cai, R., Hao, Z., Yang, X., Wen, W.: An efficient gene selection algorithm based on mutual information. Neurocomputing **72**, 991–999 (2009)
9. Raileanu, L.E., Stoffel, K.: Theoretical comparison between the gini index and information gain criteria. Ann. Math. Artif. Intell. **41**, 77–93 (2004)
10. Ding, C., Peng, H.: Minimum redundancy feature selection from microarray gene expression data. J. Bioinf. Comput. Biol. **03**, 185–205 (2005)
11. Yu, L., Liu, H.: Feature selection for high-dimensional data: a fast correlation-based filter solution. In: 20th International Conference on Machine Learning, pp. 856–863 (2003)
12. Lai, C., Reinders, M.J.T., Wessels, L.: Random subspace method for multivariate feature selection. Pattern Recogn. Lett. **27**, 1067–1076 (2006)
13. Haindl, M., Somol, P., Ververidis, D., Kotropoulos, C.: Feature selection based on mutual correlation. In: Recognition, Pattern (ed.) Image Analysis and Applications, pp. 569–577. Springer, Berlin (2006)
14. Kanan, H.R., Faez, K.: An improved feature selection method based on ant colony optimization evaluated on face recognition system. Appl. Math. Comput. **205**, 716–725 (2008)
15. Aghdam, M.H., Ghasem-Aghaee, N., Basiri, M.E.: Text feature selection using ant colony optimization. Expert Syst. Appl. **36**, 6843–6853 (2009)
16. Marinakis, Y., Marinaki, M., Doumpos, M., Zopounidis, C.: Ant colony and particle swarm optimization for financial classification problems. Expert Syst. Appl. **36**, 10604–10611 (2009)
17. Niijima, S., Okuno, Y.: Laplacian linear discriminant analysis approach to unsupervised feature selection. IEEE/ACM Trans. Comput. Biol. Bioinf. **6**(4), 605–614 (2009)
18. Kennedy, J., Eberhart, R.: Particle swarm optimization. In: Proceedings of the IEEE International Conference on Neural Networks, vol. 4, pp. 1942–1948. IEEE (1995)
19. Brandstatter, B., Baumgartner, U.: Particle swarm optimizationmass-spring system analogon. IEEE Trans. Magn. **38**(2), 997–1000 (2002)
20. Yuan, L., Zhao, Z.-D.: A modified binary particle swarm optimization algorithm for permutation flow shop problem. In: Proceedings of the International Conference on Machine Learning and Cybernetics, vol. 2, 2007, pp. 902–907
21. Marandi, A., Afshinmanesh, F., Shahabadi, M., Bahrami, F.: Boolean particle swarm optimization and its application to the design of a dual-band dual-polarized planar antenna. In: Proceedings of IEEE Congress on Evolutionary Computation, pp. 3212–3218 (2006)
22. Alba, E., Garcia-Nieto, J., Jourdan, L., Talbi, E.-G.: Gene selection in cancer classification using PSO/SVM and GA/SVM hybrid algorithms. In: Proceedings of the IEEE Congress on Evolutionary Computation, pp. 284–290 (2007)
23. Wahde, M.: Biologically Inspired Optimization Methods, 1st edn. WIT Press, Southampton (2008)
24. Chuang, L.-Y., Chang, H.-W., Tu, C.-J., Yang, C.-H.: Improved binary PSO for feature selection using gene expression data. Computat. Biol. Chemist. **32**, 29–38 (2008)
25. Reynolds, C.W.: Flocks, herds and schools: a distributed behavioral model. SIGGRAPH Comput. Graph. **21**, 25–34 (1987)
26. Lee, S., Soak, S., Oh, S., Pedrycz, W., Jeon, M.: Modified binary particle swarm optimization. Prog. Nat. Sci. **18**(9), 1161–1166 (2008)
27. Shen, Qi, Jiang, Jian-Hui, Jiao, Chen-Xu, Shen, Guo-li, Ru-Qin, Yu.: Modified particle swarm optimization algorithm for variable selection in MLR and PLS modeling: QSAR studies of antagonism of angiotensin II antagonists. Eur. J. Pharm. Sci. **22**, 145–152 (2004)

28. Susana, M. Vieira, Mendonc¸ L.F., Farinha, G.J., Sousa, J.M.C.: Modified binary PSO for feature selection using SVM applied to mortality prediction of septic patients. Appl. Soft Comput. **13**, 3494–3504 (2013)
29. Dataset Repository, Bioinformatics Research Group.: http://www.upo.es/eps/bigs/datasets.html (2014)
30. Statnikov, A., Aliferis, C.F. Tsamardinos, I.: Gems: gene expression model selector. http://www.gems-system.org/ (2005)
31. Hall, M., Frank, E., Holmes, G., Pfahringer, B., Reutemann, P., Witten, I.: The WEKA data mining software. http://www.cs.waikato.ac.nz/ml/weka
32. Tabakhi, Sina, Moradi, Parham, Akhlaghian, Fardin: An unsupervised feature selection algorithm based on ant colony optimization. Eng. Appl. Artif. Intell. **32**, 112–123 (2014)

Delay Scheduling with Reduced Workload on JobTracker in Hadoop

Krishan Kumar Sethi and Dharavath Ramesh

Abstract Job scheduling is one of the critical issues in MapReduce processing that affects the performance of Hadoop framework. Delay scheduling introduces a small delay during job scheduling to optimize the data locality. Delay scheduler may scan a job more than once before reaching a certain deadline after which the job is scheduled. This causes extra overhead on the scheduler. Moreover a higher priority job may get delayed. We propose an algorithm in which the load is distributed among the individual nodes. Our algorithm insists the scheduler to launch a high priority job on a free node. The node then executes the job locally or schedules it to some other node based on the availability of data. Experimental results show that the proposed algorithm performs better than Hadoop and records less execution time.

Keywords Mapreduce · Hadoop · Job scheduling · Data locality

1 Introduction

With the rapid growth of technology, huge amount data is generated from different sectors of Commerce, Science and Engineering. An IDC digital universe study estimates that the amount of digital data will grow 10 times from 2013 to 2020 [1]. The IDC quotes that the digital universe will reach 44 trillion by 2020. Such enormous amount of data, termed as Big Data [2–4] has the capacity to reform many fields such as Industry, healthcare, public administration and so on. This huge amount of data cannot be processed efficiently on a single machine. Therefore, the

K.K. Sethi · D. Ramesh (✉)
Department of Computer Science and Engineering, Indian School
of Mines, Dhanbad 826004, Jharkhand, India
e-mail: rameshd.ism@gmail.com

K.K. Sethi
e-mail: kksethi02@gmail.com

© Springer International Publishing Switzerland 2016
V. Snášel et al. (eds.), *Innovations in Bio-Inspired Computing and Applications*,
Advances in Intelligent Systems and Computing 424,
DOI 10.1007/978-3-319-28031-8_32

big data is analyzed and processed using different data-intensive computing environments like cloud computing [5], cluster computing. Many Clustered Computing systems like Google's MapReduce [6], Microsoft's Dryad [7] and Yahoo's Map-Reduce-Merge [8] are evolved to process Big Data.

Hadoop [9, 10] is an open-source framework which works in a distributed manner and allows to store and process huge amounts of data, i.e. big data across a cluster of commodity hardware using simple programming models. It works in master-slave fashion where master keeps the track of the data block and computing programs on the slave nodes. Data is stored in distributed fashion using Hadoop Distributed File System (HDFS) which is processed in parallel by each node. Typically the Hadoop system contains five daemon processes named as NameNode, DataNode, JobTracker, TaskTracker and Secondary NameNode. The NameNode stores the information about file system namespace and metadata of all the directories and files in HDFS. The DataNode stores the data blocks, whose information is kept with the NameNode. The JobTracker coordinates the running jobs and handles scheduling of jobs. The TaskTracker is a slave process of JobTracker that is responsible for running the job on individual node. The Secondary NameNode keeps a checkpoint for the NameNode logs to help the NameNode for better functioning. Hadoop includes MapReduce as a programming engine to provide the platform for job execution.

MapReduce incorporates the features like minimal abstraction, automatic parallel computation and fault tolerance via replication. MapReduce is composed of two methods the map() method and reduce() method [6]. The date is preprocessed and prepared for processing in map phase and intermediate output is generated. Each reduce function process the intermediate output of map to produce final output. The working model of MapReduce has been described in Fig. 1.

Job scheduling is one of the critical areas in Hadoop and responsible for better performance. Multiple users can concurrently run their jobs on a Hadoop cluster. JobTracker schedules the jobs to the nodes in the cluster. The JobTracker considers that a task should be executed on a node where data is available (local node). The JobTracker schedules a job on a local node or closer to the local node. Nonlocal data processing leads the movement of data toward processing node which consumes network bandwidth and time. Hence, Data locality is one of the important concerns to enhance the response time and throughput. Job priority is also kept into consideration during scheduling where a higher priority job should be launched first. Delay scheduling [11] dis one of the significant algorithm which adds just a small delay in job scheduling to achieve better data locality. Delay scheduling solves the problem of Head-of-line scheduling and sticky slots in fair scheduling. A job can be delayed up to D times, where D is a sensitive parameter and set to achieve desired level of locality. A problem with delay scheduling is that it skips a job until a slot (on the local node) becomes free. Hence, a job is scanned multiple times by JobTracker. Therefore, this increases the overhead on the JobTracker. Moreover, a higher priority job may be skipped due to the absence of the data block. We have developed a new algorithm that launches the job on a free node in first scan. Further, in case of unavailability of data, the node communicates with all

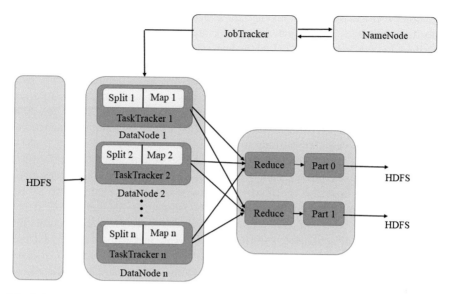

Fig. 1 Working model of MapReduce

the local nodes. At each node, finishing time of each running task is computed. A local node with smallest finishing time is chosen as right node. We also introduce a threshold which is applied at the finishing time of right node to improve the response time. Our contributions are listed as follows:

- Identified a problem with delay scheduling regarding the overhead on JobTracker.
- An algorithm to reduce the overhead on the JobTracker by distributing the load on TaskTrackers.
- Our algorithm considers finishing time computation and a variable threshold to decide for a right node.

The rest of the paper is organized as follows. Different algorithms about the scheduling in Hadoop have been discussed in 'Related Work' in Sect. 2. Section 3 describes about the problem and proposed methodology. Simulation and analysis of the proposed algorithm are covered in Sect. 4. Finally, conclusion and future work are given in Sect. 5.

2 Related Work

In the literature, several works have been performed on scheduling related activities with different methodologies. FIFO [10] is the default scheduling scheme in Hadoop Scheduler. JobTracker schedules the jobs in the order of their arrival in the system. FIFO has the limitations that, for a mixed job (both short and long), Short job may wait until a long job releases its resources. The limitation of FIFO is rectified by fair scheduler [12] of Yahoo which considers each job is equally important and needs an equal share of resources. A set of jobs has been assigned in a pool and each pool takes an equal share of resources. Typically, each pool corresponds to a single user. Fair scheduler has the limitation that it does not consider the weight of jobs. Facebook's Capacitive scheduling [10] is the most complex among the above schemes which puts the jobs in the multiple queues and assign the certain resource capacity to each queue.

Zaharia et al. [13] proposed LATE (Longest Approximate Time to End) scheduler to improve the execution of speculative tasks in heterogeneous environment. LATE finds the speculative tasks by computing finishing time for all running tasks. We used the same method of computing finishing time in our algorithm. LATE scheduler couldn't find the speculative tasks correctly because the remaining time is calculated incorrectly [14]. The problem is resolved by a Self-Adaptive MapReduce scheduler (SAMR) which computes progress of tasks dynamically and adapts to the continuously varying environment automatically.

Zaharia et al. [11] proposed some extension to address the conflict between locality and fairness in the Hadoop scheduler termed as 'Delay Scheduling' discussed in the previous section. Guo [15] proposed an algorithm to schedule multiple tasks simultaneously rather than one by one to give optimal data locality. Ibrahim et al. [16] developed a novel algorithm named LEEN: locality-aware and fairness-aware key partitioning to improve locality and reduce shuffle data. LEEN partitions all the buffered intermediate keys according to their frequencies and fairness after the shuffle phase. Nguyen [17] proposed a Hybrid Scheduler (HybS) that apply the greedy approach to assign dynamic priorities to each task based on task length, job size and job waiting time. This priority affects the choice of the task to reduce the latency for variable length jobs and maintain data locality. Another methodology on scheduling technique [18] has been developed to increase data locality of map tasks by placing tasks on nodes with seeking data blocks. Techniques in [19, 20] present data replication based scheduling techniques to enhance the data locality and performance. Different Methodologies proposed in [21–24] demonstrate the scheduling techniques to improve the data locality of reduce tasks by approaching multiple techniques of assigning the reduce tasks to the appropriate node.

3 Proposed Methodology

This section begins by introducing the problem statement and then describes the solution by incorporating the proposed method.

Table 1 Job details

Name of jobs	No. of data splits/Map tasks	Name of data splits	Name of map tasks
J_1	4	$B_{11}, B_{12}, B_{13}, B_{14}$	$J_1T_1, J_1T_2, J_1T_3, J_1T_4$
J_2	3	B_{21}, B_{22}, B_{23}	J_2T_1, J_2T_2, J_2T_3
J_3	3	B_{31}, B_{32}, B_{33}	J_3T_1, J_3T_2, J_3T_3

3.1 Problem Statement

A typical cluster can have thousands of nodes, running the jobs of multiple users. Here, we take an instance of a cluster of 7 nodes and one map slot at each node. Now, we assume that three jobs, J_1, J_2 and J_3 are submitted to JobTracker. Details of the jobs are given below in Table 1.

Data splits of all the Jobs are distributed on multiple nodes of cluster as shown in Fig. 2. Map tasks of J_1, i.e. J_1T_1, J_1T_2, J_1T_3 and J_1T_4 are running on the respective local nodes Node 1, Node 2, Node 3 and Node 4. In the same manner JobTracker schedules J_2's tasks on the free nodes in the cluster. Let's assume that Node 6 gets free slot at some time instance and it sends a heartbeat message to JobTracker. JobTracker has to schedule the high priority job J_3 on Node 6 but the corresponding data block is not available. In such a case, if delay scheduling is applied, then JobTracker may skip the job J_3 up to a certain time until a slot becomes free on any appropriate node (a node keeps the data block for the job, i.e. J_3). This situation increases the load on the JobTracker. Moreover the high priority job, i.e. J_3 will be delayed, which causes the higher job response time.

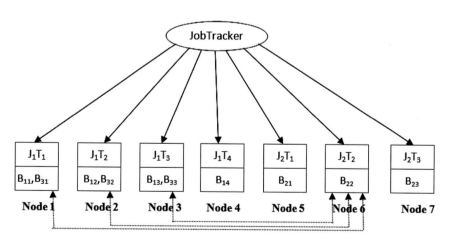

Fig. 2 An illustration to problem

3.2   Algorithmic Approach

A key question is how to schedule the jobs to reduce the overhead at JobTracker. We introduce an innovative approach to reduce the overhead by transferring the workload at TaskTracker. According to our proposed method, whenever a node gets a free slot, JobTracker launch a higher priority job irrespective of the availability of data. In case of unavailability of data, the node will further communicate with all appropriate nodes and find the right node. A right node is selected based on the following two constraints. Firstly, a right node is the one that will be free earliest among all appropriate nodes. Each appropriate node computes finishing time for each running task and estimates the remaining time to complete. A node with smallest finishing time is selected. Secondly, finishing time of the selected node should be lesser or equal to the threshold. A threshold defines the limit of finishing time for a right node to provide better response time. A threshold can be varied to get optimal results for various types of jobs. Usually, the value of the threshold is chosen according to the job size. We set the best value of threshold according the previous experience of the framework. If both the constraints are satisfied, then the task can be launched on the right node after waiting the least finishing time. Otherwise, the data will be moved from the appropriate node.

In the above example, when a heartbeat message is received from Node 6, JobTracker sorts all the jobs according to priority in ascending order. Job J_3 is a high priority job and it also has an unlaunched task. Although the data is not available on Node 6 then also the unlaunched task will be assigned on Node 6 with the details of the appropriate Nodes i.e. Node 1, Node 2 and Node 3. Node 6 communicates with all the appropriate nodes and sends a message asking for a free slot as shown in Fig. 2. In the absence of a free slot, each appropriate Node calculates the finishing time and sends this to Node 6. Finishing time is a parameter to judge for time to completion of a task. Node 6 will either transfer the task on a right node (satisfy both the constraints given above) or the data block of J_3 will be transferred on Node 6.

3.3   Algorithm Design

A free node n sends a heartbeat message to JobTracker. A higher priority Job J_i with an unlaunched task will be assigned on *node n*. In case of data unavailability, *node n* communicates with all the appropriate nodes (*AN*) and asks to compute the finishing time at each node for all running map tasks. Each appropriate node calls a method *nodeFinishTime()* that calculates the finishing time of each running tasks by the Eq. (1) and returns the minimum value of finishing time.

3.3.1 Estimating Finishing Time

Finishing time is the time left for a task to complete. We estimate the time left for a task based on the progress score as given in [13]. Hadoop computes the progress score to monitor the task progress that lies in 0 to 1. Progress score for a map is the fraction of input data read. The finishing time is given as:

$$\text{Finishing Time} = (1 - \text{Progress Score})/\text{Progress Rate} \qquad (1)$$

The progress Rate of each task is given as "Progress Score/T", where T is the amount of time the task has been running for.

Algorithm: For map tasks

Input: *JP: Pool of Unscheduled Jobs*
             T_i: Task Set of ith job
             N: Set of Nodes
             T_h: Threshold
             AN: Set of Appropriate Nodes

Output: *STP: Pool of scheduled tasks*

1. When a heartbeat message from a node n ∈ N is received:
2. Declare Integer *min_finish_time* =<biggest value>
3. **if** *n* has a free slot **then**
4. Sort jobs in increasing order of their priority
5. **for** each job J_i ∈ *JP* **do**
6.   **if** an incomplete task *t* ∈ T_i **then**
7.   Launch *t* on Node *n*
8.   **if** *t* has the data block on *n* **then**
9. $STP_i = STP_i$ ∪ *t*
10.   **else**
11.   **for** each appropriate nodes *AN* ∈ *N* **do**
12.   **if** any *AN* has a free slot **then**
13.   Launch task *t* on that *AN*
14.   break
15.   **end if**
16.   *finish_time = nodeFinishTime ()*
17.   **if** *finish_time* is lesser than *min_finish_time* **then**
18.   *Node_Index* = current node index
19.   *min_finish_time = finish_time*
20.   **endif**
21.   **end for**
22.   **end if**
23. **end if**
24. **end for**
25. **if** *min_finish_time* is lesser than T_h **then**
26. wait for the *min_finish_time* and transfer the task *t* on node with *Node_index*
27. **else**
28. Transfer the data block on node *n*
29. **end if**
30. $STP_i = STP_i$ ∪ *t*

Node *n* selects the node as the right node with the lowest value of *finish_time* i.e. *min_finish_time*. The value of min_*finish_time* is also compared with threshold T_h. Node *n* waits for the *min_finish_time* and launches the task on the node after it gets free. To enhance the response time, a local queue is maintained at node *n* that holds the task *t* locally and can free the slot to execute some other task. If there is more than one right node, one of the nodes can be chosen to launch the task. The other case, if no such a node exists, then the data block for the task is carried to *node n* to start the job execution. The data block is relocated from any *AN* which is nearest to the node *n* to minimize the data movement cost. A task is assigned to a pool of scheduled task i.e. *STP*. Every Job consists its own *STP* to accumulate all its completed tasks.

```
Method: Calculate the finishing time

function nodeFinishTime ()
  1.   Declare Integer min = <biggest value>
  2.   Constant Integer slots = 4
  3.   Declare Integer finishing_time [slots]
  4.   Declare index
  5.   for index = 0 to slots do
  6.       Calculate finishing time for each task
  7.       finishing_time [index] = calculated finishing time from step 6
  8.       if finishing_time [index] is lesser than min then
  9.           min = finishing_time [index]
 10.       end if
 11.   end for
 12.   return min
end function
```

4   Simulation and Result Analysis

To study the experimental and analytical part, we have written Java programs to simulate the basic Hadoop scheduling part. We modified the Hadoop scheduling code according to our scheduling strategy. The simulation environment contains 8 computing nodes with 2 maps and 2 reduce slots on each. We executed different kind of jobs simultaneously on the nodes and analyzed the results with multiple parameters. Moreover, we found that both FIFO and fair share scheduling affects very nominal on our algorithm. Therefore, we took single instance to consider both the scheduling algorithms in our experiment.

For analysis, we took 2 different parameters "time to completion" and "percentage local maps" as shown in Fig. 3. We took 4 instances of input jobs, i.e. 5 jobs, 10 jobs, 20 jobs and 50 jobs with 300 MB data size for each job. In basic Hadoop scheduler, locality of map tasks continuously decreases for an increasing number of jobs, i.e. 95 % for 5 jobs, 79 % for 10 jobs, 42 % for 20 jobs and 26 % for

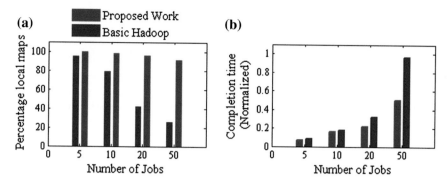

Fig. 3 Analysis between basic Hadoop and modified hadoop for different number of jobs. **a** Locality analysis **b** Time to completion analysis

Fig. 4 Locality for different threshold values

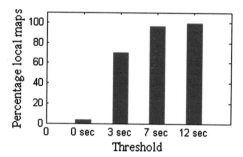

50 jobs as per the Fig. 3a. While in modified framework the locality always remains more than 90 %. In basic Hadoop, a growing number of jobs increase time to completion due to decrease in data locality as shown in Fig. 3b. In contrast, in modified framework, time to completion increases linearly with the growing number of jobs. Our algorithm has also improved the data locality. So, our algorithm improves the response time, job execution time and throughput.

Furthermore, we performed the simulation to measure the impact of different values of threshold on data locality. The experiment was executed for 30 jobs. Threshold value of 0 s doesn't allow to wait for the task to run at appropriate node. Hence, it would be same as the basic Hadoop. A small threshold value can improve the data locality by a sufficient margin as illustrated in Fig. 4.

5 Conclusion and Future Work

We analyzed the problems with delay scheduling which increase the load on the JobTracker by scanning the same job multiple times. In order to rectify this, we proposed an incremental algorithm to reduce overhead on the JobTracker by moving the load on TaskTracker. We employed the computation of finishing time on each node to select a right node. Moreover, to optimize the selection, a variable threshold is also added to optimize the selection. An appropriate value of the threshold can be considered with the previous experience of the framework. To achieve good performance, threshold can be picked smaller for small jobs while it may grow larger for larger jobs. Experimental results show that the performance of our algorithm is better than basic Hadoop. In future, we intend to compute a dynamic threshold for better results.

Acknowledgements The research work is supported by Department of Computer Science & Engineering, Indian School of Mines, Dhanbad, India.

References

1. Turner, V., et al.: The digital universe of opportunities: rich data and the increasing value of the internet of things. In: International Data Corporation, White Paper, IDC_1672 (2014)
2. Philip Chen, C.L., Zhang, Chun-Yang: Data-intensive applications, challenges, techniques and technologies: a survey on big data. Inf. Sci. **275**, 314–347 (2014)
3. Hashem, Ibrahim Abaker Targio, Yaqoob, Ibrar, Badrul Anuar, Nor, Mokhtar, Salimah, Gani, Abdullah, Ullah Khan, Samee: The rise of big data on cloud computing: review and open research issues. Inf. Syst. **47**, 98–115 (2015)
4. Kambatla, Karthik, Kollias, Giorgos, Kumar, Vipin, Grama, Ananth: Trends in big data analytics. J. Parallel Distrib. Comput. **74**(7), 2561–2573 (2014)
5. Hashem, Targio, Ibrahim Abaker, et al.: The rise of "big data" on cloud computing: review and open research issues. Inf. Syst. **47**, 98–115 (2015)
6. Dean, Jeffrey, Ghemawat, Sanjay: MapReduce: simplified data processing on large clusters. Commun. ACM **51**(1), 107–113 (2008)
7. Isard, M., Budiu, M., Yu, Y., Birrell, A., Fetterly, D.: Dryad: Distributed data-parallel programs from sequential building blocks. In: Conference Computer System (EuroSys), pp. 59–72 (2007)
8. Yang, H.C., Dasdan, A., Hsiao, R.-L., Parker, D.S.: Map-Reduce-Merge: simplified relational data processing on large clusters. In: Proceeding of ACM SIGMOD International Conference Management of Data (2007)
9. Polato, Ivanilton, et al.: A comprehensive view of Hadoop research—A systematic literature review. J. Netw. Comput. Appl. **46**, 1–25 (2014)
10. Apache Hadoop.: http://hadoop.apache.orgJune 2011
11. Zaharia, M., et al.: Delay scheduling: a simple technique for achieving locality and fairness in cluster scheduling. In: Proceedings of the 5th European Conference on Computer Systems. ACM (2010)
12. Hadoop's Fair Scheduler.: https://hadoop.apache.org/docs/r1.2.1/fair_scheduler
13. Zaharia, M., et al.: Improving MapReduce performance in heterogeneous environments. In: OSDI, vol. 8(4) (2008)
14. Chen, Q., et al.: Samr: A self-adaptive Mapreduce scheduling algorithm in heterogeneous environment. In: 2010 IEEE 10th International Conference on Computer and Information Technology (CIT). IEEE (2010)
15. Guo, Z., Fox, G., Zhou, M.: Investigation of data locality in Mapreduce. In: Proceedings of the 2012 12th IEEE/ACM International Symposium on Cluster, Cloud and Grid Computing (ccgrid 2012). IEEE Computer Society (2012)

16. Ibrahim, S., et al.: LEEN: Locality/fairness-aware key partitioning for Mapreduce in the cloud. In: IEEE Second International Conference on Cloud Computing Technology and Science (CloudCom), (2010)

17. Nguyen, P., et al.: A hybrid scheduling algorithm for data intensive workloads in a Mapreduce environment. In: Proceedings of the 2012 IEEE/ACM Fifth International Conference on Utility and Cloud Computing. IEEE Computer Society (2012)

18. He, C., Lu, Y., Swanson, D.: Matchmaking: a new Mapreduce scheduling technique. In: 2011 IEEE Third International Conference on Cloud Computing Technology and Science (CloudCom). IEEE (2011)

19. Abad, C.L., Lu, Y., Campbell, R.H.: DARE: Adaptive data replication for efficient cluster scheduling. In: 2011 IEEE International Conference on Cluster Computing (CLUSTER). IEEE (2011)

20. Ibrahim, S., et al.: Maestro: Replica-aware map scheduling for Mapreduce. In: 2012 12th IEEE/ACM International Symposium on Cluster, Cloud and Grid Computing (CCGrid). IEEE (2012)

21. Ahmad, Faraz, et al.: MapReduce with communication overlap (MaRCO). J. Parallel Distrib. Comput. **73**(5), 608–620 (2013)

22. Tang, Zhuo, et al.: A self-adaptive scheduling algorithm for reduce start time. Future Gener. Comput. Syst. **43**, 51–60 (2015)

23. Hammoud, M., Rehman, M.S., Sakr, M.F.: Center-of-gravity reduce task scheduling to lower Mapreduce network traffic. In: Cloud Computing (CLOUD). IEEE (2012)

24. Hammoud, M, Sakr, M.F.: Locality-aware reduce task scheduling for MapReduce. In: IEEE Third International Conference on Cloud Computing Technology and Science (CloudCom). IEEE (2011)

Reducing Travel Time in VANETs with Parallel Implementation of MACO (Modified ACO)

Vinita Jindal and Punam Bedi

Abstract Routing plays a major role in VANETs by helping a vehicle to reach the destination by finding an optimal path. These routing decisions are affected by the congestion on roads. Several approaches have been proposed to improve this problem of handling congestion thorough various traffic management strategies. ACO is being used in literature to provide routing in real time environment. Modified Ant Colony Optimization (MACO) algorithm is used to reduce the travel time of the journey by avoiding congested routes. This paper proposes a parallel implementation for MACO algorithm in order to further reduce the travel time due to faster computation using GPUs for the vehicles on move. Parallel implementation is done using parallel architecture on the Graphics Processing Unit (GPU) at NVIDIA GeForce 710 M using C language running CUDA (Compute Unified Device Architecture) toolkit 7.0 on Microsoft Visual Studio 2010. The obtained results for proposed parallel MACO when compared with the parallel implementation of the standard Dijkstra algorithm and that of the existing MACO algorithm on a real world North-West Delhi map with an increased number of vehicles significantly reduce the travel time.

Keywords Parallelization · Ant colony optimization · CUDA · Routing problem · GPU

1 Introduction

Vehicular Ad hoc NETworks (VANETs) comprises of vehicles that are considered as nodes and their distance from each other on underlying roads are considered as edges in forming a kind of decentralized ad hoc network [1]. Routing plays an

V. Jindal (✉) · P. Bedi
Department of Computer Science, University of Delhi, Delhi, India
e-mail: vjindal@keshav.du.ac.in

P. Bedi
e-mail: punambedi@ieee.org

© Springer International Publishing Switzerland 2016
V. Snášel et al. (eds.), *Innovations in Bio-Inspired Computing and Applications*,
Advances in Intelligent Systems and Computing 424,
DOI 10.1007/978-3-319-28031-8_33

important role in VANETs as spending less time on roads is an issue that always attracts researchers to design some mechanism for the commuters. They always search for a congestion free route in order to reach the destination however that path may be a bit larger than the shortest path. The general behavior of driver inclines to be a greedy one as they are seeking for shortest route to the destination. This can eventually cause congestion in heavy traffic conditions [2]. In real time, congestion increases if vehicles do not take optimized routes to reach their destination. The shortest route to the destination is not always the optimal one. However if all vehicles keep following their respective shortest routes, this will result in congestion on some set of routes.

Authors in [3] propose a MACO algorithm that provide the solution to the above problem by providing optimized routes with the avoidance of the congestion enroute. MACO is the variation of the classical ACO in which repulsion effect is used instead of attraction towards pheromones for avoiding congested routes and dispersing the traffic towards paths with less pheromone value. Their approach was able to significantly reduce the overall travel time of the journey. All their work in the algorithm was carried by a single processor and hence computation time is more in case of large number of vehicles.

To overcome this drawback, parallel computing was introduced in which many computations are carried out simultaneously by multiple processors. Here, a large problem is divided into multiple small sub problems and each of them can be executed simultaneously on the different processors. This reduces the computation time as compared to the serial execution of the same problem. Parallel Computing mainly uses Flynn's Taxonomy in which it uses SIMD (Single Instruction Multiple Data). In this approach the instruction remains constant, but the data changes continuously. This increases the data handling capacity of the processor and hence can reduce the computational time of the algorithm.

In the paper we are proposing the parallel implementation of MACO to further reduce the overall travel time. The MACO approach comprises of route construction, pheromone deposition and pheromone evaporation. In the literature to the best of our knowledge, only route construction path has been implemented on GPU for ACO [4, 5]. But in our proposed work we are parallelize all the function used in the MACO algorithm to reduce the overall travel time. Experimental results show that the proposed work is successfully reducing computational time with increase in congestion. For implementation, we have used CUDA 7.0 toolkit on NVIDIA GeForce 710 M. We have hardcoded a network obtained from real world map of North-West Delhi. The obtained results were compared with the parallel implementation of standard Dijkstra's algorithm and the MACO algorithm proposed by [3]. Obtained results show the significant improvement in reduction of overall travel time of the vehicles by the proposed technique under heavy traffic conditions.

1.1 Overview of CUDA (Compute Unified Device Architecture)

CUDA was first introduced by NVIDIA in 2007. It was developed for both Windows and Linux platform and later versions were also compatible with Mac OS. NVIDIA has developed CUDA especially for NVIDIA Graphic Cards environment and allows a great resource tool used to solve problems with a high degree of computational complexity [6, 7]. It is a unique programming language for the NVIDIA Graphic cards as it uses the all the cores (Graphic Process Unit, GPU) of the graphic card efficiently. To use the CUDA architecture, extended C language is used to program on NVIDIA cards. Kernel is executed in GPU which is written in C/C++ language. Kernel is executed as many times as selected by the programmer through number of threads. The building blocks of GPU are Grids. Grids are further divided into blocks and the blocks are divided into threads. There are various parallel computing platforms available as MPI (Message Passing Interface), OpenCL (Open Computing Language) and CUDA. This paper is implemented using CUDA toolkit version 7.0.

1.2 Overview of MACO (Modified ACO)

The MACO was developed to avoid the congestion in a city environment. It uses the existing Dijkstra's algorithm in low traffic conditions and modified ACO approach in heavy traffic conditions. Initially it selects the shortest path through Dijkstra's algorithm, and then uses MACO for optimization. MACO was used in a decentralized environment, where vehicles are treated as ants that leave pheromones on the trailed paths. When the accumulated pheromone trail reaches to a threshold value, the ant will be using the repulsion effect introduced in the pheromone behavior. On contrary from the classical ACO approaches, pheromone levels produce repulsion for other ants to avoid the congestion rather than an attraction [8–10]. In the absence of significant traffic, MACO behaves like the Dijkstra algorithm. In case of heavy traffic conditions, the deposition of the pheromone gradually increases due to large number of vehicles on roads. With time, the pheromone level is decreased using the evaporation process. Therefore, the working of the whole system depends on the deposition and evaporation rates of the pheromone as in each time step; every vehicle sense the updated pheromone value based on which the decision for the route is taken. We have compared the MACO algorithm [3] with our proposed parallel implementation of MACO to show the significant reduction in overall travel time for the whole journey. The parallel implementation of MACO is also compared with the parallelized standard Dijkstra's algorithm.

The rest of the paper is structured as follows: Sect. 2 presents the literature survey. The proposed parallel MACO algorithm is presented in Sect. 3. Section 4 describes the experimental setup and the obtained results. Finally, Sect. 5 concludes the paper.

2 Literature Survey

Swarm intelligence is a large field used in AI that includes the study of the behavioral patterns of living creatures like ants, bees, birds, termites and other social insects in order to model any processes. Swarms have the ability to solve complex tasks that are otherwise difficult to solve through existing computer algorithms. Ant colony optimization (ACO) is one of the most popular algorithms applied largely it networking domain to create self-organizing methods for routing related problems [11]. Ant colony optimization was proposed by Dorigo et al. in early 1990s [10].

ACO is a meta-heuristic technique motivated through the communication strategy used by the real ants to solve an optimization problem and to discover minimum cost path in a network with given constraints. They make use of a chemical substance called pheromone for the exchange of information among them. In order to communicate with other ants, the ant leaves behind pheromone on its route. The other ants detect the presence of this pheromone and follow the path where concentration of pheromone is higher. Also if a particular route is not followed for some amount of time the pheromone starts evaporating thereby reducing the significance of that route. Selection of path based on pheromone trail by ants is a pseudo random process. It plays a major role in simulation of ACO algorithm [10]. Travelling salesman problem and quadratic assignments problem were amongst initial problems making use of ant colonies [12]. Many other domains also exist in literature that makes use of ACO technique for optimization [9, 13].

In the context of traffic simulation, the idea as adopted by real ants can be implemented to find out optimal path for the vehicles and also allowing communication between vehicles using pheromone. The pheromones in this traffic simulation can be taken as the characteristic of lanes or roads that are updated by each vehicle crossing that lane. Through this characteristic, vehicles get information about the traffic on roads thereby indirect communication takes place between vehicles. The pheromone value on every lane gives the indication of whether there is congestion ahead or not. This makes the vehicles capable to check for the congestion free route.

Several approaches using ACO have also been made in the traffic area [8, 14, 15]. The authors in [16] proposed a genetic approach for traffic light control and pedestrian crossing. Another researchers in their paper [17] use ACO with link travel time prediction in order to find routes to reduce travel times. A modification to the ACO algorithm is presented by [3] to reduce the overall travel time of the journey in the MACO algorithm. In this algorithm, all the steps are executed in sequential manner. As there are large numbers of vehicle in the system with dynamic nodes, computation process of the algorithm is time consuming. Efforts have been made in our work to improve the computation time by parallel implementation of the algorithm which will be helpful in reducing the overall travel time of the vehicles on move.

Despite the fact that ACO has been implemented using the parallelization on GPU by various researchers [4, 5, 18] in their work, it was found that none of them has implement all the parts of the algorithm in parallel as per best of our knowledge. The ACO consists of route planning, pheromone deposition and pheromone updation phase. Most of the earlier implementation parallelizes only the route planning phase. In our proposed work, we are implementing all the phases of the MACO algorithm in parallel to reduce the computation time of the algorithm and hence resulting in reduction of the travel time. MACO was proposed to reduce the overall travel time in VANETs [3]. The drawback of the algorithm was that it runs on a single CPU and in VANETs, where the number of vehicles is large; it will take large time for computations. In our proposed work, parallel processing using GPU is used to overcome this problem. The proposed parallel implementation of MACO is described in next section.

3 Proposed Parallel Implementation of MACO

MACO algorithm finds the optimal path for the vehicles by introducing the repulsion effect in pheromone to avoid the congestion in VANETs. The pheromone is a characteristic of road that is updated by each vehicle while crossing that road. Through this, vehicles get information about the vehicles on the roads thereby indirect communication takes place between vehicles. The pheromone value on every lane gives the indication of whether there is congestion ahead or not. This allows the vehicles to beware in advance and allows them to change their path in order to avoid the congested route. With this, the vehicles follow a path that may have a path length greater than or equal to the shortest path but it reduces their overall travel time from source to destination. In the algorithm, deviation of path takes place only in case of congestion.

MACO algorithm runs on a single CPU. There is large number of vehicles in VANETs with dynamic topology that require large amount of computation at faster speed. To increase the speed of computations, we are proposing the parallel version of MACO algorithm that makes use of GPU programming and run the tasks in parallel, hence reducing the decision making time. It will allow the driver to take decision early and react accordingly and reduces the overall travel time as compared with its non-parallel counterpart. The parallel implementation of the algorithm is given in Figs. 1 and 2 as follows:

In the parallel implementation of MACO algorithm, we hard coded the coordinates matrix obtained from the real time North-West Delhi map obtained using Google maps. In the *CUDA_MACO* function, first of all CUDA memory will be

---

Algorithm: *CUDA_MACO* (coordinates)

1. Allocate device memory using *cudaMalloc* function.
2. Copy inputs from host to device using *cudaMemcpy* function.
3. Invoke CUDA_MACO_*KERNEL* <<<grid size, block size>>> (coordinates, distances, route)
4. Copy results back from device to host using *cudaMemcpy* function.
5. Free CUDA memory using *cudaFree* function.

Fig. 1 CUDA_MACO Algorithm

Algorithm: *CUDA_MACO_KERNEL* (coordinates, distances, route)

1. Invoke device function *Calculate_Distance* (coordinates, distances) to calculate distances among various coordinates.
2. Assign each thread for each node in the network
3. Initialize pheromone for each edge with a random value.
4. Find next node using the value of pheromone and threshold in order to avoid the congestion
5. Invoke device function *Evaluate_Route* (route, distances) to calculate the overall travel time.
6. Invoke device function *Increase_Pheromone* (route, distances).
7. Invoke device function *Evaporate_Pheromone* (route, distances)

Fig. 2 CUDA_MACO_KERNEL Algorithm

allocated to all the variables required by the device using *cudaMalloc* function. Then each of the variables will be copied in the device individually by using *cudaMemcpy* with *HostToDevice* option. Next, *CUDA_MACO_KERNEL* will be invoked on the grid with two parameters: grid size and block size. The entire tasks running on this kernel are parallelized. Next, results variables have to copied back from device to host by using *cudaMemcpy* with *DeviceToHost* option. At last, CUDA memory needs to be freed using the *cudaFree* function.

In *CUDA_MACO_KERNEL* function, we define four device functions namely: *Calculate_Distance, Evaluate_Route, Increase_Pheromone* and *Evaporate_Pheromone* that will be called by the kernel to execute the tasks in parallel. Kernel will assign a random value to the pheromones and assign each thread corresponding to each node in the network. The selection of the next node is done based on the threshold and pheromone values. The *Calculate_Distance* function will calculate the distance matrix from the coordinate matrix using the standard distance formula. *Evaluate_Route* function will calculate the overall travel time for all the vehicles in the system. *Increase_Pheromone* and *Evaporate_Pheromone* will update the value of the pheromone for each thread in parallel. After the parallelization, the speed of computation will get increased and we get results faster. This will give some time to driver for react and behave accordingly. This helps in reducing the overall travel time for the whole journey. Algorithm 1 presented in Fig. 1 runs on the CPU while Algorithm 2 presented in Fig. 2 runs on the GPU. In Algorithm 1, at line 3, GPU kernel is being invoked and running of Algorithm 2 starts parallel. Whole experimental setup with the results obtained is discussed in the next section.

4 Experimental Study and Results

The parallel implementation of MACO algorithm was implemented for the vehicular environment on a PC with 2 GB RAM, Intel core i5-3230 M 2.6 GHz processor running windows 8.1 with NVIDIA GeForce 710 M. The graphics card has 2 GB dedicated RAM. Application was written in CUDA 7.0 toolkit and C/C++ using Visual Studio 2010. NVIDIA graphic driver version 347.62 was used for CUDA compatibility. CPU applications were written in C language using its standard library. The algorithm was able to reduce the total travel time taken by vehicles to reach destination by introducing parallelization for all the functions for reducing the computation time of the MACO algorithm as discussed above. In the algorithm, the default path followed by vehicles is decided by Dijkstra's routing algorithm in the beginning. So, we are comparing the results obtained by the proposed parallel MACO with the parallel implementation of Dijkstra's algorithm along with that of the MACO algorithm. The simulations were performed for a hardcoded real world network of North-West Delhi map and obtained results were tested with different number of vehicles. In a real world map of North-West Delhi (NWD), obtained from OSM, we have 128 roads and 52 junctions. The snapshot for the network used in experiments is shown in Fig. 3.

We have implemented both standard Dijkstra shortest path algorithm in parallel and MACO algorithm for the above specified network to test the performance of the proposed algorithm. Results obtained for overall travel time by all the three

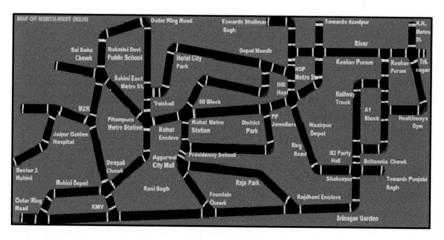

Fig. 3 Screen shot for real time North-West Delhi network

Table 1 Results obtained in real world North-West Delhi network

No. of vehicles	Overall travel time (in s)		
	Parallel Dijkstra	MACO	Parallel MACO
100	478	379	189
200	549	428	213
300	734	513	252
400	862	575	281
500	912	602	294
600	1126	742	362
700	1364	894	435
800	1448	948	458
900	1570	1025	487
1000	1798	1174	549

Fig. 4 Graphical representation of computation time in real world North-West Delhi network

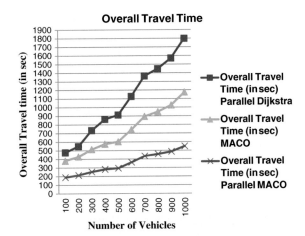

algorithms are shown in Table 1 with their graphical representation shown in Fig. 4. It is clearly evident from the results that the proposed parallel MACO algorithm shows a significant decrease in the travel time taken by vehicles as compared with both Parallel Dijkstra's and the MACO algorithm.

The results conclude that proposed parallel MACO is able to reduce overall travel time by 60–70 % when compared with the parallel implementation of the standard Dijkstra's algorithm and 50–54 % when compared with MACO algorithm. Thus, it will help commuters to reach to their destination faster by reducing the overall travel time and hence providing a solution to an important research problem in transportation.

5 Conclusion

In VANETs, large number of vehicles and limited road capacity leads to congestion, which results in increase in overall travel time. Hence, there is a need of some mechanism by which one can take decision faster to avoid the congestion and reach the destination faster by reducing overall travel time. MACO algorithm is presented in literature to reduce the congestion, but due to its serial implementation, it is comparatively slow in execution. Thus, in this paper a parallel implementation of MACO algorithm using CUDA toolkit 7.0 on NVIDIA GeForce 710 M architecture that provides us the advantage of using both CPU and GPU's processors for giving results faster in order to react quicker and hence reduce the overall travel time is presented. To validate the approach, a road network model was obtained from a real world map of North-West Delhi and hardcoded into the system. The algorithm has been simulated using CUDA toolkit 7.0 in C language on NVIDIA GPU. The results obtained by parallel implementation of the MACO algorithm were compared with the results obtained through the parallel implementation of standard Dijkstra's algorithm and MACO algorithm.

In the experiments, it was found that the use of parallel MACO algorithm reduced the overall travel time approximately by 60–70 % with the increase in number of vehicles as compared with parallel implementation of Dijkstra's algorithm. Further, comparing it with the existing MACO algorithm it was found that travel time has been reduced approximately by 50–54 % by the parallel MACO algorithm.

References

1. Bedi, P., Jindal, V., Dhankani, H., Garg, R.: ATSOT: Adaptive traffic signal using mOTes. In: 10th International Workshop on Databases in Networked Information Systems, DNIS 2015, Japan, vol. LNCS 8999, pp. 152–171 (2015)
2. Bell, M., Bonsall, P., Leaky, G., May, A., Nash, C., O'Flaherty, C.: Transport Planning and Traffic Engineering. John Wiley & Sons, NY, United State of America (1997)
3. Jindal, V., Dhankani, H., Garg, R., Bedi, P.: MACO: Modified ACO for reducing travel time in VANETs. In: Third International Symposium on Women in Computing and Informatics (WCI-2015), pp. 97–102. ACM, Kochi, India (2015)
4. Cecilia, J., Garcia, J., Ujaldon, M., Nisbet, A., Amos, M.: Parallelization strategies for ant colony optimisation on GPUs. In: IEEE International Symposium on Parallel and Distributed Processing Workshops and Phd Forum (IPDPSW), pp. 339–346. IEEE (2011)
5. Fu, J., Lei, L., Zhou, G.: A parallel ant colony optimization algorithm with GPU-acceleration based on all-in-roulette selection. In: Third International Workshop on Advanced Computational Intelligence (IWACI), pp. 260–264. IEEE (2010)
6. Sanders, J., Kandrot, E.: CUDA by Example: An Introduction to General Purpose GPU Programming. Addison-Wesley, United States (2010)
7. Kirk, D., W-m, Hwu: Programming Massively Parallel Processors: A Hands-on Approach. Morgan Kaufmann Publishers, Elsevier, USA (2010)

8. Bedi, P., Mediratta, N., Dhand, S., Sharma, R., Singhal, A.: Avoiding traffic jam using ant colony optimization—a novel approach. In: International Conference on Computational Intelligence and Multimedia Applications, vol. 1, pp. 61–67. Sivakasi, Tamil Nadu, India (2007)

9. Bell, J., McMullen, P.: Ant colony optimization techniques for the vehicle routing problem. Adv. Eng. Inform. **18**, 41–48 (2004)

10. Dorigo, M., Stützle, T.: Ant Colony Optimization. MIT Press, USA (2004)

11. Dorigo, M., Caro, G., Gambardella, L.: Ant algorithms for discrete optimization. Artif. Life **5**(2), 137–172 (1999)

12. Elloumia, W., Abeda, H., Abra, A., Alimi, A.: A comparative study of the improvement of performance using a PSOmodified by ACO applied to TSP. Appl. Soft Comput. **25**, 234–241 (2014)

13. Deneubourg, J., Aron, S., Goss, S., Pasteel, J.: The self-organizing exploratory pattern of the Argentine ant. J. Insect Behav. **3**(2), 159–168 (1990)

14. Nanda, B., Das, G.: Ant colony optimization: a computational intelligence technique. Int. J. Comput. Commmun. Technol. **2**(6), 105–110 (2011)

15. Rizzoli, A., Montemanni, R., Lucibello, E., Gambardella, L.: Ant colony optimization for real-world vehicle routing problems: from theory to applications. Swarm Intell. **1**, 135–151 (2007)

16. Turky, A., Ahmad, M., Yusoff, M.: The use of genetic algorithm for traffic light and pedestrian crossing control. Int. J. Comput. Sci. Netw. Secur. **9**(2), 88–96 (2009)

17. Claes, R., Holvoet, T.: Ant colony optimization applied to route planning using link travel time prediction. In: International Symposium on Parallel Distributed Processing, pp. 358–365 (2011)

18. Dawson, L., Stewart, I.: Improving ant colony optimization performance on the GPU using CUDA. In: IEEE Congress on Evolutionary Computation (CEC), pp. 1901–1908. IEEE, (2013)

FS-EHS: Harmony Search Based Feature Selection Algorithm for Steganalysis Using ELM

Veenu Bhasin, Punam Bedi, Neha Singh and Charu Aggarwal

Abstract The process of steganalysis involves feature extraction and classification based on feature sets. The high dimension of feature sets used for steganalysis makes classification a complex and time-consuming process, thus it becomes important to find an optimal subset of features while retaining high classification accuracy. This paper proposes a novel feature selection algorithm to select reduced feature set for multi-class steganalysis. The proposed algorithm FS-EHS is based on the swarm optimization technique Harmony Search using Extreme Learning Machine. The classification accuracy computed by Extreme Learning Machine is used as fitness criteria, thus taking advantage of the fast speed of this classifier. Steganograms for conducting the experiments are created using several common JPEG steganography techniques. Experiments were conducted on two different feature sets used for steganalysis. ELM is used as classifier for steganalysis. The experimental results show that the proposed feature selection algorithm, FS-EHS, based on Harmony Search effectively reduces the dimensionality of the features with improvement in the detection accuracy of the steganalysis process.

Keywords Steganalysis · Feature selection · Harmony search · Extreme learning machine

V. Bhasin (✉) · P. Bedi (✉) · N. Singh · C. Aggarwal
Department of Computer Science, University of Delhi, Delhi, India
e-mail: vbhasin@cs.du.ac.in

P. Bedi
e-mail: pbedi@cs.du.ac.in

N. Singh
e-mail: neha.cs.du.2013@gmail.com

C. Aggarwal
e-mail: charu.cs.du.2013@gmail.com

© Springer International Publishing Switzerland 2016
V. Snášel et al. (eds.), *Innovations in Bio-Inspired Computing and Applications*,
Advances in Intelligent Systems and Computing 424,
DOI 10.1007/978-3-319-28031-8_34

1 Introduction

Steganography is the branch of information hiding in which communication of a secret message, m, is done by hiding it in some other information (cover-object, c), thus hiding the existence of the communicated information [1]. Sender therefore changes the cover, c to a stego-object, s. The cover object (e.g. an image) should have sufficient amount of redundant data which can be replaced by secret information in order to be used for embedding message without suspicion.

Steganalysis detects the presence of messages hidden using steganography inside digital data [1]. Based on whether an image includes hidden message or not, images can be categorized into stego-images and non-stego-images. The steganalysis methods work on the principal that steganography procedures on images alter the statistical properties of the image, which can be used to detect presence of messages. Steganalysis is similar to a pattern recognition process and its main aim is to identify the presence of a covert communication by using the statistical features of the cover and stego image as clues. Framework of steganalysis involves two phases: training phase and testing phase. In former, the multidimensional feature set from the training images consisting of cover images and stego-images are extracted and this set is used to train the classifier. In the testing phase, the trained classifier is used to distinguish between stego and non-stego images.

Image steganalysis techniques have been developed with feature extraction done from several domains like spatial, DCT, and DWT etc. [2]. Several studies have led to development of different feature extraction models for steganalysis such as statistical [3], histogram, co-occurrence matrix [4], IQM [5], SPAM [6], CF moments [7], Neighboring joint density probabilities (NJD) [8] and Markov model [9, 10] etc. Pevny and Fridrich [11] have merged the features from different models, i.e. DCT and Markov model, to improve accuracy of steganalysis method. Some authors [10, 12] have used calibrated features to get better detection results.

Feature based steganalysis involves large dimensionality of feature vectors. The pattern-recognition research has indicated that training the classifier with the high-dimensional features causes difficulty and result in problems like 'over-fitting' and 'curse of dimensionality'. Thus to improve the efficiency of steganalysers, the classification process can be optimized by using reduced number of features required for classification. The individual features from the extracted high-dimensional feature set as well as their correlations need to be analyzed. It is possible that there are features which are more valuable for classification. There might be some features which together don't give performance better than their individual performance and hence some of them can be considered redundant. Thus, to increase the efficiency and practicality of steganalysis it is important to reduce the dimension of feature space by selecting features with better classification performance while distinguishing stego-images from non-stego-images.

In this paper, FS-EHS a Feature selection algorithm for steganalysis based on the swarm intelligence technique Harmony Search (HS) is proposed, that uses Extreme Learning Machine (ELM) as classifier. Harmony Search captures the coordinating nature of musicians [13]. Each musician produces a note. All the notes together generate one harmony. Different harmonies form the harmony memory. New harmonies are generated using improvisation. If a new harmony performs better than the worst harmony in the harmony memory, the worst harmony is replaced with the newly generated harmony. At the end, the best harmony is chosen. For FS-EHS, the steganalytic feature selection algorithm proposed in this paper, a harmony corresponds to a feature subset and a harmony is evaluated on the basis of computation of the classification accuracy when this feature subset is used by ELM for classification.

The steganalytic feature sets used for experiments are the Markov based Features (MB-features) proposed by Chen et al. [9], and Neighboring Joint Density based features (NJD-Features) proposed by Liu et al. [8, 14], with 486 and 338 as dimensions, respectively. FS-EHS is applied for these feature sets and experiments show that FS-EHS yields a reduced feature set that also improves the classification efficiency of the steganalyzers.

The rest of the paper is structured as follows. Section 2 gives a brief account of feature selection and related work in it. Section 3 introduces HS process. Section 4 presents FS-EHS, the proposed feature selection using HS. The experimental setup and results are given in Sect. 5, followed by the conclusion in Sect. 6.

2 Feature Selection

The feature selection is the process of electing a subset of features. It can be described as a search into a space consisting of all possible subsets of the features and finding the optimal solutions. Traversing the whole space for a full search is impractical for a large number of features, as for an n-dimensional feature set this space contains 2n different subsets. To explore the search space, a heuristic or random approach can be applied. In first case, the search can be started with an empty subset or with a full set of features and during the iterations the features are either added to or deleted from the subset for evaluation. A random approach generates random subsets within the search space and evaluates them, several bio inspired and genetic algorithms use this approach.

Every feature subset generated is evaluated. According to the method of this evaluation process, feature selection methods can be classified into two main categories: filter and wrapper. In filter methods, filtering, based on some weight values assigned to each feature, is performed before the classification process. The features that have better weight values get selected in the final subset of features; therefore, they are independent of the classification algorithm. On the other hand, wrapper approaches generate candidate subsets of features and employ accuracy predicted by a classifier to evaluate these feature subsets. Wrapper methods usually achieve superior results than filter methods, but are limited by the speed of the classifier. A fast classifier, like ELM, can be used with wrapper method to speed up the classification process.

Feature selection methods can also be categorized into univariate and multi-variate. Univariate methods examine individual features for classification abilities, whereas multivariate methods consider the correlations between features and examine subsets of features.

The comparison of performance for different types of features or analysis of the components of high-dimensional features is being generally used in research on steganalytic feature selection. Univariate method based on Bhattacharyya distance, for steganalytic feature selection, has been proposed by Xuan et al. [15]. Pevný et al. [16, 17] and Kodovský and Fridrich [18] have used comparison-based feature selection in their works. Wang et al. [19] in their univariate and sequential search based algorithms used comparison of the absolute moments of the CF (characteristic function) and the absolute moments of the PDF (probability density function) of the histogram of the wavelet coefficients for reduced feature dimensionality. Hui et al. [7] used Principal Component Analysis (PCA) for feature dimension reduction.

Avcıbas et al. [5] uses variance analysis and multiple regression analysis for reduction of feature set with features consisting of multiple values of image quality metrics. In sequential search based algorithms by Miche et al. [20], K-nearest-neighbor is used for calculating contribution of each feature to classification.

For feature selection, Mahalanobis distances and Fisher criterion has been used to evaluate the separability of each single-dimension feature by Davidson and Jalan [21] and Ji-Chang et al. [22] respectively. However, the sequential algorithms become very time-consuming as the dimensions of feature space increases. Many evolutionary algorithms have also been used for feature selection, which include genetic algorithms and swarm algorithms like Ant Colony Optimization (ACO) [23], Particle Swarm Optimization (PSO) [24], and Artificial Bee Colony [25]. Diao and Shen [26] in their stochastic method for feature selection have used a modified harmony search. Harmony search has been used for simultaneous clustering and feature selection by Sarvari et al. [27].

The proposed feature selection algorithm FS-EHS is multivariate wrapper method. It is using Harmony Search with ELM for selecting optimal feature subset. ELM being a fast classifier when used in wrapper method of feature selection provides a fast algorithm to classify image as stego or non-stego.

3 Harmony Search

Harmony Search (HS) [13] is an evolutionary algorithm inspired by the improvisation process of musicians. It follows a simple principle, requiring very few parameters. To obtain solution for an optimization problem, the process followed by musicians to create a new song is used. Each optimizing parameter is a note produced by one musician and a particular combination of these notes form a harmony, which is considered a possible solution. Different harmonies are kept in the Harmony Memory (HM). They are evaluated on the basis of some optimization criteria, new ones are created to replace the worst in the memory and the best harmony, at the end, is chosen as solution. The principal steps followed in a typical HS are:

- Initialize HM with randomly generated solutions or harmonies
- Improvise a new solution (or harmony)
- Update the HM—replace a worst member with the new solution if the new has better fitness
- Repeat the two steps of improvisation and updating for a predefined number of iterations.

The Harmony search algorithm for improvising a New Harmony (candidate solution) has to assign values to various notes either by choosing any value from HM, choosing a random value from the possible value range or choosing an adjacent value from HM. HS uses two parameters Harmony Memory Considering Rate (HMCR) and Pitch Adjustment Rate (PAR). HMCR decides which of the first two options will be chosen and PAR is the probability of choosing the third option. This improvisation method ensures the solution start converging towards the best solution.

In this paper, we propose a novel Harmony Search based feature selection algorithm using ELM, for image steganalysis. To the best of our knowledge, this is the first attempt to use Harmony Search with ELM for selecting the reduced feature set for image steganalysis.

4 FS-EHS—Harmony Search Based Feature Selection Algorithm using ELM

The principle of HS is modeled on the process followed by musicians to improvise new melodies. Harmony Search captures the coordinating nature of musicians. Each musician produces a note. All the notes together generate one harmony. Different notes leads to the production of different harmonies and all these harmonies reside in harmony memory (HM). All the harmonies have a fitness score associated with them. Depending on Harmony Memory Considering rate (HMCR), a new harmony is produced. If the newly created harmony yields better fitness than the fitness of the worst harmony in HM, then the worst harmony is replaced with the newly created harmony otherwise the harmonies in harmony memory remains the same. This process repeats for predefined number of iterations. After the completion, the harmony with best fitness is selected as solution.

In the proposed feature selection algorithm using HS, each feature corresponds to one note, thus the number of notes (and number of musicians) is equal to the number of features in original feature set. HM is a collection of binary strings, each string representing one harmony. A harmony corresponds to a feature subset and is represented as a binary string where one ('1') represents the presence of feature and zero ('0') represents its absence in that harmony. The harmonies are initially produced randomly and the number of harmonies is equal to harmony memory size (HMS). The fitness score of each harmony is the accuracy with which ELM classify images if trained using the features present in the harmony, i.e. those features corresponding to ones in a harmony. Thus, Harmony Memory is a matrix of size HMS × N, where N is the number of features in original feature set. Another vector, Fitness Score keeps track of the fitness scores of the harmonies in HM.

4.1   Parameters for the Algorithm

The parameters to be specified, for this algorithm, are:

- Harmony Memory Size (HMS) decides how many harmonies are kept in memory.
- Harmony Memory Considering rate (HMCR) specifies that whether a new harmony will be produced from the existing values stored within the harmony memory (exploiting the known good solution space) or will be a randomly produced afresh (exploring new solution space). HMCR, which varies between 0 and 1, is the probability of choosing one value for a note from the historical notes stored in the harmony memory, while (1—HMCR) is the probability of randomly generating a new value for the note.
- The number of iterations (I) specifies how many times new harmonies are generated and compared with the ones in HM.

PAR is not needed while adapting HS for feature selection, as the HM contains binary values i.e., the only values that can be assigned to HM elements is 1 or 0.

4.2   Fitness Score

For any optimization technique a Fitness criteria is required to evaluate the candidate solutions. In the case of feature selection, this fitness score should reflect how good a subset of features is in distinguishing different classes. In the proposed algorithm, FS-EHS, the accuracy of classification by Extreme Learning Machine (ELM) trained using selected features is used as Fitness Score.

ELM was proposed by Huang et al. [28] as a learning algorithm for Single-hidden Layer Feed-forward neural Networks (SLFNs). ELM randomly chooses and then fixes the input weights (i.e., weights of the connections between input layer neurons and hidden layer neurons) and the hidden neurons' biases; and analytically determines the output weights (i.e., the weights of the connections between hidden neurons and output neurons) of SLFNs. Thus, the only parameter to be determined in ELM is the number of neurons in the hidden layer. As a result ELM is much simpler and saves tremendous time.

4.3 Algorithm for FS-EHS

The algorithm for FS-EHS is given in the Algorithm 1 and in the two subroutines.

```
Algorithm 1.FS-EHS
INPUT: Feature Matrix (M) of dimension m×n where m is the number
of images in training image set and n is the number of features
in feature set (F), Group Vector (G) of dimension m, Harmony
Memory Considering rate (HMCR), Number of iterations (I)
OUTPUT: The final subset of selected features, F*.

Harmony Memory Size, HMS = n/2
Initialize Harmony Memory, HM and the Fitness Score vector, FS
(Subroutine 1).
Repeat I times
     Create a new harmony, NH
           for every note A in NH do
              Generate a random number, r
              if r ≤ HMCR, then
                  Randomly choose a harmony (h^th) from HM
                  H=HM[h], NH[A] = H[A]
              else Randomly assign value to NH[A]
        NF = Compute Fitness of NH (Subroutine 2)
        Find min such that FS[min] is minimum
        if (FS[min] < NF)
          Replace harmony with new Harmony
              HM[min] = NH, FS[min]  = NF
Select the best harmony HM[b] such that FS[b] is maximum
Return the selected feature set, F*= set of features correspond-
ing to notes with '1' in HM[b]
```

```
Subroutine 1.Initialize Harmony Memory
INPUT: Features (F), Harmony memory size(HMS)
OUTPUT: Harmony Memory, HM and Fitness Score vector, FS

for h = 1 to HMS do
     Initialize a new harmony, NH
           for every note a in NH do
              Randomly assign value to NH[a]
     F = Compute Fitness Score for NH (Subroutine 2)
     HM[h] = NH,   FS[h] = F
```

```
Subroutine 2.Compute Fitness
INPUT: Feature matrix M, Group vector G, Harmony, H
OUTPUT: Fitness Rating, R


Get the feature vectors corresponding to H from M and form new
feature matrix M'
Use ELM to compute classification accuracy with M' and G
Return R = accuracy represented as percentage
```

5 Experimental Setup and Results

Image Set: The experiments are conducted on an image dataset consisting of 2000 colored JPEG images of size 256×256. The cover images include images captured by the authors' camera and images available freely in public domain on internet. The images are of various content and textures, spanning a range of indoor and outdoor scenes. The stego-images are created using publicly available JPEG image Steganographic tools OutGuess, F5 and Model-based steganography [2]. The image dataset is divided randomly into non-overlapping training set and testing set.

Steganalytic Feature Sets: Feature extraction is a key part of Steganalysis and various steganalysis methods differ mainly in the feature sets that are extracted from images. In this paper, the experiments are conducted on two types of feature sets Markov based Features (MB-features) and Neighboring Joint Density based features (NJD-Features) with 486 and 338 as dimensions, respectively. The feature set of MB Features, proposed by Chen et al. [9], contains features extracted from the inter block as well as intra-block correlation in frequency domain using Markov Process. This Process is applied on the original image, to obtain 324 intra-blocks and 162 inter-blocks features, together forming the feature set of size 486. The feature set of NJD-Features, proposed by Liu et al. [8, 14], include features based on Intra-block Neighboring Joint Density and Inter-block Neighboring Joint Density.

Extreme Learning Machine (ELM) as classifier: In the experiments, ELM is adopted as classifier. It is used to compute classification accuracy during feature selection and also as the classifier for steganalysis. ELM with sigmoid activation function and 160 hidden neurons is used for experiments.

The feature set computed from training set is used for feature selection by FS-EHS. The reduced feature set, so obtained, is used to train ELM. The images of testing set are then classified into stego and non-stego images by this trained ELM using the selected features from testing images. The implementation is done in MATLAB.

The classification accuracy percentage and the features selected are employed to determine the performance of the proposed FS-EHS for steganalysis. For both the

Table 1 Performance of FS-EHS

Steganalytic feature set	No. of features (Original feature set)	Accuracy % using original	No. of features selected by FS-EHS	Accuracy % using selected features
NJD-features	338	78.82	166	85.19
MB-features	486	81.26	239	84.66

feature sets (NJD-features and MB-Features), the experiment has been conducted 10 times employing ten-fold cross validation. All the results are then averaged.

The experimental results, for different feature sets, as given in Table 1 indicate that: FS-EHS proposed in this paper not only can reduce the dimension greatly, but improves the steganalytic efficiency as well. These results verify the effectiveness and rationality of the proposed feature selection method, FS-EHS when it is used in steganalysis process.

6 Conclusion

This paper presented a novel feature selection method FS-EHS for blind steganalysis method of JPEG images using Harmony Search with ELM. To show the usefulness and effectiveness of FS-EHS, experiments were conducted using two types of steganalytic feature sets. The ELM was used as classifier as it is a fast learning SLFN. The experiments were conducted on JPEG images. The stego images were created using F5, Outguess and Model-based steganography techniques. The experimental results show a significant reduction in the number of features being selected with improvement in classification rate when the proposed feature selection method is used as a step in the steganalysis process.

References

1. Davidson, J.L., Jalan, J.: Feature selection for steganalysis using the Mahalanobis distance. In: SPIE Proceedings Media Forensics and Security, vol. 7541. San Jose, CA, USA (2010)
2. Abolghasemi, M., Aghainia, H., Faez, K., Mehrabi, M.A.: LSB data hiding detection based on gray level co-occurrence matrix. In: International symposium on Telecommunications, pp. 656–659. Iran, Tehran (2008)
3. Avcıbas, I., Memon, N., Sankur, B.: Steganalysis using image quality metrics. IEEE Trans. Image Process. **12**(2), 221–229 (2003)
4. Bhasin, V., Bedi, P.: Steganalysis for JPEG images using extreme learning Machine. In: IEEE International Conference on Systems, Man, and Cybernetics, pp. 1361–1366. IEEE, Manchester, UK (2013)
5. Chen, C., Shi, Y.Q.: JPEG image steganalysis utilizing both intrablock and interblock correlations. In: IEEE ISCAS, International Symposium on Circuits and Systems, pp. 3029–3032. Seattle, Washington, USA (2008)

6. Chen, G., Chen, Q., Zhang, D., Zhu, W.: Particle swarm optimization feature selection for image steganalysis. In: Fourth International Conference on Digital Home (ICDH), pp. 304–308. Guangzhou (2012)
7. Diao, R., Shen, Q.: Feature selection with harmony search. IEEE Trans. Syst. Man Cybern. B Cybern. **42**(6), 1509–1523 (2012)
8. Fridrich, J.: Feature-based steganalysis for JPEG images and its implications for future design of steganographic schemes. In: 6th Information Hiding Workshop. Toronto, ON, Canada (2004)
9. Fridrich, J., Kodovsky, J.: Calibration revisited. In: 11th ACM Multimedia & Security Workshop, pp. 63–74. Princeton, NJ (2009)
10. Fridrich, J., Pevny, T.: Merging Markov and DCT features for multiclass JPEG steganalysis. In: Proceeding SPIE Electronic Imaging, Security, Steganography, and Watermarking of Multimedia Contents IX, pp. 650503-1–650503-13. San Jose, CA (2007)
11. Huang, G.-B., Zhu, Q.-Y., Siew, C.-K.: Extreme learning machine: a new learning scheme of feedforward neural networks. International Joint Conference on Neural Networks (IJCNN), 2, pp. 985–990. Budapest, Hungary (2004)
12. Hui, L., Ziwen, S., Zhiping, Z.: An image steganalysis method based on characteristic function moments and PCA. In: Proceeding of the 30th Chinese Control Conference. Yantai, China (2011)
13. Kabir, M.M., Shahjahan, M., Murase, K.: An efficient feature selection using ant colony optimization algorithm. In: Neural Information Processing—16th International Conference, ICONIP 2009, pp. 242–252. Bangkok, Thailand: Springer, Berlin (2009)
14. Katzenbeisser, S., Petitcolas, F.: Information Hiding Techniques for Steganography and Digital Watermarking. Artech House, Boston (2000)
15. Kharrazi, M., Sencar, H., Memon, N.: Benchmarking steganographic and steganalysis techniques. Electron. Imaging **2005**, 252–263 (2005)
16. Kodovsky, J., Fridrich, J.: Rich models for steganalysis of digital images. IEEE Trans. Inf. Forensics Secur. **7**(3), 868–882 (2012)
17. Liu, Q., Sung, A.H., Qiao, M.: Improved detection and evaluation for JPEG steganalysis. 17th ACM International Conference on Multimedia, pp. 873–876. ACM, Beijing, China (2009)
18. Liu, Q., Sung, A.H., Qiao, M.: Neighboring joint density-based JPEG steganalysis. ACM Trans. Intell. Syst. Technol. (TIST) **2**(2), 16:1–16:16 (2011)
19. Lu, J.-C., Liu, F.-L., Luo, X.-Y.: Selection of image features for steganalysis based on the Fisher criterion. Digit. Invest. **11**(1), 57–66 (2014)
20. Miche, Y., Roue, B., Lendasse, A., Bas, P.: A feature selection methodology for steganalysis. In: Content, Multimedia (ed.) Representation, Classification and Security, pp. 49–56. Springer, Berlin Heidelberg (2006)
21. Pedrini, M.S.: Data feature selection based on artificial bee colony algorithm. EURASIP J. Image Video Process. 1–8 (2013)
22. Pevny, T., Fridrich, J.: Multiclass detector of current steganographic methods for JPEG format. IEEE Trans. Inf. Forensics Secur. **4**(3), 635–650 (2008)
23. Pevný, T., Bas, P., Fridrich, J.: Steganalysis by subtractive pixel adjacency matrix. IEEE Trans. Inf. Forensics Secur. **5**(2), 215–224 (2010)
24. Pevny, T., Filler, T., Bas, P.: Using high-dimensional image models to perform highly undetectable steganography. 12th International workshop on information hiding. Calgary, Canada (2010)
25. Sarvari, H., Khairdoost, N., Fetanat, A.: Harmony search algorithm for simultaneous clustering and feature selection. International Conference of Soft Computing and Pattern Recognition (SoCPaR), (pp. 202–207). Paris (2010)
26. Wang, X., Gao, X.-Z., Zenger, K.: An Introduction to Harmony Search Optimization Method. Springer International Publishing, Berlin (2015)
27. Wang, Y., Moulin, P.: Optimized feature extraction for learning-based image steganalysis. IEEE Trans. Inf. Forensics Secur. **2**(1), 31–45 (2007)
28. Xuan, G., Zhu, X., Chai, P., Zhang, Z., Shi, Y. Q., Fu, D.: Feature selection based on the Bhattacharyya distance. In: The 18th International Conference on Pattern Recognition (ICPR'06). Hong Kong (2006)

A Hybrid Dimension Reduction Technique for Document Clustering

Cynthia Marea Nebu and Sumy Joseph

Abstract The paper proposes a hybrid approach to reduce dimension in text classification problems, to overcome the issue of Curse of Dimensionality. This hybrid approach is a combination of Feature Selection (FS) and Feature Extraction (FE) methods, considering different aspects of feature relevance, to effectively reduce the dimension in large text datasets. It prevents feature selection biased in favor of a particular FS method. Many FS methods like Term Variance, Document Frequency, Information Gain, Shannons Entropy measure, Mean-Median and Mean Absolute Difference, were implemented and a comparative study was made on their performance when implemented in a hybrid system. The features selected by the individual FS methods are merged using three approaches, namely, Union, Intersection and Modified Union. The sub lists of features further undergo Feature Extraction by PCA, and the reduced feature sub list is clustered with k-means. Finally, the sentiment-score of the individual clusters are calculated using SentiWordNet database which gives the polarity of the data. The experiments were conducted on the benchmark datasets namely Reuters-21,578 and Classic4. The performance evaluation of the system made using the measures like precision, recall, f-score and accuracy shows that the proposed method has improved performance compared to its competitive methods.

Keywords Dimensionality reduction · Feature selection · Feature extraction · Text categorization · Sentiment analysis

C.M. Nebu (✉) · S. Joseph
Amal Jyothi College of Engineering, Koovappally, Kerala, India
e-mail: cynthia.nebu@gmail.com

S. Joseph
e-mail: sumyjoseph@amaljyothi.ac.in

© Springer International Publishing Switzerland 2016
V. Snášel et al. (eds.), *Innovations in Bio-Inspired Computing and Applications*,
Advances in Intelligent Systems and Computing 424,
DOI 10.1007/978-3-319-28031-8_35

1 Introduction

The phenomenal growth of the Internet technology has resulted in the enormous number of digital information making it difficult to manage manually. It is a necessary task to arrange and organize these documents so that users can query and use them as needed without much effort. Manual organization of this enormous collection of documents is impractical, costly and very time consuming. Hence modern approaches like Machine Learning and Natural Language Processing has been employed to automatically classify the documents based on the similarity between them.

The main challenge faced in text mining is the extremely large number of features that may appear in a document makes its dimensionality. This issue is called Curse of dimensionality and it affects the performance of clustering algorithm. One approach to simplify the unmanageable amount of data in the text documents is to reduce the dimensionality of them by applying some Dimension Reduction (DR) techniques. These techniques tend to take input data and convert them into a much smaller number of dimensions, while preserving important characteristics of the original data. These methods remove all the redundant, irrelevant and noisy features and keep only those relevant features which are useful in discriminating between the documents. Dimensionality reduction can be done in two ways. First type of reduction technique is Feature Selection (FS) in which a subset of the original features is selected based on some criteria. Second type of dimension reduction technique is Feature Extraction (FE). In Feature Extraction the high dimensional feature set is transformed into a lower dimensional feature subspace using some calculations.

With the growing use of social media, blogs, newsgroups, people express their opinion on every piece of news or information. Customers like to know the review on certain products before buying things; producers like to study market trends from the feedback from customers and so on. Hence Sentiment Analysis (or Opinion Mining) has become an area of active research and interest.

The rest of the paper is structured as follows. Section 2 presents the related works on areas dimension reduction. Section 3 contains the description of proposed method. Section 4 summarizes the experimental evaluations and results. Section 5 gives the conclusion of the paper.

2 Related Works

2.1 Dimension Reduction

Dimension reduction is a mandatory step in Text Mining due to its high dimensionality problem called Curse of Dimensionality. Dimension reduction methods can be broadly classified into two, Feature Extraction (FE) methods and Feature Selection (FS) methods. Traditional Feature Extraction methods include Principal Component

Analysis [1], Latent Semantic Indexing [2], and Independent Component Analysis. These methods are called feature construction methods as they produce combination of features in they reduces feature subset. Feature Selection methods like Term Variance [3], Document Frequency [4], Mean Absolute Difference [5], Mean Median [6], Information Gain [7], Shannons Entropy [8] selects the relevant subset of features from amongst the initial set features.

Single dimension reduction techniques consider only one aspect of the feature during feature selection. The recent studies have been focused on hybrid methods which incorporate more than one feature reduction technique considering different aspects of the features. Uguz et al. [9] introduces a FS-FS method (Information Gain—Genetic Algorithm) and FS-FE method (Information Gain—Principal Component Analysis) to transform a high dimensional data into a lower dimensional subspace. Bharti et al. [6] proposes a three stage dimensionality reduction technique involving two FS and one FE methods (FS-FS-FE). The methods employed were Mean Absolute Difference (MAD) to remove irrelevant features, Absolute Cosine (AC) to remove the redundant features and finally Principal Component Analysis (PCA) to remove the noisy features.

The hybrid dimension reduction techniques used a union approach [10] or intersection approach [9, 10] to merge the feature sub list. The union approach increases the total number of features selected after dimension reduction whereas, intersection approach losses those features that achieve highest significance score with respect to only one feature selection method. Bharti et al. proposes a feature merging approach named Modified Union which is an extension of the previous merging methods. It applies union approach on a defined number of top ranked features and intersection approach on the rest of the features.

2.2 Sentiment Analysis

Sentiment Analysis gives an idea of how the latest news influences important entities. Kim et al. calculates the sentiment of an entity using WordNet [11] and lists the positive and negative words. They assume that the synonyms of a word have the same polarity and the antonyms have opposite polarity. The drawback of this method is that the synonym set coherence can weaken the distance. To overcome this, Godbole et al. [12] proposes an algorithm sentiment lexicon generation through path analysis. This algorithm does not use the dictionary WordNet for querying synonyms. Pahyung et al. [13] employs SentiWordNet [14], an opinion lexicon developed from WordNet, to analyze the polarity of a word. Each word is associated with three scores with their values ranging from 0 to 1. Theses scores represent the positive, negative and neutral sentiment of the word. Kerstin et al. [15] does multi-domain sentiment analysis based on SentiWordNet as lexical resource. The sentiment classification was performed in two ways, rule based classification and machine-learning based classification.

3 Proposed System

The system implemented is a hybrid dimension reduction technique used for text document clustering (Fig. 1). The various modules in the system are as follows:

3.1 Document Preprocessing

Preprocessing of the text will remove the useless or non-informational words from the documents. This step consists of stop word elimination [16] and stemming [17] of the terms. Also processing of text data is difficult as they do not have proper standard structure or a numeric format to apply the statistical machine learning algorithms and hence term weighting technique $tf - idf$ is applied.

3.2 Feature Selection

The system uses two different features selection methods at a time. The comparative study was made with the following measures.

3.2.1 Task-Free Measures

It includes those feature selection measures that does not use the category information present in the training set.

Term Variance (TV): Term Variance assigns a relevance score to each feature based on its deviation from its mean value. The method is based on the fact that features that are non-uniformly distributed over all the documents are comparatively more descriptive than the uniformly distributed features. The TV of each term can be calculated with the equation below,

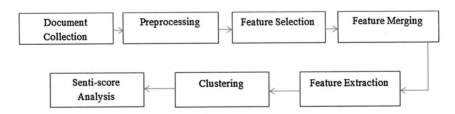

Fig. 1   The flow chart of the proposed methodology

$$TV_i = \frac{1}{n} \sum_{j=1}^{n} (x_{ij} - \overline{x_i})^2 \tag{1}$$

where, x_{ij} is the value of the term x_i in the document d_j and $\overline{x_i}$ its mean value.

Document Frequency (DF): It is the simplest feature selection measure and it measures the number of documents in which the term appears. The term with DF score greater than a predefines threshold are chosen as discriminative features. DF measure can easily scale to very large datasets and are also computationally less complex.

Mean Absolute Difference (MAD): It assigns a relevance score to each feature by calculation its difference from the mean value.

$$MAD_i = \frac{1}{n} \sum_{j=1}^{n} |X_{ij} - \overline{X}_i| \tag{2}$$

where X_{ij} is the value of the feature i with respect to the document j and $\overline{X}_i$ is the mean of the feature i given as follows:

$$\overline{X}_i = \frac{1}{n} \sum_{j=1}^{n} X_{ij} \tag{3}$$

Mean-Median (MM): It is a simplified form of skewness. It assigns a relevance score to each feature according to the absolute difference between its mean and median of $\overline{X}_i$.

$$MM_i = |\overline{X}_i - median(X_i)| \tag{4}$$

3.2.2 Task-Sensitive Measures

It includes feature selection measures that use the category information present in the training set for the selection of representative features for each category.

Shannons Entropy (E): It gives the expected value of the information contained within a term. The entropy measures the amount of randomness. The entropy value increases when the term is closer to random and entropy decreases when the term is less random. Shannons Entropy $E(t_j)$ of the term t_j is given by the formula,

$$E(t_j) = - \sum_{k=}^{r} (tf_j^k) \times \log_2(tf_j^k) \tag{5}$$

where tf_j^k frequency of the term t_j in the category c_k. If the Entropy value is equal to 0, then the entropy is minimal and if the term t_j appears only in one category. Then the term t_j is considered to have good discriminative power.

Information Gain (IG): In machine learning, Information gain is the most popular feature selection method. It is a measure based on the entropy of each feature. IG selects those features which contain most information about a category.

$$IG(t_j, c_k) = P(t_j, c_k) log \frac{P(t_j, c_k)}{P(t_j)P(c_k)} + P(\overline{t_j}, c_k) log \frac{P(\overline{t_j}, c_k)}{P(\overline{t_j})P(c_k)} \tag{6}$$

with $P(t_j, c_k)$ the probability of observing the word t_j in a document belonging to the category c_k, and $P(t_j, \overline{c_k})$ the probability of observing the word t_j in a document belonging to other categories. $P(t_j)$ is the probability that the term t_j occurs and $P(c_k)$ is the probability that category c_k occurs.

3.3   Feature Merging

Feature Merging combines the sub list of features selected by the two different feature Selection methods. Let FS_1 is the sub list of features selected with the feature section model M_1. It consists of q number of features. FS_2 is the sub list of features selected with the feature section model M_2. It consists of l number of features.

3.3.1   Union

To create a feature sub lit FS_3 using the union approach, merge all the features in the sub lists FS_1 and FS_2.

$$FS_3 = FS_1 \cup FS_2 \tag{7}$$

The newly created FS_3 feature sub list will contain f' number of features, where $f' \geq \{q, l\}$.

3.3.2   Intersection

To create a feature sub lit FS_4 using the intersection approach, merge the common features of sub lists FS_1 and FS_2.

$$FS_4 = FS_1 \cap FS_2 \tag{8}$$

The newly created FS_4 feature sub list will contain f'' number of features, where $f'' \leq \{q, l\}$.

3.3.3 Modified Union

Initially the features in both the sub lists are arranged in their decreasing order of relevance, so that the most relevant feature appears in the top of the list and the least relevant feature appears in the bottom of the list. To select the high scored features from both the list, a union approach is applies on those top ranked features, and then an intersection approach is applied on the rest of the list. Here the union approach ensures that the top ranked features are not lost and on the other hand the intersection approach ensures that the total number of selected features is not too high. System applies union approach on the top $C_1\%$ of features and then intersection approach on the remaining $C_2\%$ of features.

$$FS_5 = \{C_1\%\{FS_1\} \cup C_1\%\{FS_2\}\} \cup \{C_2\%\{FS_2\} \cap \{C_2\%\{FS_2\}\}\} \qquad (9)$$

This produces a feature sub list containing f''' number of features, where, $f''' \geq \{q, l\}$.

The merged feature set undergoes further dimension reduction by PCA algorithm and the final feature subset is clustered using k-means algorithm.

3.4 Sentiment Analysis

Sentiment analysis aims at determining the attitude, emotions and opinion of people on a certain piece of information rather than the fact involved. In other words, it analyses the reaction towards a fact.

SentiWordNet is a lexical resource based on WordNet synsets, a lexical database for English with emphasis on synonyms. In SentiWordNet, each synset is assigned three sentiment scores: positivity, negativity and objectivity. This will help to determine the polarity of the text under analysis. The value of the three attributes lie between 0 to 1 such that their sum is always 1. The more the value for positivity, the polarity of the word is considered to be positive, or if the value of negativity is more, then the polarity of the word is considered to be negative and if the objectivity score is more, then the word is said to be neutral. Thus the polarity of each cluster of data is evaluated.

4 Experimental Analysis

4.1 Dataset

The real-world text datasets Reuters-21,578 [18], Classic4 [19] and 20NewsGroups (20NG) [20] were used for the experiments conducted. Both are text documents under different class labels. Selected number of documents were used from both the datasets to conduct the experiments.

4.2 Performance Metric

The most widely used performance metrics in text categorization problems are Precision (P), Recall (R), F-score (F) and Accuracy (ACC). Precision gives the ratio right positive predictions to the entire positive predictions.

$$P_i = \frac{TP_i}{TP_i + FP_i} \tag{10}$$

Recall gives the ratio right positive predictions to the entire positive documents.

$$R_i = \frac{TP_i}{TP_i + FN_i} \tag{11}$$

F-score gives the harmonic mean of the precision and recall measures. Accuracy gives the truthfulness of the clustering algorithm.

$$ACC = \frac{\sum TP + \sum TN}{\sum TotalPopulation} \tag{12}$$

4.3 Performance Evaluation

4.3.1 Experiment 1: Impact of Dimension Reduction

The Table 1 proves the necessity of documents preprocessing. The Dimensionality Reduction Technique used here is Shannon's Entropy on dataset—Classic4.

Table 1 Improvement in clustering accuracy with document preprocessing

#docs	Dimen.Reduction	Accuracy
100	–	0.6201
100	SE	0.7550
200	–	0.6100
200	SE	0.7425

Fig. 2 Comparison of accuracy of different dimension reduction techniques on the datasets-Reuters-21,578, Classic4 and 20NG

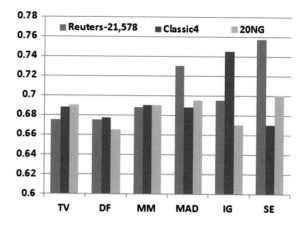

4.3.2 Experiment 2: Impact of Single Dimension Reduction Technique

Many dimension reduction techniques were employed one at a time to study the impact of dimensionality reduction on text categorization. Table 2 and Fig. 2 summarizes the experimental results of the effect of each technique on the three different datasets.

4.3.3 Experiment 3: Impact of Hybrid Dimension Reduction Technique

The system proposes a hybrid approach with more than one FS and FE methods. This is because the feature selection will be biased by the intrinsic characteristics of the technique employed. And so a single feature selection method may lose features at are proficient in some other aspects. Hence various combinations of the FS methods were studied in combination with the FE method PCA. The experimental results summarized in Table 3 and Fig. 3.

4.3.4 Experiment 4: Impact of Feature Merging Techniques

Feature merging was done sing three different ways, union, intersection, or modified union approach. The experimental study shows that modified union approach gives better accuracy when compared to the traditional union and intersection approaches of feature merging. The C1 and C2 parameters of modified union approach are fixed at 20 % and 80 % respectively for their best results proved empirically. Figure 4 shows the experimental results of the comparison between the feature merging techniques on the three different datasets Reuters-21,578, Classic4 and 20NewsGroups.

Table 2 Performance evaluation of dimension reduction techniques

Technique	Reuters-21,578			Classic4			20NG		
	Precision	Recall	F-score	Precision	Recall	F-score	Precision	Recall	F-score
TV	0.4016	0.3500	0.3740	0.4051	0.3750	0.3895	0.3752	0.3800	0.3776
DF	0.3956	0.3500	0.3714	0.4234	0.3550	0.3862	0.2988	0.3300	0.3136
MM	0.4051	0.3750	0.3895	0.3667	0.3800	0.3732	0.3667	0.3800	0.3732
MAD	0.5026	0.4600	0.4804	0.4052	0.3750	0.3895	0.3680	0.3900	0.3787
IG	0.4688	0.3900	0.4258	0.4817	0.4900	0.4858	0.4175	0.3400	0.3748
SE	0.8350	0.5150	0.6371	0.4175	0.3400	0.3748	0.4058	0.4000	0.4030

Table 3 Performance evaluation of hybrid dimension reduction approach—combination of two FS methods with PCA

Technique	Reuters-21,578			Classic4			20NG		
	Precision	Recall	F-score	Precision	Recall	F-score	Precision	Recall	F-score
TV-DF	0..3859	0.4200	0.4022	0.5645	0.5500	0.5571	0.4356	0.4300	0.4327
TV-MAD	0.5080	0.4800	0.4936	0.3413	0.4900	0.3918	0.3680	0.3900	0.3787
TV-SE	0.8345	0.5100	0.6331	0.8725	0.7400	0.8008	0.5438	0.5400	0.5419
MM-DF	0.4869	0.5200	0.5029	0.4435	0.5200	0.4787	0.4375	0.4500	0.4437
MM-MAD	0.4748	0.4900	0.4823	0.6153	0.6400	0.6274	0.3814	0.4000	0.3905
MM-SE	0.8345	0.5100	0.6331	0.8726	0.7400	0.8008	0.8333	0.5000	0.6250
IG-DF	0.4600	0.4900	0.4745	0.4644	0.4700	0.4672	0.5088	0.4200	0.4602
IG-MAD	0.4881	0.4400	0.4628	0.5946	0.5800	0.5872	0.3900	0.3900	0.3900
IG-SE	0.8345	0.5100	0.6331	0.8357	0.7700	0.8015	0.8333	0.5000	0.6250

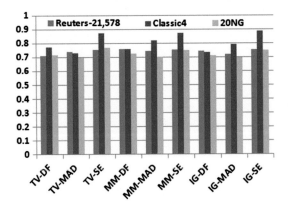

Fig. 3 Comparison of accuracy of different hybrid combinations of dimension reduction techniques on the datasets- Reuters-21,578, Classic4 and 20NG

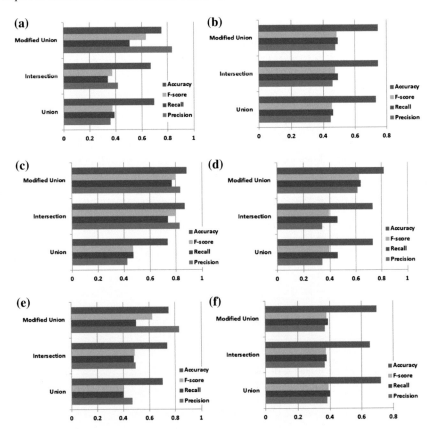

Fig. 4 Comparison of feature merging (FM) approaches. **a** with IG-SE-PCA combination in Reuters-21,578. **b** with MM-MAD-PCA combination in Reuters-21,578. **c** with IG-SE-PCA combination in Classic4. **d** with MM-MAD-PCA combination in Classic4. **e** with IG-SE-PCA combination in 20NG. **f** with MM-MAD-PCA combination in 20NG

5 Conclusion

The system puts forward the concept of hybrid dimension reduction technique which has more than one dimensionality reduction method implemented. This is because dimension reduction by a technique will always have a bias on the features chosen depending upon its intrinsic characteristics. A hybrid approach involving combination of FS and FE methods preserves those features which are prominent in different aspects. Many feature selection methods were implemented in the system and a comparative study of their performance was made.

The sub list of features selected by each feature selection technique as merged in three different ways, Union, Intersection and Modified Union approaches. The Modified Union gave better performance as it included the top features of both the feature selection technique as well as the common features of both. The sub list of features from the feature selection then undergoes further reduction by PCA to further reduce the dimensionality. The final subset of features is fed into the k-means clustering algorithm. The sentiment-score of each cluster is calculated to understand the polarity of the cluster. The performance evaluation of the system made shows that the proposed method has improved performance when compared to competitive methods. The future research will focus on n-gram tokenization of the features and impact of hybrid dimension reduction along with n-gram tokenization on the performance of clustering.

References

1. Pearson, K.: LIII. On lines and planes of closest fit to systems of points in space. Lond., Edinb., Dublin Phil. Mag. J. Sci. **2**(11), 559–572 (1901)
2. Deerwester, S.: Improving information retrieval with latent semantic indexing. In: Proceedings of the 51st Annual Meeting of the American Society for Information Science, no. 25, pp. 3640 (1988)
3. Yang, Y., Pedersen, J.O.: A comparative study on feature selection in text categorization.In: Proceedings of the 14th International Conference on Machine Learning, pp. 412–420
4. Xu, Y.: A comparative study on feature selection in unbalance text classification. In: 2012 International Symposium on Information Science and Engineering (ISISE), pp. 44–47. IEEE (2012)
5. Ferreira, A.J., Figueiredo, M.A.T.: Efficient feature selection filters for high-dimensional data. Pattern Recognit. Lett. **33**(13), 1794–1804 (2012)
6. Bharti, K.K., Singh, P.K.: A three-stage unsupervised dimension reduction method for text clustering. J. Comput. Sci. **5**(2), 156–169 (2014)
7. Patil, L.H., Atique, M.: A novel feature selection based on information gain using WordNet. In: Science and Information Conference (SAI), pp. 625–629. IEEE (2013)
8. Largeron, C., Moulin, C., Gry, M.: Entropy based feature selection for text categorization. In: Proceedings of the 2011 ACM Symposium on Applied Computing, pp. 924–928. ACM (2011)
9. Uguz, H.: A two-stage feature selection method for text categorization by using information gain, principal component analysis and genetic algorithm. Knowl.-Based Syst. **24**(7), 1024–1032 (2011)

10. Tsai, C.-F., Hsiao, Y.-C.: Combining multiple feature selection methods for stock prediction: union, intersection, and multi-intersection approaches. Decis. Support Syst. **50**(1), 258–269 (2010)
11. Miller, G.A.: WordNet: a lexical database for English. Commun. ACM **38**(11), 39–41 (1995)
12. Godbole, N., Srinivasaiah, M., Skiena, S.: Large-scale sentiment analysis for news and blogs. ICWSM **7**, 21 (2007)
13. Meesad, P., Li, J.: Stock trend prediction relying on text mining and sentiment analysis with tweets. In: 2014 Fourth World Congress on Information and Communication Technologies (WICT), pp. 257–262. IEEE (2014)
14. Esuli, A., Sebastiani, F.: Sentiwordnet: a publicly available lexical resource for opinion mining. Proc. LREC **6**, 417–422 (2006)
15. Denecke, K.: Are SentiWordNet scores suited for multi-domain sentiment classification? In: Fourth International Conference on Digital Information Management. ICDIM 2009, pp. 1–6. IEEE (2009)
16. http://jmlr.org/papers/volume5/lewis04a/a11-smart-stop-list/english.stop
17. http://tartarus.org/martin/PorterStemmer/
18. https://archive.ics.uci.edu/ml/datasets/Reuters-21578+Text+Categorization+Collection
19. http://www.dataminingresearch.com/index.php/2010/09/classic3-classic4-datasets/
20. https://kdd.ics.uci.edu/databases/20newsgroups/20newsgroups.html

2D Image Reconstruction After Removal of Detected Salient Regions Using Exemplar-Based Image Inpainting

Hima Anns Roy and V. Jayakrishna

Abstract Salient region detection is useful for applications like image segmentation, adaptive compression, and object recognition. In this paper, a novel approach is proposed to detect salient region which combines image pyramid and region property. The proposed salient region detection approach contains the three principal steps, multi-scale image abstraction, salient region detection in a single scale, saliency map fusion under multiple scales. Then image reconstruction is done after removal of detected salient regions using exemplar-based image inpainting. The results of this method were evaluated on the two publicly available databases, including MSRA-1000 and CMU Cornell iCoseg datasets. The experimental results shows that our method consistently outperforms two existing salient object detection methods, yielding better precision and recall rates. Also, better structural similarity index is also obtained in our proposed exemplar-based image inpainting.

Keywords Inpainting · Saliency map · Region segmentation · Image reconstruction

1 Introduction

Salient region detection mainly aims to locate the important region or object in images. Saliency detection plays an important role in a variety of applications including salient object detection, salient object segmentation, image resizing, and content-aware image retargeting, etc. Humans are able to detect visually salient regions effortlessly and rapidly. Visual saliency captures the most noticeable part in a scene. Saliency detection can be classified into two groups: top down and bottom up. In top down approach it requires prior knowledge about the target object. But bottom

H.A. Roy (✉) · V. Jayakrishna
Amal Jyothi College of Engineering, Kerala, India
e-mail: himaannsroy1991@gmail.com

V. Jayakrishna
e-mail: jayakrishna.vb@gmail.com

© Springer International Publishing Switzerland 2016
V. Snášel et al. (eds.), *Innovations in Bio-Inspired Computing and Applications*,
Advances in Intelligent Systems and Computing 424,
DOI 10.1007/978-3-319-28031-8_36

up approach do not have the prior knowledge of the region to be detected. The goal of preliminary bottom-up saliency detection research is to detect salient region of image.

Saliency detection aims to find the most significant object from an image. Many salient detection methods like Spatiotemporal cues, Spectral Residual approach (SR), Histogram based Contrast (HC), Region based Contrast (RC), Frequency Tuned (FT), Global Contrast (GC), cluster based method are available for capturing salient objects in an image. In most of the existing methods the input images are represented in a pixel-grid manner. In pixel-grid representation of input images, all images with large salient regions have poor performance. All color, contrast, and orientation features affect human attention and they are mostly attracted by objects rather than individual pixels.

So, the goal of our work is to reconstruct an image after removal of detected salient region using exemplar-based image inpainting. Our method employs an integration approach based on region color contrast and histogram contrast method to generate final saliency map. The advantage of our method compared to existing is that it is simple and efficient.

2 Related Works

Visual saliency aims to find the most significant object from an image. Various saliency detection methods are also available. Region based saliency detection method proposed by [1] is used to detect saliency of each pixel in an image. According to this model, the input image is partitioned into several regions. For each region, visual metrics such as locality, visual saliency, size and border count are computed. Saliency for each region is computed based on energy of each region. Finally, the calculated importances of the regions are used to construct saliency map.

Fan et al. [2] is one of the most relevant method for single image saliency detection. Three main factors are considered in this saliency detection model: isolation, distribution and location prior. Isolation deals with the feature difference between pixel and background. Distribution deals with the feature distribution of region in image and location prior explores location of salient region. Image feature based saliency map is obtained by combining superpixel isolation and distribution. Then location prior map and image feature based saliency map is combined to get the final saliency map.

Wang et al. [3] proposes a new method for saliency detection based on region descriptors and prior knowledge. Most of the existing methods lack prior knowledge about the salient regions. So to overcome this situation, prior knowledge about salient regions are also taken into consideration.

According to Ming-Ming Cheng [4] this method is used to define saliency value for image pixels based on color statistics of the input image. Pixel with similar colors are assigned with same saliency value and a color histogram is computed. But this method is computationally expensive. Region based Contrast method is an improvement over histogram contrast based method. Here the input image is divided into regions and saliency value is computed as weighted sum of regions contrast to all other region in the image [4].

Achanta et al. [5] propose a method named frequency-tuned saliency. It is obtained as a result of frequency analysis of images. The main advantage over existing methods is that it is computationally efficient. Wagh et al. [6] proposes a new framework for text detection and removal from images. This removal of unwanted text from images is done using an inpainting technique. Image inpainting techinque is used to remove partially damaged and corrupted image.

The rest of our paper is organized as follows. The main techniques for salient region detection and image reconstruction mentioned in the proposed method are introduced in the next section. Section 4 sums up the performance analysis of our scheme in terms of accuracy, F-measure and SSIM index on MSRA-1000 and CMU Cornell iCoseg dataset. Conclusions are provided in Sect. 5.

3   Proposed System Architecture

The proposed algorithm mainly consists of 3 stages. Stage 1: Salient region detection and saliency map fusion, Stage 2: Salient region removal, Stage 3: Image reconstruction using exemplar-based image inpainting.

3.1   Stage 1: Salient Region Detection and Saliency Map Fusion

Salient region detection finds the most prominent region in an image. An image pyramid is constructed and two different salient features based on region color contrast and histogram-based contrast are calculated. Saliency map is obtained by performing feature integration in each layer of image pyramid. Stage 1 goes through the following sub-stages.

3.1.1   Image Pyramid Construction

An image level muti-scale image representation is used in this method. Image pyramid is exploited to extract image layer. In each layer of image pyramid, the given input image is divided into non-overlapped regions so that saliency analysis can be done based on these regions. Given an input image I, and multi-scale image abstraction is defined as $Q = \{Q_1, Q_2, Q_3, \ldots Q_N\}$ where N denotes number of image pyramid. Q_1 denotes the first layer of image pyramid and the number of region is maximum in this layer. Q_N represents the top layer of image pyramid and here the number of region is minimum (Fig. 1).

Fig. 1 Block diagram of
proposed system

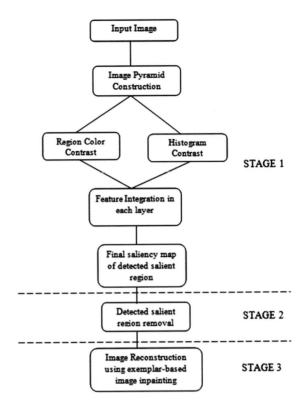

3.1.2   Region Color Contrast

Region color contrast and histogram based approach are the two important methods used for saliency analysis. Region color contrast considers spatial relationship
between regions. In region contrast based method, the input image is first segmented
into regions using efficient graph based image segmentation. Three important parameters in this algorithm are set to sigma $= 0.7$, $K = 100$ and minsize $= 20$. K denotes
threshold function and sigma is used to smooth the input image before segment-
ing it. The segmentation results in each layer of image is defined as $Q_1 = \{R_1^1, R_2^1,$
$R_3^1, \dots R_{N_1}^1\}$.

Three factors are used for region color contrast calculation. These three factors
include spatial distance between two regions, color distance between two regions and
region size. For computing region contrast, distance between two region is a major
factor. Color distance between two region $\varphi_1(R_j^n, R_k^n)$ is obtained by:

$$\varphi_1(R_j^n, R_k^n) = \| F(R_j^n - R_k^n) \| \tag{1}$$

Spatial distance between two region is obtained by:

$$\varphi_2(R_j^n, R_k^n) = 1 - DistSpatial(R_j^n, R_k^n) \tag{2}$$

where Dist Spatial is the spatial distance between two regions. Then final region contrast for each layer of image pyramid is obtained by integrating color distance, spatial distance between two regions and region color.

$$S_{region}(R_j^n) = \sum_{k=1, K \neq j}^{k_n} \varphi_1(R_j^n, R_k^n) \varphi_2(R_j^n, R_k^n) Size(R_k^n) \tag{3}$$

3.1.3 Histogram-Based Contrast

Histogram based method is used for efficient processing and here saliency value for image pixels are computed based on color statistics of input image. Given an input image, then compute color histogram of the corresponding input image. All pixels with same color value have same saliency and those same color values are grouped together to calculate saliency of each color. In natural images, number of colors are reduced by ignoring less frequently occurring colors. Saliency value of each color is calculated by weighted average of saliency values of similar colors. Saliency value of each color is defined as:

$$S_{histogram}(c) = \frac{1}{(m-1)T} \sum_{j=1}^{m} (T - Dist(c, c_j)) S(c_j) \tag{4}$$

where $T = \sum_{j=1}^{m}(T - Dist(c, c_j))$ denotes the sum of distances between color c and its m nearest neighbors c_j.

3.1.4 Feature Integration

Saliency map under different scales are obtained based on these image pyramid by integration of region color contrast and histogram based approach such that it can highlight salient object and also reduce the influence caused by complex texture background. Salient feature integration is performed in each layer of image pyramid by:

$$Saliency = S_{region}(R_j^n).S_{histogram}(c) \tag{5}$$

where η is set to 0.3.

3.1.5   Saliency Map Fusion

Final saliency map is generated by fusing integration result obtained from each layer
of image pyramid.

$$Saliency(I) = \oplus_{m=1}^{M} Saliency(I^m) \qquad (6)$$

where I denotes the input image, I^m is the mth layer of image pyramid. $Saliency(I)$
is the final saliency map obtained by integration.

3.2   Stage 2: Salient Region Removal

After detection of salient region, a segmentation map is constructed for an input
image. An efficient graph-based segmentation is used in which it divides the input
image into several regions. The output of salient region detection is selected as the
target region and the salient region is removed from the input image.

3.3   Stage 3: Image Reconstruction

After removal of salient regions, image reconstruction is done using exemplar based
image inpainting. Here Difference of Gaussians (DoG) is used for selecting appropriate parameter values and uses exemplar based image inpainting that fills the target
region with most similar patch in source region S. Inorder to calculate patch priority
$P(p)$, confidence term $C(p)$ and data term $D(p)$ are considered. This priority value is
obtained as a product of confidence and data term. Confidence term $C(p)$ is calculated as:

$$C(p) = \frac{\sum_{q \in \Psi_p \cap S} C(q)}{|\Psi_p|} \qquad (7)$$

where number of pixels are represented using Ψ_p. Data term $D(p)$ is calculated as:

$$D(p) = \frac{|\nabla I_p^{\perp}.n_p|}{255} \qquad (8)$$

where $\perp$ represents normal component and n_p is a normal vector at pixel p. An exemplar based image inpainting is used in which the target region is filled with most similar patch in the source image. The output of salient region detection is selected as
the target region and target region is marked as green. Here the filling order is done
based on priority functions and data and confidence terms are calculated. Then find
a patch with maximum priority Ψ_q and search the source region to find a patch with
minimum distance from the patch Ψ_p. Confidence term is updated.

4   Experimental Results and Analysis

The data used in the experiments are obtained from publicaly available databases MSRA-1000 [7], CMU Cornell iCoSeg datasets [8]. Each of these dataset has its ground truth in the form of human marked labels. The proposed algorithm was implemented in a MATLAB environment. Figure 2 shows some image samples taken from MSRA-1000 and CMU Cornell iCoseg dataset.

4.1   Accuracy Analysis

Accuracy is calculated as sum of foreground pixels detected in binary image to sum of foreground pixels in ground truth. The column named 'Avg' in the table represents the average score on all categories. The quality comparison of these methods was performed by calculating the accuracy value.

$$Accuracy = \left(\frac{B}{G}\right) * 100 \qquad (9)$$

where B is the sum of foreground pixels detected in binary image and G denotes the sum of foreground pixels in ground truth.

Figure 3 shows the inpainting result using our saliency map. Detected salient regions are removed using exemplar-based image inpainting.

(a)   **(b)**   **(c)**   **(d)**

Fig. 2   Image samples taken from different Image databases. **a, b** Shows MSRA-1000 dataset images. **c, d** Belong to iCoseg dataset

(a)   **(b)**   **(c)**

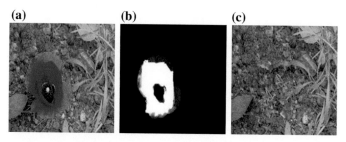

Fig. 3   Inpainting result of proposed saliency method using MSRA-1000 dataset. **a** Input image, **b** Saliency map and **c** Inpainting result

Table 1 Accuracy results of MSRA-1000 dataset

Method	Apple	Post box	Flower	Flag	Car	Avg
Our method	86.4	83.5	90.3	78.49	83.2	84.37
[9]	73.8	74.4	81.6	68.3	84.3	76.48
[10]	63.8	70.4	78.6	72.3	82.5	73.52

Table 1 shows the accuracy results on MSRA-1000 dataset. Our method obtains better accuracy results when compared with [9] and [10].

4.2 Precision-Recall Analysis

The precision and recall measures are calculated using the binary ground truths as reference masks. These measures allow us to evaluate all maps of all images in the database for each threshold independently. Performance of proposed method is measured based on precision recall analysis. The precision and recall is calculated using:

$$Precision = \frac{\sum_x g_x s_x}{\sum_x s_x} \tag{10}$$

$$Recall = \frac{\sum_x g_x s_x}{\sum_x g_x} \tag{11}$$

where g_x is the ground truth of human marked labels in xth image, s_x is the salient region detected in xth image. In addition to precision and recall, F measure is calculated as:

$$F = \frac{(1 + \beta^2).Precision.Recall}{\beta^2.Precision + Recall} \tag{12}$$

where $\beta^2 = 0.3$. The proposed method is compared with two saliency detection methods namely, context aware saliency [9] and spectral residual saliency method [10]. The performance evaluation is done using precision-recall analysis.

Figure 4 shows performance comparison of proposed method on MSRA-1000 and CMU Cornell iCoseg dataset. The proposed method on MSRA-1000 dataset obtained an F-measure of 0.8254.

4.3 SSIM Index Analysis

Another important measure used for comparing inpainting results is the structural similarity (SSIM) index. Table 2 shows the comparison of SSIM index on

(a)   **(b)**

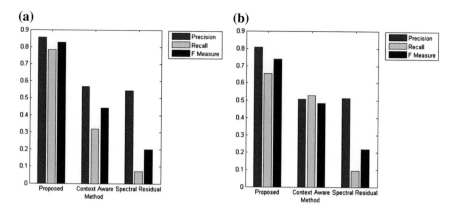

Fig. 4 Performance of the proposed method compared with two saliency detection methods on **a** MSRA-1000 dataset and **b** CMU Cornell iCoseg dataset

Table 2 Comparison of SSIM index on MSRA-1000 dataset

Image	Proposed method	Inpaint NaNs
Bird	0.9367	0.2530
Parachute	0.9159	0.3678
Fish	0.7444	0.5612
Leaf	0.7424	0.2352
Tree	0.7876	0.0111
Flower	0.8788	0.4512
Flag	0.8314	0.4673

MSRA-1000 Dataset. Better structural similarity index is obtained in our proposed exemplar-based image inpainting as shown in Table 2.

5 Conclusion

A novel approach is proposed to detect salient region which combines image pyramid and region property. Firstly, an image pyramid is constructed and then calculate the region color contrast and histogram based contrast of each layer of the image pyramid. The two salient features in each layer of image pyramid are then combined by a simple non-linear approach and saliency map of single scale can be obtained. Finally, fuse the saliency map of multiple levels to detect salient region in an image. Image reconstruction is done after removal of detected salient region using exemplar-based image inpainting. Both structure and texture information are utilized in this exemplar-based inpainting technique to automatically select parameter values and to select appropriate patch size. On analysing the performance statistics with similar techniques like context aware and spectral residual saliency detection that uses MSRA-1000 and CMU Cornell iCoseg datasets, our proposed method gives promising results on precision-recall rates.

Future research will focus on the hole-filling application in depth image based rendering(DIBR) using segmentation and depth-based inpainting technique for three-dimensional video systems.

References

1. Manipoonchelvi, P., Muneeswaran, K.: Region-based saliency detection. IET Image Process. **8**(9), 519–527 (2013)
2. Fan, Q., Qi, C.: Two-stage salient region detection by exploiting multiple priors. J. Vis. Commun. Image R. **25** (2014)
3. Wang, W., Cai, D., Xu, X., Liew, A.W.-C.: Visual saliency detection based on region descriptors and prior knowledge. Signal Process.: Image Commun. **29** (2014)
4. Cheng, M., Zhang, G., Mitra, N.J., Huang, X., Hu, S.: Global contrast based salient region detection. In: Proceedings of Computer Vision Pattern Recognition (2011)
5. Achanta, R., Hemami, S.S., Estrada, F.J., Ssstrunk, S.: Frequency tuned salient region detection. In: IEEE Computer Society Conference on Computer Vision Pattern Recognition (CVPR) (2009)
6. Wagh, P.D., Patil, D.R.: Text detection and removal from image using inpainting with smoothing. In: International Conference on Pervasive Computing (ICPC) (2015)
7. http://mmcheng.net/msra10k/
8. http://chenlab.ece.cornell.edu/projects/touch-coseg/index.html
9. Goferman, S., Zelnik-Manor, L., Tal, A.: Context-aware saliency detection. In: Proceedings of IEEE Transactions on Computer Vision and Pattern Recognition (2012)
10. Hou, X., Zhang, L.: Saliency detection: a spectral residual approach. In: Proceedings of IEEE Computer Society Conference on Computer Vision and Pattern Recognition (CVPR) (2009)

Solution to Constrained Test Problems Using Cohort Intelligence Algorithm

Apoorva S. Shastri, Priya S. Jadhav, Anand J. Kulkarni and Ajith Abraham

Abstract Most of the real world problems are inherently constrained in nature. There are several nature inspired algorithms being developed; however their performance degenerate when applied solving constrained problems. This paper proposes Cohort Intelligence (CI) approach in which a probability based constrained handling approach is incorporated. This approach is tested by solving four well known test problems. The performance is compared and discussed with regard to the robustness, computational cost, standard deviation and rate of convergence etc. The constrained CI approach is used here to solve few inequality based constrained problems. The solution to these problems indicates that the CI approach can be further efficiently applied to solve a variety of practical/real world problems.

Keywords Cohort intelligence · Constrained optimization · Constrained test problems

1 Introduction

There are several nature inspired algorithms have been devised since past few years such as Particle Swarm algorithm (PSO) [1], Evolutionary strategy (ES) [2], Differential Evolution [3], Genetic Algorithm [4], Similar to these methods, the

A.S. Shastri (✉) · P.S. Jadhav · A.J. Kulkarni
Symbiosis Institute of Technology, Symbiosis International University,
Pune 412115, India
e-mail: apoorva.shastri@sitpune.edu.in

P.S. Jadhav
e-mail: Priya.jadhav@sitpune.edu.in

A.J. Kulkarni
e-mail: anand.kulkarni@sitpune.edu.in

A. Abraham
Machine Intelligence Research Labs (MIR Labs), Scientific Network
for Innovation and Research Excellence, WA, USA
e-mail: ajith.abraham@ieee.org

© Springer International Publishing Switzerland 2016
V. Snášel et al. (eds.), *Innovations in Bio-Inspired Computing and Applications*,
Advances in Intelligent Systems and Computing 424,
DOI 10.1007/978-3-319-28031-8_37

427

immerging technique referred as Cohort Intelligence (CI) [5–7] also suitable and efficient for solving unconstrained problems and when applied for constrained problems its performance may degenerate. To solve real world problems which are constrained in nature suitable constrained handling technique needs to be devised. There are few methods being used such as Penalty function method, Feasibility based method and repair based method.

The Penalty function methods are sensitive to penalty parameters as well as the performance highly depends on problem at hand. The feasibility based methods require the objective function to be calculated and then the solution needs to be tested for feasibility. Also a mechanism needs to be incorporated to push the solution towards feasible region. This considerably increases computational cost. In the repair approach the solution needs to be significantly modified to make it feasible. This approach may quickly find a feasible solution however such methods are problem specific and needs exclusive knowledge about the problem. More over repair approach may work for small size and simple problems however as problem size increases the approach may become tedious with excessive computational efforts. In this paper we proposed a problem based constrained handling approach in which a problem distribution is devised for every individual constraint. In this approach a window is created in which the lower limit of the window lies in the feasible region and upper limit lies in the infeasible region. The lower and upper limits are chosen based on the preliminary trials of the algorithm and are problem specific.

2   Constrained Cohort Intelligence

Consider a general constrained optimization problem as follows:

$$\text{Minimize } f(\mathbf{X}) = f(x_1, \ldots x_i, \ldots, x_N) \tag{1}$$

Subject to:

$$\begin{cases} g_j(\mathbf{X}) \leq 0 & \text{for } j = 1, \ldots, m \\ \psi_i^{lower} \leq x_i \leq \psi_i^{upper} & \text{for } j = 1, \ldots, N \end{cases} \tag{2}$$

In the context of CI [8] the objective function $f(\mathbf{X})$ is considered as the behavior of an individual candidate in the cohort with associated set of qualities = $(x_1, \ldots x_i, \ldots, x_N)$.

Consider a cohort with number of candidates C, where every individual candidate $c(c = 1, \ldots, C)$ belongs to a set of qualities $\mathbf{X}^c = (x_1^c, \ldots x_i^c, \ldots x_N^c)$. The individual behavior of each candidate c is generally being observed by itself and every other candidate (c) in the cohort. This naturally urges every candidate (c) to follow the behavior better than its current behavior. The procedure begins with initialization of no of cohort candidates C, sampling interval $\left[\psi_i^{lower}, \psi_i^{upper}\right]$ for each quality $x_i, i = 1, \ldots, N$, learning attempt counter $n = 1$, the sampling interval reduction factor $r \in [0, 1]$ and convergence parameter ε. The values of C and r are chosen based on preliminary trials of the algorithm.

Step 1: Every candidate $c(c = 1, \ldots, C)$ randomly generates qualities $\mathbf{X}^c = \left(x_1^c, \ldots x_i^c, \ldots, x_N^c\right)$ from within the associated sampling interval $\left[\psi_i^{lower}, \psi_i^{upper}\right] i = 1, \ldots, N$.

Step 2: Every candidate $c(c = 1, \ldots, C)$ evaluates associated constrain value

$$\forall c\ g_j^c(\mathbf{X}), j = 1, \ldots, m \qquad (3)$$

Step 3: For every constraint $g_j^c(\mathbf{X}), c(c = 1, \ldots, C), j = 1, \ldots, m$ a probability distribution is developed (refer to Fig. 1) with predefined lower and upper limits for probability calculation as $k_{l,j}^c$ and $k_{u,j}^c$ respectively. In addition, any constrained value if lower than associated $k_{l,j}^c$ or exceeding $k_{u,j}^c$ is assigned a probability value 0.00001.

Kulkarni and Shabir (2014) proposed a modified approach to the CI method for solving knapsack problems. This approach makes use of probability distributions for handling constraints. This approach is adopted here. For inequality constraint, probability distribution is developed (refer to Fig. 1) and probability is calculated based on following rules:

1. If $k_{l,j}^c \leq g_j^c(\mathbf{X}) \leq 0$, then based on the probability distribution presented in Fig. 1

$$p = 1 - (slope_{1, k_{l,j}^c} \times g_j^c(\mathbf{X})) \qquad (4)$$

2. If $0 < g_j^c(\mathbf{X}) \leq k_{u,j}^c$, then based on the probability distribution presented in Fig. 1

$$p = 1 - (slope_{2, k_{u,j}^c} \times g_j^c(\mathbf{X})) \qquad (5)$$

Fig. 1 Probability distribution for constrained handling

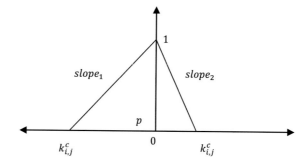

3. And If

$$k_{l,j}^c \langle g_j^c(\mathbf{X}) \rangle k_{u,j}^c, \text{ then } p = 0.00001 \tag{6}$$

Step 4: After probability for individual candidate, probability score for each candidate is calculated as follows:

$$p^c = \sum_{j=1}^m p\big(g_j(\mathbf{X})\big) \quad c = 1, \ldots, C \tag{7}$$

$$p_{score}^c = \frac{p^c}{\sum_{c=1}^C p^c(c)} \quad c = 1, \ldots, C \tag{8}$$

p_{score}^c is score of each individual candidate in the cohort.

Step 5: If all the m constraints for each individual candidate $c = 1, \ldots, C$ are satisfied i.e. $g_j^c(\mathbf{X}) \le 0, (j = 1, \ldots, m)$, then objective function $f^c(\mathbf{X})$ is evaluated and the value β is added in p_{score}^c of respective candidate. If above condition is not satisfied then value of objective function $f^c(\mathbf{X})$ is assigned as 0.0009 for respective candidate not satisfying the condition.

Step 6: $p_{score}^c (c = 1, \ldots, C)$ and the roulette wheel mechanism [5] is used to make candidates follow one another. It is important to mention that 'following' here refers to shrinking. Every candidate $c(c = 1, \ldots, C)$ shrinks the sampling interval ψ_i^c, associated with every variable $x_i^c, i = 1, \ldots, N$ to its local neighborhood as follows:

$$\psi_i^c \in \left[x_i^c - (\|\psi_i\|/2), x_i^c + (\|\psi_i\|/2) \right] \tag{9}$$

where $\psi_i = (\|\psi_i\|) \times r$.

3 Result and Discussion

The proposed CI algorithm was coded in MATLAB (R2013a) and simulations were run on a Windows platform using Intel Core 2 duo, 3 GHz processor speed with 2 GB RAM. The continuous constraint test problems: G01, G04, G06, G08 have been solved. The characteristics of these problems are listed in Table 1. Every problem was solved 20 times. The best, mean solutions and SD for each test problem is listed in Table 2. The associated standard deviation (SD), number of function evaluations (FE), computational time are listed in Table 3. For every test problem, the parameters such as the reduction factor r and numbers of candidates

Table 1 Characteristics of test problem

Problem	Nature	Problem type	N	LI	NI	NE	α
G01	Minimization	Quadratic	13	9	0	0	6
G04	Minimization	Quadratic	5	9	0	0	2
G06	Minimization	Cubic	2	0	2	0	2
G08	Maximization	Non-linear	2	0	2	0	0

NE Nonlinear Equality, NI Nonlinear Inequality, LI Linear Inequality, α number of active constraints at optimum, N No of variables

Table 2 Comparison of CI solutions with existing algorithms

Problem		DE	GA	ES	PSO	CI
G01	Best	−15.00	−14.064	−14.972	−14.935	−15
	Mean	−15.00	−13.982	−14.932	−14.851	−15.0002541
	SD	0.00E+00	4.60E-02	3.00E-02	4.30E-02	1.77E-04
G04	Best	−30665.539	−30654.531	−30665.539	−30637.414	−30631.0504
	Mean	−30665.539	−30582.522	−30665.539	−30613.192	−30631.0518
	SD	0	4.00E+01	0	1.10E+01	0.007761769
G06	Best	−6961.814	−6846.993	−6961.814	−6959.643	−6961.31323
	Mean	−6961.814	−6303.951	−6961.814	−6932.836	−6961.31323
	SD	0	285.186118	0	1.70E+01	7.34983E-12
G08	Best	−0.095825	−0.095825	−0.095825	−.095823	−0.09571196
	Mean	−0.095825	−0.095825	−0.095825	−0.095727	−0.09551298
	SD	0	1.00E-06	0	8.00E-05	0.006430892

Table 3 Performance of CI

Problem	FE	SD	Time (s)	Parameters (C, r)
G01	5000	1.77E-04	0.12	5, 0.939
G04	3500	7.76E-03	0.01546535	5, 0.939
G06	1500	7.35E-12	0.078152478	5, 0.890
G08	1500	6.43E-03	0.006430892	5, 0.830

are listed in Table 3. The solution convergence plot for each the problem is presented in Fig. 2a–d. It validated the self supervised learning behavior of each candidate solving the constrained problems (Figs. 3, 4, 5 and 6).

The results are compared with other the contemporary optimization methods such as DE, GA, ES, PSO. The best, mean solutions of the CI with other method is compared (Refer Table 2). The standard deviation (SD), number of function

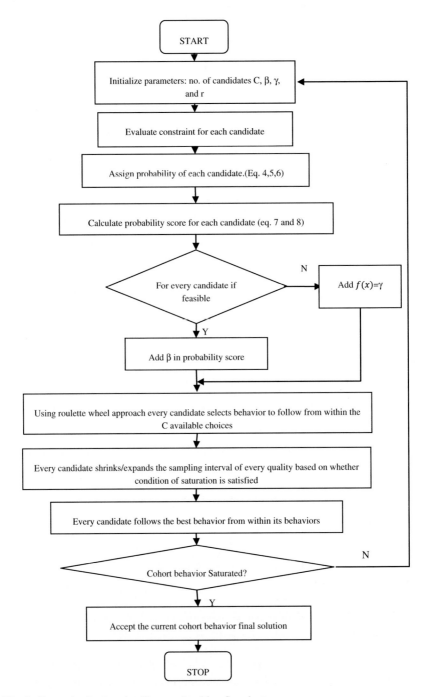

Fig. 2 Constrained cohort intelligence algorithm flowchart

Fig. 3 Convergence history of G01

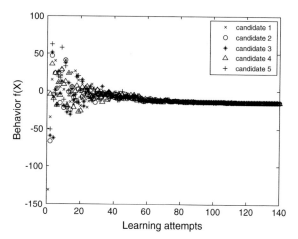

Fig. 4 Convergence history of G04

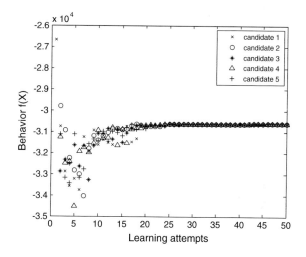

evaluations (FE) and the computational time are presented in Table 3. It is observed that CI is more robust as compared to the algorithms being compared. The obtained solutions for function G01 and G06 are better than other methods with considerably less computational time, function evaluations and standard deviation. However, the comparable function values are obtained for G04 and G08.

Fig. 5 Convergence history
of G06

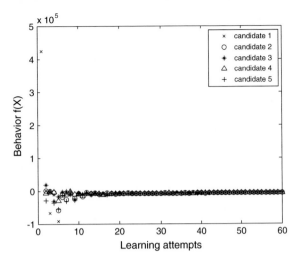

Fig. 6 Convergence history
of G08

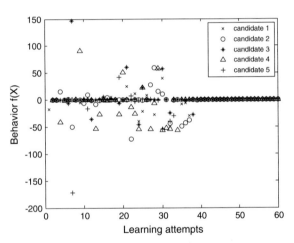

4   Conclusions and Future Directions

These results are reasonably robust with acceptable computational cost. In addition
as compared to other algorithms, the solution quality was reasonable good. The
method of constrained handling based on probability distribution worked well for
solving the continuous constrained test problems. In the future, constrained CI
algorithm can be improved in order to make this approach more generalized to
solve problems having with equality constraints. Also in order to avoid preliminary
runs of the algorithm a parameter fine tuning method needs to be devised. In
addition, a rigorous analysis of the constrained CI approach is required to be
conducted.

References

1. Kennedy, J., Eberhart, R.: Particle swarm optimization. In: Proceedings of IEEE International Conference on Neural Networks. pp. 1942–1948 (1995)
2. Coello, C.A.C.: Theoretical and numerical constraint handling techniques used with evolutionary algorithms: a survey of the state of the art. Comput. Methods Appl. Mech. Eng. **191**(11–12), 1245–1287 (2002)
3. Price, K.V., Storn, R.M., Lampinen, J.A.: Differential Evolution. A Practical Approach to Global Optimization. Springer, New York (2005)
4. Deb, K.: An efficient constraint handling method for genetic algorithms. Comput. Methods Appl. Mech. Eng. **186**(2/4), 311–338 (2000)
5. Kulkarni, A.J., Durugkar, I.P., Kumar, M.: Cohort intelligence: a self supervised learning behaviour. In: IEEE International Conference on Systems, Man, and Cybernetics pp. 1396–1400 (2013)
6. Kulkarni, A.J., Shabir, H.: Solving 0-1 knapsack problem using cohort intelligence algorithm. J. Mach. Learn. Cybern. (in Press)
7. Kulkarni, A.J., Tai, K.: A probability collectives approach with a feasibility-based rule for constrained optimization. Appl. Comput. Intell. Soft Comput. **2011**, Article ID 980216
8. Kulkarni, A.J., Tai, K.: Probability collectives: a distributed optimization approach for constrained problems. In: Proceedings of IEEE World Congress on Computational Intelligence, pp. 3844–3851 (2010)

Placement Strategies for Faulty Cells in Module Relocation Based BISR Approach

Madhuri Elsa Eapen, C. Pradeep, Anila Ann Varghese and Jisha M. Nair

Abstract Field programmable gate arrays are used as a core component in many safety and mission critical applications. In most cases these systems will be continuously exposed to radiations and change in temperature and pressure. This can result in defects within the IC which leads to malfunctioning or total system failure before mission completion. Traditionally, fault tolerance in FPGA is achieved by using spare cells to replace a faulty cell. Higher fault coverage demands more number of spares. So there is a need to develop a repair strategy with minimal hardware overhead capable to respond to defects without bothering system performance. The paper discusses a Built-in-Self-Repair (BISR) approach for FPGA based reconfigurable systems with limited number of spares, reduced area overhead, routing complexity and maximum resource utilization. The key concept used in this BISR is relocation of reconfigurable modules using dynamic runtime partial reconfiguration. This is an on-line repair method which can be used to handle multiple faults without affecting system functioning and throughput. The efficient placement of relocated module helps handle more number of faults with least area overhead and routing complexity. According to the required lifetime of the system the designer can flexibly use this method to maintain fault coverage until mission completion.

Keywords Relocation · Self-repair · FPGA · Placement · Dynamic runtime partial reconfiguration · Multiple faults

M.E. Eapen (✉) · C. Pradeep (✉) · A.A. Varghese (✉) · J.M. Nair (✉)
Department of Electronics and Communication, SAINTGITS College
of Engineering, Kerala, India
e-mail: madhurilseapen@gmail.com

C. Pradeep
e-mail: pradeepcee@gmail.com

A.A. Varghese
e-mail: anilaann7@gmail.com

J.M. Nair
e-mail: jishamnr@gmail.com

© Springer International Publishing Switzerland 2016
V. Snášel et al. (eds.), *Innovations in Bio-Inspired Computing and Applications*,
Advances in Intelligent Systems and Computing 424,
DOI 10.1007/978-3-319-28031-8_38

1 Introduction

Arrival of FPGAs into the electronic industry has revolutionized VLSI based digital systems [1]. FPGAs have now become one of the most commonly used IC in majority of the VLSI based high computing mission critical applications. For systems used in space crafts, avionic applications, orbital satellites etc. the core component used is FPGA. Due to the adverse climatic conditions, these FPGAs are highly prone to defects. A defect causes change to the implemented design within the system which leads to malfunctioning or at times mission failure. Failure of such systems can lead to huge yield loss and sometimes may turn dangerous to human life or property loss. To avoid fault is not possible, so FPGAs must have the capability to function normally even after a fault has occurred. To achieve fault tolerance, care must be taken from design time itself.

Ability to reconfigure is the key feature exhibited by FPGAs which makes it the best choice to implement self adaptive systems. A self adaptive system has the capability to adapt to diverse faulty conditions throughout its life time without human intervention. The main challenge for reconfiguration in earlier days was its time duration. To make a small change to the implementation, the entire FPGA had to undergo reconfiguration, which created a huge time delay. But, as part of technology advancement, modern FPGAs are designed with a state of art reconfigurability called dynamic run time partial reconfiguration. This is one of the most advanced level of reconfiguration ever since. This allows changing a part of the implemented design during run time, without affecting the rest of the system [2]. Researchers use this feature to design and implement self repairing systems on FPGAs. The main vendors of such FPGAs are Xilinx, Altera etc. and the most commonly used among them for implementing reconfigurable based systems are Virtex 5, 6, 7 series.

The concept of self healing or self repairing was first inspired from biology and was first proposed in 1990s [3]. Since then design of fault tolerant systems has gained increasing attention. More than a million faults are repaired during the life time of an organism, with or without its knowledge, which permits its normal operation. Researchers are trying to achieve this level of fault repair or healing into the electronic systems as well. Most of the repairing techniques are done by replacing the faulty module with the spare module allocated inside the system. The main challenges of the existing fault tolerant techniques are requirement of more number of spares, area overhead, routing complexity, reconfiguration time, and reduced throughput due to repair.

The paper deals with a new self repairing strategy for FPGA based reconfigurable systems using dynamic run-time partial reconfiguration. The main concept is to relocate the faulty cell from the fault location and two approaches to determine this new location is discussed and demonstrated. To eliminate the effect of the fault, fine grain isolation is done which enables best resource utilization.

The paper layout is as follows; Sect. 2 discusses some of the interesting literatures published in the field of FPGA self-repairing. Section 3 describes the proposed work. Section 4 concludes the paper with future scopes.

2 Previous Works

After understanding the importance of developing fault tolerant systems on FPGA many literatures have been published, proposing different self-repairing schemes, architectures and algorithm. In [4], Kim, S et al. discuss a fast fault recovery method by use of both redundant cell and spare cell associated with a single working unit; also a detection unit is allotted for every cell. A self-repairing architecture for sixteen working cells is demonstrated in the paper. Each working cell is allotted with a redundant cell; and for a group of four working cells, a spare is allotted. Due to the presence of spare and redundant cells fast fault recovery is possible. But the area overhead is very high and an extra decision making circuitry is necessary as the inputs to the spares and working cell are already pre-routed; the decision making circuitry decides which among the two has to function at a particular time.

The most commonly used method of fault tolerance in FPGA is by using hardware redundancy. Often redundancy measures like duplication or triplication of subsystems is done. In TMR, three multiples of the same sub-unit is implemented and the output is taken simultaneously via a majority gate; it can handle only a single fault but requires no fault detection unit separately [5]. Other diversities of TMR were also proposed like TMR with alternative computing and partial TMR. All of them were proposed to reduce the overhead caused by regular TMR. TMR with alternative computing uses three multiples of the hardware but each one is implemented in different way internally (one using gates other using multiplexers). Partial TMR was proposed mainly to limit the huge area overhead produced in TMR [6]. In this case, the only units which demands higher fault coverage or more importance are implemented using TMR. All the others are left out normally. TMR as a repair method has mainly two disadvantages; higher area overhead and power consumption. Also it can only handle limited number of faults, mostly only one.

Lala proposed another method of repair inspired from human immune system [7]. In this method a spare cell is allotted for 4 working cells and routing cells are arranged in between them. The cells are arranged such that every working cell has two spare cells always present adjacent to them. When a fault is detected in any of the working cell, any one of the adjacent spare is taken to replace it; which will be according to the decision router cells.

Pradeep et al. [8] presents a fault recovery algorithm using king spare allocation of the spare cell. King spare allocation means eight working cells with one spare cell located at the centre. And this spare cell is used by any of the 8 working cells when they get faulty. In the absence of a spare unit in that tile, Dijkstra's algorithm is used to locate the nearest spare and cells are shifted to bring the spare more close to the faulty cell location. The shifting is very time consuming as it requires that much more number of reconfigurations. A self-repairing algorithm which acts on the configuration of selected LUTs inside the FPGA is discussed in [9]; based on the faulty condition, LUT-level repairing is done by mutation and cross-over process. A novel approach for FPGA self-repair systems using dynamic partial reconfiguration called reconfigurable adaptive redundancy systems is discussed in [10]. It is completely based on evolutionary genetic algorithms.

3 Proposed Work

The paper deals with a self repairing technique for FPGA based reconfigurable systems. The key feature used is relocation of reconfigurable modules using dynamic run-time partial reconfiguration. Relocation of reconfigurable modules can be done by either software (PARBIT) [11] or hardware (REPLICA) [12] technique. Efficient relocation of reconfigurable modules inside an FPGA can improve the efficiency of the system [13]. For a system consisting of heterogeneous modules, self repairing property can be easily achieved by efficient relocation of the faulty modules. The efficiency is mainly determined by the location to which the cell is relocated. The paper also discusses two approaches for the placement of the relocated modules. Relocation with best placement can help the system to attain self-repairing capability.

FPGA basically consists of Configurable Logic Blocks (CLBs) connected through programmable interconnects [14]. The FPGA fabric area is divided into static and reconfigurable regions. There are two ways to perform partial reconfiguration: difference based and module based [15]. In case of module based design flow, the system is implemented as different modules in the FPGA. Among which, the modules within the reconfigurable area can only be changed or reconfigured. To achieve self repairing capability, the system has to be designed with some fault tolerant methods. The object is to handle and repair multiple finite number of faults with minimum number of spares. In a system with finite number of reconfigurable cells, some will be high priority cells and others will be low priority cells. If the high priority cell gets faulty, it must be repaired fast; so a spare cell can be reserved for this and also for transient fault repair. This spare cell should be of such dimension that it can be used to replace any of the reconfigurable cells. Now to repair low priority cells, relocation of the faulty cell can be effected (Table 1).

When a defect occurs in a FPGA, it can be expressed in different level of resolution. Basically it affects a location inside the IC; but it can be expressed in different levels of resolution namely, a transistor is faulty, or a CLB is faulty or the module affected by this defect is faulty or the entire FPGA is faulty. Each level of resolution has considerable effect on the efficiency of the repair and its subsequent overhead. To get the maximum resource utilization, fine grain resolution need to be used. For obtaining the results below, the resolution is in CLB-level. So when a fault is detected, the faulty module is initially isolated and priority of the cell is

Table 1 Simulation parameters

Parameters	Values
Fault_info	1
FPGA area	50 × 40
Reconfigurable area	40 × 30
No. of tiles	1
No. of cells	9
No. of working cells	8
No. of spares	1
Working cells	1,2,3,4,5,6,7,8
High priority cells	1, 2, 3
Spare cell	9
Faulty cell	8
Total no. of CLBs	300
Occupied CLBs	194

checked. For low priority cells, effort is made to relocate the faulty cell to a non-defective location and the same is done for high priority cells in cases of unavailable spare cell. After successful repair, the isolated faulty module undergoes self test and only the faulty CLB is isolated and the rest of the healthy CLBs are made unoccupied for future relocation processes. This fine-grain resolution help maximize the resource utilization. And the single spare cell reduces the area overhead compared to other repairing techniques which uses more number of spares for repairing multiple faults.

Entire reconfigurable area of the FPGA is modeled as a two dimensional array of CLBs. And the reconfigurable modules are located within it. Each reconfigurable area will consist of finite number of CLBs. The Fig. 1 shows eight reconfigurable modules located arbitrarily within the reconfigurable area of the FPGA. The 9th cell is the spare cell and is of dimension capable of replacing any of the modules around. A finite number of CLBs is reserved during the design time for the initial

Fig. 1 Initial placement of the reconfigurable modules

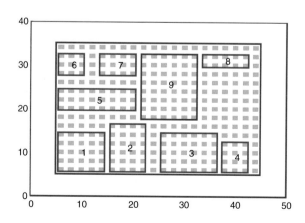

relocation. Say if i_1, i_2, i_3, ..., i_n represents the number of CLBs in n-reconfigurable modules within the reconfigurable area of the FPGA. Then, i is the number of CLBs that has to be reserved for relocation, which is given by Eq. (1).

$$i = \{i_1 + i_2 + i_3 \cdots + i_n\}/n \tag{1}$$

For easy demonstration, let us consider that a fault has occurred on cell number 8. It is a low priority cell, so repair is done using relocation. To nullify the effect of the defect, the faulty cell is isolated and is denoted "FAULTY" in Figs. 2a and 3a. Number of CLBs used in this particular module is considered and a search is done to find out, that much number of consecutive unoccupied CLBs within the reconfigurable area. The faulty cell is relocated to this new location. As we can see in Fig. 2a, enough number of such locations is available but the new location must be chosen with great care, because a careless placement of the relocated cell can lead to fragmentation [16] or high routing overhead. Two placement approaches for relocated faulty module is discussed below.

Fig. 2 Repairing with best resource utilization

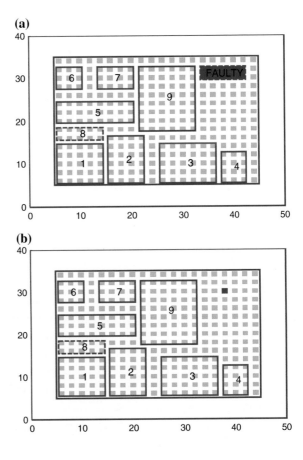

First one is the placement approach for the best area utilization so that maximum area is available for future relocation. In this case, the placement location is considered from any one among the four corners of the reconfigurable area. After the fault is detected, a search is done by considering one of these corners as the origin. The first apt location with the required number of unoccupied CLBs is utilized for relocating the faulty module. Figure 2 shows the relocation of faulty module-8 by this approach. Here, left bottom corner is considered as the origin and module-8 is relocated to the first set of consecutive unoccupied CLBs by searching. After successful relocation, the faulty cell undergoes a self-test and only the faulty CLB will be isolated. The rest of the healthy CLBs are made un-occupied so that they can be useful for future.

The next approach of placement is to reduce routing complexity with the static module. The main idea is to relocate the faulty module close to its earlier location, so that the routing complexity is not increased. In Fig. 3, when module-8 is detected faulty, the search starts from the current location of module-8. The first consecutive set of unoccupied CLBs of the required number (total number of CLBs needed to regenerate the faulty cell) is considered as the new location.

Fig. 3 Repairing with least routing complexity with static module

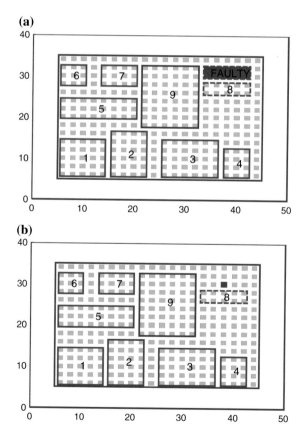

Table 2 Comparison with previous techniques

Approach	Resolution	Total no. of cells	No. of working cells	No. of spare cells per working cell	No. of redundant cells per working cell	Area overhead	Resource utilization	Routing complexity
Paralogous genes [4]	Module level (low resolution)	6	4	1/4	1/1	High	Low	–
TMR [5]	Module level (low resolution)	3	3	0/1	2/1	High	Low	–
Proposed method	CLB level (high resolution)	9	8	1/8	0/1	Low	High	Low

According to the user needs and the life time requirement of the system, the designer can flexibly use any one among the above placement. The compact placement technique for maximum resource utilization can handle maximum number of faults while the second one will give least routing complexity. In either case, repairing using relocation can be considered as a better option to achieve fault tolerance in FPGA based reconfigurable systems.

Table 2 demonstrates the comparison of the proposed work with two of the previous existing works. In case of repair method inspired from paralogous genes, each working cell will be allotted with a redundant cell and for a group of four working cells, a spare is reserved. When a fault is detected in any working cell, an immediate repair is done with the help of redundant cell; only in the unavailability of the redundant cell, the spare will be utilized. This approach demands both spare and redundant block for a single working cell, which causes high area-overhead [4]. TMR (Triple Modular redundancy) is a hardware redundancy technique and it uses two redundant blocks for a single sub-unit and output is taken via a majority gate. This makes a total of three cells for a single sub-unit; all three working simultaneously, which cause both area overhead and power wastage. The system functions normally up-to a single fault. After which the working cell along with the two redundant cells will be purposeless [5]. In both the methods, the faulty cell undergoes apoptosis on fault detection, which causes resource wastage. In the proposed approach, only limited number of spares are required and there is no need of any redundant hardware, which limits the area overhead. Also the fine-grain isolation and relocation of the faulty module of the proposed method helps to achieve high resource utilization and improved fault handling capacity. The second placement approach discussed in the paper can reduce the routing complexity with the static module to a great extent.

4 Conclusion and Future Scope

An efficient FPGA based self repairing technique of faulty module relocation and two approaches to determine the best location of the relocated cells according to user need is discussed and demonstrated in this paper. The method has the capability to recover from multiple fault events with minimum overheads. This strategy can be effectively implemented with minimum number of spares and does not require any extra hardware. Self healing capability will be adhered to the system with least overheads. As dynamic run-time partial reconfiguration is used for repair, the system can be repaired online and the time taken will be less compared to an entire FPGA-reconfiguration process. The power consumption is also less as the method will not consume unnecessary power like in case of modular redundancy techniques because repairing action will be stimulated only on the event of a fault. This approach can be flexibly used under different design requirements namely, design for maximum fault coverage and design for the least routing complexity. As for future scope, these techniques can be absorbed to develop a novel self repairing algorithm for reconfigurable systems implemented using FPGA.

References

1. Akoglu, A., Sreeramareddy, A., Josiah, J.G.: FPGA based distributed self healing architecture for reusable systems. Cluster Comput. **12**(3), 269–284 (2009)
2. Hagemeyer, J., Boris K., Markus, K., Mario, P.: Design of homogeneous communication infrastructures for partially reconfigurable FPGAs. In: Proceedings of the International Conference on Engineering of Reconfigurable Systems and Algorithms (ERSA'07) (2007)
3. Mange, D., Sanchez, E., Stauffer, A., Tempesti, G., M, P., Piguet, C.: Embryonics: a new methodology for designing field-programmable gate arrays with self-repair and self-replicating properties. IEEE Trans Very Large Scale Integr. (VLSI) Syst. **6**(3), 387–399 (1998)
4. Kim, S., Chu, H., Yang, I., Hong, S., Jung, S.H., Cho, K.-H.: A hierarchical self-repairing architecture for fast fault recovery of digital systems inspired from paralogous gene regulatory circuits. IEEE Trans Very Large Scale Integr. (VLSI) Syst. **20**(12), 2315–2328 (2012)
5. Lyons, R.E., Vanderkulk, W.: The use of triple-modular redundancy to improve computer reliability. IBM J. Res. Dev. **6**(2), 200–209 (1962)
6. Pratt, B., Caffrey, M., Graham, P., Morgan, K., Wirthlin, M.: Improving FPGA design robustness with partial TMR. In: 44th Annual Reliability Physics Symposium Proceedings, 2006. IEEE International, pp. 226–232. IEEE (2006)
7. Lala Parag, K., Kiran Kumar, B., Patrick Parkerson, J.: On self-healing digital system design. Microelectron. J. **37**(4), 353–362 (2006)
8. Pradeep, C., Radhakrishnan, R., Samuel, P.: Fault recovery algorithm using king spare allocation and shortest path shifting for reconfigurable systems. J. Theoret. Appl. Inf. Technol. **61**(2) (2014)
9. Ashraf, R., DeMara, R.F.: Scalable FPGA refurbishment using netlist-driven evolutionary algorithms. IEEE Trans. Comput. **62**(8), 1526–1541 (2013)
10. Al-Haddad, R., Oreifej, R., Ashraf, R.A., DeMara, R.F.: Sustainable modular adaptive redundancy technique emphasizing partial reconfiguration for reduced power consumption. Int. J. Reconfig. Comput. **2011** (2011)
11. Horta, E.L., Lockwood, J.W.: Automated method to generate bitstream intellectual property cores for Virtex FPGAs. In Field Programmable Logic and Application, pp. 975–979. Springer, Berlin, Heidelberg (2004)
12. Kalte, H., Lee, G., Porrmann, M., Rückert, U.: Replica: a bitstream manipulation filter for module relocation in partial reconfigurable systems. In: Parallel and Distributed Processing Symposium, 2005. Proceedings. 19th IEEE International, pp. 151b–151b. IEEE (2005)
13. Compton, K., Li, Z., Cooley, J., Knol, S., Hauck, S.: Configuration relocation and defragmentation for run-time reconfigurable computing. IEEE Trans. Very Large Scale Integr. (VLSI) Syst. **10**(3), 209–220 (2002)
14. Farooq, U., Marrakchi, Z., Mehrez, H.: FPGA architectures: an overview. In: Tree-based Heterogeneous FPGA Architectures, pp. 7–48. Springer, New York (2012)
15. Nath, Sasamal Trailokya, Prasad, Rajendra: Module based and difference based implementation of partial reconfiguration on FPGA: A review. International Journal of Engineering Research and Applications (IJERA) **1**(4), 1898–1903 (2011)
16. Joseph, S., Baskaran, K.: Performance analysis of various fragmentation techniques in runtime partially reconfigurable FPGA. Int. J. Comput. Appl. **94**(8) (2014)

Fetal Heart Rate Variability: Multiple Regression Models Using Autoregressive Analysis and Fast Fourier Transform

Manoj S. Sankhe and Kamalakar D. Desai

Abstract The system is designed to measure the fetal heart rate variability for the evaluation of autonomic nervous system (ANS) indices in the normal and abnormal fetus using Doppler ultrasound method. We have tested the hypothesis that a LF/HF ratio [Parametric autoregressive (AR) and nonparametric fast Fourier transform (FFT) Based] as an index of fetal sympathetic activity is a function of ten variables, age, gestation week, body mass index, CVRR %, HR Mean, HR Std, RMSSD, NN50, pNN 50 and non linear index SD1/SD2 ratio, a multiple regression analysis was performed. The overall model explained 46.47 and 36.12 % of the variation in LF/HF ratio as an index of fetal sympathetic activity using AR and FFT based methods respectively. Age, CVRR %, HR mean, HR Std, and RMSSD are significant predictors (or significantly related to) of LF/HF ratio as an index of fetal sympathetic activity using AR based method and age, CVRR %, HR mean, HR Std and RMSSD are significant predictors using FFT based method. The standardized beta tells us the strength and direction of the relationships (interpreted like correlation coefficients). CVRR % is positively related to LF/HF ratio as an index of fetal sympathetic activity using both the methods.

Keywords Autonomic nervous system · Doppler ultrasound · Heart rate variability · Multiple regressions

M.S. Sankhe (✉)
Department of Electronics & Telecommunication Engineering, SVKM's NMIMS University M.P.S.T.M.E., Bhakti Vedanta Swami Marg, J.V.P.D, Vile Parle (West), Maharashtra 400056, Mumbai, India
e-mail: manojsankhe1@gmail.com

K.D. Desai
VPM's Maharshi Parshuram College of Engineering, Mumbai University, Hedavi - Guhagar Road, Tal-Guhagar Dist. Ratnagiri, Velneshwar 415729, Maharashtra, India
e-mail: kddesai@hotmail.com

© Springer International Publishing Switzerland 2016
V. Snášel et al. (eds.), *Innovations in Bio-Inspired Computing and Applications*,
Advances in Intelligent Systems and Computing 424,
DOI 10.1007/978-3-319-28031-8_39

1 Introduction

Fetal heart starts pulsating at around 250 beats per minutes (BPM) at 12th weeks of gestation period. It decreases down to around 120–160 BPM at 36 weeks (9th month). The average heart rate and heart rate variation are related to development of the fetal nervous system and development of different body organs. The most common method for fetal monitoring is recording of fetal heart rate and analysis of fetal heart rate variability (fHRV). The fHRV analysis has a physiological significance as the changes in fetal heart rate (FHR) are responsible for fetal well-being. Congenital heart defects can be detected during gestation period if we measure the heart rate of the fetus during its growth. The defect may be so slight that the baby appears healthy for many years after birth, or so severe that its life is in immediate danger. Congenital heart defects originate in early stages of pregnancy when the heart is forming and they can affect any of the parts or functions of the heart [1–6].

Heart Rate Variability (HRV) was first used clinically in 1965 when Hon and Lee noted that fetal distress was accompanied by changes in beat-to-beat variation of the fetal heart rate, even before there was detectable change in heart rate. HRV refers to the beat-to-beat alterations in heart rate. Stress, certain cardiac diseases, and other pathologic states affect on HRV. Here we talk about HRV; we actually mean variability of RR intervals. HRV measurements analyze how these RR intervals, which show the variation between consecutive heartbeats, change over time.

Analyses based on the time and frequency domains of heart rate variability using Doppler ultrasound method enable an evaluation of fetal ANS diagnostic indices. These diagnostic indices derived from fetal heart rate data can be utilized to predict the fetal future life growth and can be utilized for preventive measure. Our design system not only measures heart rate variation but also heart rate power spectrum which can be utilized for determining diagnostics indices helpful for the medical community.

Cortez et al. proposes a FHRV analysis based on the evaluation of time domain parameters (statistic measures); frequency domain parameters; and the short and long term variability obtained from the Poincare plot. A normal distribution is presumed for each parameter and a normality criterion is proposed. Specific and overall classifications are proposed to help improve the fetal conditions interpretation, expanding the conventional FHR analysis [7].

A method of estimation of a fetus condition includes abdominal ECG registration, correlation processing of the received data, fetal R-R intervals allocation, estimation of distribution parameters and diagnostic index calculation, describing activity of sympathetic nervous system of fetus. This technique is used in real-time mode and serves as an approach to the problem of fetal stress diagnostics by means of maternal abdominal ECG processing [8].

An analysis based on heart rate variability in normal subjects of various age groups using the various time domain, frequency domain and nonlinear parameters show that, with aging the heart rate variability decreases [9].

Jezwski et al. compared Doppler ultrasound and direct electrocardiography acquisition techniques for quantification of fetal heart rate variability, and showed that evaluation of the acquisition technique influence on fetal well-being assessment cannot be accomplished basing on direct measurements of heartbeats only. The more relevant is the estimation of accuracy of the variability indices, since analysis of their changes can significantly increase predictability of fetal distress [10].

An estimation of fetal autonomic state by time-frequency analysis of fetal heart rate variability confirmed that there is a neural organization during the last trimester of the pregnancy, and the sympathovagal balance is reduced with the gestational age [11].

Time-domain and frequency domains analysis of heart rate variability using fetal magnecardiography enable an evaluation of fetal autonomic nervous system (ANS) activity. The result show that sympathetic nervous activity increased with gestational age in the normal pregnancy group [12].

A Heart rate variability non-invasive monitoring of autonomic nervous system function special measurements, based on time and frequency domain analysis was introduced [13]. The results show that, heart rate variability gives many parameters that are related to the functioning of two branches of autonomous nervous system: sympathetic and parasympathetic system.

The HRV indexes are obtained by analyzing the intervals between consecutive R waves, which can be captured by instruments such as electrocardiographer, digital-to-analog converter and the cardio-frequency meter, from external sensors placed at specific points of the body. The results show that, changes in the HRV patterns provide a sensible and advanced indicator of health involvements [14].

A group of experiments performed to investigate whether anxiety during pregnancy can be linked with the autonomic nervous system via different heart rate variability parameters, confirmed that the ANS modulation is slightly influenced by the anxiety level, but not as strongly as hypothesized before [15].

A novel technique for fetal heart rate estimation from Doppler ultrasound signal on a beat-to-beat basis offers a high accuracy of the heart interval measurement enabling reliable quantitative assessment of the FHR variability, at the same time reducing the number of invalid cardiac cycle measurement [16].

The cardiovascular indices in pregnant women are significantly altered in comparison to non-pregnant women, thus highlighting the importance of cardiovascular monitoring during pregnancy [17].

In this study we have tested the hypothesis that a LF/HF ratio as an index of fetal sympathetic activity is a function of ten variables, age, gestation week, body mass index, CVRR %, HR Mean, HR Std, RMSSD, NN50, pNN 50 and non linear index SD1/SD2 ratio, a multiple regression analysis was performed.

2 Method

Through the ongoing safe passage study at Brihan Mumbai Municipal Corporation (BMC) Hospital, Mumbai. The proposed system is tested using real time Doppler ultrasound fetal data acquisition system. Subjects enrolled in the present study were pregnancies (n = 41) at 26–39 weeks of gestation whose body mass index (BMI) ranging from 19.2 to 36.6 that visiting hospital as either outpatients or inpatients.

We got the special permission from BMC with approved patient protocol to get 200 female subjects for measurement of Doppler ultrasound fetal signal. Written informed consent was obtained from all subjects after being briefed about the clinical study, which was approved by the Ethics Committee of the Brihan Mumbai Municipal Corporation (BMC) Hospital. Doppler ultrasound signal is recorded from the abdominal transducer placed on mothers abdomen. The recording time was 5 min, although some fragments in which either Doppler ultrasound transducer lost the heart signal, was marked as signal loss and removed. The LF/HF ratio in a supine resting posture has been suggested for the evaluation of autonomic nervous system (ANS) activities. The methods of ultrasonography and cardiotocography, which are incapable of measuring CVRR, LF/IIF ratio, and various fetal heart-rate variability analyses, can be improved upon with Doppler ultrasound, thereby enabling these indices to be determined. In the present study, we evaluated the significance of heart rate variability as an actual autonomic nervous system development of normal and abnormal fetuses at 26–39 weeks of gestation using Doppler ultrasound method.

R-R interval indicates instantaneous heart rate which is (1/T*60) beats per minute. This can be derived from electrocardiogram (ECG) by measuring the time between two consecutive QRS complexes. Same results are expected by measuring the time between the two consecutive movements of the same part of the fetal heart. So by focussing the Doppler ultrasound signal on fixed part of the fetal heart, the waveforms generated are proportional to the velocity of the movements of that part which can be used for detection of the same event in the consecutive cardiac cycle. So detected waveform can be correlated to QRS complexes of the consecutive cardiac cycles. This can be validated by measuring time between two QRS complexes and at the same time measuring time between two detected events from Doppler ultrasound transducer. Timing diagram of direct electrocardiography and Doppler ultrasound method for HRV signal detection is shown in Fig. 1. If practically T1 = T2 for the duration of the test procedure then it can be assumed that the HRV signal produced either by direct electrocardiography or Doppler ultrasound can be similar. Hence analysis will be similar.

Fetal heart rate signals are recorded using Dipel make Doppler ultrasound (DFM-051) machine. Figure 2 shows a real time abdominal Doppler ultrasound recording setup in hospital and abdominal fetal ECG and Doppler electrodes placed on mother abdomen. The monitor is equipped with ultrasound transducer which continuously emits (with repetition frequency of 3 kHz) 2 MHz ultrasound wave of

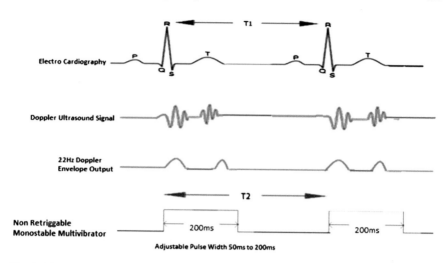

Fig. 1 Timing diagram of direct electrocardiography and Doppler ultrasound method for HRV signal detection

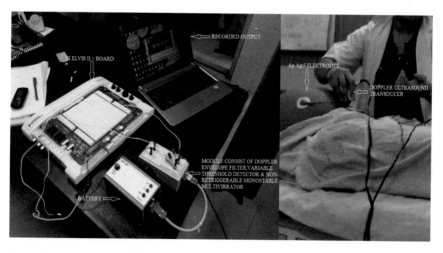

Fig. 2 Real time abdominal Doppler ultrasound recording setup in hospital

a very low power 1.5 mw/cm^2. The wave reflected from moving parts of fetal heart (walls or valves) returns to the transducer, which has receiving elements. Frequency shift between emitted and reflected waves is caused by the Doppler effect and provides information on the speed of moving object on which the ultrasound beam is focused.

Doppler ultrasound transducer is held on patient abdomen in the direction such that ultrasound waves emitted will pass the fetal heart movement. The reflected waves from moving fetal heart rate are received by receiving element in the transducer. This signal is fed to the RF amplifier (2 MHz) and FM demodulator to detect the movement of the fetal heart.

Demodulated detected waveform has definite events relating to contraction and relaxations of fetal heart. Each event is a combination of different frequency components relating to motion of fetal heart and angle of incidence of the ultrasound wave on it. This signal is then passed through envelope filter (Band Pass Filter 22 Hz) with centre frequency of 22 Hz which results in generating two simple peaks per cardiac events. This signal is then passed through a variable threshold detector where threshold is kept at half the peak value of incoming signal.

Two separately detected pulses then pass through a non retriggerable monostable multivibrator for avoiding double triggering of a single cardiac event. The adjustable pulse width for this monostable multivibrator is 50–200 ms giving fetal heart rate range up to 300 BPM. This output is given to National instruments ELVIS II+ board to personal computer USB port for HRV analysis. At the same time, Doppler signal related to heart movements and contained in the audio frequency range (from 0.2 to 1 kHz) is fed to the speaker, which helps in correct positioning of transducer on maternal abdomen. The maternal and per abdomen ECG is also monitored during the process for separate filtering and evaluation studies.

Measurement station has been based on a laptop PC with the ELVIS II+ (National Instruments) data acquisition board. This ELVIS II+ board has eight differential, sixteen single ended analog inputs and 16 bits resolution analog-to-digital (A/D) converter which can operate with the maximum sampling rate of 1.25 MS/s. Battery power supply and patient's electrical barrier ensures full standard safety for a patient, and minimizes power line interferences. All procedures for acquisition and processing of the signals have been developed in LABVIEW Version 10.0 (32 bit) environment (National Instruments). Figure 3 shows Conceptual diagram of the real time hardware for Doppler ultrasound signal analysis.

The R–R interval variability which shows the variation between consecutive heart beats, change over time was eventually adopted to calculate for time-domain, frequency domain and non linear analysis. Based on frequency analysis, the ranges of the LF and HF domains were defined as 0.04–0.15 and 0.15–0.4 Hz, respectively. The result of Doppler ultrasound monitoring which demonstrates the convenience and simplicity of performing HRV analysis by using Lab VIEW. As Doppler ultrasound signal is more spread out in time, making its timing more

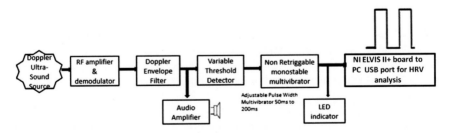

Fig. 3 Conceptual diagram of the real time hardware for Doppler ultrasound signal analysis

difficult to measure, and it begins before the ECG complex. We have converted this Doppler ultrasound fetal signal into output pulse for quantification of heart rate variability.

HRV analysis methods can be divided into time-domain, frequency-domain, and nonlinear methods. Denotations and definitions for HRV parameters in this work and in the developed software follow the guidelines given in [18–20].

The system is designed to measure the fetal heart rate variability for the evaluation of autonomic nervous system (ANS) indices. The system is used to differentiate the autonomic nervous system diagnostic indices of normal and abnormal fetus using Doppler ultrasound method. We have taken coefficient of variance (CVRR) of a patient's normal RR intervals as an index of parasympathetic activity which is defined as ratio of standard deviation of normal RR intervals value to mean of such intervals and a low frequency/high frequency (LF/HF) ratio as a sympathetic activity index. The relationships among CVRR, LF/HF, and gestational age in each group were analyzed by linear regression, while inter-group changes in CVRR, LF/HF over the gestational period in each group were verified by one-way ANOVA. The pregnancy group was divided into three groups for classifying one-way ANOVA analysis of CVRR and LF/HF as follows: Group A, 26–29 ± 1 wk (7th month pregnancy); Group B, 30–33 ± 1 wk (8th month pregnancy); and Group C, 34 onwards ± 1 wk (9th month pregnancy).

3 Statistical Analysis

The purpose of multiple regressions is to predict a single variable from one or more independent variables. Multiple regression is an extension of simple linear regression. It is used when we want to predict the value of a variable based on the value of two or more other variables. The variable we want to predict is called the dependent variable (or sometimes, the outcome, target or criterion variable). The variables we are using to predict the value of the dependent variable are called the independent variables (or sometimes, the predictor, explanatory or regressor variables) [21].

In single regression we have been concerned with predicting the value of a response on the basis of the value of a single input variable. However, in many situations the response is dependent on a multitude of input variables. Suppose that we are interested in predicting the response value Y on the basis of the values of the k input variables $x_1, x_2, \ldots, x_k$. The multiple linear regression model supposes that the response Y is related to the input values x_i, $i = 1, \ldots, k$, through the relationship

$$Y = \beta_0 + \beta_1 x_1 + \beta_2 x_2 + \cdots + \beta_k x_k + e \tag{1}$$

We have taken all time-domain, frequency domain and non linear parameters together to perform multiple regression analysis.

3.1 LF/HF Ratio [Parametric (AR) Based] as an Index of Sympathetic Activity

We could use multiple regression to understand whether LF/HF ratio [Parametric (AR) Based] as an index of fetal sympathetic activity can be predicted based on age, gestation week, body mass index, CVRR, HR Mean, HR Std, RMSSD, NN50, pNN 50) and nonlinear (SD1/SD2 index) parameters.

Multiple regression was conducted to examine whether LF/HF ratio [Parametric (AR) Based] as an index of fetal sympathetic activity is a function of ten variables, age, gestation week, body mass index, CVRR, HR Mean, HR Std, RMSSD, NN50, pNN 50, non linear index SD1/SD2 ratio.

There are three general tables that must be interpreted in the write-up of the regression analysis. The summary output table is shown in Table 1. The information that needs to taken from this table is the R-square (0.4647). The R-square is the proportion of variation in the dependent variable (LF/HF ratio as an index of fetal sympathetic activity) [Parametric (AR) Based] as an index of fetal sympathetic activity that is explained by the ten independent variables. It is expressed as a percentage. So 46.47 % of the variation in dependent variable LF/HF ratio as an index of fetal sympathetic activity can be explained by ten independent variables in the model.

The ANOVA table (Table 2) shows whether the proportion of variance explained in the first table is significant or not significant. It also tells whether the overall effect of the ten independent variables on dependent variable (LF/HF ratio as an index of fetal sympathetic activity) is significant or not significant. The significant (or p-value) is 0.02075856 which is below the 0.05 level; hence, we conclude that the overall model is statistically significant, or that the variables have significant combined effect on the dependent variable $F (10, 30) = 2.60503, p < 0.1$

Look at the sig. (p-values) first in Table 3. We can see that age (p-value 0.02500), CVRR % (p-value 0.00722), HR Std (p-value 0.036755), and RMSSD (p-value 0.023961) are significant predictors (or significantly related to) of LF/HF ratio [Parametric (AR) Based] as an index of fetal sympathetic activity.

Table 1 Multiple regression summary output table

Summary output	
Regression statistics	
Multiple R	0.681738142
R Square	0.464766895
Adjusted R Square	0.28635586
Standard error	0.176641621
Observation	41

Predictors (Constants), Age, Gestation Week, Body Mass Index, CVRR, HR Mean, HR Std, RMSSD, NN50, pNN 50, Non linear Index SD1/SD2 ratio

Table 2 Multiple Regression ANOVA table

ANOVA					
	df	SS	MS	F	Significance F
Regression	10	0.81288689	0.081283	2.605034	0.020758756
Residue	20	0.936067872	0.031202		
Total	40	1.774897561			

[a]*Predictors* (Constants), Age, Gestation Week, Body Mass Index, CVRR, HR Mean, HR Std, RMSSD, NN50, pNN 50, Non linear Index SD1/SD2 ratio
[b]*Dependant variable* LF/HF ratio [Nonparametric Fast Fourier Transform (FFT) Based] as an index of fetal sympathetic activity

The standardized beta tells us the strength and direction of the relationships (interpreted like correlation coefficients) CVRR (Beta = 0.082968, $p < 0.1$) is positively related to LF/HF ratio [Parametric (AR) Based] as an index of fetal sympathetic activity. Gestation week (Beta = -0.003506, $p > 0.1$), Body mass index (Beta = -0.00056, $p > 0.1$), HR Mean (Beta = -0.00360, $p > 0.1$), NN50 (Beta = -0.00015, $p > 0.1$), pNN50 (Beta = -0.00279, $p > 0.1$) and SD1/SD2 index (Beta = -0.228919, $p > 0.1$) is not a significant predictor of LF/HF ratio [Parametric (AR) Based] as an index of fetal sympathetic activity.

3.2 LF/HF Ratio [Nonparametric Fast Fourier Transform (FFT) Based] as an Index of Sympathetic Activity

Multiple regression was conducted to examine whether LF/HF ratio [Nonparametric Fast Fourier Transform (FFT) Based] as an index of fetal sympathetic activity is a function of ten variables, age, gestation week, body mass index, CVRR, HR Mean, HR Std, RMSSD, NN50, pNN 50 and non linear index SD1/SD2 ratio.

There are three general tables that must be interpreted in the write-up of the regression analysis. The summary output table is shown in Table 4. The information that needs to taken from this table is the R-square (0.361287). The R-square is the proportion of variation in the dependent variable (LF/HF ratio as an index of fetal sympathetic activity) that is explained by the ten independent variables. It is expressed as a percentage. So 36.12 % of the variation in LF/HF ratio as an index of fetal sympathetic activity can be explained by ten independent variables in the model.

The ANOVA table (Table 5) shows whether the proportion of variance explained in the first table is significant. It also tells whether the overall effect of the ten independent variables on dependent variable (LF/HF ratio as an index of fetal sympathetic activity) is significant or not significant. The p-value is 0.127661516

Table 3 Multiple regression coefficients

	Coefficients	Standard error	t Stat	p-Value	Lower 95 %	Upper 95 %	Lower 95.0 %	Upper 95.0 %
Intercept (LF/HF parametric, AR models)	2.018021029	0.902547723	2.235916	0.032947	0.17477268	3.8612694	0.174477268	3.86126938
X variable 1 (age)	−0.019553897	0.008287578	−2.35942	0.025008	−0.036479389	−0.0026284	−0.03647939	−0.0026284
X variable 2 (gestation week + 1 week)	−0.003506203	0.008249585	−0.42502	0.673858	−0.020354103	0.0133417	−0.0203541	0.0133417
X variable 3 (body mass index)	0.000563387	0.007529613	0.074823	0.940852	−0.014814133	0.0159409	−0.01481413	0.01594091
X variable 4 (CVRR (%) = RR Std/RR mean)	0.082968899	0.028782846	2.882581	0.007225	0.024186485	0.1417513	0.02418649	0.14175131
X variable 5 (HR mean)	−0.003607079	0.004614993	−0.77572	0.443989	−0.013103631	0.0058895	-0.01310363	0.00588947
X variable 6 (HR Std)	−0.023254621	0.01069807	−2.18598	0.036755	−0.044980459	−0.0058895	−0.04489046	−0.0015288
X variable 7 (RMSSD)	−0.009040398	0.003801223	−2.37829	0.023961	−0.016803532	−0.0012773	−0.01680353	−0.0012773

(continued)

Table 3 (continued)

	Coefficients	Standard error	t Stat	p-Value	Lower 95 %	Upper 95 %	Lower 95.0 %	Upper 95.0 %
X variable 8 (NN50)	−0.0001494	0.000848472	−0.18261	0.856332	−0.00188775	0.0015779	−0.00188775	0.00157787
X variable 9 (pNN50)	−0.002794079	0.006176262	−0.45239	0.654243	−0.015407689	0.0098195	−0.01540769	0.00981953
X variable 10 (SD1/SD2 index)	−0.228919728	0.408595175	−0.56026	0.579462	−1.063382396	0.6055429	−1.0633824	0.60554294

The effect of individual independents variable on dependent variable

Dependent variable LF/HF ratio [Parametric (AR) Based] as an index of fetal sympathetic activity

Table 4 Multiple regression summary output table

Summary output	
Regression statistics	
Multiple R	0.6010716
R Square	0.361287
Adjusted R Square	0.1483127
Standard error	0.2601175
Observations	41

Predictors (Constants), Age, Gestation Week, Body Mass Index, CVRR, HR Mean, HR Std, RMSSD, NN50, pNN 50, Non linear Index SD1/SD2 ratio

Table 5 Multiple Regression ANOVA table

ANOVA	df	ss	MS	F	Significance F
Regression	10	1.148171878	01148172	1.696945332	0.127661516
Residue	20	2.029833	00676611		
Total	40	3.178004878			

Predictors (Constants), Age, Gestation Week, Body Mass Index, CVRR, HR Mean, HR Std, RMSSD, NN50, pNN 50, Non linear Index SD1/SD2 ratio

which is not below the 0.1 level; hence, we conclude that the overall model is statistically not significant, or that the variables have a not significant combined effect on the dependent variable F $(10, 30) = 1.69694$, $p > 0.1$

Look at the sig. (p-values) first in Table 6. We can see that age (p-value 0.11350), CVRR % (p-value 0.00460), HR mean (p-value 0.007017), HR Std (p-value 0.08827) and RMSSD (p-value 0.00782) are significant predictors (or significantly related to) of LF/HF ratio [Nonparametric Fast Fourier Transform (FFT) Based] as an index of fetal sympathetic activity.

Gestation week (Beta $= -0.0033438$, $p > 0.1$), Body mass index (Beta $= -0.00526$, $p > 0.1$), NN50 (Beta $= -0.00063$, $p > 0.1$), pNN50 (Beta $= 0.00474$, $p > 0.1$) and SD1/SD2 index (Beta $= 0.08643$, $p > 0.1$) is not a significant predictor of LF/HF ratio [Nonparametric Fast Fourier Transform (FFT) Based] as an index of fetal sympathetic activity.

The standardised beta tells us the strength and direction of the relationships (interpreted like correlation coefficients). CVRR (Beta $= 0.1297954$, $p < 0.1$) is positively related to LF/HF ratio [Nonparametric Fast Fourier Transform (FFT) Based] as an index of fetal sympathetic activity.

Table 6 Multiple regression coefficients

	Coefficients	Standard error	t Stat	p-Value	Lower 95 %	Upper 95 %	Lower 95.0.%	Upper 95.0.%
Intercept (LF/HF parametric, FFT models)	3.1230098	1.320966341	2.3497772	0.25558872	−0.408694223	5.837325	0.408694227	5.83732537
X variable 1 (age)	−0.0198941	0.012204054	−1.630124	0.113531766	−0.04481812	0.00503	−0.04481812	0.00502989
X variable 2 (gestation week ± 1 week)	−0.0033438	0.012148107	−0.275254	0.785008433	−0.02815355	0.021466	−0.02815355	0.02146593
X variable 3 (body mass index)	−0.0052691	0.011087895	−0.475209	0.638079713	−0.02791357	0.017375	−0.02791357	0.01737543
X variable 4 (CVRR (%) = RR Std/RR mean)	0.12979537	0.042384808	3.0623089	0.004605144	0.043234049	0.216357	0.043234049	0.2163567
X variable 5 (HR mean)	−0.0128574	0.006847449	−1.877687	0.070176766	−0.02684172	0.001127	−0.02684172	0.00112699
X variable 6 (HR Std)	−0.0276006	0.015665323	−1.76189	0.08827659	−0.05959344	0.004392	−0.05959344	0.00439228
X variable 7 (RMSSD)	−0.0159561	0.005597574	−2.850531	0.007820346	−0.02738783	−0.004524	−0.02738783	−0.0045243

(continued)

Table 6 (continued)

	Coefficients	Standard error	t Stat	p-Value	Lower 95 %	Upper 95 %	Lower 95.0.%	Upper 95.0. %
X variable 8 (NN50)	−0.0006355	0.001249435	−0.508591	0.614759117	−0.00318714	0.001916	−0.00318714	0.00191624
X variable 9 (pNN50)	0.00474963	0.00909499	0.5222246	0.605349078	−0.01382482	0.023324	−0.01382482	0.02332407
X variable 10 (SD1/SD2 index)	0.08643047	0.601685739	0.1436472	0.886739653	−1.14237574	1.315237	−1.14237574	1.31523667

The effect of individual independents variable on dependent variable Multiple Regression coefficients. The effect of individual independents variable on dependent variable

Dependent variable LF/HF ratio [Nonparametric Fast Fourier Transform (FFT) Based] as an index of fetal sympathetic activity

4 Conclusion

Analyses based on the time and frequency domains of heart rate variability using Doppler ultrasound method enable an evaluation of fetal ANS diagnostic indices. We have checked the fetal sympathetic activity by two methods. The result of the methods is comparable. The overall model explained 46.47 and 36.12 % of the variation in LF/HF ratio as an index of fetal sympathetic activity using AR and FFT based methods respectively. We conclude that age, CVRR %, HR Std, and RMSSD are significant predictors (or significantly related to) of LF/HF ratio as an index of fetal sympathetic activity in both the methods. CVRR % is positively related to LF/HF ratio as an index of fetal sympathetic activity in both the methods.

References

1. Desai, K.D., Sankhe, M.S.: A Real-time fetal ECG feature extraction using multiscale discrete wavelet transform, In: 5th International Conference Biomedical Engineering and Informatics, BMEI, Chongqing, China (2012) (Appeared in IEEE Explorer)
2. Desai, K.D., Jadhav, S.D., Sankhe, M.S.: A comparison and quantification of fetal heart rate variability using Doppler ultrasound and direct electrocardiography techniques. In: International Conference on Advances in Technology, ICATE, Mumbai, India (2013) (Appeared in IEEE Explorer)
3. Desai, K.D., Sankhe, M.S.: Heart rate variability power spectrum as health signature, techno-path. J. Sci. Technol. Manage. 3(2), 46–53 (2011)
4. Desai, K.D.: A system for the detection of diabetic myocardial infarction. Ph. D. thesis V.J.T. I., Bombay (1996)
5. Sankhe, M.S., Desai, K.D., Gadam, M.A.: Assessment of fetal autonomic nervous system activity by abdominal Doppler ultrasound recordings. In: Health Tech Innovations—2015 Conference, organised by Society Applied Microwave Electronics Engineering & Research (SAMEER), IIT Bombay under the aegis of Department of Electronics and Information Technology, Government of India in technical collaboration with National Health Systems Resource Center (NHSRC) & Indian Council of Medical Research (ICMR), awarded as Best Concept note under the theme Technology Innovations in Diagnostic/Prognostic
6. Desai, K.D., Sankhe, M.S.: Correlations of Fetal Cardiac Sympathetic Activity with Maternal Body Mass Index. Indicon, IIT, Powai, Mumbai (2013)
7. Cortez, P.C., Madeiro, J.P.V., Schlindwein, F.S.: Classification system for fetal heart rate variability measures based on cardiotocographies. J. Life Sci. Technol. 184–189 (2013)
8. Kalakutskij, V., Konyukhov, V., Manelis, E.: Estimation of fetal heart rate using abdominal ECG recordings. In: IFMBE Proscedddings, 2nd European Medical and Biological Engineering Conference, Vienna (2002)
9. Acharya, U.R., Kannathal, N., Sing, O.W., Ping, L.Y., Chua, T.: Heart rate analysis in normal subjects of various age groups. Biomed. Eng. Online 24, 1–8 (2004)
10. Jezwski, J., Wrobel, J., Horoba, K.: Comparision of Doppler ultrasound and direct electrocardiography acquisition techniques for quntification of heart rate variability. IEEE Trans. Biomed. Eng. 53(5), 855–863 (2006)
11. David, M., Hirsch, M., Karin, J., Toledo, E., Akselrod, S.: An estimate of fetal autonomic state by time-frequency analysis of fetal heart rate variability. J. Appl. Physiol. 102, 1057–1064 (2007)

12. Fukushima, A., Nakai, K., Itoh, M., Horigome, H., Suwabe, A., Tohyma, K., Kobayashi, K., Yoshizawa, M.: Assessment of fetal autonomic nervous system activity by fetal magnetocardiograph. Clin. Med.: Cardiol. **2**, 33–39 (2008)
13. Omerbegovic, M.: Heart rate variability—noninvasive monitoring of autonomic nervous system function. Professional Paper, pp. 53–58 (2009)
14. Vanderlei, L.C.M., Pastre, C.M., Hoshi, R.A., de Carvalho, T.D., de Godoy, M.F.: Basic notions of heart rate variability and its clinical applicability. Rev Bras Cir Cardiovasc. **24**(2), 205–217 (2009)
15. Taelman, J., Vandeput, S., Widjaja, D., Braeken, M.A.K.A., Otte, R.A., Van den Bergh, B.R. H., Van Huffel, S.: Stress during pregnancy: Is the autonomic nervous system influenced by anxiety? Comput. Cardiol. **37**, 725–728 (2010)
16. Jezewski, J., Roj, D., Wrobel, J., Horoba, K.: A novel technique for fetal heart rate estimation from Doppler ultrasound signal. Biomed. Eng. Online **10**(92), 1–17 (2011)
17. Panja, S., Bhowmick, K., Annamalai, N., Gudi, S.: A study of cardiovascular autonomic function in normal pregnancy. Al Ameen J Med. Sci. **6**(2), 170–175 (2013)
18. Task Force of The European Society of Cardiology, The North American Society of Pacing and Electrophysiology: Heart rate variability standards of measurement, physiological interpretation, and clinical use. Eur. Heart J. **17**, 354–381 (1996)
19. Acharya, U.R., Paul Joseph, K., Kannathal, N., Lim, C.M., Suri, J.S.: Heart rate variability: a review. Med. Bio. Eng. Comput. **44**, 1031–1051 (2006)
20. Niskanen, J.-P., Tarvainen, M.P., Ranta-aho, P.O., Karjalainen, P.A.: Software for advanced HRV analysis. Comput. Methods Programs Biomed. (2002)
21. Ross, S.M.: Introductory Statistics, 2nd edn. Academic Press, New York (2006)

Restricted Boltzmann Machine Based Energy Efficient Cognitive Network

P. MohanaPriya, S. Mercy Shalinie and Tulika Pandey

Abstract Current network technology is statically configured and it is difficult to self-adjust changes on demand. Existing protocols react for situations but it cannot take intelligent decisions. Emerging cognitive network plays a key role in networking environment because of its unique features namely reasoning and decision making. Energy efficiency is highly desirable for effective data communication network. In this paper, energy aware routing protocols and trust based metrics for improving energy efficiency is addressed. The proposed method uses Restricted Boltzmann Machine to stabilize the energy level of the network during routing. RBM based routing is comparatively better than conventional Boltzmann Machine based routing in terms of self-learning the trust metrics. The performance graph shows that the proposed RBM based routing has achieved lesser energy level consumption and higher trust values which ensures effective cognitive approach.

Keywords Cognitive network · Energy efficiency · Trust metrics · Restricted Boltzmann Machine · Self-learning

1 Introduction

Conventional networks has security challenges in different aspects namely energy consumption, link failure, trust management among participating nodes, bandwidth consumption and security attacks. To overcome these limitations, cognitive function-

P. MohanaPriya (✉) · S.M. Shalinie (✉)
Department of Computer Science and Engineering, Thiagarajar College of Engineering,
Madurai 625 015, Tamilnadu, India
e-mail: mohanapriyatce@gmail.com; pmptce@gmail.com

S.M. Shalinie
e-mail: shalinie@tce.edu

T. Pandey (✉)
Department of Electronics and Information Technology (DeitY),
Ministry of Communications & IT, Government of India, New Delhi, India
e-mail: 2likapandey@gmail.com

© Springer International Publishing Switzerland 2016
V. Snášel et al. (eds.), *Innovations in Bio-Inspired Computing and Applications*,
Advances in Intelligent Systems and Computing 424,
DOI 10.1007/978-3-319-28031-8_40

ality is added into the network which brings the intelligence into the network. Cognitive Network (CN) is the data communication network which consists of intelligent devices that make them aware of everything happening inside the device and the network they are connected to. This network learns from every situation it encounters and uses that information for future reference and is known as proactive network. Network becomes cognitive when all the statically configured components are replaced with self-aware and self-adjustable components.

Ryan et al. [1] discusses about a new form of adaptive network technology, the CN and its Software Adaptable Network (SAN). Wang et al. [2] proposed a novel Cognitive Network Framework (CNF) which provides the detailed description of the components belonging to the cognitive network.

Li et al. [3] gives a new definition of a cognitive network and provide the clear line between cognitive and other types of networks. Mihailovic et al. [4] discuss the aspects for developing cognitive functionalities in self-managed networks according to the vision of the Future Internet networks being developed in the Self-NET project. Niandong Du et al. proposed the cognitive network model [5] with an integration of knowledge base which is a primary part of cognitive networks to optimize the network performance.

The statically configured conventional network structure motivates the researchers community to establish the intelligent cognitive network which can self-adjust to dynamic changes in the network behavior. This research paper provides the energy-efficient cognitive network with the help of Restricted Boltzmann Machine (RBM) for the first time. It also achieves an effective routing strategy when compared to the existing routing methods.

The paper is organized as follows. In Sect. 2, energy aware routing protocols are summarized. In Sect. 3, trust metrics to improve energy efficiency are addressed. In Sect. 4, an overview of the stochastic learning method, DSR protocol and the proposed RBM based DSR method is discussed. Results and its discussions are summarized in Sect. 5. Finally, Sect. 6 is summarized with conclusion and the directions of future work.

2 Energy Efficiency in Cognitive Network

An infrastructure-less wireless network is highly dependent on battery power, which becomes useless when the battery power gets depleted. Umang et al. proposed Enhanced Intrusion Detection Systems (EIDS) for detecting malicious nodes and minimize energy consumption of a node in MANET. In [6, 7], the authors have evaluated the performance of Destination Sequenced Distance Vector (DSDV), Dynamic Source Routing (DSR) and Ad hoc On-Demand Distance Vector (AODV) routing protocols with respect to energy consumption of a node. Many energy-efficient techniques are proposed and they are classified into two major categories such as minimum energy routing protocol and maximum network lifetime [8] routing protocols.

Minimum energy routing protocols [9, 10] finds the energy-efficient path in the network and Maximum network lifetime protocol balance the battery power of each node in the case for searching for a path.

In [11], some of the energy efficient techniques are proposed to reduce the energy consumption in MANET and they are power control technique and topology control technique. In power control technique, next hop node is chosen based on the power level of the node which conserves the energy level of a node. Maleki et al. [12] have proposed a Power-aware Source Routing (PSR) protocol for MANET to increase the lifetime of the network and it is also responsible for preventing node mobility and energy depletion of node. Sahoo et al. [13] proposed a distribution transmitted power control protocol which is used to construct the power saving hierarchy topologies and it maintains the network on account of changing the transmission power.

Doshi et al. [14] proposed Minimum Energy Routing (MER) which minimizes the total transmission energy of the network by avoiding low energy nodes. It requires the perspective about the cost of a network link in terms of energy, information about minimum energy routes. Conditional Max-Min Battery Capacity (CMMBCR) by Toh [15] prefers the route with minimal transmission power to effectively reduce the energy consumption of the node. In [16], Sleep/Power mode protocol is discussed which reduces energy consumption of a node during an idle state. Song et al. [17] have presented the minimal achievable broadcast energy consumption scheme to save energy in the network. In [18], the authors have found the packets responsible for increasing energy consumption with the use of various routing protocols on different traffic models. Dimokas et al. [19] proposed an energy-efficient distributed clustering model for improving the energy conservation in MANET.

3 Trust Based Metrics to Improve Energy Efficiency

Energy Aware Deterioration Routing (EADR) scheme is discussed in [20], which deals with efficient utilization of energy resources and helps addressing some of the issues of energy consumption of a node. Zhu and Wang [21] proposed a PEER protocol to reduce the energy consumption in the case of mobility. Node Alarming Mechanism (NOAL) by [22], in which the node having lower energy would alarm its status to the other nodes in the network. In [20], the proposed system has four modules to detect the malicious node by using trust value parameter. In [23], hexagonal clustering based mechanism is used for energy consumption in Wireless Sensor Network (WSN). TRACE [24] method is based on centralized competence technique and trust based energy efficiency. The sink (or) Base Station (BS) in WSN is more intelligent which involves the maintenance of trust and reliability. The framework [25] provides energy efficiency in terms of number of hops, transmission and transmission delay. Almasri et al. [26] proposed a Trusted and Energy Efficient Routing Protocol (TERP) to maximize the network lifetime. It also reduces the power consumption by using the trust factor. The higher the trust factor, lesser is the energy consumption as it involves lesser number of encryptions.

4   Proposed Methodology

The proposed methodology focuses on designing the stochastic neural network method to self-learn the characteristics of dynamic network infrastructure. This method is a protocol-dependent method which concentrates on the energy efficiency of the network through the trust metrics.

4.1   Design of Stochastic Neural Network Method for Learning the Trust Metrics

The proposed research work is divided into three stages which stabilize the energy level of a node using machine learning algorithm along with QoS trust parameters. In the first stage, the common categories of trust and their metrics are studied which includes social and QoS trust in order to find the trust value of a node in the network infrastructure. The trust metrics used to calculate the social trust properties include the frequency of communication of a node, malign or benign behavior of a node and quality of reputation about a node. Some of the trust metrics of QoS trust of a node includes competence, reliability and dependability. In the second stage, various learning methods like reinforcement, stochastic and unsupervised methods are studied. In the third stage, an intelligent stochastic learning method named Restricted Boltzmann Machine (RBM)is proposed and it is applied to the scenario of Dynamic Source Routing (DSR) protocol to fine tune the energy parameter in order to provide the better energy efficient trust based cognitive (or) intelligent network to its intended users.

According to the standards of RFC 3561, 3684 and 3626 the self-configuring nature of MANET is supported by three routing protocols namely Ad-hoc On Demand Distance Vector (AODV), Open Link State Routing (OLSR) and Dynamic Source Routing (DSR). In this paper, DSR is taken into account which is energy less protocol by its birth.

4.2   DSR Routing Protocol

DSR is an on-demand reactive routing protocol that adapts to quickly routing changes in the network environment [27]. It is a simple and effective routing protocol which is composed of route discovery and route maintenance. It provides a loop-free routing platform as the information of the node is maintained in the route cache for the future use. Some of the advantages of DSR includes reduced overhead of route maintenance, route caching mechanism reduces the route discovery process. The disadvantage of this protocol are,

(1) The size of the packet header grows with route length due to source routing,
(2) RREQ packets may potentially flood all over the network,
(3) No energy saving as every node needs to turn its receiver and
(4) Route reply messages using a stale cache by an intermediate node pollutes other nodes cache will result in mess up routing and forwarding.

4.3 Restricted Boltzmann Machine Based Routing

Restricted Boltzmann Machine (RBM) based routing in DSR protocol assigns a global energy state in which the initial configuration of all the nodes has to be done in such a way that they are able to minimize the global energy. At some time interval, some of the nodes clamp their states to accept the incoming information such as Route Request (RREQ) message from the node and it maintains the route information of the node in its route cache. RBM gathers the input vectors rather than the training samples from the solution space and hence it belongs to unsupervised learning method. RBM is a stochastic neural network learning method, with the restriction that their neurons must form a bipartite graph. It does not support intra layer connection with visible and hidden units. This characteristic leads to efficient learning when compared to general Boltzmann Machine (BM). The structure of RBM consists of binary valued "m" visible units $V = (v_1, \ldots, v_m)$ to represent observable data and "n" hidden units $H = (h_1, \ldots, h_n)$ to capture dependencies between observable data. It consists of a matrix of weights $W = (w_{m,n})$ associated with the connection between hidden unit "h_n" and visible unit "v_m" and the bias units as "a_m" for the visible units and "b_n" for the hidden units. Here, the energy for the configuration (m, n) is defined as,

$$E(V,H) = \sum_{i=1}^{m} \sum_{j=1}^{n} w_{m,n} h_n v_m - \sum_{i=1}^{m} a_i v_m - \sum_{i=1}^{n} b_j h_n \qquad (1)$$

where,

$E(V,H)$ Total energy of the RBM configuration
$w_{m,n}$ Strength of the connection between two nodes m and n in the network
a_i, b_j bias units of visible and hidden layers
v_m visible unit
h_n hidden unit of the network.

The probability distributions over hidden and visible vectors are defined in terms of the following energy function,

$$P(V,H) = \frac{1}{Z} e^- E(v,h) \qquad (2)$$

where,

$P(V,H)$ Probability distribution over hidden and visible vectors.

The probability of visible vectors of booleans is the sum of overall possible hidden layer configurations as follows,

$$P(V) = \frac{1}{Z} \sum_h e^- E(v, h) \qquad (3)$$

where,

$P(V)$ Probability distribution over visible vectors.

The bias units such as a_i and b_j are the minimum threshold values of visible nodes and hidden nodes. In DSR, the route discovered from the source to the destination attached in the packet header is stored in the route cache. Based on this information, RBM detects the selfish nodes, compromised nodes by their energy consumption in the global state energy of the hidden visible network. It also uses simulated annealing procedure to reach thermal equilibrium state at temperature ?t.

The traces of selfish nodes can be detected using,

$$SN = \frac{1}{Z} e^- E(v, h)/t \, [REM_{energy}] \qquad (4)$$

where,

SN Selfish Node.
REM_{energy} Remaining energy.

The traces of compromised nodes can be detected based on the reception of node information in the network and it is given by,

$$CN = \frac{1}{Z} e^- E(v, h)/t \, [RECV_{energy}] \qquad (5)$$

where,

CN Compromised node.
$RECV_{energy}$ Receiving energy.

RBM is applied in the routing from source to destination in order to stabilize the overall energy of the network and to identify the selfish and compromised nodes present in the network scenario. It self-learns the network based on the patterns generated in the network and finds the nodes with higher energy consumption. It moves in an independent manner in which it learns the internal mechanism of the network which has the ability to modify the connection strengths of the nodes. The connection link of node with higher energy depletion gets break and it is splited into the hidden layer nodes. The network becomes self-adaptable which provides a cognitive network environment. Three types of nodes such as legitimate, selfish and malicious nodes can be detected based on energy and trust threshold parameters. The nodes whose energy consumption value is below the energy threshold value is known as selfish nodes and nodes whose energy consumption value is above the energy threshold value is known as malicious nodes. Nodes whose energy consumption value is equal in all the stages are detected as legitimate nodes. Energy level parameter is indirectly proportional to the trust threshold parameter. The computed trust value seems to be average for selfish nodes, lesser for malicious nodes and greater for the legitimate nodes.

5   Results and Discussions

The network structure is simulated with 32 number of mobile nodes using Network Simulator 2 (NS2). These nodes are also called as control nodes as it transfers the status of neighboring nodes to all the participating nodes in the network scenario. These nodes are configured with DSR protocol and Distributed Denial of Service (DDoS) attack has been carried out for 10 min. The trace file produced at the end of simulation is kept for the deep analysis by RBM.

RBM is implemented using python and tracefile is fed as an input to detect the node with higher energy level consumption. In addition, RBM based python script has two input matrices for energy level and trust level threshold values. Based on this information, RBM detects three kinds of nodes namely legitimate, selfish and malicious characteristics.

Table 1 infers that the node information before the generation of DDoS attack. The nodes participating in the scenario are node 12, node 14, node 16, node 17, node 18, node 29 and node 30. The data routed from node 14 to node 30 consumes lesser energy as 26 J. Table 2 infers the node information after the generation of DDoS attack. Here, the data routed from node 14 to node 30 consumes higher energy as 130 J. The energy level consumption depends on the trust level of all the participating nodes.

Table 1   Node information before attack

Nodes	Member nodes	Routing from	Routing to	Energy consumption (J)
1	0,2,3,4,6,9,10,25	10	25	50
11	12,14,16,17,18,29,30	14	30	26
22	24,25,26,27,28,31,32	32	16	30

Table 2   Node information after attack

Nodes	Member nodes	Routing from	Routing to	Energy consumption (J)
1	0,2,3,4,6,9,10,25	10	25	100
11	12,14,16,17,18,29,30	14	30	130
22	24,25,26,27,28,31,32	32	16	110

Fig. 1 Energy level
performance comparison

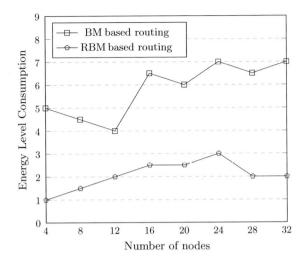

Fig. 2 Trust level
performance comparison

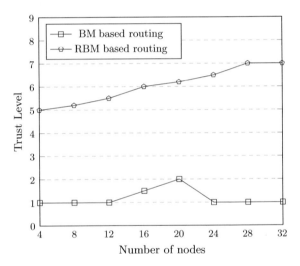

Figure 1 shows that RBM based routing consumes energy level below the defined average energy threshold value 5. This proves that RBM based routing is comparatively better than the BM based routing method. RBM, by its cognitive nature detects the malicious node 29 and disconnects from the network. Figure 2 shows that RBM based routing allows the accurate trusted nodes into the network. This proves that RBM based routing is highly trusted energy-efficient method when compared to BM based routing method.

6 Conclusion

In this paper, the RBM based routing is analyzed which provides a cognitive network environment in terms of energy level and trust value of a node. It stabilizes the energy level of the network during the occurrence of malicious traffic inside the network. It effectively identifies the selfish nodes, legitimate nodes and malicious nodes by its self-learning capability. From the performance graph, it is proved that the proposed RBM based routing technique is comparatively better than the conventional method. The future work is to implement in real time as a protocol independent method for energy efficiency parameter along with context-aware trust based security methods.

References

1. Thomas, R.W., Friend, D.H., DaSilva, L.A., MacKenzie, A.B.: Cognitive networks, pp. 17–41. Springer, Netherlands (2007)
2. Wang, Z., Wang, H., Feng, G., Li, B., Chen, X.: Cognitive networks and its layered cognitive architecture. In: 5th IEEE International Conference on Internet Computing for Science and Engineering, pp. 145–148 (2010)
3. Li, Q., Quax, P., Luyten, K., Lamotte, W.: A cognitive network for intelligent environments. In: 6th IEEE International Conference on Innovative Mobile and Internet Services in Ubiquitous Computing, pp. 317–322 (2012)
4. Mihailovic, A., Nguengang, G., Borgel, J., Alonistioti, N.: Building knowledge lifecycle and situation awareness in self-managed cognitive future internet networks. In: 1st IEEE International Conference on Emerging Network Intelligence, pp. 3–8 (2009)
5. Du, N., Bai, Y., Luo, L., Wu, W., Guo, J.: Building the knowledge base through Bayesian network for cognitive wireless networks. In: 17th IEEE International Conference on Parallel and Distributed Systems, pp. 412–419 (2011)
6. Kafhali, S.E., Haqiq, A.: Effect of mobility and traffic models on the energy consumption in MANET routing protocols. Int. J. Soft Comput. Eng. 2231–2307 (2013)
7. Yu, C., Lee, B., Youn, H.Y.: Energy efficient routing protocols for mobile ad hoc networks. Wirel. Commun. Mobile Comput. 8, 959–997 (2003)
8. Misra, A., Banerjee, S.: MRPC: Maximizing network lifetime for reliable routing in wireless environments. In: IEEE International Conference on Wireless Communications and Networking Conference, vol. 2, pp. 800–806 (2002)
9. Hussain, M.A., Ravi Sankar, M., Vijaya Kumar, V., Srinivasa Rao, Y., Nalla, L.: Energy conservation techniques in Ad hoc networks. Int. J. Comput. Sci. Inf. Technol. 2(3), 1182–1186 (2011)
10. Bonatti, P., Duma, C., Olmedilla, D., Shahmehri, N.: An integration of reputation-based and policy-based trust management. Networks 2(14) (2007)
11. Gupta, H.P., Rao, S.V.: DBET: Demand Based Energy Efficient Topology for MANETs. In: International Conference on Devices and Communications, pp. 1–5 (2011)
12. Maleki, M., Dantu, K., Pedram, M.: Power-aware source routing protocol for mobile Ad Hoc networks. In: International Symposium on Low Power Electronics and Design, pp. 72–75 (2002)
13. Sahoo, P.K., Sheu, J.P., Hsieh, K.Y.: Power control based topology construction for the distributed wireless sensor networks. Comput. Commun. 30, 2774–2785 (2007)
14. Doshi, S., Bhandare, S., Brown, T.X.: An on-demand minimum energy routing protocol for a wireless ad hoc network. ACM SIGMOBILE Mobile Comput. Commun. Rev. 6(3), 50–66 (2002)

15. Lee, E., Kim, M., Yu, C., Kim, M.: NOAL: Node Alarming Mechanism For Energy Balancing in Mobile Ad hoc Networks (2002)
16. Ray, N.K., Turuk, A.K.: Energy efficient techniques for wireless Ad Hoc network. In: International Joint Conference on Information and Communication Technology, pp. 105–111 (2010)
17. Wang, Y., Song, W., Wang, W., Li, X.-Y., Dahlberg, T.A.: LEARN: Localized Energy Aware Restricted Neighbourhood routing for ad-hoc networks. In: 3rd Annual IEEE Communications Society Conference on Sensor, Mesh and Ad-hoc Communications, vol. 2, pp. 502–517 (2006)
18. Zhu, J., Qiao, C., Wang, X.: A comprehensive minimum energy routing protocol for wireless adhoc networks. IEEE INFOCOM (2004)
19. Dimokas, N., Katsaros, D., Manolopoulos, Y.: Energy-efficient distributed clustering in wireless sensor networks. J. Parallel Distrib. Comput. 70(4), 371–383 (2010)
20. Patel, D., Patel, Y.: Intrusion detection systems for trust based routing in Ad-Hoc networks. Int. J. Comput. Sci. Inf. Technol. Secur. 2(6), 1160–1165 (2012)
21. Zhu, J., Wang, X.: Model and protocol for energy-efficient routing over mobile ad hoc networks. IEEE Trans. Mobile Comput. 10(11), 1546–1557 (2011)
22. Toh, C.K., Cobb, H., Scott, D.: Performance evaluation of battery-life-aware routing schemes for wireless Ad hoc networks. In: IEEE International Conference on Communication (2001)
23. Wang, D., Xu, L., Peng, J., Robila, S.: Subdividing hexagon-clustered wireless sensor networks for power-efficiency. In: IEEE International Conference on Communications and Mobile Computing, vol. 2, pp. 454–458 (2009)
24. Tajeddine, A., Kayssi, A., Chehab, A.: TRACE: A centralized Trust And Competence-based Energy-efficient routing scheme for wireless sensor networks. In: 7th IEEE International Conference on Wireless Communications and Mobile Computing, pp. 953–958 (2011)
25. Bade, S., Sawant, H.K.: A comparative analysis for detecting uncertain deterioration of node energy in MANET through trust based solution. Global J. Comput. Sci. Technol. 12(8), 41–48 (2012)
26. Almasri, M., Elleithy, K.,Bushang, A., Alshinina, R.: TERP: a trusted and energy efficient routing protocol for wireless sensor networks. In: 17th IEEE/ACM International Symposium on Distributed Simulation and Real Time Applications IEEE Computer Society, pp. 207–214 (2013)
27. Leung, R., et al.: MP-DSR: a QoS-aware multi-path dynamic source routing protocol for wireless ad-hoc networks. In: IEEE International Conference on Computer Networks (2001)

An Overview of Computational Intelligence Technique in Drug Molecular Structure Identification

Yee Ching Saw and Azah Kamilah Muda

Abstract This review focus on the use of computational intelligence techniques for the identification of drugs via its molecular structure. Automation identification of drugs has been a challenging problem in bioinformatics. Hence, this has successfully drawn the attention from the researchers. In the past decade, computational intelligence has been widely applied in several fields such as electrical engineering, computer science, business, and electronic and communication engineering. However, recently there also many researchers who apply these techniques in the Bioinformatics field. There are various techniques that have been applied in the drug identification over the past few years. In this paper, we will present a theoretical overview of computing techniques for drug molecular structure identification, a brief description of the problem and issues involved in it, is first discussed. Then, we will discuss the application of computational intelligence in the drug identification based on the previous work.

1 Introduction

Over the past decade, there are broad increasing in the illicit manufacturing of new and unfamiliar synthetic drugs that release to the market. The producers continuously invent new drugs by altering the chemical composition of the drugs substances and distribute illegally to generate huge profits. Although there are many

Y.C. Saw (✉) · A.K. Muda
Faculty of Information and Communication Technology,
Universiti Teknikal Malaysia Melaka, Hang Tuah Jaya,
76100 Durian Tunggal, Melaka, Malaysia
e-mail: ycsaw@hotmail.com

A.K. Muda
e-mail: azah@utem.edu.my

© Springer International Publishing Switzerland 2016
V. Snášel et al. (eds.), *Innovations in Bio-Inspired Computing and Applications*,
Advances in Intelligent Systems and Computing 424,
DOI 10.1007/978-3-319-28031-8_41

473

illicit drugs can be used for medical purpose (treating a disease or a symptom), but consumption of illegal drugs without the assistance of medical professionals can lead to drug abuse. Drug abuse can also be known as drug addiction is a phenomenon where people become dependent to the drugs and they will start craving for it. Most of the drug abuse are targeting the human's brain central nervous system (CNS), which will cause high impart to the human body due to the seizure of illegal drug substance. Because of these risks, the scientist of drug chemistry also known as forensic drug chemistry will undergoes a series of processes performed in the laboratory in order to analyze and identify the presence or absence of controlled substance in the drugs that submitted by the law enforcement. Basically, there are two categories of forensic test used to identify the drug substances: presumptive test (such as color tests) and confirmatory tests (such as gas chromatography/mass spectrometry). Presumptive tests are less precise as compare to confirmatory tests and only give an indication to which type of substance is present. Confirmatory tests are more accurate and specific in determining the type of substances present in the drug specimen using the instrumental analysis [1]. However, the experimental studies of confirmatory tests for drug identification involved multi-step process and expensive due to its high cost of R&D and time-consuming. An overview of the chemical analysis process of controlled substances that consists of many steps are shown in Appendix A.

Therefore, to overcome this drawback, numerous developments in informatics and computer science offer new opportunities. Generally, the principle of forensic drug chemistry in the illegal drug identification follows the chemistry discipline, which examine how the molecule in the drugs are bonded to each other. Hence, there are numerous researches that apply computational intelligence techniques for the molecular identification in the biomedical area. In order to present the literature review, the well-established and most representative work is classified into three main categories of computational intelligence techniques, which is Genetic Algorithm, Artificial Neural Network, and Fuzzy Logic.

The rest of the paper is organized as follows: in the next section, different computational intelligence techniques like neural networks, genetic algorithms, and fuzzy logic are discussed along with their application in the molecular structure's application. Subsequently, a theoretical analysis of these techniques is presented. Finally a brief summary about this paper are given.

2   Overview of Computational Intelligence Techniques

In this section, a comprehensive review of various feature selection techniques used in wide areas of molecular structure applications, which including Genetic Algorithm, Artificial Neural Network, and Fuzzy Logic. In addition, a brief discussion of hybrid techniques are presented.

2.1 Genetic Algorithm (GA)

Genetic algorithm (GA) is heuristic search that proposed by John Holland in 1975. It is a heuristic search technique that used to solve optimization problem for the problem at hand. The basic concept of GA is based on the Charles Darwin, the theory of evolution, the principle of "survival of the fittest" in the natural biological evolution in The Origin of Species [2]. GA is used to search the best population by following the basic step in Fig. 1. Initially, a population is generated randomly. Next, a fitness function is applied to evaluate the fitness of each individual in the population. The fitness score will be passed to the breeding function to produce new and better parent population through selection, crossover and mutation process. This cycle process will be repeated until a convergence criterion is met.

In [3], David Fogel pointed out that there are several advantages of using evolutionary computation such as genetic algorithm techniques over traditional techniques. The advantages include conceptual simplicity, broad applicability, better performance than classic methods on real problems, potential to use knowledge and hybridize with other methods, parallelism, robustness to dynamic changes, capability for self-optimization, and the ability to solve problems that have no known solutions.

Solomon et al. [4], applied genetic algorithm in parameterization of interatomic potentials for Metal oxides. In this study, they proved that genetic algorithm based methodology are capable to reproduce and energize the structure, capture the basic structural, mechanical, and thermal properties of the BTO (a member of an important class of metal oxide) and predict the two crystalline phases of the BTO.

Fig. 1 Flow chart of the cycle of the genetic algorithm [24]

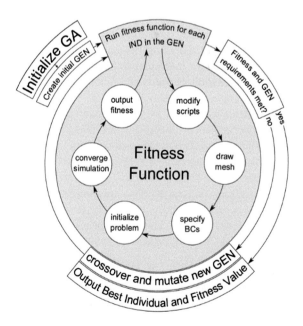

Wałejko et al. [5], developed a system using genetic algorithm (GA) to perform structural analysis of Phenyl galactopyranosides using C MAS NMR spectroscopy and conformational analysis. Conformational analysis was performed using genetic algorithm-assisted grid search method (GAAGS), which is a techniques that using search technique and genetic algorithm approach to increase the conformational analysis performance.

Li et al. [6], used the genetic algorithm with the multi-population evolution and entropy-based searching technique with narrowing down space to solve the optimization model for molecular docking problem. This proposed approach has evaluated based on RMSD value and it has achieved a rate of 41 % excellent docking solutions, 38 % of good docking solutions and 13 % of poor docking solutions.

Paul and Iba [7] developed Probabilistic Model Building Genetic Algorithm (PMBGA) for the identification of informative genes for molecular classification and present the unbiased experimental results on three bench-mark data sets. According to [8] genetic algorithm provides a better percentage of accuracy in identifying patterns between HIV-1 protease against organic leads and FDA approved inhibitors of HIV-1 protease.

2.2 Artificial Neural Network (ANN)

ANN is inspired by biological neural network of the structure and functional aspect of human being nervous system especially the human's brain [9]. ANN are popular today because it works by simulating the human intuition in decision making and can tolerate with complex, noisy and incomplete data. They are useful to learn complex relationships or patterns hidden in large scale dataset.

Lorenz et al. [10] used artificial neural network to successfully identify mosquito species based on the molecular mosquito DNA samples. It is proved that ANN have better classification performance than traditional discriminant analysis with high accuracy ranging from 85.70 to 100 %.

According to Li et al. [11], it is proven that artificial neural network model have high performance in identifying complex, multi-dimensional and non-linear patterns than response surface methodology in optimization of controlled release nanoparticle formulation of verapamil hydrochloride.

Laosiritaworn [12], proved that ANN have a success performance in modeling spin-transition behavior in ultra-thinfilm molecular magnet. In [13], the artificial neural network have been applied to predict the solubility of sulfur dioxide in different ionic liquids. The study proved that ANN can be a new alternative approach to predict different thermodynamic properties (solubility in this case) of the materials with the low error rate about 2 %.

Ghaedi [14] applied artificial neural network to predict the density, viscosity, thermal expansion coefficient, molar volume and viscosity deviation of ethylene glycol monoethyl ether (EGMEE) solution. In his study, he reveals that the ANN model can be an excellent alternative for simultaneous prediction of the thermo-dynamic properties of the aqueous solution of EGMEE with mean square error (MSE b 0.0051 %) and high coefficient of determination (R2 ≥ 0.9913).

2.3 Fuzzy Logic

Fuzzy logic concept was first introduced by Zadesh in 1965 [15]. It provides mathematical tool for dealing with linguistic variables to describe the relationship between the system variables. The membership function in fuzzy logic can range within the interval [0–1]. Hence, fuzzy logic is suitable to apply in many fields due to its specialty in dealing with uncertainty and complexities data. The Fig. 2 shows the membership function for temperature i.e. low, medium and high.

Saravanan and Lakshmi [16], applied a novel fuzzy rule based system for assessing the protein for allergenicity and evaluate the quality of the query protein. This system is helpful for distinguishing the allergen-like non-allergens from allergens and providing a uniform examination and interpretation of the results.

Sun et al. [17], applied the dynamic fuzzy modelling approach for modelling genetic regulatory networks from gene expression data. This approach is useful in utilized the structural knowledge on gene network and the nonlinear dynamic property embedded in the gene network can be well collected. Hence, this approach can achieve a faster convergence in the identification process and the fuzzy gene network will have a better performance in prediction.

Umoja et al. [18] used the fuzzy logic and shape definition approaches to reduce the search size and predict the possible molecular docking location. In [19], it have applied a fuzzy-logic based system for both diagnosis and control of a convention process of ammonia into nitrogen gas by different microbial groups. In addition, the fuzzy logic control module was tested and evaluated using dynamic simulations and has shown to achieve high and stable nitrogen removal efficiency (around 90 %).

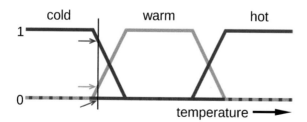

Fig. 2 Fuzzy logic representation of temperature [25]

2.4 Hybrid Techniques

In a lot of others cases, a combination of different techniques is better than any single one.

Theofilatos et al. [20], has successfully combined genetic algorithms with Extended Kalman Filters, and applied the combination for protein interaction prediction and classification. This combination is proven to be useful and effective if the search parameter is big and complex or it does not have a valid mathematical analysis of the problem.

Hamzehie and Najibi [21] used artificial neural network and Deshmukh-Mather model to predict the solubility of carbon dioxide (CO_2) in amino acid salt solutions.

In [22], the electron conformational–genetic algorithm (EC-GA) method is presented as a novel hybrid 4D-QSAR approach for pharmacophore identification and bioactivity prediction using the best subset of parameters. Besides, [22] stated that the genetic algorithm manages to show the importance of thermodynamic, electronic and geometric parameters.

3 Conclusion

The molecular structure, as a complex structure, is akin to be fuzzy and evolutionary. In this paper, we have provided a brief overview of computational intelligence techniques and their application in the area of molecular structure application. Each computational intelligence technique is summarized and its utility to molecular structure application is analyzed. The paper consists of four main aspects: artificial neural network can learn complex nonlinear input-output relationships to improve the learning capability of the molecular structure; fuzzy logic summarize the domain knowledge with the use of symbol to deal with vagueness and uncertainty in molecular structure; genetic algorithm provides an efficient search methodology to deal with the vastness and tractability issues of molecular structure data; artificial neural network to improve the learning capability of the molecular structure; hybrid techniques combine one or more different techniques can give a robust solution to the problem at hand. We expect that this review will be of some assistance in the task of choosing the proper computational intelligence techniques for the application which involve complex drug molecular structure.

Acknowledgments This work was supported by Collaborative Research Programme (CRP)—ICGEB Research Grant (CRP/MYS13-03) from International Centre for Genetic Engineering and Biotechnology (ICGEB), Italy and University Technical Malaysia Melaka under GLuar/2014/FTMK(1)/A00004.

Appendix A

An overview of chemical analysis process of controlled substances [23].

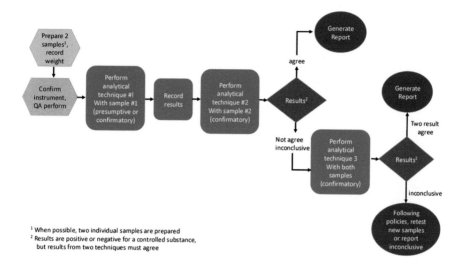

¹ When possible, two individual samples are prepared
² Results are positive or negative for a controlled substance, but results from two techniques must agree

References

1. Unodoc—United Nations Office on Drugs and Crime.: Recommended methods for the identification and analysis of cocaine in seized materials. Unodoc (2012)
2. Darwin C.: On the origin of species by means of natural selection
3. Fogel, D.B.: The advantages of evolutionary computation. Bcec. **1995**, 1–11 (1997)
4. Solomon, J., Chung, P., Srivastava, D., Darve, E.: Method and advantages of genetic algorithms in parameterization of interatomic potentials: metal oxides. Comput. Mater. Sci. **81**, 453–465 (2014)
5. Wałejko, P., Paradowska, K., Bukowicki, J., Witkowski, S., Wawer, I.: Phenyl galactopyranosides—13C CPMAS NMR and conformational analysis using genetic algorithm. Chem. Phys. **457**, 43–50 (2015)
6. Li, Z., Gu, J., Zhuang, H., Kang, L., Zhao, X., Guo, Q.: Adaptive molecular docking method based on information entropy genetic algorithm. Appl. Soft Comput. **26**, 299–302 (2015)
7. Paul, T., Iba, H.: Identification of informative genes for molecular classification using probabilistic model building genetic algorithm. Genet. Evol. Comput. Conf. **3102**, 414–425 (2004)
8. Harishchander, A., Senapati, S., Anand, D.A.: Analysis of drug resistance to HIV-1 protease using fitness function in genetic algorithm. **12**(Suppl 1), 2334 (2012)
9. Haykin, S.: Neural networks: a comprehensive foundation (1999)
10. Lorenz, C., Ferraudo, A.S., Suesdek, L.: Artificial Neural Network applied as a methodology of mosquito species identification. Acta Trop. **152**, 165–169 (2015)

11. Li, Y., Abbaspour, M.R., Grootendorst, P.V., Rauth, A.M., Wu, X.Y.: Optimization of controlled release nanoparticle formulation of verapamil hydrochloride using artificial neural networks with genetic algorithm and response surface methodology. Eur. J. Pharm. Biopharm. **94**, 170–179 (2015)
12. Laosiritaworn, W., Laosiritaworn, Y.: Artificial neural network modeling of spin-transition behavior in two-dimensional molecular magnet: The learning by experiences analysis. Polyhedron **66**, 108–115 (2013)
13. Bahmani, A.R., Sabzi, F., Bahmani, M.: Prediction of solubility of sulfur dioxide in ionic liquids using artificial neural network. J. Mol. Liq. **211**, 395–400 (2015)
14. Ghaedi, A.: Simultaneous prediction of the thermodynamic properties of aqueous solution of ethylene glycol monoethyl ether using artificial neural network. J. Mol. Liq. **207**, 327–333 (2015)
15. Zadeh, L.A.: Fuzzy Sets. Inf. Control **8**(3), 338–353 (1965)
16. Vijayakumar, S., Lakshmi, P.T.V.: A fuzzy inference system for predicting allergenicity and allergic cross-reactivity in proteins. In: Proceedings of the 2013 IEEE International Conference Bioinformatics and Biomedicine IEEE BIBM 2013, no. Mlc, pp. 49–52 (2013)
17. Sun, Y.S.Y., Feng, G.F.G., Cao, J.C.J.: A new approach to dynamic fuzzy modeling of genetic regulatory networks. IEEE Trans. Nanobiosci. **9**(4), 263–272 (2010)
18. Umoja, C., Durham, E.A., Rosen, A., Harrison, R.W.: A novel approach to determine docking locations using fuzzy logic and shape determination. 31–33 (2014)
19. Boiocchi, R., Mauricio-Iglesias, M., Vangsgaard, A.K., Gernaey, K.V., Sin, G.: Aeration control by monitoring the microbiological activity using fuzzy logic diagnosis and control. application to a complete autotrophic nitrogen removal reactor. J. Process Control **30**, 22–33 (2014)
20. Theofilatos, S.D.K.A., Dimitrakopoulos, C.M., Tsakalidis, A.K., Likothanassis, S.P.M., Papadimitriou, S.T. (2010) A new hybrid method for predicting protein interactions using genetic algorithms and extended kalman filters. 0–3 (2010)
21. Hamzehie, M.E., Najibi, H.: Prediction of carbon dioxide loading capacity in amino acid salt solutions as new absorbents using artificial neural network and Deshmukh–Mather models. J. Nat. Gas Sci. Eng. (2015)
22. Geçen, N., SarIpInar, E., Yanmaz, E., Şahin, K.: Application of electron conformational-genetic algorithm approach to 1, 4-dihydropyridines as calcium channel antagonists: Pharmacophore identification and bioactivity prediction. J. Mol. Model. **18**(1), 65–82 (2012)
23. Principles of Forensic Drug Chemistry: Forensic Drug Chemistry: Principles. N.p., n.d. Web. 30 Nov 2015
24. Genetic Algorithm. Digital image. UD FCRL: Research: Flow Geometry Optimization. University of Delaware, n.d. Web. 30 Sep 2015
25. Fuzzy Logic—Wikipedia, the Free Encyclopedia: Wikipedia. Wikimedia Foundation, n.d. Web. 30 Nov 2015

An Effective Bio-Inspired Methodology for Optimal Estimation and Forecasting of CO_2 Emission in India

A. Sangeetha and T. Amudha

Abstract The exponential increase in Greenhouse Gases (GHGs) emission due to the oxidation of fossil fuels is an alarming ecological problem all over the world. Such harmful emissions affect the air quality, thereby creating serious concern to the human health, natural life and agriculture. Consequently, the intercontinental community has developed air quality standards to observe and control pollution rates worldwide. Among the GHGs, carbon-di-oxide (CO_2) plays a major role in polluting the air heavily, and hence the estimation and control of CO_2 emission has become the need of the hour. The main objective of this research is to study, estimate and forecast CO_2 emission in India from various sources of energy consumption. The estimation of CO_2 emission and analysis are done through Multiple Linear Regression (MLR) model and Particle Swarm Optimization (PSO) algorithm. In our research, PSO could obtain a more accurate estimation compared to MLR and hence PSO estimation was used for future forecasting of CO_2 emission in India. The performance of MLR and PSO were also measured using statistical quality parameters and the results have proven the effectiveness of PSO algorithm.

Keywords Air pollution · Bio-inspired computing · CO_2 emission estimation · CO_2 emission forecasting · Particle swarm optimization

1 Introduction

Biologically inspired computing (BIC) is a field of study dedicated to deal with complex problems using computational methods which are modelled after the principles encountered in nature. BIC is a problem solving technique usually

A. Sangeetha (✉) · T. Amudha (✉)
Department of Computer Applications, Bharathiar University, Coimbatore, TamilNadu, India
e-mail: sangekarthi2505@gmail.com

T. Amudha
e-mail: amudha.swamynathan@gmail.com

© Springer International Publishing Switzerland 2016
V. Snášel et al. (eds.), *Innovations in Bio-Inspired Computing and Applications*,
Advances in Intelligent Systems and Computing 424,
DOI 10.1007/978-3-319-28031-8_42

481

referred as bottom-up, decentralised approaches that specifies a set of rules and conditions to solve a complex problem by iteratively applying these rules. Biological system heavily relies on individual components of the system [4]. BIC algorithms are generally divided into evolutionary computation and swarm intelligence algorithms, which are derived from the analogy of natural evolution and through biological activities. Evolutionary computation is a term used to describe algorithm which was stirred by 'survival of fittest', whereas swarm intelligence is a term used to illustrate the algorithms which was inspired by the cooperative intelligence of insect colonies and other animal societies.

Climate change due to the rise in air contamination presents a major ecological problem, which are caused by the emission of various GHGs into the environment. Carbon-di-oxide, methane, nitrous oxide, F-gases are the GHGs which directly pollutes our air. CO_2 emission is considered as a largest contributor to air pollution. CO_2 emission from different sources directly pollutes 50 % of atmosphere, 26 % of land and 24 % of ocean. In 2011, Energy Information Agency (Department of Energy), estimates CO_2 emissions from all sources globally in which India ranks fourth with about 5 % of global CO_2 emission (1725.76 million metric tonne) [5]. Globally CO_2 emission from fossil fuel combustion and from industrial process [cement and metal production] increased in 2013 to the new record of 35.3 billion tonnes (Giga tonnes-Gt) CO_2, which is 0.7 Gt higher than preceding year record, in which India shares about 7.7 % of CO_2 emission in worldwide [10]. The outlook of GHGs emission shows the importance of the need for CO_2 emission modeling [8]. The main factors of CO_2 emission are termed as thermal power plant, transportation, deforestation, residential and commercial buildings, cement and ceramics industry etc.

Ant Colony Optimization, Bee Colony Optimization and Genetic Algorithms are nature inspired algorithms which were applied for estimating and predicting the CO_2 emission worldwide [1, 2, 7, 9]. These algorithms are robust and effective. This paper presents the application of PSO method to estimate and forecast the CO_2 emission in India based on the socioeconomic indicators, coal, oil, natural gas, and primary energy consumption. The CO_2 estimation of PSO is compared with multiple linear regression model and it was identified that PSO could make better estimates. By estimating the CO_2 emission in India, the air quality in our environment can be managed and their harmful effects can be reduced. Air quality forecasting models will be much helpful in the planning of cleaner environment.

2 Estimation of CO_2 Emission

Coal, oil, natural gas (NG), and primary energy (PE) are used as socio economic indicator variables for CO_2 emission estimation in this research. A linear model is applied for estimation of CO_2 emission in India. The linear form of equation for the estimation model is written as,

$$Y_{linear} = w_1 X_1 + w_2 X_2 + w_3 X_3 + w_4 X_4 + w_5 \qquad (1)$$

where, w_i represents the corresponding weighting factors, X_1 specifies coal consumption, X_2 specifies oil consumption, X_3 specifies natural gas (NG) consumption, X_4 specifies primary energy (PE) consumption, and Y_{linear} represents the predicted value of CO_2 emission.

The objective function is to minimize the function F(x), and it is formulated as,

$$F(x) = \sum_{i=1}^{n} |A_i - P_i| \qquad (2)$$

where, A_i is actual value, P_i is predicted value of CO_2 emission and n is the number of observations.

Estimation of CO_2 emission is performed using PSO algorithm and Multiple Linear Regression (MLR) model. The accuracy in obtaining closer estimate is analysed by using statistical measures RMSE (Root Mean Square Error), MAPE (Mean Absolute Percentage Error), VAF (Variance Accounted For) and ED (Euclidian Distance), given in Eqs. 3, 4, 5 and 6.

$$RMSE = \sqrt{\frac{\sum_{i=1}^{n}(A_i - P_i)}{n}} \qquad (3)$$

$$MAPE = \left(\frac{1}{n} \sum_{i=1}^{n} \left| \frac{(P_i - A_i)}{A_i} \right| \right) * 100 \qquad (4)$$

$$VAF = \left[1 - \frac{var(A_i - P_i)}{var(A_i)} \right] * 100 \qquad (5)$$

$$ED = \sqrt{\sum_{i=1}^{n} (A_i - P_i)^2} \qquad (6)$$

where, A_i is the actual value, P_i is the predicted value, n is the number of observations and var is the variance of the actual and predicted value.

3 Particle Swarm Optimization

Particle Swarm Optimization is a population-based metaheuristic technique, developed by James Kennedy and Russell Eberhart in 1995 [6]. This algorithm is inspired from the social behavior of bird flocking, where a group of birds randomly search for food in an area by following the nearest bird to the food. It combines exploration (local search) methods with exploitation (global search) methods in the search space. The individuals in a PSO have its own position and velocity and are referred as "particles". Each particle is treated as a point in a multi-dimensional search space. The social interaction between the swarm is to locate the best position achieved so far. The main aim of PSO algorithm is to obtain the number of particles that comprises a group moving around the search space and looking for the best solution [8]. For every particle, the fitness is evaluated and if the particle fitness is better than the best-so-far fitness, then update is performed on the global best fitness. The new velocity and position for each particle will be evaluated and this process will be repeatedly done for a number of iterations. The basic PSO steps are represented in Algorithm 1.

Algorithm 1. Basic PSO Technique [1]

```
1        Begin PSO
2        Randomly initialize the position and velocity of the particles
3        While (termination condition not reached) do
4            for i=1 to number of particles
5                Evaluate the fitness F(xᵢ)
6                Update particle best pᵢ and global best gᵢ
7                Update velocity of the particle Vᵢ
8                Update position of the particle Xᵢ
9            Next For
10       End While
11       End PSO
```

The PSO equation is formulated as follows,

$$v_{id}^{new} = v_{id}^{old} + \varphi_1 * c_1 * (p_{id} - x_{id}) + \varphi_2 * c_2 * (p_{gd} - x_{id}) \tag{7}$$

$$x_{id}^{new} = x_{id}^{old} + v_{id}^{new} \tag{8}$$

where, v_{id} represents the velocity of particle i in dimension d, x_{id} represents the position of particle i in dimension d, φ_1, φ_2 are acceleration constants, c_1, c_2 are random numbers, p_{id} specifies the best position of the particle reached so far, and p_{gd} specifies the global best position reached so far. The two components in the velocity update equation are influenced by cognitive component and social component. $(p_{id} - x_{id})$ in the equation represent the cognitive component, and $(p_{gd} - x_{id})$ in the equation represent the social component.

4   Implementation Results and Discussion

This section presents the CO_2 emission estimation results obtained through PSO and MLR model. The figures correlated to coal, oil, natural gas, and primary energy consumption are obtained from British Petroleum Statistical Review of World Energy [3]. These four energy commodities are considered as the highest emission factors of CO_2 and based on these energy consumption the long-term forecasting was done. The related data from 1965 to 2013 are used, partly for training the model (1965–2002) and partly for testing the model (2003–2013). The algorithm is developed in Java language and executed for 100 runs to identify the optimal weighting factors. By applying PSO algorithm, the linear form of equation coefficients for the estimation model in Eq. (1) becomes,

$$Y_{linear} = 3.1705X_1 + 3.7935X_2 + 2.804X_3 + 0.4380X_4 + 3.1946 \qquad (9)$$

Using SPSS package, Multiple Linear Regression is applied to identify the weighting factors and by applying MLR, the linear form of equation coefficients for the estimation model in Eq. (1) becomes,

$$Y_{linear} = 3.983X_1 + 2.797X_2 + 3.032X_3 + 0.039X_4 + 1.316 \qquad (10)$$

The parameter setting used in the implementation of PSO algorithm to estimate the linear model is represented in Table 1. The algorithm gave its best performance with 5 particles and 100 runs, beyond which it converged.

The observed and estimated value of CO_2 emission in India is represented in Table 2.

It can be understandable from Fig. 1 that PSO has resulted in better estimate of CO_2 emission value than MLR. CO_2 emission estimate obtained through PSO and MLR are analysed using the statistical performance measures tabulated in Table 3. The performance measures have shown that error rates are less in PSO than MLR, which has proven the accuracy of PSO.

The validation of the models specifies that PSO linear model is in good agreement with the observed data. Correlation coefficient (R^2) for developed model is about 99.3 %. Paired t-test is also performed in SPSS to prove the reliability of PSO estimation model, and the results are tabulated in Table 4.

The significance obtained through t-test is <5 %. As the results obtained through PSO model are proven to be highly accurate and reliable, forecasting of CO_2 emission from the year 2014–2030 is done by using PSO results and the corresponding polynomial trend line is shown in Fig. 2.

Table 1 Tuning parameters for linear PSO model

Variables	Values/Range
No. of particles	5
No. of runs	100
Acceleration constant	[0.2, 0.3]
Search space	[−10, 10]
Random values	[0, 1]

Table 2 CO_2 emission—observed [3] and estimated—PSO & MLR

Year	Coal (Mtoe)[a]	Oil (Mtoe)[a]	NG (Mtoe)[a]	PE (Mtoe)[a]	Actual-CO_2 Emission (Mt)[b]	PSO-CO_2 Estimation (Mt)[b]	MLR-CO_2 Estimation (Mt)[b]
1965	35.5	12.6	0.2	52.7	180.1	187.5	231.6
1966	35.5	14.1	0.2	54.3	184.3	193.7	237.1
1967	36.1	14.6	0.3	56.0	188.2	198.2	242.6
1968	37.3	16.3	0.3	59.8	198.6	210.5	256.4
1969	39.6	19.6	0.4	66.3	218.0	233.3	281.5
1970	37.6	19.5	0.6	64.8	210.0	226.3	272.1
1971	38.1	20.5	0.6	67.0	215.1	232.7	279.2
1972	40.2	22.1	0.7	70.5	228.7	247.3	295.9
1973	39.7	23.3	0.7	71.9	230.3	250.9	298.6
1974	44.8	22.8	0.8	76.3	249.3	267.4	322.3
1975	48.1	23.3	1.0	81.9	264.7	283.1	343.4
1976	50.1	24.6	1.2	85.9	276.9	296.5	359.4
1977	52.5	26.4	1.3	90.8	291.9	313.2	379.0
1978	50.8	28.7	1.4	94.2	292.7	318.4	382.5
1979	54.0	31.0	1.8	99.5	313.1	340.5	407.9
1980	56.7	31.6	1.1	102.5	324.2	350.9	421.4
1981	63.2	34.0	1.8	113.1	359.3	387.6	467.1
1982	63.1	35.4	2.4	112.7	364.1	393.6	471.6
1983	66.2	37.2	2.9	117.8	383.0	413.9	495.6
1984	69.5	39.9	3.3	125.8	405.4	439.3	525.5
1985	72.5	43.3	4.0	132.7	429.8	467.0	556.4
1986	78.0	45.5	5.6	142.2	462.0	501.4	598.8
1987	85.9	47.0	6.5	135.1	499.8	531.3	629.9
1988	91.7	51.5	7.6	139.2	539.1	571.6	672.9
1989	100.0	55.8	9.1	143.3	588.4	620.0	726.3
1990	95.5	57.9	10.8	147.4	581.4	620.6	723.9
1991	101.8	58.9	12.1	151.5	612.2	649.5	759.5
1992	108.2	62.1	13.5	155.6	650.8	687.8	802.5
1993	112.5	62.7	13.7	159.7	670.1	706.0	826.0
1994	115.8	67.4	14.8	163.7	700.4	739.4	859.9
1995	125.0	75.2	16.9	167.8	765.5	805.7	928.6
1996	134.4	81.1	18.5	171.9	824.4	863.9	991.2
1997	135.9	86.5	20.1	176.0	850.8	895.6	1021.4
1998	136.1	92.5	22.0	180.1	874.6	926.2	1048.9
1999	135.8	100.3	22.6	184.2	898.5	958.0	1075.2
2000	144.2	106.1	23.7	188.3	952.8	1012.2	1132.9
2001	145.2	107.0	23.8	192.4	959.2	1020.2	1143.3
2002	151.8	113.2	24.8	196.4	1007.2	1069.9	1194.6

(continued)

Table 2 (continued)

Year	Coal (Mtoe)[a]	Oil (Mtoe)[a]	NG (Mtoe)[a]	PE (Mtoe)[a]	Actual-CO_2 Emission (Mt)[b]	PSO-CO_2 Estimation (Mt)[b]	MLR-CO_2 Estimation (Mt)[b]
2003	156.8	116.5	26.6	200.5	1040.9	1104.5	1232.8
2004	172.3	119.5	28.7	204.6	1116.3	1172.6	1313.2
2005	184.4	121.9	32.1	208.7	1180.0	1231.8	1383.0
2006	195.4	130.5	33.5	212.8	1246.5	1305.2	1459.3
2007	210.3	135.3	36.1	216.9	1341.2	1379.4	1543.6
2008	230.4	140.2	37.2	221.0	1443.9	1466.5	1644.8
2009	250.3	145.0	46.7	225.1	1569.4	1576.6	1770.7
2010	260.2	149.9	56.7	229.1	1640.7	1656.1	1858.0
2011	270.1	163.0	55.3	233.2	1699.8	1735.0	1933.8
2012	302.3	173.6	52.9	237.3	1854.2	1872.3	2088.5
2013	324.3	175.2	46.3	241.4	1931.1	1931.8	2165.0

(Mtoe)[a]: Million tonne oil equivalent
(Mt)[b]: Million tonne

Fig. 1 Comparison observed and estimated of CO_2 emission in India

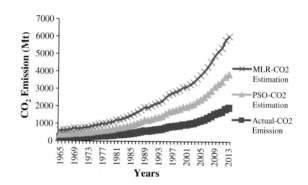

Table 3 Comparison of PSO & MLR using statistical performance measures

Performance measures	MLR	PSO
RMSE	11.73436	5.632097
MAPE	23.75685	5.865665
VAF	98.69	99.88
ED	1039.8030	249.4347

Table 4 Comparison of PSO & MLR using paired t-test

	Prediction error-MLR	Prediction error-PSO
Mean	137.6953206	31.7205201
Variance	3169.797867	269.047415

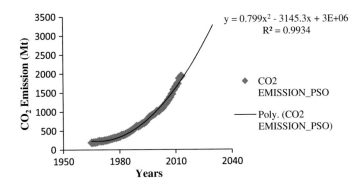

Fig. 2 Forecast of CO_2 emission in India [2014–2030]

5 Conclusion

This research work was carried out with the aim of creating awareness on the tremendous increase of CO_2 emission in India, which will have a huge impact on humanity and environment. In this paper, carbon-di-oxide emission estimation was done based on the socioeconomic indicators coal, oil, natural gas and primary energy using PSO algorithm and Multiple Linear Regression model. A related data set of 49 years (1965–2013) was used for estimating the parameters of the linear model in both the training and testing cases. Statistical validation of the results specifies that there is a good agreement between the observed data and the estimated data by PSO. Based on the PSO estimation, forecasting of CO_2 emission is done from the year 2014–2030. As a future work, it is planned to forecast the CO_2 emission for the forthcoming years using bio-inspired algorithms, which will create an awareness towards prevention and control of the hazardous CO_2 emission. Exponential and quadratic models for estimation of CO_2 emission will also be developed and a validation among those models will be performed.

References

1. Alaa, A., Sahar, Mokhtar, A., Alaa, S., Basma, S.: Forecast global carbon dioxide emission using swarm intelligence. Int. J. Comput. Appl. **77**(12) (2013)
2. Behrang, M.A., Assareh, E., Assari, M.R., Ghanbarzadeh, A.: Using bees algorithm and artificial neural network to forecast world carbon dioxide emission. Energy Sour. Part A: Recovery Utilization Environ. Eff. **33**(19), 1747–1759 (2011)
3. British Petroleum (BP) Statistical review of world energy 2014: http://www.bp.com/statistical review. (2014)
4. Harini, C., Pomil, Bachan, Proch., Roshini, R., Chandrasekharan, K.: Bio inspired approach as a problem solving technique. Netw. Complex Syst. **2**(2), 14 (2012)
5. International Energy Agency (IEA).: World Energy Outlook 2010. http://www.iea.org (2010)

6. James, K., Russell, E.: Particle swarm optimization. In Proceedings of the IEEE International Conference on Neural Networks, vol. IV, pp. 1942–1948, Piscataway, NJ. IEEE Press (1995)
7. Kavoosi, H., Saidi, M.H., Kavoosi, M., Behrang, M.: Forecast global carbon dioxide emission by use of genetic algorithm. Int. J. Comput. Sci. Issues (IJCSI), **9**(5), 1 (2012)
8. Particle Swarm Optimization.: http://www.swarmintelligence.org
9. Reza, S.: Application of ant colony optimization (ACO) to forecast CO_2 emission in Iran. Bull. Env. Pharmacol. Life Sci. **2**(6), 95–99 (2013)
10. Trends in Global CO_2 Emissions 2014 Report: http://edgar.jrc.ec.europa.eu/news_docs/jrc-2014-trends-in-global-co2-emissions-2014-report-93171.pdf (2014)

Conceptual Voice Based Querying Support Model for Relevant Document Retrieval

Olufade F.W. Onifade and Ayodeji O.J. Ibitoye

Abstract To increase the precision and recall values of retrieved documents while using a voice information retrieval system for documents (VIRD), is to develop a model that best captures users' objectives in query speech. An approach to achieving this objective is by establishing conceptual relationship between keywords in user's query before search. Here, the research proposed a Universal Fuzzy Concept Network Language (UFCNL) that represents' the users speech with a declarative formal language, and establishes an associated degree of relationship between conceptual relations (auxiliaries and determiners) and concepts. The essence is to produce new sets of conceptual queries with potentials to retrieve documents which are more relevant to the information seeker unlike the ordinary keyword extraction by spoken term detection in voice based retrieval or the conventional text based retrieval system.

Keywords Universal network language · Fuzzy concept network · Conceptual structure · Voice information retrieval system for documents · Information retrieval · Concept

1 Introduction

Communication as a process of sending and exchanging information is a core requirement whose output either in speaking or writing determines the input either through listening or reading. For communication to take place effectively, at least

O.F.W. Onifade
Department of Computer Science, University of Ibadan, Ibadan, Nigeria
e-mail: fadowilly@yahoo.com

A.O.J. Ibitoye (✉)
Department of Computer Science and Information Technology,
Bowen University, Iwo, Nigeria
e-mail: ibitoye_ayodeji@yahoo.com

© Springer International Publishing Switzerland 2016
V. Snášel et al. (eds.), *Innovations in Bio-Inspired Computing and Applications*,
Advances in Intelligent Systems and Computing 424,
DOI 10.1007/978-3-319-28031-8_43

two systems must be put in place. One of such system must put something "out" while the other system must take something "in". This is called "output" and "input" processes [1]. The "out", "in" scenario of communication best describe what transpired between an information seeker who outputs his/her intention and an information retrieval system which take in the query for processing. Hence, for an effective result to be obtained by IR systems, effective communication must take place between the searcher and the information retrieval system from the start of problem definition to the end of interpreting the retrieved document. Over the years, information retrieval systems via text have played a long vital role in the structuring, analysis, organization, storage, searching and retrieval of information [2]. But her values have not been fully optimised. More success is to be discovered and achieve especially on the aspect of retrieving of relevant document. Since, its conventional querying mechanism is through writing. Currently, voice is becoming more and more a preferred medium of interaction between people and the World Wide Web [3]. However, due to existing technical difficulties and limitations with natural language processing, grammar generation, voice recognition, result representation, etc. a very little attention has been given to document information retrieval using voice. In this research, we present a model to establish conceptual relationship in spoken text detected from user's speech in order to increase the chances of retrieving relevant documents on a semantic structured data corpus.

In the remaining sections of this work, we examined related works in the realm of VIRD and their qualities in Sect. 2. Section 3 explores the proposed model by detailing the processes of the universal fuzzy concept network language and the roles of universal network language. Also, the roles of fuzzy concept network (FCN) in engendering new sets of user query that is fit to retrieve relevant documents was discussed. In Sect. 4, we used the model in 3 to experiment 10 distinct user queries on a clustered data set of 3200 to test for the precision rate, recall rate and degree of cohesion to compare existing methods like keyword spotting and query expansion. We conclude the research in Sect. 5.

2   Overview of Voice Information Retrieval for Document

In voice information retrieval for document, query is given as voice command to the system; the system has to generate the information based on the spoken word to retrieve the needed documents [4]. Several methods have been used to implement a voice information retrieval system for documents (VIRD). From the telephony system of interaction with the indexed database to VoiceXML and to using a large vocabulary continuous speech recognition (LVCSR) [5] are approaches that help by using its language information to search for contents that include speech in the database. A more recent approach for VIRD is the Automatic Speech Recognition (ASR) mechanism which is used together with language processing for the purpose of producing transcripts of spoken words which are useful for document retrieval. The ASR approach seems good because speech recognition allows the user to query the system by voice, making the interaction direct, easy to use and more immediate. However, unlike text retrieval, the basic challenge of VIRD is dealing with speech recognition errors and out-of-vocabulary (OOV) words document processing [6]. OOV terms are words missing in the automatic speech recognition (ASR) system's vocabulary and are replaced in the output transcript by alternative words which are probable, given the language models of the ASR. With the OOV present in query search, no retrieved document will be relevant to users query. In resolving the OOV problem, several IR technique of query expansion have also been introduced to tackle OOV. Some of this technique involves finding synonyms of query terms from the speech to finding all the various morphological and adding additional words by using techniques such as relevance feedback [7]. Many other studies have also focused on dealing with OOV and misrecognized words. Such of this methods include the example of using sub word units as the recognition result or search unit by using the multiple recognition candidates [8]. Tomoko [9] also proposed a method that incorporates Spoken Text Document (STD) into Spoken Content Retrieval (SCR) process to deal with OOV and miss-recognized words. However, reviewed VIRD approach mostly use keyword search from speech recognition, the option give existing VIRD systems a low precision and recall value [10]. Also, the level of accuracy in speech transcription, particularly on terms of interest such as named entities and content words are issues with retrieving information from spoken data [11]. Here, the research proposed a formal language approach to be used alongside the fuzzy concept network technique for generating and establishing a User-IR model that engenders conceptual relationship between the user queries before search is conducted.

3   The Proposed Conceptual Querying Support Model

Concept-based IR represents both documents and queries using semantic concepts in place of keywords extracted from user query in a concept space. The essence is to establish a conceptual knowledge of the user's query. Hence, the conceptual query support model is designed as an information retrieval system that allows users to communicate their intentions through voice command. The model is designed using the UNL and FCN approaches. These methods act dynamic roles for the purpose of minimizing noise, and increasing the level of cohesion by finding and establishing the conceptual structures for a speech based user query before performing a search. This system is trained to listen through voice communication on a speech recognition device that can identify words and convert the user's speech to text format that is needed to build a conceptual relationship for the captured query. Here, the Universal Fuzzy Concept Network Language is used for the generation of conceptual query vectors by utilizing the conceptual relations and Universal Words that are extracted from the user speech to describe the objectivity of the captured sentences. Figure 1 is a detailed diagram that illustrates the flow of the proposed UFCNL.

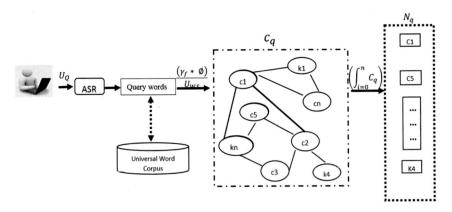

Fig. 1 Diagram illustrating the conceptual structure of the user query

From Fig. 1, U_Q is the user query, U_{Ws} is the Universal word corpus, γ_f is the universal network language, $\varnothing$ is the fuzzy concept network, C_q is the conceptual query and N_q is the new set of ranked concepts from the conceptual queries.

The Universal Network Language γ_f in Fig. 1 is used on the captured text in the user speech U_Q in order to establish a conceptual relations between function words and concepts. γ_f is a digital meta-language for describing, summarizing, refining, storing and disseminating information in a machine independent and human language neutral form. Since UNL is used as a language for knowledge representation in Information Retrieval, here, the meta-language focuses to express meanings in a group of word retrieved by the ASR. In UFCNL, the Universal Words (UWs) are the main concepts or words that constitute the vocabulary which are always represented by a node in a hypergraph while function words such as auxiliaries and determiners are attributes to UWs that provides additional information to the sentence These words are loaned from English and disambiguated by their positioning in a knowledge base of conceptual hierarchies. Hence, UFCNL uses the UNL to establish relationship between the auxiliaries or/and determiners and the available concepts in the user query in order to define the semantic links wherein high-level concepts can be related to lower-level ones through the ontological relations "equ" (= is equal to), "icl" (= is a kind of), and "iof" (= is an instance of), logical relations ("and", "or") and thermal relations; such as "tim" = time, "plc" = place, etc.

In line with the objective to increase the conceptual cohesion between concepts, the Fuzzy concept network $\varnothing$ as indicated in Fig. 1 is used in UFCNL to build relationships and increase the degree of cohesion between concepts being multiple defined in order to extend the generality of the knowledge base architecture. It utilizes the universal word corpus to generate and build a wide range of concept with fuzzy relative degree of association. In the UFCNL, FCN is modeled by a relation matrix and a relevance matrix, where the elements in a relation matrix represent the fuzzy relationships between concepts, and the elements in a relevance matrix indicate the degrees of relevance between concepts. The concepts are explicitly linked by one of the four fuzzy relationships at a time, i.e., fuzzy positive, negative, generalization and fuzzy specialization association relationship. The essence is to provide a more powerful knowledge representation method which is more appropriate than the Universal Network language in the information retrieval environment. Instead of establishing binary relation between concepts, based on available concepts in a sentence (user query), FCN is used to broaden the scope of the universal words by utilizing the available words present in the Universal words corpus as illustrated in Fig. 1. Then fuzzy relations with degree of relationship within the range of 0 and 1 are established for each pair of concept. Hence the FCN in UFCNL allows the users to perform positive queries, negative queries, generalization queries, and specialization queries in vector space. From this, the new conceptual query C_q can then be rank based on degree value of association to form the new set of queries required to perform a search using the function $\left(\int_{i=0}^{n} C_q \right)$. This process earns a better recall rate, precision value, and high degree of cohesion in user query.

4 Experiments and Evaluation

The work examined 10 distinct spoken queries on a dataset of 3200 clustered documents. The experiment was conducted to reveal the percentage increase by which the proposed model outperforms existing approaches like keyword spotting and query expansion in precision and recall values. We used the mean average precision, mean average recall, and mean average cohesion values of user queries in relation to document clustered and document retrieved to showcase this percentage increase. In Table 1 below, we present the overall precision and recall values on the user query.

Table 1 Precision and recall values for ten user queries

Queries	Precision value			Recall value		
	Keyword spotting	Query expansion	UFCNL	Keyword spotting	Query expansion	UFCNL
Q_1	45.53	53.54	59.54	62.44	69.44	74.17
Q_2	43.88	54.28	60.46	60.68	68.51	72.45
Q_3	43.65	53.17	59.68	60.78	68.12	74.76
Q_4	45.83	53.34	60.74	60.74	69.23	75.62
Q_5	46.68	50.45	58.72	63.98	69.45	75.67
Q_6	46.81	51.26	58.16	61.72	70.52	76.75
Q_7	44.41	52.33	60.14	60.49	69.1	75.21
Q_8	46.55	52.41	59.24	59.14	70.34	76.43
Q_9	45.42	51.34	59.86	61.62	69.35	75.66
Q_{10}	45.46	50.88	61.88	62.85	69.62	75.39

Table 2 Summarized experiment and result

Approach	Mean average precision (%)	Mean average recall (%)	Mean average cohesion (%)
Keywords spotting	45.42	61.44	54.03
Query expansion	52.30	69. 37	60.22
UFCNL	59.84	75.21	83.5

From Table 1, we evaluate the mean average recall (MAR) value and mean average precision (MAP) value. The MAR tested for the average fraction of relevant documents that are retrieved by the system out of the entire document collection, while the MAP tested for average fraction of documents retrieved by the system that are relevant. In the precision test, the proposed model outperformed query expansion and keyword spotting algorithms by 7.5 and 14.4 % respectively. While evaluating the recall values, UFCNL engendered 5.8 % relevant document when compared to query expansion and 13.8 % when compared to keyword spotting algorithm. In like manner, we also tested for the mean average degree of cohesion that existed in the user query. The cohesive test is necessary to evaluate the degree to which documents in a cluster are related to each other as compared to the traditional keyword spotting approach, the query expansion technique where such is minimal. By using a threshold value of 0.5, the total mean average results from these experiments are summarized in Table 2.

From Table 2, the degree of relationship between clustered words in UFCL approach is 23.3 and 31.3 % better to query expansion and keyword spotting approach respective. Thus, the obtained results indicates that the proposed UFCNL outperforms the existing keyword spotting and Query expansion technique in all experiment conducted. This approach in no doubt has helped to enhance the retrieval of relevant document from the data corpus based on the users query in voice information retrieval system for documents.

5   Conclusion

The main given contribution in this research is establishing a conceptual knowledge of the user's query on recognition after reviewing the semantic structure of the database and later evaluate the impact of the former on the latter. The essence of achieving a conceptual knowledge in user query is channeled towards modeling the information seekers subject, order than just selecting the presented keywords for document search. Here, the model can intelligently learn and extract concept from the user's search query which will further provide relevant search results for each differentiated end-users from the inferred user query concept.

References

1. Sugimoto, K., Nishizaki, H., Sekiguchi, Y.: Effect of document expansion using web documents for spoken documents retrieval. In: Proceedings of the 2nd Asia-Pacific Signal and Information Processing Association Annual Summit and Conference, pp. 526–529 (2010)
2. Fayolle, J., Saraclar, M., Moreau, F., Raymond, C., Gravier, G.: Lexical-phonetic automata for spoken utterance indexing and retrieval. In: Proceedings of International Conference on Speech Communication and Technology (2012)
3. Parlak, S., Saraclar, M.: Spoken term detection for Turkish broadcast news. In: submitted to ICASSP (2008)
4. Hazen, T., Shen, W., White, C.: Query-by-example spoken term detection using phonetic posteriogram templates. In: Proceedings of the IEEE Workshop on Automatic Speech Recognition and Understanding. Merano, Italy (2009)
5. Mamou, J., Ramabhadran, B., Siohan, O.: Vocabulary independent spoken term detection. In: Proceeding of SIGIR (2007)
6. Katsurada, K., Sawada, S., Teshima, S., Iribe, Y., Nitta, T.: Evaluation of fast spoken term detection using a suffix array. In: Proceedings of International Conference on Speech Communication and Technology, pp. 909–912 (2011)
7. Itoh, Y., Iwata, K., Ishigame, M., Tanaka, K., Shi-wook, L.: Spoken term detection results using plural subword models by estimating detection performance for each query. In: Proceedings of International Conference on Speech Communication and Technology, pp. 2117–2120 (2011)
8. Akiba, T., Honda, K.: Effects of query expansion for spoken documnet passage retrieval. In: Proceedings of International Conference on Speech Communication and Technology, pp. 2137–2140 (2011)
9. Tomoko, T, Akiba, T.: Open vocabulary spoken content retrieval by front ending with spoken text detection. In: Signal and Information Processing Association Annual Summit and Conference (APSIPA), 2013 Asia-Pacific, pp. 1–6. IEEE (2013)
10. Mamou J., Ramabhadran B., Siohan, O.: Phonetic query expansion for spoken term detection. In: Proceedings of Interspeech (2008)
11. Toth, B., Hakkani-Tur, D., Yaman, S.: Summarization and learning based approaches to information distillation. In: Proceedings of ICASSP (2010)

Online Pairwise Ranking Based on Graph Edge–Connectivity

Carlos Quintero, Reinaldo Uribe, Juan Calderón and Fernando Lozano

Abstract We propose a novel ranking algorithm that takes into account specific properties of the graph that represents the items and the user votes in pairwise comparison scenarios. The algorithm models the scoring relationships between instances as local edges among vertices in a corresponding graph and use such properties to find scores for each instance. We have compared the performance of the algorithm with other widely known information retrieval techniques tasked with ranking a set of movies. As a baseline implementation, we have used the topological ordering of the acyclic subgraph with maximum weight, by solving an approximated version of the maximum acyclic subgraph problem. The results show accurate ranking lists for the movie dataset.

Keywords Online ranking · MASP · Vertex–connectivity · Machine learning

1 Introduction

The rapid growth of the world wide web has allowed the easy generation of large amounts of available data collected as opinions from different users all over the world. Data mining researchers have turned their attention to developing new methods related to ranking a set of instances. Dozens of applications such as web search engines, image sharing websites, document search, social network graphs, video

C. Quintero (✉) · J. Calderón (✉)
Universidad Santo Tomás, Bogotá, Colombia
e-mail: carlosquinterop@usantotomas.edu.co

J. Calderón
e-mail: juancalderon@usantotomas.edu.co; juan_mch@yahoo.com

R. Uribe · F. Lozano
Universidad de Los Andes, Bogotá, Colombia
e-mail: r-uribe@uniandes.edu.co

F. Lozano
e-mail: flozano@uniandes.edu.co

© Springer International Publishing Switzerland 2016
V. Snášel et al. (eds.), *Innovations in Bio-Inspired Computing and Applications*,
Advances in Intelligent Systems and Computing 424,
DOI 10.1007/978-3-319-28031-8_44

postings, comments and movie recommendations have directed recent research in this field. We consider the problem of learning an order of objects based on a preference function provided by a human in the form of pairwise comparisons between the objects. It is well known that it is difficult for humans to create an ordered list of more than 5 to 7 instances according to their preferences [1]. How ever, humans are easily capable of defining a preference relationship between pairs of items. The challenge then becomes how to create an ordered list based on pairwise instance comparisons.

Two main approaches are common in the problem of ranking a set of instances. On one hand, there are independent scoring methods where each instance is graded by the user on some scale. On the other hand, there are pairwise–based methods, where two instances are shown to the user and she must select her preferred one. Sreenivas et al. [2] has shown that (under certain assumptions) creating ordered lists using pairwise–voting methods usually converges faster to accurate rankings than using independent score methods. Furthermore, in some contexts such as sport tournaments, only pairwise comparisons are allowed, making pairwise comparison methods natural for ranking data [3].

An interesting example comes from rating items in forums on the web such as Youtube videos, Facebook comments or posts on twitter forums. Sreenivas et al. [2] proposed a mechanism to rank items in twitter–like forums and contrasted the pairwise–comparison and independent score methods. They have shown theoretically that for some probability models there is no bounded number of reviews per item that a user can perform using thumb (or star) based ranking (score–based methods) to achieve high accurate rankings. However, bounded cost pairwise–based methods that produce highly accurate rankings can indeed be found. These "learning to rank" methods can be combined with the usual approaches from the field of statistical learning, such as SVM–Rank [4] and RankBoost [5] in order to create recommendation systems for different applications based on the lists of multiple users.

The main contributions of this paper are twofold:

- First, we proposed an heuristic algorithm based on breadth first search (BFS) to approximately solve the maximum acyclic subgraph problem (MASP). It assumes the votes of a fairly consistent user as the source of the graph and exploits the structure of the ranking problem to reduce memory storage and hasten computation.
- Finally, we proposed an algorithm that takes advantage of explicit properties of the graph that represents the problem, in order to incrementally create an ordered list of user preferences from pairwise comparisons. To the best of our knowledge, this is the first approach that uses explicit graph properties in order to create accurate preference lists.

2   The Maximum Acyclic Subgraph Problem in Ranking

A natural way of modeling the pairwise ranking problem is to consider each instance as a vertex in a directed graph, where there is an edge between the preferred and the dismissed instances in a vote. Formally, each instance is projected into a vertex

$v \in V$ of the directed graph $G = (V, E)$ and $(i, j) \in E$ only if instance i is preferred to instance j (denoted $i \succ j$). Assuming that the user preference function is *transitive*, that is, when $i \succ j$ and $j \succ k$ implies $i \succ k \; \forall i j, \; k \in V$, then the corresponding graph will be acyclic and the indexes of its topological ordering correspond to the positions of the items in the preference list.

In typical applications, however, the graph G is constructed from the user's pairwise preferences through a voting procedure, and even for relatively small sets of instances, it will contain cycles. Nevertheless, for a fairly consistent user one can argue that the number of cycles in such situation will remain low. In this context, the removal of all such conflicting edges would make the graph acyclic, allowing a topological ordering of the resulting subgraph and thus finding a solution to the ranking problem. The task of removing the minimum number of edges such that the remaining subgraph is acyclic is equivalent to the well known maximum acyclic subgraph problem.

Definition 1 (*MASP*) Given a directed graph $G = (V, E)$, $|V| = n$ with arc weights $w_{ij} > 0$, the *maximum acyclic subgraph problem* is to find a subset $E' \in E$ such that $G' = (V, E')$ is acyclic and $w(E') = \sum_{(i,j) \in E'} w_{ij}$ is maximized.

The solution found by solving an instance of MASP is the one with maximum total weight. However, MASP is known to be NP–hard [6].

2.1 Related Work

MASP belongs to the family of edge deletion problems and it is known to be polynomially solvable when G is planar [7]. The easiest approach consists on what is known as the 0.5 approximation. It basically consists on randomly choosing two subgraphs $E_1 = \{(i, j) \in E | i < j\}$ and $E_2 = \{(i, j) \in E | i > j\}$ and returning the one with maximum weight $\max\{w(E_1), w(E_2)\} \geq 0.5 w(E)$. One of them will have weight larger than half of the total weight of the graph.

Many bounds have been presented in the last decades trying to improve the 0.5 approximation to the MASP [8–11]. However, improving this approximation still a long standing open problem [8]. For example [9] proposed two algorithms to maintain a topological ordering of a directed acyclic graph as arcs are added and to detect a cycle when one is created. Alternatively, [8] proposed an algorithm that is able to find a permutation of the vertices that induces an acyclic subgraph with a small polynomial gain over random. This approximation solves the problem of finding the cut norm of a skew–symmetric matrix due to [12] by solving a relaxed semidefinite problem and a novel rounding technique. For more information on the bounds related to the advantage over random for the MASP, see [11].

2.2 Solving an Instance of the MASP

Most of the techniques mentioned above guarantee bounds on the MASP that are not far from the 0.5 approximations and are applicable to general graphs. However, these techniques might not be appropriate for the problem of ranking a set of objects. In the ranking problem the user is usually consistent, i.e., most of the times her votes will reflect transitivity among objects in the set, leading to a graph with few cycles. In other words, the amount of edges that need to be removed from the graph to obtain an acyclic graph should remain low.

Using the previous statement, one can try to solve an instance of the MASP by directly searching the elementary cycles in the graph. This strategy consists on building a breadth first search tree starting from one chosen vertex (the root) and storing all the paths to all other vertices in the graph. On each step, each new vertex is checked so that if the reached vertex corresponds to the current root then a new cycle has been found. Running the algorithm for each vertex as the root will find all the cycles in the graph. Then, we can keep a count of the edges that appear in most cycles and remove them until no cycles are left.

The procedure mentioned above has some implementation issues. The number of possible paths and the number of cycles for a given root may grow very rapidly, bringing storage problems. This growth can be kept limited by running small instances of the problem and keeping the most popular edges out of the graph for larger instances. We have implemented a strategy that allowed us to solve[1] an instance of the MASP with $|V| = 100$ vertices and $|E| = 4950$ edges within a reasonable time. This set (G, E) represents the pairwise preferences of a user's 100 favorite movies. The strategy is shown next:

1. Initialize the set of instances with a low number of vertices (i.e., 10 vertices)
2. Build the BFS tree and store all the elementary cycles in the graph
3. If maximum storage capacity is reached, decrease the number of vertices and go back to step 2, otherwise go to step 4
4. Find the most popular edges for the current instance (i.e., the edges that appear most frequently in the elementary cycles)
5. If the number of cycles related to a specific edge in the graph doubles or increases beyond a given threshold between one iteration and the next, remove it and go to step 2, otherwise go to step 6.
6. Increase the size of the set of instances and go to step 2

The acyclic subgraph found by our strategy had $|E'| = 4869$ edges, meaning that only 81 edges were removed from the original graph. After that, we were able to compute a topological ordering of the subgraph that allowed us to obtain an ordered list of object preferences. This order will be used as a baseline for our experiments.

---

[1] Actually, any solution of a MASP instance found by using the strategy proposed in this section will not be optimal since the removal of critical edges before building the BFS tree in step 5 changes the original problem.

3 Connectivity–based Ranking

In this section we present an algorithm that creates a user preference list that is based on specific properties of the graph represented by the user votes. The idea, as stated before, consists on mapping the set of instances as the set of vertices V and user votes as a set of edges E in a directed graph. In the graph, $e_{ij} \in E$ if the user prefers instance i over instance j. Figure 1a shows a graph that represents a small instance of a ranking problem. Each directed edge (i, j) represents a preference $i > j$ and is assumed to have unity weight. In this case, vertex 0 is clearly better than all other vertices and also that vertex 4 is the worst of the whole set.

In Fig. 1a, it is easy to see that vertex 0 is better than vertex 4, not only because the $(0, 4)$ edge is in the graph, but also because there are several other possible directed paths through which 4 could be reached from 0 This suggests that an interesting question would be how and *how much* better is 0 than 4. In order to answer such a question we can take a look at the *number of different paths* that we can use to go from 0 to 4. These paths represent all the ways in which instance 0 could defeat instance 4 indirectly. Figure 1b–f show these paths highlighted. In this way, we can say that we would need to *remove* 5 edges of the graph so that instance 0 could not defeat instance 4 (in order to disconnect vertex 0 from vertex 4). This is closely related to a very well known property of the graph, its edge–connectivity.

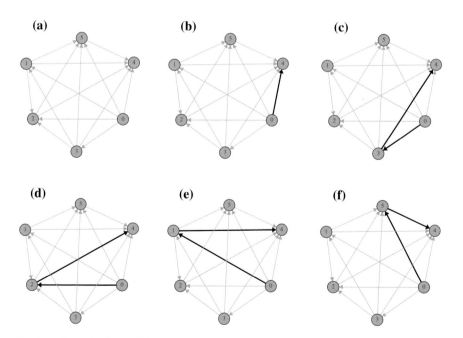

Fig. 1 **a** Example of a small instance in a ranking problem. **b–f** Paths from vertex 0 to 4

The local edge connectivity $\lambda_G(a, b)$ from vertex a to vertex b is defined as the number of edges–disjoint paths between a and b [13]. Intuitively, it can be seen as the minimum number of edges that are required to be removed in order to disconnect a from b. The usual procedure to compute $\lambda_G(a, b)$ when G is a directed graph consists on assigning a value of 1 to all edges in the graph and computing the maximum flow from source a to sink b. The matrix containing all local edge–connectivities is called Λ.

The local connectivity $\lambda_G(a, b)$ may give an idea of how good is item a with respect to item b. We can easily generalize this in order to know how good is item a with respect to all other items in the set by simply summing over all local connectivities,

$$ s_a = \sum_{i=1, i \neq a}^{n} \lambda_G(a, i) . \tag{1} $$

Finally, the list of preferences can be found by sorting the vector of scores S,

$$ \text{Ranking} = \text{sort}([s_1, s_2, \ldots, s_n]) . \tag{2} $$

The full ranking method based on edge–connectivity would first compute the complete local connectivity matrix Λ_G between all pairs of instances, then it would sum over the rows and sort the resulting vector. However, this would need the graph to be completed (for a problem with n instances, the number of pairwise comparisons is $O(n^2)$). This is usually not true for real ranking applications such as movie ranking, where only a small percentage of the votes are actually available. Besides, the complexity of the all–pairs–local–connectivity algorithm for a graph with $|V| = n$ vertices and $|E| = m$ is $O(n^2 m \min\{m^{1/2}, n^{2/3}\})$ []. Taking into account that $m = O(n^2)$, the total complexity is $O(n^5)$. This technique must be carefully used for large matrices and may not be suitable for online pairwise comparison.

In the context of online pairwise comparison, one should take into account that both, only one or none of the two instances being compared at a certain time could be new vertices. Additionally, the local connectivity between all pairs of movies does not need to be recomputed between one iteration and the next, and one can use the previously available information to update the matrix of connectivities given that only one edge is added at each step (every user vote is represented as a single edge in the corresponding directed graph).

Assuming that instances p_i and p_j are chosen to be compared, the possible scenarios are:

- Both instances are new, $p_i \notin V$, $p_j \notin V$ and $p_i \succ p_j$ as shown in Fig. 2a. Then two rows and two columns should be added to the connectivity matrix and the connectivity between the old vertices in the graph remain the same.

  - $\lambda_G^{(t)}(p_i, p_j) = 1$
  - $\lambda_G^{(t)}(p_j, p_k) = \lambda_G^{(t)}(p_i, p_k) = 0, \forall k \in V$
  - $\lambda_G^{(t)}(p_k, p_j) = \lambda_G^{(t)}(p_k, p_i) = 0, \forall k \in V$

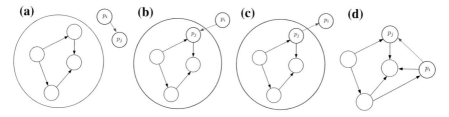

Fig. 2 **a** Instances p_i and p_j do not belong to the current list and $p_i \succ p_j$. **b** Instance p_i does not belong to the current list but p_j does and $p_i \succ p_j$. **c** Instance p_i does not belong to the current list but p_j does and $p_j \succ p_i$. **d** Instances p_i and p_j belong to the current list and $p_i \succ p_j$

- One instance is new and the other one is old, $p_i \notin V, p_j \in V$ and $p_i \succ p_j$ as shown in Fig. 2b. Then one new row and one column should be added to the connectivity matrix. The connectivity between every vertex and p_i is zero (since there are no incoming edges), the connectivity between p_i and all other vertices is the same as between p_j and all other vertices and the connectivities between all other vertices remain the same.

  - $\lambda_G^{(t)}(p_i, p_j) = 1$
  - $\lambda_G^{(t)}(p_k, p_i) = 0, \forall k \in V$
  - $\lambda_G^{(t)}(p_i, p_k) = \lambda_G^{(t-1)}(p_j, p_k), \forall k \in V$

- One instance is new and the other one is old, $p_i \notin V, p_j \in V$ and $p_j \succ p_i$ as shown in Fig. 2c. Then one new row and column should be added to the connectivity matrix. The connectivity between p_i and every other vertex is zero (since it has no outgoing edges), the connectivity between every vertex and p_i is the same as the connectivity between every vertex and p_j and the connectivities between all other vertices remain the same.

  - $\lambda_G^{(t)}(p_j, p_i) = 1$
  - $\lambda_G^{(t)}(p_i, p_k) = 0, \forall k \in V$
  - $\lambda_G^{(t)}(p_k, p_i) = \lambda_G^{(t-1)}(p_k, p_j), \forall k \in V$

- Both instances are already in the list, $p_i \in V, p_j \in V$ and $p_i \succ p_j$ as shown in Fig. 2d. Then the connectivity between p_i and p_j is increased by one, the connectivity between any vertex that is not connected with p_i and p_j remains the same as does the connectivity between p_i and any vertex that has no connection from p_j. The connectivity between any vertex and all others also remains the same if the connectivity between such vertex and any other is equal to its out–degree d^+. In the same way, the connectivity between any vertex and all others remains the same if the connectivity between any vertex and such vertex is equal to its in–degree d^-.

  - $\lambda_G^{(t)}(p_i, p_j) = \lambda_G^{(t-1)}(p_i, p_j) + 1$
  - $\lambda_G^{(t)}(p_k, p_j) = \lambda_G^{(t-1)}(p_k, p_j), \forall k$ such that $\lambda_G^{(t-1)}(p_k, p_i) = 0$ or $\lambda_G^{(t-1)}(p_k, \cdot) = d^+(p_k)$

- $\lambda_G^{(t)}(p_i, p_k) = \lambda_G^{(t-1)}(p_i, p_k), \forall k$ such that $\lambda_G^{(t-1)}(p_j, p_k) = 0$ or $\lambda_G^{(t-1)}(\cdot, p_k) = d^-(p_k)$
- The connectivity between all other vertices must be recomputed using the maximum flow method

The procedure mentioned before gives rise to the online pairwise connectivity algorithm that creates a preference list based on user pairwise voting. The algorithm is shown in Algorithm 1.

Algorithm 1 Pairwise edge connectivity algorithm

1: **for** $t = 1, 2, \dots$ **do**
2:     Pick p_i and p_j randomly
3:     **if** $p_i \notin V$ and $p_j \notin V$ **then**
4:         Update V as $V \cup \{p_i, p_j\}$ and E as $E \cup \{e_{ij}\}$ or $E \cup \{e_{ji}\}$
5:         Update Λ by adding two rows and columns of zeroes and $\lambda_G(p_i, p_j) \leftarrow 1$
6:     **else if** $p_i \notin V$ and $p_j \in V$ **then**
7:         **if** $p_i \succ p_j$ **then**
8:             Replicate row $\Lambda_G(p_j, :)$ in new row $\Lambda_G(p_i, :)$
9:             $\lambda_G(p_i, p_j) \leftarrow 1$
10:        **else**
11:            Replicate column $\Lambda_G(:, p_j)$ in new column $\Lambda_G(:, p_i)$
12:            $\lambda_G(p_j, p_i) \leftarrow 1$
13:        **end if**
14:     **else**
15:         $\lambda_G(p_i, p_j) \leftarrow \lambda_G(p_i, p_j) + 1$
16:         Compute $\lambda_G(p_k, p_j) = \lambda_G(p_k, p_j), \forall k$ such that $\lambda_G(p_k, p_i) = 0$ or $\lambda_G(p_k, \cdot) = d^+(p_k)$
17:         Compute $\lambda_G(p_i, p_k) = \lambda_G(p_i, p_k), \forall k$ such that $\lambda_G(p_j, p_k) = 0$ or $\lambda_G(\cdot, p_k) = d^-(p_k)$
18:         For the rest, compute λ_G using the max flow routine
19:     **end if**
20: **end for**

---

4 Experiments and Discussion

We have performed experiments to show the behavior of our connectivity-based online ranking algorithm and compare it with simple and widely known techniques. The dataset used in the experiments was created by ourselves and it consists on 100 movies and 4950 votes (all possible pairwise votes). The procedure has the simple mechanics of selecting two randomly chosen movies from the database and updating the user preference list at each step, according to each algorithm's policy. As stated in Sect. 2, we used the preference list obtained by approximately solving the MASP as reference in order to assess the performance of all other algorithms.

We have implemented five ranking algorithms that are based on user's pairwise comparisons and compared their performance for the task at hand. Initially, we have implemented two simple greedy algorithms that try to find a good ordering of the instances. **Findrank** finds in the adjacency graph the index of the first instance that is preferred over instance *i* and rearranges the rows and columns of the adjacency matrix in that ordering iteratively. **Stumprank** finds the index in each row of the adjacency matrix that minimizes the error separating zeroes to the left and ones to the right of the row.

From a different perspective, Flickchart's algorithm (http://www.flickchart.com) has been used in recent years to rank movies by using pairwise comparisons in the web, unlike other movie ranking sites that assign stars or tomatoes to each movie. The idea consists on maintaining a top subset of winners and a bottom subset of losers. Every new pair of movies enters the list at some threshold value that separates winners from losers after the user's vote, potentially changing the order of the instances. Secondly, the ELO rating system, widely used to rate chess players and sports teams. The idea is to model the skill of each player as a logistic distributed random variable. The expected score of a player is updated with the results of every game, either winning, losing or drawing. Finally, the EigenRank algorithm assumes that the importance of some element is proportional to the sum of importances of all elements that are worse than it. The mathematical model is written in such a way that the vector of preferences is given by the entries of one eigenvector of a matrix that contains the user's voting information. Many information retrieval methods, such as the well–known Page Rank algorithm [14], are based on this approximation.

We have used two evaluation measures in order to assess the performance of each algorithm in the movie ranking task. On one hand, the mean reciprocal rank (MRR) is the mean of the multiplicative inverse of the position for the actual top movie. The MRR is updated at iteration *t* as $MRR = \frac{1}{t} \sum_{i=1}^{t} \frac{1}{R_i}$ where R_i is the position of the top movie in iteration *i*. On the other hand, the disagreement [5] is the fraction of distinct pairs of movies that are disordered with respect to the correct list.

4.1 Results

We have performed the experiments of online pairwise ranking for a total of 19800 iterations to make sure that all possible comparisons are made and for each iteration we have updated 4 different preferences lists given by the 4 algorithms previously described. For each iteration, the MRR and Disagreement have been calculated.

Figure 3a shows the mean reciprocal rank measure for every algorithm. It can be seen that Flickchart and Eigenrank are able to find the top item faster than the rest of the algorithms followed by connectivity–rank. ELO is the only algorithm that is incapable of finding the top movie most of the times.

On the other hand, Fig. 3b shows the Disagreement for all the mentioned algorithms. This is a measure that indicates how accurate is the list created by some algorithm with respect to the base user list. All algorithms seem to approach a low disagreement at the end of the iterations.

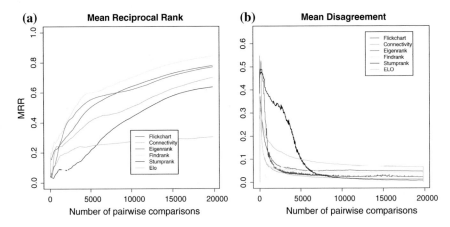

Fig. 3 **a** Disagreement measure for each step iteration found by each algorithm. **b** Disagreement measure for each step iteration found by each algorithm

5 Conclusions

In this paper we have shown the development of a framework that allowed us to introduce a new online pairwise ranking algorithm that uses specific properties of the graph. The algorithm models the problem in terms of local edge–connectivities between instances to compute the relative importance between them. By examining certain aspects of the problem and properties of the graph, we have also provided a version that does not require the calculation of the entire connectivity matrix every time a new vote is performed by the user. We have created a dataset to create an ordered list of movies and used it to compare the performance of the proposed algorithm to other widely known techniques. The initial results proved experimentally that the algorithm is capable of finding a fairly accurate list without the need for too many user votes, as desired.

References

1. Jiang, X., Lim, L.-H., Yao, Y., Ye, Y.: Statistical ranking and combinatorial hodge theory. In: Mathematical Programming (2011)
2. Sreenivas, G., Rina, P., Sarma, A.D., Atish, D.S.: Ranking mechanisms in twitter-like forums. In: Proceedings of the Third ACM International Conference on Web Search and Data Mining, pp. 21–30, 2010
3. Hochbaum, Dorit S., Levin, Asaf: Methodologies and algorithms for group-rankings decision. Manage. Sci. **52**, 1394–1408 (2006)
4. Large Margin Rank Boundaries for Ordinal Regression, chapter 7, pp. 115–132. MIT Press (2000)
5. Freund, Yoav, Iyer, Raj, Schapire, Robert E., Singer, Yoram: An efficient boosting algorithm for combining preferences. J. Mach. Learn. Res. **4**, 933–969 (2003)

6. Karp, R.M.: Reducibility among combinatorial problems. Complexity Comput. Comput. (1972)
7. Lucchesi, C.L.: A minimax equality for directed graphs. Ph.D. thesis, University of Waterloo (1976)
8. Charikar, M., Makarychev, K., Makarychev, Y.: On the advantage over random for maximum acyclic subgraph. In: 48th Annual IEEE Symposium on Foundations of Computer Science, pp. 625–633 (2007)
9. Haeupler, B., Kavitha, T., Mathew, R., Sen, S., Tarjan, R.E.: Faster algorithms for incremental topological ordering. In: Proceedings of the 35th international colloquium on Automata, Languages and Programming, Part I, pp. 421–433. Springer, Berlin (2008)
10. Hassin, Refael, Rubinstein, Shlomi: Approximations for the maximum acyclic subgraph problem. Inf. Process. Lett. **51**, 133–140 (1994)
11. Khot, S., O'Donnell, R.: Sdp gaps and ugc-hardness for maxcutgain. In: Proceedings of the 47th Annual IEEE Symposium on Foundations of Computer Science, pp. 217–226. IEEE Computer Society, Washington, DC, USA (2006)
12. Noga, A., Assaf, N.: Approximating the cut-norm via grothendieck's inequality. In: Proceedings of the Thirty-sixth Annual ACM Symposium on Theory of Computing, STOC '04, pp. 72–80. ACM, New York, NY, USA, (2004)
13. Ulrok B., Thomas, E.: Network Analysis. Springer (2005)
14. Lawrence, P., Sergey, B., Rajeev, M., Winograd, T.: Bringing order to the web. Stanford Digital Library project, talk, the pagerank citation ranking (1999)

Fuzzy Variable Stiffness in Landing Phase for Jumping Robot

Juan M. Calderón, Wilfrido Moreno and Alfredo Weitzenfeld

Abstract Some important applications of humanoid robots in the nearest future are elder care, search and rescue of human victims in disaster zones and human machine interaction. Humanoid robots require a variety of motions and appropriate control strategies to accomplish those applications. This work focuses on vertical jump movements with soft landing. The principal objective is to perform soft contact allowing the displacement of the Center of Mass (CoM) in the landing phase. This is achieved by affecting the nominal value of the constant parameter P in the PID controller of the knee and ankle motors. During the vertical jump phases, computed torque control is applied. Additionally, in the landing phase, a fuzzy system is used to compute a suitable value for P, allowing the robot to reduce the impact through CoM displacement. The strategy is executed on a gait robot of three Degrees of Freedom (DoF). The effect of the impact reduction is estimated with the calculations of the CoM displacement and the impact force average during the landing phase.

1 Introduction

The development of humanoid robots has increased exponentially in the last few years. Many organizations around the world, such as RoboCup and DARPA [1], are involved in establishing guidelines for humanoid robotics development in various

J.M. Calderón (✉) · W. Moreno
Department of Electrical Engineering, University of South Florida, Tampa, FL, USA
e-mail: juancalderon@mail.usf.edu

W. Moreno
e-mail: wmoreno@usf.edu

A. Weitzenfeld
Department of Computer Science and Engineering, University of South Florida,
Tampa, FL, USA
e-mail: aweitzenfeld@usf.edu

J.M. Calderón
Department of Electronic Engineering, Universidad Santo Tomás, Bogotá, Colombia

© Springer International Publishing Switzerland 2016   511
V. Snášel et al. (eds.), *Innovations in Bio-Inspired Computing and Applications*,
Advances in Intelligent Systems and Computing 424,
DOI 10.1007/978-3-319-28031-8_45

application domains. The robotics road map for USA [2] and Europe has mentioned some important applications of humanoid robots. Some examples include: rescuing human victims in disaster zones, taking care of the elderly population, and any other application that involves risky actions for human beings. For these to succeed, humanoid robots need to develop advanced locomotion capabilities analogous to humans, such as walking, jumping and running. These movement capabilities are natural for humans, but they are significantly difficult for robots. Any humanoid movement requires special attention to aspects such as equilibrium, stability, soft contact, and low energy consumption. The work presented in this paper focuses on the process of vertical jumping and soft contact during the landing phase. The rest of the paper is composed of Sect. 2—Related Work, Sect. 3—Humanoid Jumping Process, Sect. 4—Model and Jumper Robot, Sect. 5—Control of Vertical Jump, Sect. 6—Experiment and Results, Sect. 7—Conclusions.

2 Related Work

The work presented in this paper extends the original ideas exposed by Kajita et al. in [3] where jumping is used as a first attempt to run, and by Raibert et al. in [4] where a study about balance and control of a legged hopping robot that is able to jump and run was presented. Additional related work includes different aspects of jumping and running models. Sakka and Yokoi explain in [5] how to use Ground Reaction Force (GRF), and robot inertia to optimize jump height using a virtual version of a HRP robot. They also propose a motion pattern generation method for vertical jumping in a humanoid robot [6]. They present a special policy of movement in the landing phase that reduces the impact force. This approach, however, does not use compliant actuators and test was performed in simulation only. A robot with compliant motor capabilities is described by Missura and Behnke [7], proposing an algorithm to generate an open-loop walking motion in a bipedal robot. While their work focused on the walking movement, they used comparable motors as in this work to generate compliant actions in the robot movements. A Fuzzy logic control approach is presented by [8–10] in a real robot called KURMET. This robot is a five-link planar biped, and it uses Unidirectional Series-Elastic Actuators (USEAs). The actuator consists of a DC brushless motor, and a planetary gearhead in series with a spiral torsion spring. This robotics platform uses a fuzzy control approach to perform jumping and running movements. Other types of actuators have been used to perform compliant movements. Such is the case of [11, 12] where they used pneumatic actuators to create an artificial muscle with similar mechanical properties to human muscles. These papers describe different features of the proposed artificial muscle, including the basic control system that is employed to develop humanoid robots with compliant movements. Similarly, [13] proposed a trajectory generation method to perform jumping and walking movements in a LUCY robot. This robot was developed using artificial pneumatic muscles as explained previously. In [14, 15] are presented studies about responsibility of gastrocnemius muscles and elastic tendons in the jumping process. The researchers evaluated the effect of muscles, tendons, and their stiffness in the height of vertical jump. These works introduced a robot dynamic model including the stiffness effect concluding that there is a dependency between the stiffness of the muscle and the ability to jump. This work focuses in the vertical jump process control and variable stiffness generation. Computed Torque Control to track the desired trajectories according to the ZMP and CoM conditions were used. The variable stiffness capability is generated without the use of any spring or dampers. All the actuators in the robot are DC Motors and the stiffness is achieved by recalculating the P and D values of the PID controller for each motor. A fuzzy system estimates the new value of P in the landing phase accordingly to the impact velocity and the desired landing displacement, i.e. soft or hard contact.

3 Humanoid Jumping Process

Vertical jumping is the action executed by human beings when the Center of Mass (CoM) is raised over the normal standup human position. This movement has to be performed solely by the muscles actions without the help of any external device. The main criteria to evaluate vertical jump efficiency is the maximum height reached in the flight phase. The vertical jump is normally divided in four phases: preparatory, take-off, flight and landing phase depicted in Fig. 1.

Preparatory Phase: When the CoM is moved to a lower position and, the potential energy decreases. The hips and knees are flexed, and the ankles are dorsiflexed [16].

Take-off Phase: When the CoM is moved to a higher position and, the potential energy increases. The hips and knees are extended, and the ankle plantars are flexed. During this phase the feet stay in contact with the ground. This phase finishes when the feet are no longer touching the ground.

Flight Phase: When the body is in the air. It starts when the feet are no longer in contact with the ground, and it finishes when they touch the ground again. The height

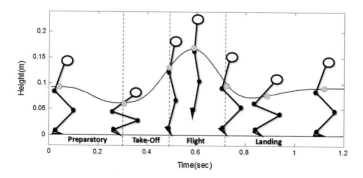

Fig. 1 Vertical jump phases and CoM trajectory

of the jump depends on the velocity reached by the CoM at the beginning of the phase. During flight, the body loses control of the rotation and trajectory, thus the position of the body is determined by the trajectory, velocity, and acceleration of the previous phase. Acceleration, velocity, position, and maximum height in the flight phase are described by (1) through (4), respectively.

$$\ddot{y}_{com}(t) = -g \tag{1}$$

$$\dot{y}_{com}(t) = -gt + V_{to} \tag{2}$$

$$y_{com}(t) = -\frac{1}{2}gt^2 + V_{to}t + y_{to} \tag{3}$$

$$y_{comMax}(t) = -\frac{V_{to}^2}{2g} \tag{4}$$

where g is gravity force, V_{to} and y_{to} are the velocity and position at the end of the take-off phase respectively.

Landing Phase: When the feet touch the ground again. In this phase the lower body tries to absorb and reduce the impact force exerted by the floor. Assuming that the robot velocity before the impact (landing velocity) is known, the impact force can be estimated as:

$$F_{i-avg} = \frac{\frac{1}{2}mV_l^2}{d} \tag{5}$$

where m is the mass of the body, V_l is the landing velocity, which can be computed using (3), and d represents the distance traveled by the object after the impact. Figure 1 shows the position of a human CoM during a vertical jump movement. The trajectory shows how the CoM is going down in the landing phase, when the legs are trying to damp the impact exerted by the ground. A similar approach is employed by this work, where the landing impact is naturally reduced with the use of low stiffness in the electrical motors driving the ankle and knee joints.

4 Model and Jumper Robot

In the human vertical jump, both legs are doing the same movements. Using this idea, a reduced planar model with three Degrees of Freedom (DoF) is proposed. The proposed model has three joints and four links. The joints are the ankle, the knee, and the hip. The links are the foot, the shank, the thigh, and the trunk.

The kinematic model is used to calculate the position of every link, and to estimate the velocity, acceleration, and position of whole robot's CoM, as depicted in Fig. 2a. L_i denotes the length of link i, θ_i is the absolute rotation of joint i, q_i is the relative

Fig. 2 Robot model and jumper robot in sagittal plane. **a** Robot Model. **b** Jumper Robot

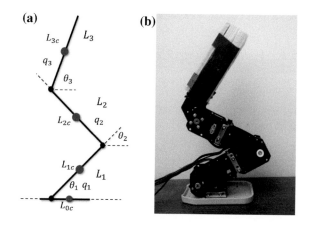

rotation of joint i, and L_{ic} shows the position of CoM for the corresponding link i. The dynamic model is given by the Lagrange-Euler formulation as shown in (6), where $q(t) \in \mathbb{R}^n$ is the joint variable vector, $D(q)$ is a 3x3 inertial matrix, $H(q, \dot{q})$ is a 3x1 vector of Coriolis and centrifugal forces, $G(q)$ is a 3x1 gravitational forces matrix, τ represents the control torque input, and τ_d disturbance.

$$D(q)\ddot{q} + H(q, \dot{q})\dot{q} + G(q) + \tau_d = \tau \qquad (6)$$

The jumper robot is a three DoF robot. The actuator of every joint is a MX-28 motor from Robotis Inc. The actuator's weight is 72 gms, and it can provide a maximum torque of 3.1 Nm. The motor runs in "endless turn" and in torque control mode. The torque control input signal can be adjusted at a 0.1 % resolution of the maximum torque allowed by the current voltage supply. The motor provides feedback about the angular position with a 0.088° resolution, the velocity, and the joint torque. Additionally, the motor has internally a micro controller unit Cortex-M3 of 32 bits, where a programmable PID controller is implemented. The Controller is implemented using a CM-2 board with an Atmega128 CPU. This board communicates with the PC via the RS232 protocol, and with the motors via the RS485 protocols. Table 1 shows the robot link parameters used in the kinematic and dynamic model.

Table 1 Parameters of robot

	Length $L_i(m)$	Mass $m_i(Kg)$	Center of mass $L_{ic}(m)$
Foot	0.010	0.051	0.045
Shank	0.098	0.051	0.072
Thing	0.098	0.051	0.072
Trunk	0.080	0.153	0.063

5 Control of Vertical Jump

5.1 Vertical Jump Conditions

Using a similar approach from Babič et al. in [17], two primary conditions to achieve vertical jump is considered. The first one, the CoM has to move upward and the displacement in the horizontal axes has to be minimum or close to zero; also, the CoM must stay into the support polygon, foot of the robot. The second one is related to ZMP [18], where the components in the horizontal axes have to be equal to zero.

The CoM position is defined by (7), where x_{com} and y_{com} are the horizontal and vertical position of the robot's CoM, m_i is the mass of the ith link, x_i and y_i are the coordinates of the CoM of the ith link.

$$x_{com} = \frac{\sum_{i=1}^{n} m_i x_i}{\sum_{i=1}^{n} m_i}, y_{com} = \frac{\sum_{i=1}^{n} m_i y_i}{\sum_{i=1}^{n} m_i} \tag{7}$$

The ZMP is defined by (8) where θ_i is the angular velocity of the ith link, and $\mathbf{I}_i$ is the inertial tensor of the ith link around the CoM.

$$x_{zmp} = \frac{\sum_{i=1}^{n} m_i x_i (\ddot{y}_i + g) - \sum_{i=1}^{n} m_i y_i \ddot{x}_i + \sum_{i=1}^{n} (\mathbf{I}_i \dot{\omega}_i + \omega_i \times \mathbf{I}_i \omega_i)}{\sum_{i=1}^{n} m_i (\ddot{y}_i + g)} \tag{8}$$

Using (7), the second derivate of x_{com} and y_{com} for control purposes can be formulated, and x_{zmp} can be computed as shown in the following equations:

$$\ddot{x}_{com} = \alpha_1 \ddot{q}_1 + \alpha_2 \ddot{q}_2 + \alpha_3 \ddot{q}_3 + d_1 \tag{9}$$

$$\ddot{y}_{com} = \beta_1 \ddot{q}_1 + \beta_2 \ddot{q}_2 + \beta_3 \ddot{q}_3 + d_2 \tag{10}$$

$$x_{zmp} = \gamma_1 \ddot{q}_1 + \gamma_2 \ddot{q}_2 + \gamma_3 \ddot{q}_3 + d_3 \tag{11}$$

where α_i, β_i, γ_i, and d_i are functions of joint angles (q_i)

$$\ddot{q}_d = \begin{bmatrix} \ddot{q}_1 \\ \ddot{q}_2 \\ \ddot{q}_3 \end{bmatrix} = \begin{bmatrix} \alpha_1 & \alpha_2 & \alpha_3 \\ \beta_1 & \beta_2 & \beta_3 \\ \gamma_1 & \gamma_2 & \gamma_3 \end{bmatrix}^{-1} \left(\begin{bmatrix} \ddot{x}_{com} \\ \ddot{y}_{com} \\ 0 \end{bmatrix} - \begin{bmatrix} d_1 \\ d_2 \\ d_3 \end{bmatrix} \right) \tag{12}$$

5.2 Computed Torque Control

Computed Torque Control is a widely used control strategy based on two special approaches [19, 20]. The first one uses feedback linearization of nonlinear systems.

The second one is based on a computation of the robot's required torque by the use of the nonlinear feedback control law [21]. This kind of control is based on the concept that there is a desired tracking signal and the system tries to follow it. The main idea here is to reduce the error through a feedback linearization of the system.

The error is defined as the difference between the desired trajectory and the actual joint position as shown in equation (13), and $\dot{e}(t)$ and $\ddot{e}(t)$ can be defined using a similar equation.

$$e(t) = q_d(t) - q(t) \tag{13}$$

where $q(t)$ is the current position of the actuator (14), and it is defined from dynamic robot model given by equation (6).

$$\ddot{q} = D^{-1}(q)(H(q, \dot{q})\dot{q} + G(q) + \tau_d - \tau) \tag{14}$$

Now by back substitution (14) into (13), the second derivate of error is obtained as shown in (15)

$$\ddot{e}(t) = \ddot{q}_d(t) + D^{-1}(q)(H(q, \dot{q})\dot{q} + G(q) + \tau_d - \tau) \tag{15}$$

By defining $u(t)$ as the control input function, and $w(t)$ as the disturbance function as shown below.

$$u(t) = \ddot{q}_d(t) + D^{-1}(q)(H(q, \dot{q})\dot{q} + G(q) - \tau) \tag{16}$$

$$w(t) = D^{-1}(q)\tau_d \tag{17}$$

The feedback linearization of (16) can be inverted to yield τ as given in (18).

$$\tau = D(\ddot{q}_d(t) - u(t)) + H(q, \dot{q})\dot{q} + G(q) \tag{18}$$

where $u(t)$ is then selected as the PID feedback loop control signal,

$$u(t) = -k_v\dot{e} - k_p e - k_i\varepsilon, \quad \text{where } \dot{\varepsilon} = e \tag{19}$$

To conclude, (19) in back substitute into (18), which yield the final form for τ as shown in (20). Figure 3 depicts the Computed Torque Control schema with outer PID loop feedback.

$$\tau = D(\ddot{q}_d(t) + k_v\dot{e} + k_p e + k_i\varepsilon) + H(q, \dot{q})\dot{q} + G(q) \tag{20}$$

The Computed Torque Control is applied during all four jumping phases. The system tracks the desired trajectory $(\ddot{q}_d, \dot{q}_d, q_d)$ on the take-off phase. For the flight and landing phases, the control system tries to keep the stand-up position and reject the disturbance produced by the ground impact.

Fig. 3 Computed Torque
Control schema with outer
PID loop

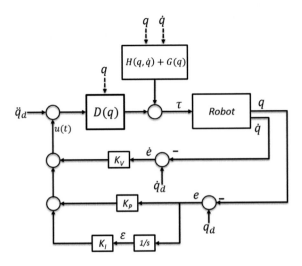

5.3 Landing Phase with Variable Stiffness

One of the principal aims of this work is the generation of soft landing in vertical
jumping movement for a real robot. The approach used to accomplish this objective
is the generation of a variable stiffness in the ankle and knee joints. The stiffness
capability is produced through the variation of the P gain of the PID controller. A P
gain value below of the nominal designed value implies low stiffness in the motor.
The D value of the PID controller is calculated assuming a critical damping response
where $D = 2\sqrt{P}$.

The impact force is proportional to the squared velocity reached by the robot
at contact, in accordance with (5). Based on the P gain effect on stiffness and the
impact force definition given above, a fuzzy system is proposed in order to estimate
the adequate P value to implement a soft contact capability.

The proposed fuzzy system is composed by two inputs and one output. The inputs
are Desired-Landing and Impact-Velocity. The output is the P gain value. The fuzzy
system is explained in terms of fuzzifier, set of rules and defuzzifier next.

Fuzzifier: The system has two inputs. The first one is the Desired Landing. It is relative to displacement of the center of mass from the initial landing position. This
input has three membership functions called soft, medium and hard. Soft means low
stiffness and it generates a large displacement. Hard means big stiffness and generates small displacement. The second input is called Impact Velocity. This variable is
calculated by (2) and (3) immediately after the take-off phase using the value of the
take-off velocity. This input has three memberships function called low, medium and
high. Finally, the output estimates the P gain value and it is classified in three memberships called: small, medium and large. The inputs, output and P value estimation
surface are shown by Fig. 4.

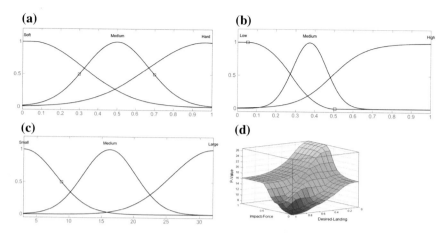

Fig. 4 Fuzzy variable stiffness estimator. **a** Desired-Landing input. **b** Impact-Velocity input. **c** P-Value output. **d** P-Value estimation surface

Table 2 Set of rules

		Impact-Velocity		
		Low	Medium	High
Desired-Landing	Soft	Medium	Medium	Large
	Medium	Small	Medium	Medium
	Hard	Small	Small	Medium

Set of rules: The set of rules are composed by nine rules that define the relation between inputs and output. This set of rules is depicted in Table 2.
Defuzzifier: The centroid is employed to estimate the final value of the P gain value.

6 Experiment and Results

In order to perform validation experiments, for vertical jumping movements in the robotics platform, the Computed Torque Control and fuzzy variable stiffness were implemented in a CM-2 board and external PC.

The experiment consists on executing several vertical jumps with different desired landing values. The objective of this experiment is to check the impact force reduction through variation of motor stiffness in the landing phase. The impact force is estimated using (5), where d is assumed as the different between touchdown position and lowest position reached by CoM in the landing phase. Large values of d, displacement in y axis of CoM in landing phase correspond to soft landing and low impact force, whereas small values of d correspond to hard landing and high impact force. Landing velocity is calculated using (3) and (4), since takeoff velocity and

maximum height are known. For every attempt the inputs of the fuzzy variable stiffness system were established according to the desired landing and impact velocity
estimated as previously explained. The results are shown in Figs. 5 and 6 and Table 3.

Figure 5a depicts CoM position during the whole vertical jump process. Every
phase of the vertical jump is clearly recognized, especially the landing phase. The
red line is the CoM position in a vertical jump with full stiffness. The blue line is

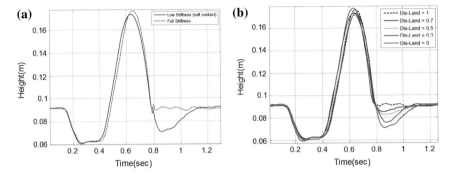

Fig. 5 Center of Mass position for different values of Disared-Landing parameter. **a** Low and full
stiffness. **b** Several levels of stiffness

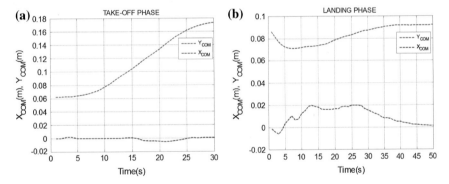

Fig. 6 X and Y position of Center of mass. **a** Take-O phase. **b** Landing phase

Table 3 Parameters of Fig. 5b

Desired-Landing	Max. high (cm)	Impact velocity (m/s)	d (cm)	Impact force average ($Kg * m/s2$)
1	2.66	1.298307	N.A.	N.A.
0.7	2.34	1.273923	0.6	41.38344
0.5	2.16	1.260041	0.855	28.40968
0.3	2.79	1.308083	1.528	17.13320
0	2.35	1.274692	2.009	12.37434

when the fuzzy stiffness system is activated and soft landing is reached. Figure 5b shows several trials with different desired landing values. According with Fig. 5b, lower values of desired landing increase the distance d, which means the impact force average was reduced. The results show impact force average, impact velocity, maximum high reached and desired landing are shown in Table 3. Table 3 shows how the impact force is reduced according to desired landing level with an impact velocity almost constant. Additionally, results show that the Computed Torque Control is allowing experiment repeatability, as the values of landing velocity and maximum height have low variance between trials. Other important aspect to analyze is the CoM displacement in takeoff and landing phases. Figure 6 shows x_{com} and y_{com} in the time for takeoff and landing phases. The maximum deviation of the CoM in the take-off phase is 0.4 cm. That means the Computed Torque Control has a good performance following the original trajectory. But in the landing phase the standard deviation of CoM is 0.75 cm and maximum deviation is 2.1 cm. Although the robot keeps balance, the movement of CoM far from the zero point can affect the robots equilibrium. We are planning to implement other control approaches such as impedance control using fuzzy estimators to improve this aspect of the landing phase.

7 Conclusions

A Fuzzy Variable Stiffness System was proposed to generate compliance capabilities in the landing phase of a jumper robot. The proposed system uses a trajectory generator based on CoM and ZMP. The control model uses a Computed Torque Control approach and the stiffness capability is generated using a fuzzy estimator. The fuzzy system estimates the P gain value of a PID controller for the landing phase allowing the displacement of the CoM from the stand up position. The estimation of the P value was done using information about Impact Velocity and Desired Landing. The proposed model was tested in a real robot having three DoF. The robot uses DC motor actuators without any dampers or springs to generate compliance capabilities. The proposed control system was evaluated running several experiments where Desired Landing was varied to get different levels of compliance. The results show a reduction of impact force according with Desired Landing where the implemented Computed Torque Control prevents the robot from falling. Different values of compliance were achieved depending on the Impact Velocity and Desired Landing values. The displacement of the x coordinate of CoM during the landing phase suggests the need for an additional control approach to improve the robot balance. Future work includes the use of compliance control with a fuzzy estimator to improve the balance of the robot.

Acknowledgments This work is funded by NSF IIS Robust Intelligence research collaboration grant #1117303 at USF and U. Arizona entitled "Investigations of the Role of Dorsal versus Ventral Place and Grid Cells during Multi-Scale Spatial Navigation in Rats and Robots"and "Octava Convocatoria Interna de Proyectos de Investigacion FODEIN 2015 # 049" at Universidad Santo Tomás, Colombia.

References

1. DARPA, DARPA Robotics Challenge, DARPA. http://www.theroboticschallenge.org/ (2015). Accessed Jan 2015
2. Georgia, I.M.I.: A Roadmap for U.S. Robotics From Internet to Robotics. Robotics in The United State of America (2013)
3. Kajita, S., Nagasaki, T., Kaneko, K.,Yokoi, K.: A hop towards running humanoid biped. In: Proceedings of ICRA'04. 2004 IEEE International Conference on Robotics and Automation (2004)
4. Raibert, M.H., Brown, B.H., Chepponis, M.: Experiments in balance with a 3D one-legged hopping machine. Int. J. Robot. Res. (1984)
5. Sakka, S., Yokoi, K.: Humanoid vertical jumping based on force feedback and inertial forces optimization. In: IEEE International Conference on Robotics and Automation (2005)
6. Sakka, S., Sian, N.E., Yokoi, K.: Motion pattern for the landing phase of a vertical jump for humanoid robots. In: International Conference on Intelligent Robots and Systems, 2006 IEEE/RSJ (2006)
7. Missura, M., Behnke, S.: Self-stable omnidirectional walking with compliant joints. In: Proceedings of 8th Workshop on Humanoid Soccer Robots, IEEE International Conference on Humanoid Robots, Atlanta, USA (2013)
8. Yiping, L., Wensing, P.M., Orin, D.E., Schmiedeler, J.P.: Fuzzy controlled hopping in a biped robot. In: International Conference on Robotics and Automation (ICRA), IEEE (2011)
9. Hester, M., Wensing, P.M., Schmiedeler, J.P., Orin, D.E.: Fuzzy control of vertical jumping with a planar biped. In: ASME 2010 International Design Engineering Technical Conferences and Computers and Information in Engineering Conference (2010)
10. Hester, M.S.: Stable Control of Jumping in a Planar Biped Robot. The Ohio State University, Ohio (2009)
11. Daerden, F.: Conception and Realization of Pleated Pneumatic Artificial Muscles and Their Use as Compliant Actuation Elements. Vrije Universiteit Brussel, Belgium (1999)
12. Beyl, P., Vanderborght, B., Van Ham, R., Van Damme, M., Versluys, R., Lefeber, D.: Compliant actuation in new robotic applications. In: NCTAM067th National Congress on Theoretical and Applied Mechanics (2006)
13. Vermeulen, J.: Trajectory Generation for Planar Hopping and Walking Robots: An Objective Parameter and Angular Momentum Approach. Vrije Universiteit Brussel, Brussel (2004)
14. Babič, J., Lenarčič, J.: Vertical jump: biomechanical analysis and simulation study, New Devel. Humanoid Robot. (2007)
15. Babič, J., Lenarčič, J.: Optimization of biarticular gastrocnemius muscle in humanoid jumping robot simulation. Int. J. Humanoid Robot. 02, 218–234 (2006)
16. Umberger, B.: Mechanics of the vertical jump and two-joint muscles: implications for training. Strength Cond. J. 20(5) (1998)
17. Babič, J., Damir, O., Lenarčič, J.: Balance and control of human inspired jumping robot. In: Advances in Robot Kinematics: Mechanisms and Motion (2006)
18. Vukobratović, M., Borovac, B.: Zero moment point thirty five years of its life. Int. J. Humanoid Robot. 157–173 (2004)
19. Hunt, L.R., Renjeng, S., Meyer, G.: Global transformations of nonlinear systems. IEEE Trans. Autom. Control, 2431 (1983)
20. Gilbert, E.G., Joong, H.: An approach to nonlinear feedback control with aplications to robotics. IEEE Trans. Syst. Man Cybern. 879–884 (1984)
21. Piltan, F., Hossein, M., Shamsodini, M., Mazlomian, E., Ho, A.: PUMA-560 robot manipulator position computed torque control methods using matlab/simulink and their integration into graduate nonlinear control and MATLAB courses. Int. J. Robot. Autom. 167–191 (2012)

An Extended Study on the Association Between Elicitation Issues and Software Project Performance: A Theoretical Model

S. Neetu Kumari and S. Pillai Anitha

Abstract Requirements elicitation issues are categorized under problems of scope, problems of volatility and problems of understanding. This study is proposed as an extension to two critical research work related to requirements elicitation issues and their impacts on overall project performance. One study determines factors that categorically represent focus areas in elicitation. The cause and effects of elicitation issues under each of these factors are identified and relevance is drawn with reference to the general categories of elicitation issues. Another study demonstrates an empirical association between categories of elicitation issues and overall project performance. A theoretical model is proposed to provide an in-depth analysis and perspective on the association between the factors influencing elicitation issues and overall project performance. This provides a framework to extend the empirical analysis to deduce focused quantitative metrics for priority-setting and decision-making to address elicitation issues.

Keywords Requirements elicitation issues · Software project performance · Contingency model

1 Introduction

Research in software engineering has laid significant focus on requirement engineering and particularly requirements elicitation (RE). RE influences the overall project performance and determines the project success or failure. In the world of

S.N. Kumari (✉) · S.P. Anitha
School of Computing Sciences, Hindustan University, Chennai, Tamil Nadu, India
e-mail: neetu_sethia@yahoo.com

S.P. Anitha
e-mail: anithasp@hindustanuniv.ac.in

© Springer International Publishing Switzerland 2016
V. Snášel et al. (eds.), *Innovations in Bio-Inspired Computing and Applications*,
Advances in Intelligent Systems and Computing 424,
DOI 10.1007/978-3-319-28031-8_46

global software development, it becomes imperative to remain constantly improvise existing processes to ensure an effective elicitation. RE issues are categorized as problems of scope, problems of volatility and problems of understanding. Studies [1] have identified factors that influence the elicitation activity. These factors require being addressed in a certain fashion owing to the project complexity, volume and the framework (for example, GSD framework) under which the system is built. The cause and effects of the issues that can surface under each of these factors have also been identified and documented in literature. These studies [2, 3] draw analyst attention to the various aspects of elicitation activity and the potential threats to effective elicitation. There have been empirical studies that determine the association between elicitation factors and overall project performance [4]. While some of the studies empirically confirm the positive or negative association between the elicitation factors such as scope, human factors, communication, etc. and overall project performance; few other studies encompass all of the elicitation factors and determine their overall association with the overall project performance. Such studies [4, 5] have enabled analysts and requirements engineers to understand the levels of influence that can be caused by the elicitation issues and relook at the requirements engineering process handbooks to effectively manage the elicitation activities.

This study is an extension [4–6] of the research work in-progress and provides a framework for related future studies in the field of requirements engineering, with particular focus on requirements elicitation.

2   Background

2.1   Factors Influencing Requirements Elicitation

A recent systematic literature review reported in [4, 6] concluded nine classifications of factors that influenced the overall elicitation activity. These classifications are related to changes, communication, human factors, knowledge, scope, social and organizational, stakeholder, tools, techniques and methods and requirements. Critical factors grouped within these classifications determined the success of failure of these factors and in turn determined their impacts on elicitation. A detailed description of the classifications and their factors are discussed in [4, 6]. To generalize these classifications in terms of elicitation issue categories, these factors are grouped under the generalized categories of elicitation issues such as problems of scope, problems of volatility and problems of understanding. These are depicted in Figs. 1, 2 and 3.

Problems of scope. The *Problems of scope (PoS)* are those in which the requirements may address too little or too much information [7]. Based on the systematic literature review conducted recently to capture and formally consolidate various details of elicitation issues, there are three key factors that influence the

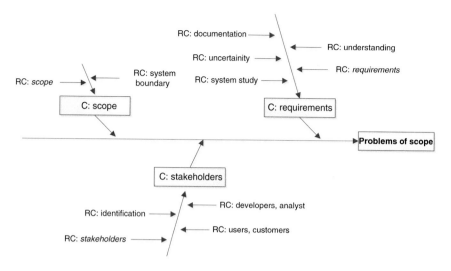

Fig. 1 Problems of scope (as adapted from [4, 6])

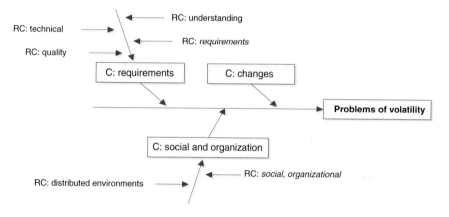

Fig. 2 Problems of volatility (as adapted from [4, 6])

problems of scope. Careful consideration and execution of these factors will eliminate or to a large extent, minimize the effects of problems of scope in elicitation. These are requirements, scope and stakeholders. The causes that can significantly impact the overall scope is also determined, as shown in Fig. 1.

Problems of volatility. The *Problems of volatility (PoV)* is the extent of changes that the requirements undergo during the project life cycle [7]. As an outcome of the systematic literature review conducted on the elicitation issues and their cause and effects to project performance, problems of volatility surfaced three key factors that requires focus. These are changes, requirements and social and organizational factors. A broad understanding of the causes that can influence these factors

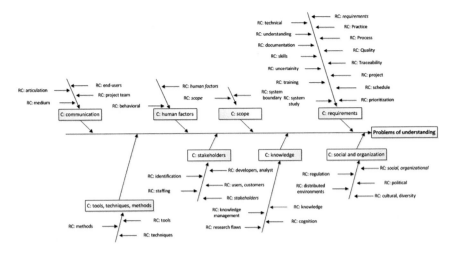

Fig. 3 Problems of understanding (as adapted from [4, 6])

positively or negatively is also depicted in the fishbone diagram as shown in Fig. 2. The current market conditions, nature of business and demands from the customers make business requirements dynamic in nature. Volatility issues can strongly influence the project performance and hence, these factors must be carefully addressed until project completion.

Problems of understanding. The *Problems of understanding* (*PoU*) is the degree of requirements understanding absorbed as part of the elicitation process. This describes the extent of ambiguity and communication challenges that can result in poor elicitation [7]. While problems of scope and problems of volatility are important aspects that influence elicitation activities, most of the challenges encountered are related to "*understanding*" of requirements. There are multiple factors that contribute to problems of understanding. The systematic literature review determines eight causes of problems of understanding. These are requirements, scope, human factors, communication, tool, techniques and methods, stakeholders, knowledge and social and organizational factors that influence the problems of understanding as depicted in Fig. 3. Each of these causes have been drilled down to determine further causes that influence the overall elicitation.

The consolidation of the root cause and effects on each of the generalized categories of elicitation issues and challenges have made it simpler for business analysts and requirements engineers to streamline their focus on the critical aspects that can adversely impact the elicitation and overall project performance. The detailed view of elicitation issues described above will be applied as an extension to the causal model [4]. This extension would enable construction of a new theoretical model that will provide a deeper insight into the elicitation issues, their cause and effects and the ones that will significantly influence the overall project performance.

2.2 Empirical Association Between Elicitation Issues and Project Performance

Residual performance risk and uncertainty coping mechanisms are critical to determine the overall project performance. They have a profound role to play in the overall causal relationship and in determining the impacts to project performance. Residual performance risk is the extent of difficulty in estimating the performance-related outcomes of the project, regardless of a specific estimation technique used [2]. Uncertainty coping mechanism is related to the coordination activities amongst multiple groups in the organization that are involved in the system build. There are two types of coordination activities that govern the level of coordination amongst the groups—horizontal and vertical coordination [2]. The levels of such coordination determines the uncertainty that can be managed during project execution.

Figure 4 depicts the structural model that describes the association of elicitation issues and project performance with residual performance risk and uncertainty coping mechanism as intervening variables [4]. The empirical analysis [4] on the causal model deduced the following

- Higher levels of elicitation issues will lead to higher levels of residual performance risk.
- Higher levels of residual performance risk will lead to lower levels of project performance.

An empirical analysis [3] of the association between the elicitation issues and project performance not only deduced the direct influence of elicitation issues on

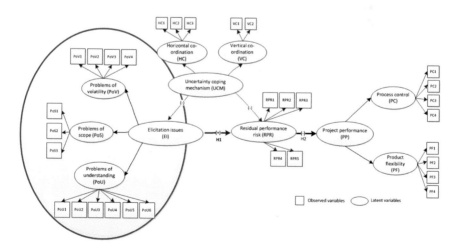

Fig. 4 Structural model of the association between elicitation issues and project performance

overall project performance, but also statistically demonstrated problems that can be prioritized while dealing with elicitation. The study [3] confirmed that problems of volatility contributed highest to elicitation issues followed by problems of understanding and problems of scope. This insight served as a motivating factor to extend this study and derive a model that can further drill down on the issues related to problems of scope, problems of volatility and problems of understanding.

2.3   A Drilldown of the Elicitation Issue Categories

For decades, elicitation issues has been the prime focus for researchers and industry professionals. Literature witnesses several studies that focus on aspects of requirements elicitation to bring to light the issues and challenges that persist in this field and discuss approaches to overcome them. These studies are based on the nature of projects, complexity, volume and global framework in which the projects are executed. While recent studies have attempted to consolidate the elicitation issue details such as cause and effects, recommendations, practices, classifications, etc. there are minimal or no evidences in literature that empirically confirms focus areas to minimize elicitation challenges. Critical observations have been made in the study [3] that contributions to elicitation issues were in the order of problems of volatility, problems of understanding and problems of scope. Given the critical observations made in the study [3] while deducing the association between elicitation issues and project performance, they have been a key motivating factor to further understand if such statistical analysis can help priority-setting on the detailed factors (represented in Figs. 1, 2 and 3) that cause problems of scope, problems of volatility and problems of understanding and in turn influence elicitation issues and overall project performance. This forms the basis for the theoretical model to be proposed. Confirmatory factor analysis will be required to be performed on the model like done in other causal model to seek outcomes and discuss results.

The construction of survey instrument, data collection and statistical analysis of this model is not part of the proposed study. This will be done as part of author's upcoming research work.

3   Proposed Structural Model

This section discusses the structural model (a theoretical view) of the extension of elicitation issues and their influence on project performance.

3.1 Constructs

Elicitation Issues. Two critical research work discussed in the previous section forms the basis of the proposed structural model. Figure 5 represents the extended model with focus on the elicitation issues. The remaining variables such as uncertainty coping mechanism, residual performance risk and project performance will be related to elicitation issues as depicted in the original model (Fig. 4).

Elicitation Issue Categories. The factors that influence the problems of scope, problems of volatility and problems of understanding are adapted from [4, 6] and applied to this model. The objective is to enable a detailed analysis of the impacts of these factors on the elicitation issue categories and in turn the overall project performance. The description of each of these categories with reference to the new model is discussed below.

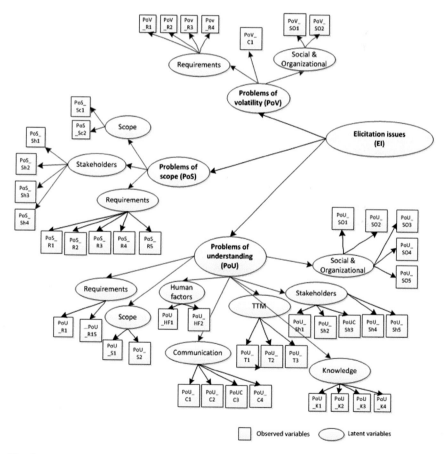

Fig. 5 Proposed elicitation issue construct

Problems of Scope. There are three factors such as scope, stakeholders and requirements identified under the problems of scope based on systematic literature review. These are identified as third-order latent variables in the model (Fig. 5). Observed variables for each of these latent variables are also adapted from the systematic literature review [4, 6]. While the actual observed variables will require to be constructed carefully based on nature of responses expected from the survey respondents, Table 1 represents the aspect on which the observed variables require to be constructed.

Problems of Volatility. There are three factors such as changes, requirements and social and organizational that represent the problems of volatility based on [4, 6]. The observed variables are constructed based on the causes determined in the Fig. 2. We may encounter overlaps in the issue category factors, but they have to be carefully studied to confirm if the same understanding is shared between the issue categories. Table 1 depicts the basis for observed variables for volatility issues.

Problems of Understanding. There are eight factors such as requirements, scope, human factors, communication, tools, techniques and methods, stakeholders, knowledge and social and organizational that represent the problems of understanding based on [4, 6]. The observed variable constructs for this issue category

Table 1 Basis for observed variable constructs for elictation issue category

Issue category	Issue category factors	Observed variables related to
Problems of Scope	Scope	[PoS_Sc1] System boundary
		[PoS_Sc2] Scope
	Stakeholders	[PoS_Sh1] Identification
		[PoS_Sh2] Stakeholders
		[PoS_Sh3] Users, customers
		[PoS_Sh4] Developers, analysts
	Requirements	[PoS_R1] Requirements
		[PoS_R2] Understanding
		[PoS_R3] Documentation
		[PoS_R4] Uncertainty
		[PoS_R5] System study
Problems of volatility	Changes	[PoV_C1] *variable to be constructed based on changes in the project*
	Requirements	[PoV_R1] Understanding
		[PoV_R2] Requirements
		[PoV_R3] Technical
		[PoV_R4] Quality
	Social and organizational	[PoV_SO1] Distributed environments
		[PoV_SO2] Social and organizational

(continued)

Table 1 (continued)

Issue category	Issue category factors	Observed variables related to
Problems of understanding	Requirements	[PoU_R1] Requirements
		[PoU_R2] Practice
		[PoU_R3] Process
		[PoU_R4] Quality
		[PoU_R5] Traceability
		[PoU_R6] Project
		[PoU_R7] Schedule
		[PoU_R8] Prioritization
		[PoU_R9] Technical
		[PoU_R10] Understanding
		[PoU_R11] Documentation
		[PoU_R12] Skills
		[PoU_R13] Uncertainty
		[PoU_R14] Training
		[PoU_R15] System study
	Scope	[PoU_S1] System boundary
		[PoU_S2] Scope
	Human factors	[PoU_HF1] Human factors
		[PoU_HF2] Behavioral
	Communication	[PoU_C1] Project team
		[PoU_C2] End-users
		[PoU_C3] Articulation
		[PoU_C4] Medium
	Tools, techniques, methods	[PoU_T1] Tools
		[PoU_T2] Techniques
		[PoU_T3] Methods
	Stakeholders	[PoU_Sh1] Identification
		[PoU_Sh2] Staffing
		[PoU_Sh3] Users, customers
		[PoU_Sh4] Developers, analysts
		[PoU_Sh5] Stakeholders
	Knowledge	[PoU_K1] Knowledge management
		[PoU_K2] Research flaws
		[PoU_K3] Knowledge
		[PoU_K4] Cognition
	Social and organizational	[PoU_SO1] Regulation
		[PoU_SO2] Distributed environments
		[PoU_SO3] Social, organizational
		[PoU_SO4] Political
		[PoU_SO5] Cultural, diversity

needs to be carefully framed based on the understanding captured in the systematic literature review [4, 6]. Table 1 consolidates the basis for observed variable constructs for understanding issues as well.

3.2 Next Steps

The empirical validation of the model will support in priority-setting and decision making when considering the dimensions of elicitation for a system development. The empirical outcomes of this model can determine areas of prime focus to prevent elicitation to negatively influence project performance. In order to empirically validate the proposed theoretical model, the following steps can be considered. SPSS Statistics and SPSS AMOS is recommended to perform the statistical analysis.

- Construct the survey instrument to capture responses on the observed variables
- Identify the population from whom the responses require to be collected
- Determine a (sizable) sample size.
- Collect survey responses from the identified population
- Perform Confirmatory Factor Analysis
- Determine the outcomes based on the results

4 Conclusion

The objective of this proposed study is to determine the level of influence the identified factors that have the elicitation issue categories such as problems of scope, problems of volatility and problems of understanding. The outcomes of the confirmatory factor analysis is expected to determine the level of impacts of impacts that the factors can have on the issue categories and on overall elicitation. This will in-turn determine the degree of influence on the project performance. The insights that the industry practitioners obtained based on the outcome of the empirical analysis is valuable [6]. In a similar manner, this study will streamline the focus of elicitation and will attempt to control and manage issues that surface through these activities. There are limitations to this proposed model—Constant research is needed to capture the elicitation issues that occur and requires to be reported back to literature. This is important to continue validating the focus areas required for conducting the elicitation activity.

References

1. Neetu Kumari, S.; Pillai, A.S.: A survey on global requirements elicitation issues and proposed research framework. In: Software Engineering and Service Science (ICSESS), pp. 554, 557. IEEE Xplore, 23–25 May 2013
2. Nidumolu, Sarma: The effect of coordination and uncertainty on software project performance: residual performance risk as an intervening variable. Inf. Syst. Res. **6**(3), 191–219 (1995)
3. Neetu Kumari, S.; Pillai, A.S.: The Effects of Requirements Elicitation Issues on Software Project Performance: An Empirical Analysis, LNCS, vol. 8396, pp 285–300. Springer (2014)
4. Neetu Kumari, S., Pillai, A.S.: A study on the software requirements elicitation issues—its causes and effects. In: World Congress on Information and Communication Technologies (WICT), pp. 15–18, Dec 2013
5. Neetu Kumari, S.; Pillai, A.S.: Requirements elicitation issues and project performance: a test of a contingency model. In: Science and Information Conference, pp. 889–896. IEEE Xplore (2015)
6. Neetu Kumari, S.; Pillai, A.S.: A systematic literature review on the software requirements elicitation issues. J. Netw. Innov. Comput. **2**, 283–308 (2014)
7. Christel, M.G., Kang, K.C.: Issues in requirements elicitation. Technical report no. CMU/SEI-92- TR-12), Software Engineering Institute (1992)

Compact Design of Rectangular Patch Antenna with Symmetrical U Slots on Partial Ground for UWB Applications

Sandeep Toshniwal, Somesh Sharma, Sanyog Rawat, Pushpendra Singh and Kanad Ray

Abstract This paper presents a design of rectangular microstrip patch antenna with slotted finite ground plane. Impedance bandwidth 53.6 % is achieved with stable pattern characteristics, within its bandwidth. The antenna has an operating impedance bandwidth of 3.34 GHz (4.57–7.91 GHz). VSWR is <2 over the operating frequency range. This antenna is designed on Ansoft HFSS 11 software. Details of the simulated results are presented and discussed.

Keywords Slotted microstrip patch antenna · Impedance bandwidth · Radiation pattern · Return loss

1 Introduction

In the past few years, ultra-wideband (UWB) systems have been used in multiple kinds of applications, due to their inherent features such as small size, high data transmission rate with short-range, larger bandwidth, simple hardware configuration, low power consumption and omni-directional radiation pattern. UWB is a high data rate and short range wireless technology, which utilizes the unlicensed spectrum ranging from 3.1 GHz to 10.6 GHz allocated by the Federal Communications Commission (FCC). In this range, several other kinds of licensed narrowband systems exist for which the UWB system causes the potential interference.

S. Toshniwal · S. Sharma
Department of Electronics and Communication, Kautilya Institute of Engineering and Technology, Jaipur, India

S. Rawat · P. Singh · K. Ray (✉)
Department of Electronics and Communication, Amity University Rajasthan, Jaipur, India
e-mail: kanadray00@gmail.com

© Springer International Publishing Switzerland 2016
V. Snášel et al. (eds.), *Innovations in Bio-Inspired Computing and Applications*,
Advances in Intelligent Systems and Computing 424,
DOI 10.1007/978-3-319-28031-8_47

These narrow band systems are namely, IEEE 802.11a Wireless Local Area Network (WLAN) in the frequency band of 5.15 GHz to 5.35 GHz and 5.725 GHz to 5.825 GHz, and HYPERLAN/2 in the frequency band of 5.45 GHz to 5.725 GHz. In order to avoid these interferences, a filter circuit is required which adds up to the system complexity and cost as well. An antenna capable of filtering these frequency bands is a more ideal solution. There are number of modified antenna design schemes reported in recent times. The most common way to get a band-notched characteristic in a printed monopole antenna is cutting the slots on the patch or in the ground plane such as cutting a C-shaped slot, cutting a fractal slot etc. [1–6]. Recently several methods have been investigated to obtain ultra wideband behavior of the antennas such as different shaped patches, employing finite ground, use of stacked elements, parasitic patches and different feeding techniques [7–14]. In this paper, rectangular microstrip patch antenna with identical U-slot on the ground plane is proposed. The patch is mounted on FR4 substrate of thickness equivalent to 1.59 mm. Both the patch and the ground plane are of the thickness of about 0.03 mm.

This paper is structured in four sections. Antenna design has been presented in Sect. 2. Results and discussion constitute Sect. 3 and conclusion is mentioned in Sect. 4.

2 Antenna Design

The geometry of a rectangular patch antenna for wideband application with U-slotted ground is shown in Figs. 1 and 2. The proposed antenna is printed on the glass epoxy FR-4 dielectric substrate with substrate thickness 'h' = 1.59 mm,

Fig. 1 Geometry of proposed microstrip patch antenna with top view

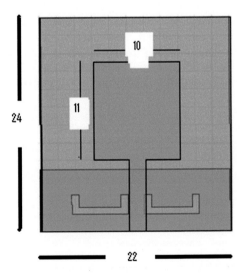

Fig. 2 Geometry of proposed
microstrip patch antenna with
back and side view

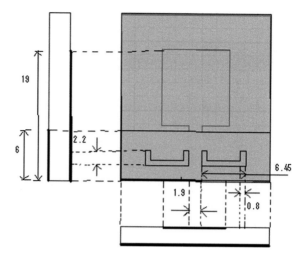

relative permittivity $ir = 4.4$ and loss tangent $tan\ i = 0.02$. A rectangular patch
(10 mm × 11 mm) is printed on the top side of the dielectric substrate. A rectangular feed line (1.9 mm × 8 mm) is printed on the same side of the substrate. The
bandwidth is increased by cutting two symmetrical U-shaped slots on the ground
plane. Due to this a bandwidth of 3.4 GHz (Approx.) is achieved. The overall
dimensions of this antenna are 22 mm × 24 mm. The performance of this structure
can be varied by varying the dimensions of the rectangular patch and the slots that
have been cut on the ground plane.

It should also be noted that higher bandwidth can also be achieved by increasing
the height of the substrate but due to this surface waves are introduced which
usually are not desirable because they extract power from the total available for
direct radiation (space waves). The surface waves travel within the substrate and
they are scattered at bends and surface discontinuities, such as the truncation of the
dielectric and ground plane, and degrade the antenna pattern and polarization
characteristics. Surface waves can be eliminated, while maintaining large band-
widths, by using cavities. This is why slotted microstrip patch antennas came into
existance.

3   Results and Discussion

In this section, predicted results of the rectangular microstrip patch antenna with
high bandwidth are presented. Figure 3 shows the return loss with frequency, curve
for the proposed antenna. The range of frequency falling below −10 dB is from
4.57–7.91 GHz. Due to this a wideband of 3.34 GHz is achieved and antenna is

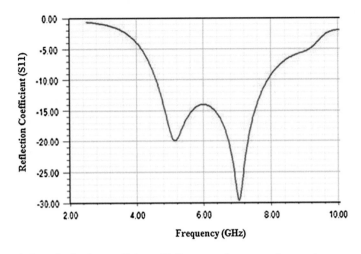

Fig. 3 Variation of reflection coefficient with frequency for proposed geometry

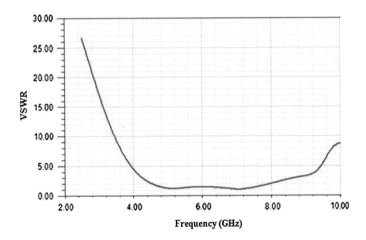

Fig. 4 Variation of reflection coefficient with frequency for proposed geometry

operating at resonant frequencies of 5.23 and 7.18 GHz. The central frequency of 6.24 GHz is obtained. Therefore a bandwidth as high as 53.57 % is achieved.

Figure 4 shows the VSWR with frequency graph of the proposed antenna. The VSWR falls below 2 for the proposed geometry under the desired band.

The simulated E and H plane elevation radiation pattern of antenna for resonant frequency of 7.18 GHz within the impedance bandwidth region is shown in Fig. 5. It may be seen that the radiation patterns are normal to the patch geometry and has radiation in both directions. The Fig. 6 clearly indicated that gain of as high as

Fig. 5 Simulated radiation
pattern of the antenna.
a E-Plane. **b** H-plane

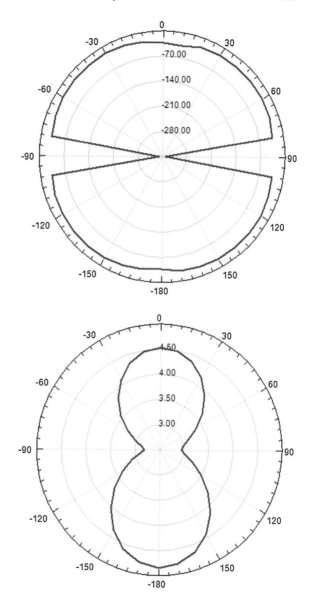

3.96 dB is achieved for this resonant frequency which is shown by red colour. The measured input impedance corresponding to two resonance frequencies are (51.36 + j3.45) ohm and (50.87 + j1.489) ohm respectively as shown in 6 which are quite close to 50 ohm impedance of the feed line (Fig. 7).

Fig. 6 3D radiation pattern
of proposed antenna

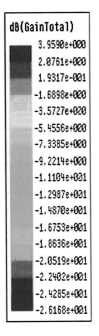

Fig. 7 Variation of input
impedance for proposed
geometry

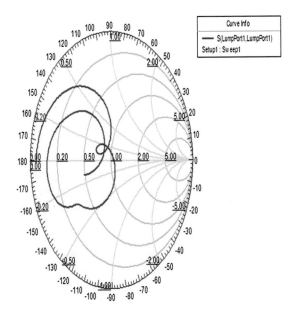

4 Conclusion

The design has demonstrated a high bandwidth microstrip patch antenna can be designed by embedding rectangular patch on the substrate and by cutting two symmetrical U-Shaped slots on the ground. A bandwidth of around 3.34 GHz is achieved at the central frequency of 6.24 GHz. Due to this a bandwidth of as high as 53.57 % is achieved. By varying the parameters some better results can be obtained. The future scope of this paper is that we can create band-notched characteristics by cutting slots and embedding parasitic elements on the ground plane for filtering some licensed bands so as to make this antenna a wideband antenna with band-notched characteristic. Now-a-days, modern communication systems require antennas with broadband and/or multi-frequency operation modes. These goals have been accomplished by employing slotted patch for the radiating element, with the aim to preserve compactness requirements and to maintain the overall layout as simple as possible and keeping the realization cost very low.

References

1. Balanis, C.A.: Antenna Theory Analysis and Design, 3rd edn. Wiley, London
2. Cho, Y.J., Kim, K.H., Choi, D.H., Lee, S.S., Park, S.-O.:A miniature UWB planar monopole antenna with 5-GHz band-rejection filter and the time-domain characteristics. IEEE Trans. Antenna Propag. 1453–1460 (2006)
3. Ma, T.G., Wu, S.J.: Ultrawideband band-notched folded strip monopole antenna. IEEE Trans. Antenna Propag. 2473–2479 (2007)
4. Hong, C.Y., ling, C.W., Tarn, I.Y. Chung, S.J.: Design of a planar ultra wideband antenna with a new band-notched structure. IEEE Trans. Antenna Propag. 3391–3396 (2007)
5. Wu, Q., Jin, R., Geng, J. and ding, M.: Printed omni-Directional UWB monopole antenna with very compact size. IEEE Trans. Antenna Propag., 896–899 (2008)
6. Sadat, S., Fardis, M., Geran, F., Dadashzadeh, G.: A compact microstrip square-ring antenna for UWB applications. Prog. Electromagn. Res. **67**, 173–179 (2007)
7. Moeikham, P., Akkaraekthalin, P.: A compact ultra wide band monopole antenna with tapered CPW feed and slot stubs. In: Proceedings of the 8th International Conference on Electrical Engineering/Electronics, Computer, Telecommunications and Information Technology, Khon Kaen, pp. 180–183 Thailand (2011)
8. Cappelletti, G., Caratelli, D., Cicchetti, R., Simeoni, M.: A low-profile printed drop-shaped dipole antenna for wideband wireless applications. IEEE Trans. Antennas Propag. **59**, 3526–3535 (2011)
9. Kang, C.H.; Wu, S.J.; Tarng, J.H.: A novel folded UWB antenna for wireless body area network. IEEE Trans. Antennas Propag. 1139–1142 (2012)
10. Rawat, S., Sharma K.K.: Stacked elliptical patches for circularly polarized broadband performance. In: International Conference on Signal Propagation and Computer Technology (ICSPCT 2014), pp. 232–235 (2014)
11. Rawat, S and Sharma K K.: A compact broadband microstrip patch antenna with defected ground structure for C-band applications. Central European Journal of Engineering, Springer, pp. 287–292 (2014)

12. Puri, M., Dhanik, S.S., Mishra, P.K., Khubchandani, H.: Design and simulation of double ridged horn antenna operating for UWB application. In: IEEE India Conference (2013)
13. Rawat, S., Sharma, K.K.: Stacked configuration of rectangular and hexagonal patches with shorting pin for circularly polarized wideband performance. Central Eur. J. Eng. **4**, 20–26 (2014)
14. Rawat, S., Sharma, K.K.: Annular ring microstrip patch antenna with finite ground plane for ultra-wideband applications. Int. J. Microwave Wireless Technol. 179–184 (2015)

Big Data Challenges and Solutions in Healthcare: A Survey

Prabha Susy Mathew and Anitha S. Pillai

Abstract The digitization of medical data, field of genomics and use of wearable sensors to monitor patient health are some of the factors that have dramatically increased the growth of Big Data in Health Care/Biomedicine. Big data in healthcare actually refers to electronic health data sets which are large and complex that is very difficult to manage with traditional/conventional data management tools and techniques. Big data analytics in healthcare is cumbersome not just because of its volume but also because of the diversity of data types and the speed at which it is generated and must be managed/analyzed. Rapid progress is to be made for analyzing this data and for gleaning new insights for making better informed decisions. There are unprecedented opportunities to use big data. The Health Care Industry should find methods to properly analyze this Big HealthCare Data generated and stored around the world each seconds in order to discover associations, understand the patterns and trends which will provide significant opportunities for real-time tracking of diseases, predicting disease outbreaks, to improve care, save lives and lower costs. Extraction, integration and analysis of heterogeneous, enormous and complex HealthCare data captured from various Electronic Health Care sources are a major challenge. New methods, applications and tools that are used by Healthcare industries, practitioners and researchers to tackle the big data challenges are discussed in this paper.

Keywords Big data · Big data analytics · Biomedicine · Electronic Medical Record (EMR) · Healthcare · Genomics

P.S. Mathew (✉) · A.S. Pillai (✉)
MCA, School of Computing Sciences, Hindustan University, Chennai, India
e-mail: prabhasm@hindustanuniv.ac.in

A.S. Pillai
e-mail: anithasp@hindustanuniv.ac.in

© Springer International Publishing Switzerland 2016
V. Snášel et al. (eds.), *Innovations in Bio-Inspired Computing and Applications*,
Advances in Intelligent Systems and Computing 424,
DOI 10.1007/978-3-319-28031-8_48

1   Background

In this 'Digital Age' there is tremendous growth in data in terms of volume, variety and velocity. According to Gartner, Big Data is "high-volume, high-velocity, and high-variety information assets that demand cost-effective, innovative forms of information processing for enhanced insight and decision making" [1].

Recent years have seen a remarkable change in how data has been handled across industries. Healthcare industry in specific generates a vast set of data. As the Healthcare industry is information intensive [2] it is even more important to get the right inference from the data to make effective decision [3, 4]. The Healthcare data these days are originated from multiple sources including mobile devices, sensors attached to patient bed, wearable sensors, social media, Internet of Things (IoT), Electronic Medical Records (EMR), claim data, medical images, Radio Frequency Identification Device (RFID) monitoring/tracking devices etc [5, 6]. Such 'Data Explosions' have led to challenges in managing and analyzing these data. To do a timely analysis of this massive patient dataset in real time is a challenge. If the big data is synthesized and analyzed properly to identify associations, patterns and trends then healthcare providers and other stakeholders in the healthcare system can get better insightful diagnoses and treatments, which in turn would result in higher quality care at lower costs. For Example the increase in the search term in Google such as 'Flu symptoms' or 'Flu Treatment' or availability of public data sets available in Canada such as FluWatch can be used to predict the outbreak of flu in a particular region thereby keeping the medical facilities ready for treating the patients [7].

In this paper systematic review of literature related to healthcare, the Big Data problems faced, as well as solutions adopted by healthcare industry to combat these issues are discussed.

2   Related Work

The traditional approaches used to store and analyze the data no longer provide effective solution for handling Big Data. There are several solutions like Hadoop, High Performance Computing, In-Memory Computing, and Cloud Computing that can effectively handle the massive datasets. The Open-source software has proven to be quite instrumental in handling the big data challenges. Hadoop and MapReduce play a significant role in processing large clinical data [7].

According to experimental studies conducted [8] the authors have identified that the conventional relational database does not support all operations required for interactive visualization and analytics. NoSQL as a data management solution provides more flexibility to model different complex EMR data and is scalable for handling large data as it can be used with distributed computing architecture.

Panahiazar et al. [9] investigates, Hortonworks Data Platform (HDP) in Mayo's health care systems that uses Hadoop-MapReduce framework to predict survival score of each heart failure patients using Pig to translate queries to a sequence of MapReduce jobs. A test ran to compare Pig with other tools like SQL revealed that SQL took more time while Pig based alternatives took considerably less time to run a query.

Raghupathi and Raghupathi have described the use of big data analytics and applications in healthcare to gain valuable insights from the clinical data, but mercurial advances in big data platforms and tools can accelerate their maturing process [4].

Bioinformatics study increasingly relies on high-performance computation and large-scale data storage. The datasets are often complex, heterogeneous and incomplete. Considering these aspect, visual techniques plays a vital role in bioinformatics. There are many powerful scientific toolsets available ranging from software libraries such as SciPy, Chimera, Taverna, Galaxy, and the Visualization Toolkit (VTK). Most of them are designed for small, local datasets and cannot handle recent advances in data generation and acquisition.

To resolve these big data challenges, Steven et al. developed a visual analytics software framework named DIVE—Data Intensive Visualization Engine. It is a data-agnostic, ontologically-expressive software framework capable of streaming large datasets at interactive speeds. The platform provides parallelized operations, high-throughput and structured data streaming [10].

3 Big Data Challenges in Healthcare

The Healthcare industry leverages Big Data and Analytics to increase effectiveness in terms of better clinical care, reduced operational cost and providing personalized patient care. However there are growing complexities due to the presence of huge amount of diverse Healthcare data [11, 12].

3.1 V's of Big Data Analytics

The V's of Big Data is predominant even in healthcare mainly Volume, Variety and Velocity and Veracity. Huge volume of healthcare data comes from Electronic medical records, clinical images, Diagnosis data and health claim data, etc. in structured, semi-structured and unstructured form in real time. Capturing and analyzing streaming data is a challenge. Sensors attached to the patient's bedside to continually track patient vitals produce huge chunks of data that the traditional systems were incapable of storing and analyzing effectively [13]. Changes in pattern in these vital signs are alerted to a team of doctors and assistants. All this was successfully achieved using Hadoop ecosystem components [4]. Real time processing of monitoring data can mean the difference between life and death for a patient. Mitsui Knowledge Industry MKI uses SAP HANA and Hadoop for providing personalized cancer treatment based on analysis of one's DNA. The solution uses Hadoop to align the patient's DNA sequence with the normal sequence, as the data is in a semi-structured and parallelization across multiple machines is possible. Identifying the mutations and predicting the best treatment requires a lot of highly iterative analysis which is achieved by SAP HANA. As a result the analysis which took them 2–3 days earlier has reduced to 20 min now [6]. Advancement in data management, Virtualization, cloud computing and Big Data Ecosystem is facilitating easier and effective means to capture, store and manipulate large volumes of data. Today, enterprises are adopting NoSQL (Not Only SQL) technology as it overcomes the limitations of the traditional Relational Database technology. NoSQL provides a more flexible schema less data model which can efficiently store unstructured and semi structured data. It provides an easier and cost effective approach to database scaling [14]. Several Big Data challenges are as represented in the Fig. 1. In this paper some of the major big data challenges and approaches will be discussed.

Fig. 1 Big data challenges

3.2 Data Storage

The Dominance of Relational database is slowly fading away with the rise of new type of databases. The relational database which uses strict schema based approach cannot efficiently store unstructured and semi structured data. The ever increasing volume of data causes the relational database technology performance to decline as it does not provide a scalable solution for handling the "big data" [14]. Dealing with Massive data sets requires large storage space. Two approaches to deal with it are Compression and Sampling [15]. By using Compression techniques time taken is generally more but less space will be utilized. While using sampling, information is lost by takes less space. Feldman et al. [3] in his work uses coresets which are small sets that approximate the original data for a given problem to reduce the complexity of Big Data with merge-and-reduce approach for problems such as K-means, PCA and Projective clustering. Cloud Computing can be used as a powerful assistance to store and process large amount of healthcare data. One of the commonly used solutions to handle this problem of big data is dimensionality reduction. Linear mapping methods, such as principal component analysis (PCA), singular value decomposition, as well as non-linear mapping methods, such as Sammon's mapping, laplacian eigenmaps and kernel principal component analysis have been used for dimensionality reduction [16].

3.3 Data Processing

Some optimal mechanism is required as near real time processing of information is the need of the hour. The In-Database processing and In-Memory computing technologies can be adopted by organizations to improve their processing speed. Many organizations are leveraging on hybrid transactional/analytical processing (HTAP) allowing transactions and processing to reside in the same in-memory database. Analytics with HTAP is much faster compared to the solutions already available [17].

Analyzing genomic data is a computationally intensive task and combining them with standard clinical data adds additional layers of complexity. This sort of data explosion has led to complexity in handling and analyzing data with respect to increasing volume, velocity and variety. The solution provided by Hadoop is extensively used to deal the big data problem, but Hadoop being a batch processing framework it does not cope with the need for real time analytics [18]. To deal with issues of the fundamental architecture, Nathan Marz came up with Lambda Architecture (LA) paradigm which is a scalable and fault tolerant data processing architecture. Lambda Architecture consists of batch layer, serving layer and speed layer that compute real time analytics to compensate for the slow batch layer. The LA is able to serve a wide range of workloads in which low-latency is required

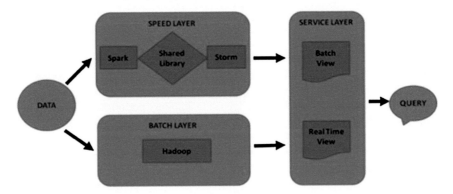

Fig. 2 Lambda architecture. (Based on Nathan Marz design)

[19]. Lambda Architecture is an approach that handles both batch and real time processing to achieve best of both world (Fig. 2).

Spark has the ability to address both batch and near real-time data processing and provides high performance using in-memory computing. It is a cluster computing framework that guarantees up to 100 times faster performance making it best suited for machine learning and graph data processing. It gives a comprehensive, unified framework to manage big data processing requirements with a variety of data sets that are diverse in nature (text data, graph data etc.) as well as the source of data (streaming data) [20]. Like Spark, Storm is also fault- tolerant, does distributed computation and has the ability to handle arbitrarily complex computation [21, 22].

3.4 Existing Algorithms Redesigned to Perform Big Data Mining

Many data mining techniques are not capable to carry out distributed mining. Data is ever evolving so the data mining techniques must be able to adapt to the changing needs. Efficient parallel algorithms and implementation techniques can greatly improve and address the performance and scalability requirements.

For example SVM classification is only directed to two-class tasks, so to solve multi-class tasks, algorithms that deal with reduction to several binary problems have to be applied. It suffers from scalability issues in both memory use and computational time, to overcome this issue Parallel SVM (PSVM) has been used [5]. Ke Xu et al. proposed PSVM based on iterative MapReduce for e-mail classification. They have used ontology based semantics to improve the accuracy of PSVM based on MapReduce. The use of MapReduce framework significantly reduces the training time. However, for data intensive mining purpose MapReduce in cloud computing proves to be effective than MapReduce in Hadoop [23].

Association Rule Mining algorithms are often used to find association between items in a dataset. The Apriori algorithm uses an iterative approach where results of previous iterations are used to find the frequent item sets for the current iteration. Main drawback with this approach is the candidate sets become large if the dataset is too huge and multiple data scans are required. Eclat algorithm deals with the issue of Apriori algorithm's multiple scan by making a single pass on the dataset there by making it a faster algorithm when compared to Apriori. However Eclat does not deal with the huge candidate set generated by Apriori, which is effectively handled by FP-Growth. FP-Growth eliminates the candidate generation step. It is faster than Apriori algorithm but as the size of data increases the efficiency of the algorithm drops. Thus to handle the issues relating to huge datasets parallel algorithms are introduced. Hence in order to scale across multiple machines fault tolerant framework for distributed application such as MapReduce is used. MapReduce based on Hadoop implementation of Apriori algorithm handle the large datasets in HDFS, for each iteration result is to be sourced from the HDFS which requires high I/O time leading to a reduced performance. Yet another frequent itemset Mining (YAFIM) Algorithm based on Spark RDD (Resilient Distributed Dataset) framework resolves the issues. The Apriori algorithm's implementation on spark platform speeds up to nearly 18 times on an average over different benchmarks. Results generated on real world healthcare data are observed to be many times faster when compared to MapReduce framework. Thus in an attempt to further improve the algorithm, Rathee et al. proposed Reduced-Apriori (R-Apriori) based on Spark RDD framework, it eliminates the time consuming part of the YAFIM algorithms second pass to further reduce the number of computations required thereby improving the speed [24].

Existing Clustering algorithms do not provide scalable solutions to handle the Big Data. A clustering method Fuzzy C-means based on the Google's MapReduce paradigm was proposed in [25]. The Experimental evidence of MapReduce-based Fuzzy C-Means Algorithm (MR-FCM) demonstrated that the algorithm scales well with increasing data sets. Another comparison between the two Mahout algorithms KMeans and Fuzzy KMeans (FKM) with MR-FCM for data set sizes varying showed that though FKM and MR-FCM are computationally quite similar, the Mahout FKM algorithm scales better than the MR-FCM algorithm.

High dimensionality data clustering methods are designed to handle data with hundreds of attributes, including DFT and MAFIA [26].

Algorithms such as SPRINT (Scalable Parallelizable Induction of Decision Tree algorithm) and SLIQ (Supervised Learning In Quest) are highly scalable, and has no storage constraint on larger data sets [26]. The drawback associated with SLIQ algorithm is that it requires computation of large number of Gini indices at each node of the decision tree to decide which attribute is to be split at each node. This computation is carried out for all attributes and for each successive pair of values. For huge datasets, lot of such computation is required. A performance enhancement to SLIQ algorithm, CC-SLIQ (Cascading Clustering and Supervised Learning In Quest) was proposed by Prasad et al. [27]. The CC-SLIQ algorithm constructs the binary decision tree by cascading two machine learning algorithms: k-means clustering and SLIQ decision tree learning. The CC-SLIQ approach results in decision trees that have smaller sizes, fewer rules. It is proved to be useful in noisy environments when compared to standard methods.

A parallel version of the random forest using MapReduce programming framework on top of Hadoop has been developed [28] for large-scale population genetic association studies involving multivariate traits. The algorithm has been applied to a genome-wide association study on Alzheimer disease (AD) in which the quantitative characteristic consists of a high-dimensional neuroimaging phenotype describing longitudinal changes in human brain structure. Remarkable speed-ups in the processing were observed.

There are some limitations with the traditional association rule mining algorithms for large-scale data. The FP-Growth algorithm's success is limited by internal memory size because mining process is on the base of large tree-form data structure, which results in high computation time. The algorithm that constructs an Optimum pattern Tree with the node as the data item of the transaction has been proposed. This algorithm is implemented on Hadoop to reduce the computation cost and for handling the large data and processes them parallely to improve efficiency [29].

Traditional machine-learning frameworks such as Weka and Rapidminer do not scale to big data. Apache's Mahout is a framework which is open source and is a distributed framework to deal with big data. Mahout is a scalable machine-learning library used on top of Hadoop. Two approaches for performing machine learning algorithm on distributed framework using MapReduce are Mahout and Spark. In case of Mahout all iteration results are written to and read from disk while for Spark all iterations can be stored in memory. As processing of data can be done directly from memory there is a significant performance improvement when using Spark as opposed to Mahout [30].

3.5   Comparison Between MapReduce and Spark

Apache Spark is an open source framework for big data. It is a cluster computing framework that guarantees up to 100 times faster performance making it best suited for machine learning and graph data processing. It gives a comprehensive, unified framework to manage big data processing requirements with a variety of data sets that are diverse in nature (text data, graph data etc.) as well as the source of data (streaming data). The lack of speed and absence of in-memory queuing are considered to be the biggest drawback plaguing MapReduce. Apache Spark allows processing of data streams unlike MapReduce which processes data in batches causing queuing delays which are not acceptable in several real time applications. Real time studies conducted have proved Spark to sort 100 TB of data in just 23 min when compared to the 72 min taken by Hadoop to accomplish the same using a number of Amazon Elastic Cloud machines. Spark runs on Hadoop just the way MapReduce does but with the exception that MapReduce runs only on Hadoop

Table 1 MapReduce versus Spark

	MapReduce	Spark
Performance	Relatively slow	10 to 100 times faster than MapReduce engine
Processing	Majorly for batch processing	Streaming and batch processing
Compatibility	Hadoop	Mesos, YARN and Hadoop
Data store	Stores data on disk	Stores data in-memory
Implementation	Implemented using java	Using API's for Java, Scala and Python
Ease of use	Code is lengthy	Less word count in code compared to MapReduce
Failure tolerance	Slightly more failure tolerant	Less tolerant when compared to MapReduce
Security	More security features	Security aspects still in Infancy

while Spark on the other hand has the ability to run and exist without Hadoop. Spark also runs on Mesos, standalone, or in the cloud, making it the next big thing in data analytics [31].

Spark stores data in-memory whereas Hadoop stores data on disk. Hadoop uses replication to achieve fault tolerance whereas Spark uses different data storage model, resilient distributed datasets (RDD) that minimizes network I/O. It can access diverse data sources including HDFS, Cassandra, HBase, and S3 [20]. Spark permits writing applications in Java, Scala, or Python. It comes with a built-in set of over 80 high-level operators.

Some of the differences between MapReduce and Spark are highlighted in the table mentioned below [32, 33] (Table 1).

4 Conclusion

In Healthcare industry the medical professionals and patients are generating huge data for monitoring and improving patient. The biggest challenge faced by data scientists is how to integrate these diverse data and make use of this massive data effectively.

To overcome the Big Data challenge several new methods and technologies have been devised to resolve storage, processing and security related issues. Till date, the Hadoop ecosystem has proven to be the most mature framework for handling big data, but is restricted to batch processing. New technologies such as Storm, Spark, and Mahout are emerging to provide flexibility, support real-time processing and to run adhoc queries on large data sets. A major challenge for the next couple of years is to come up with robust implementations of analytical methods that are required for the biomedical and health domain, within these frameworks. This will require transforming the existing analytical algorithms to fit into the distributed processing model.

Organizations must understand that big data solutions are not a replacement to the existing solutions but a complement to it for solving big data problems. It is essential for them to take an end to end solution leveraging from multiple big data and traditional solutions in order to obtain the desired Healthcare Outcome.

References

1. http://www.gartner.com/it-glossary/big-data
2. Chen, C.L.P, Zhang, C.-Y.: Data-intensive applications, challenges, techniques and technologies: a survey on Big Data. Inf. Sci. Elsevier (2014)
3. Feldman, D., Schmidt, M., Sohler, C.: Turning Big data into tiny data: Constant-size coresets for k-means, PCA and projective clustering. In: SODA '13 Proceedings of the Twenty-Fourth Annual ACM-SIAM Symposium on Discrete Algorithms, pp. 1434–1453 (2013)
4. Raghupathi, W., Raghupathi, V.: Big data analytics in healthcare: promise and potential. Health Inf. Sci. Sys. http://www.hissjournal.com/content/2/1/3 (2014)
5. Yadav,C., Wang, S., Kumar, M.: Algorithm and approaches to handle large data—a survey. IJCSN Int. J. Comput. Sci. Netw. 2(3) 2013. ISSN (Online):2237-5420
6. Jonker, D.: https://blogs.saphana.com/2013/05/09/saps-hadoop-strategy/
7. Wang, W., Krishnan, E.: Big Data and Clinicians: A Review on the State of the Science, vol. 2, no. 1 (2014)
8. Archenaa, J.: Big Data analytics for health care using hadoop. Int. J. Appl. Eng. Res. 9(18), 3301–3308 (2014). Research India Publications, ISSN:0926-4513
9. Panahiazar, M., Taslimitehrani, V., Jadhav, A., Pathak, J.: Empowering personalized medicine with Big Data and semantic web technology: promises, challenges, and use cases. In: Proceedings of IEEE International Conference on Big Data, pp. 790–795, Oct 2014. doi:10.2409/BigData.2014.7004307
10. Kumar, V., et al.: Exploring clinical care processes using visual and data analytics: challenges and opportunities. http://dssg.uchicago.edu/kddworkshop/papers/kumar.pdf
11. Pratt, M.K.: No Quick Cure for Healthcare Systems Computerization is Slowly Improving the Healthcare System, But it's a Long Way From Living up to Expectations. Computer World, Healthcare IT
12. Mathew, P.S., Pillai, A.S.: Big Data Solutions in Healthcare: Problems and Perspectives. IEEE Xplore (2015). doi:10.2409/ICIIECS.2015.2293224
13. Banaee, H., Ahmed, M.U. Loutfi, A.: Data mining for wearable sensors in health monitoring systems: a review of recent trends and challenges. Sensors 13(12), 12245–12900 (2013). doi:10.3390/s131212245
14. Sadalage, P.: NoSQL databases: an overview. https://www.thoughtworks.com/insights/blog/nosql-databases-overview (2014)
15. http://albertbifet.com/big-data-mining-tools
16. Hirak, K., Afzal, A.H., Nazrul, H., Swarup, R., Kumar, B.D.: Big Data analytics in bioinformatics: a machine learning perspective. J. Latex Class Files 13(9) (2014)
17. http://www.computerworld.com/article/2690856/big-data/8-big-trends-in-big-data-analytics.html
18. Hausenblas, M., Bijnens, N., inspired by Marz, N.: Lambda Architecture (2015)
19. Laurent Bride.: Hadoop Summit 2015 Takeaway: The Lambda Architecture (2015)
20. Mohammed, J.: Is apache spark going to replace hadoop. http://aptuz.com/blog/is-apache-spark-going-to-replace-hadoop/ (2015)

21. Giamas, A.: Spark, Storm and Real Time Analytics (2014)
22. https://storm.apache.org/
23. Xu, K., Wen, C., Yuan, Q., He, X., Tie, J.: A MapReduce based parallel SVM for email classification. J. Netw. 9(6), (2014)
24. Rathee, S., Kaul, M., Kashyap, A.: R-Apriori: an efficient apriori based algorithm on spark. In: PIKM'15, Melbourne, VIC, Australia. ACM, Oct 19 2015. ISBN:978-1-4503-3782-3/15/10. doi:http://dx.doi.org/10.1145/2809890.2809893
25. Ludwig, S.A.: MapReduce-Based Fuzzy C-Means Clustering Algorithm: Implementation and Scalability
26. Chen, F., Deng, P., Wan, J., Zhang, D., Vasilakos, A.V., Rong, X.: Data mining for the internet of things: literature review and challenges. Int. J. Distrib. Sens. Netw. 2015, Article ID 431047, 14 (2015). http://dx.doi.org/10.1155/2015/431047
27. Narasimha Prasad LV., Naidu, M.M.: CC-SLIQ: performance enhancement with 2 K split points in SLIQ decision tree algorithm. IAENG Int. J. Comput. Sci. 41(3), IJCS_41_3_02
28. Mohammed, E.A., Far, B.H., Naugler, C.: Applications of the MapReduce programming framework to clinical big data analysis: current landscape and future trend. BioData Min. 7, 22 (2014). doi:10.1186/1756-0381-7-22
29. Shah, A.H., Patel, P.A.: Optimum frequent pattern approach for efficient incremental mining on large databases using MapReduce. Int. J. Comput. Appl. (0975–8887) 120(4) (2015)
30. Aydin, G., Hallac, I.R., Karakus, B.: Architecture and implementation of a scalable sensor data storage and analysis system using cloud computing and Big Data technologies. J. Sens. 2015, Article ID 834217, 11 (2015). http://dx.doi.org/10.1155/2015/834217
31. Suresh, R.: Apache spark and the future of big data analytics. http://suyati.com/apache-spark-and-the-future-of-big-data-analytics/ (2015)
32. https://www.xplenty.com/blog/2014/11/apache-spark-vs-hadoop-mapreduce/
33. http://www.dezyre.com/article/hadoop-mapreduce-vs-apache-spark-who-wins-the-battle/83

Experimental Study on Bound Handling Techniques for Multi-objective Particle Swarm Optimization

Devang Agarwal and Deepak Sharma

Abstract Many real world optimization scenarios impose certain limitations, in terms of constraints and bounds, on various factors affecting the problem. In this paper we formulate several methods for bound handling of decision variables involved in solving a multi-objective optimization problem using particle swarm optimization algorithm. We further compare the performance of these methods on different 2-objective test problems.

Keywords Bound handling · Multi-objective optimization · Particle swarm optimization

1 Introduction

Particle swarm optimization (PSO) is a stochastic, population-based evolutionary algorithm inspired by the collective behavior of flocks and uses swarm intelligence to perform the task of search and optimization [1]. The process starts with random particles (points lying between the specified upper and lower bounds) being generated in the search space. The particles move in the search space guided by their own experiences and the acquired knowledge of the swarm, i.e., they have personal and global guides that guide them to better regions (solutions) in the solution space. These guides are the best solutions discovered so far by the particle itself, in the case of the personal guide, and the best solutions discovered so far collectively by the group in the case of the global guides. In the latter case each particle is assigned a global guide based on some criteria. When we say the particles move we

D. Agarwal (✉) · D. Sharma (✉)
Department of Mechanical Engineering, Indian Institute of Technology Guwahati, Guwahati 781039, Assam, India
e-mail: a.devang@iitg.ernet.in; devangagarwal93@gmail.com

D. Sharma
e-mail: dsharma@iitg.ernet.in

© Springer International Publishing Switzerland 2016
V. Snášel et al. (eds.), *Innovations in Bio-Inspired Computing and Applications*,
Advances in Intelligent Systems and Computing 424,
DOI 10.1007/978-3-319-28031-8_49

essentially mean to say that the position of the particle is changed or updated after every iteration. A velocity vector is added to the current position vector to move it to a new position. The personal guides and global guides play a major role in the formulation of the velocity vector. One of the biggest problems faced with PSO is that of maintaining the swarm within the feasible region, especially when swarm goes out the bound [2, 3]. The particles get large velocities. The velocity vector adds momentum to the particle and forces the particle out of the feasible region. Empirical analysis and theoretical proofs show that particles leave search boundaries very early during the optimization process [4, 5]. This results in inefficient performance of the algorithm, and poor convergence of the swarm, as even after millions of function evaluations, the swarm fails to converge, wasting a lot of time. As a result, in most cases, we do not get any feasible solution at all.

Over the past, several methods have been proposed in the literature to extend PSO to deal with multi-objective optimization problems, which are known as Multi-Objective Particle Swarm Optimization (MOPSO) techniques [6–8]. However so far, little attention has been drawn for bound handling to MOPSO [9]. A task of respecting the bounds is even more difficult with MOPSO when many particles are likely to converge to the Pareto-optimal front and get out of the bounds. Helwig et al. in [2] have suggested several methods for bound handling on a flat landscape. Padhye et al. [3] suggested inverse parabolic distribution strategy and compared the results with other techniques for single-objective optimization. In this paper we implement the various methods of [2] for the bound handling in multi-objective problems. These methods can help to not only achieve solutions faster but also can help improving the efficiency of PSO. We test the methods on benchmark problems including ZDT1, ZDT2, ZDT3 and ZDT6 [10]. It is observed that the swarm fails to converge to the Pareto-optimal front in the absence of coupling with one of the bound handling methods mentioned in [2].

The outline of the paper is as follows. In Sect. 2 we describe the general scheme of the PSO algorithm. In Sect. 3 we describe the bound handling methods that we have coupled with MOPSO. In Sect. 4 we present the results of various bound handling methods with MOPSO for ZDT1, ZDT2, ZDT3 and ZDT6 benchmark multi-objective problems. Finally in Sect. 5 we draw our conclusion.

2 Generalized MOPSO Algorithm

The generalized MOPSO algorithm is given in Table 1. It is a simple MOPSO algorithm coupled with a boundary handling technique. The fitness is assigned to each particle using non-dominated sorting and crowding distance operators [11]. The following equations illustrate the position, velocity, personal guide and global guide updates respectively.

Table 1 MOPSO coupled with boundary handling technique

1. Set the iteration counter, t = 0, maximum allowed iterations = T, c1 = 1.7, c2 = 1.6 and w = 0.9
2. Initialize position and velocity of the particles randomly
3. Evaluate particles and assign fitness
4. while t < T do
a. Calculate $p_{i,lb}^t$ using (3)
b. Calculate $p_{i,gb}^t$ using (4)
c. Update velocity for each particle using (2)
d. Update position for each particle using (1)
e. Check If the position of the particle has exceeded the bounds
If yes
Using a boundary handling technique described in Sect. 3, update the position and velocity for each particle
Else
Keep the same position and velocity as calculated in 4(b) and 4(c)
f. Evaluate particles and assign fitness
g. t = t + 1
5. end while

(1) Position of the particle (*i*) is adjusted as $x_i^{t+1} = x_i^t + v_i^{t+1}$

(2) Velocity of particle (*i*) is updated as

$$v_i^{t+1} = wv_i^t + c_1 r_1 \left(p_{i,lb}^t - x_i^t \right) + c_2 r_2 \left(p_{i,gb}^t - x_i^t \right)$$

where *i* is the *i*th particle, t is the iteration number, v_i^0 is set randomly, *w* adds to inertia of particle, *n* is the number of particles in the swarm, c_1 and c_2 are acceleration coefficients, and r_1 and r_2 are random numbers $\in$ [0, 1].

(3) $p_{i,lb}^t$ is the personal guide of the *i*th particle at *t*th iteration which is calculated as

If $f\left(x_i^{t+1}\right)$ has better fitness than $f\left(p_{i,lb}^t\right)$ then $p_{i,lb}^{t+1} = x_i^{t+1}$
Else $p_{i,lb}^{t+1} = p_{i,lb}^t$

(4) $p_{i,gb}^t$ is the global guide of the *i*th particle at *t*th iteration, calculated as

$$p_{gb}^t \in \{p_{1,lb}^t, p_{2,lb}^t, \ldots, p_{n,lb}^t\} \text{ if } \left(p_{gb}^t\right) = best\{f\left(p_{1,lb}^t\right), f\left(p_{2,lb}^t\right), \ldots, f\left(p_{n,lb}^t\right)\}.$$

The general MOPSO scheme can be modified to produce improved results and faster convergence of the swarm, by introducing the concept of bound handling while updating the position of the particle. This maintains all the particles within the feasible search region always during the search. These bound handling methods are described in the following section.

3 Bound Handling Methods

In this section we discuss various bound handling methods which are coupled with MOPSO.

1. **Mutation**: Mutation based method that perturbs the unbounded position by a small amount to bring the particle back into the feasible region.
2. **Reflect Methods**: In the reflect method we reflect all the violating variables along the closer bound and repeat the process unless the variable is within the bounds. There are three variations to alter the velocity.

    i. **Reflect-Adjust**: $V_{new} = X_{new,bounded} - X_{old,unbounded}$
    ii. **Reflect Unmodified**: Velocity is unmodified.
    iii. **Reflect-Zero**: Velocity is set to zero.

3. **Nearest Methods**: Put the particle on the nearer bounds with four variations to alter the velocity.

    i. **Nearest–Zero**: Velocity is set to zero.
    ii. **Nearest–Unmodified**: Velocity is unmodified.
    iii. **Nearest with deterministic back**: $V_{new} = -0.5 * V_{old}$
    iv. **Nearest with random back**: $V_{new} = -lambda * V_{old}$, where lambda is a random real number between 0 and 1.

4. **Random Methods**: Randomly locate the particle within the bounds with two variations to change the velocity.

    i. **Random-Zero**: Velocity is set to zero.
    ii. **Random-Unmodified**: Velocity is unmodified.

5. **Hyperbolic Method**: In this method the particle is not allowed to leave the feasible area at all. The new velocity is normalized before updating the position as,

$$\begin{aligned} \text{If } V_i^{t+1} > 0 \quad & V_i^{t+1} = V_i^{t+1} / \left(1 + \left|V_i^{t+1} / (X_{i,max} - X_i^t)\right|\right) \\ \text{Else} \quad & V_i^{t+1} = V_i^{t+1} / \left(1 + \left|V_i^{t+1} / (X_i^t - X_{i,min})\right|\right) \end{aligned}$$

The normalized velocity ensures that the new position of the particle is always within the bounds.

The methods described in this section help to handle variable bounds during the execution of the algorithm. These methods restrict the movement of the particle thereby keeping them in the feasible region. Hence they help the algorithm to converge faster and find better solutions. We now describe the benchmark problems over which we have tested the MOPSO algorithm coupled with these methods.

4 Results and Discussion

MOPSO with various bound handling methods is tested on four ZDT benchmark problems [10] in multi-objective optimization. The mathematical description of the problems is given in Table 2 with their nature of Pareto-front. Their characteristic features of convexity, non-convexity, discreteness and non-uniformity respectively are known to cause difficulty to an EA in converging to the Pareto optimal front. We run our code 20 times with different initial particle positions using the algorithm described in Table 1. The performance of bound handling methods are assessed by using inverse generalized distance. The best, median and worst inversed generalized distance (IGD) values [12] from our runs are tabulated in Table 3.

For solving problems with convex Pareto fronts, the best performance is shown by the Hyperbolic method, but it is not consistent, hence Nearest Z serves as the best method for such problems as shown by its consistent IGD values for ZDT1 problem. Other versions of the Nearest method can also prove to be useful.

Solving problems having a non-convex front like ZDT2 proves to be difficult using MOPSO even after the deployment of bound handling methods. Although Nearest Z shows maximum convergence in the best case in ZDT2 problem, but in several other runs it fails to live up to its best performance. Similar behavior is shown by Nearest U and Nearest DB methods. However Nearest RB shows incredible consistency in solving ZDT2 problem as evident from the consistent best, median and worst IGD values for ZDT2 problem. Hence for solving problems with a previous knowledge of it having a non-convex Pareto front, Nearest RB method could prove to be extremely useful for an improved bound handling and a better convergence to the true Pareto using MOPSO. None of the other methods solve the ZDT2 problem.

When there are several non-contiguous convex parts in the Pareto front, both Nearest U and Nearest Z perform the best and equally well as in the case of ZDT3 problem. Hyperbolic and other versions of Nearest method could also be used as they show consistent IGD values for all cases. Reflect Z and A also may be used, while other methods result in weak Pareto front.

In the case of ZDT6 problem which has a non-uniformly distributed, non-convex Pareto-optimal front, satisfactory performance is shown by Hyperbolic, all Nearest methods and Reflect A, Z methods. Nearest RB method is consistent and shows the best performance in the worst case.

The analysis of Table 3 and the plots in Fig. 1 give a clear idea that out of all the methods, Random-U, Random-Z and Reflect U are the poorest and generate weak Pareto fronts in all cases. The reason for the poor performance of Reflect U is that even after being reflected back into the feasible region, the particle is forced out again into infeasibility because of the unmodified velocity vector. The Random methods also fail because they randomly put the violating variable in the feasible region. In both the above cases even though feasibility is maintained, the swarm loses out on the areas of better fitness. Hence the particles fail to converge to the Pareto front. The other two variants of Reflect method, Reflect-A and Reflect-Z

Table 2 Details of ZDT benchmark problems [10]

	n	Variable bounds	Objective functions	Optimal solutions	Nature of Pareto front
ZDT1	30	[0, 1]	$f_1(x_1) = x_1$ $g(x_2,\ldots,x_m) = 1 + 9 \cdot \sum_{i=2}^{m} x_i/m - 1$ $h(f_1,g) = 1 - (f_1/g)^2$	$x_1 \in [0,1]$, $x_i = 0$, $i = 2,\ldots,n$	Convex continuous
ZDT2	30	[0, 1]	$f_1(x_1) = x_1$ $g(x_2,\ldots,x_m) = 1 + 9 \cdot \sum_{i=2}^{m} x_i/m - 1$ $h(f_1,g) = 1 - \sqrt{f_1/g}$	$x_1 \in [0,1]$, $x_i = 0$, $i = 2,\ldots,n$	Nonconvex continuous
ZDT3	30	[0, 1]	$f_1(x_1) = x_1$ $g(x_2,\ldots,x_m) = 1 + 9 \cdot \sum_{i=2}^{m} x_i/m - 1$ $h(f_1,g) = 1 - \sqrt{f_1/g} - (f_1/g)\sin(10\Pi f_1)$	$x_1 \in [0,1]$, $x_i = 0$, $i = 2,\ldots,n$	Convex disconnected
ZDT6	10	[0, 1]	$f_1(x_1) = 1 - \exp(-4x_1)\,sin^6(6\Pi x_1)$ $g(x_2,\ldots,x_m) = 1 + 9 \cdot \left(\left(\sum_{i=2}^{m} x_i\right)/(m-1)\right)^{0.25}$ $h(f_1,g) = 1 - (f_1/g)^2$	$x_1 \in [0,1]$, $x_i = 0$, $i = 2,\ldots,n$	Nonconvex non uniformly spaced

All the problems are minimized

Table 3 Best, Median and Worst IGD values obtained for MOPSO

	Mutation	Hyperbolic	Nearest-DB	Nearest-RB	Nearest-U	Nearest-Z	Random-U	Random-Z	Reflect-A	Reflect-U	Reflect-Z
ZDT1	5.414e-03	**4.427e-04**	6.083e-04	5.656e-04	4.502e-04	4.795e-04	1.311e-01	7.272e-02	8.593e-04	9.585e-02	7.337e-04
	1.251e-02	5.382e-04	6.681e-04	6.577e-04	**4.913e-04**	5.404e-04	1.489e-01	8.971e-02	1.170e-03	1.109e-01	9.854e-04
	2.238e-02	7.458e-02	8.004e-04	7.284e-04	8.695e-04	**5.795e-04**	1.692e-01	9.654e-02	3.095e-03	1.244e-01	6.538e-03
ZDT2	4.838e-03	3.409e-02	4.876e-04	4.518e-04	4.223e-04	**4.162e-04**	2.285e-01	1.508e-01	1.776e-02	1.686e-01	4.395e-02
	1.398e-02	7.231e-02	5.736e-04	**5.531e-04**	1.647e-03	5.462e-02	2.468e-01	1.944e-01	7.231e-02	2.013e-01	7.231e-02
	2.036e-02	8.918e-02	7.231e-02	**6.085e-04**	7.231e-02	7.231e-02	2.636e-01	2.099e-01	8.918e-02	2.173e-01	7.332e-02
ZDT3	1.886e-02	3.697e-04	3.933e-04	3.587e-04	3.447e-04	**3.423e-04**	7.916e-02	5.120e-02	6.716e-04	6.136e-02	5.431e-04
	2.005e-02	4.124e-04	4.252e-04	4.202e-04	**3.856e-04**	4.106e-04	8.515e-02	5.493e-02	1.471e-03	6.876e-02	9.870e-04
	2.822e-02	6.612e-04	4.579e-04	5.301e-04	4.680e-04	**4.461e-04**	9.164e-02	6.162e-02	2.917e-03	7.519e-02	3.744e-03
ZDT6	3.508e-03	3.174e-03	**3.165e-03**	3.194e-03	3.498e-03	3.503e-03	5.367e-01	4.848e-01	3.301e-03	5.075e-01	3.197e-03
	2.271e-01	3.489e-03	3.523e-03	**3.476e-03**	3.506e-03	3.509e-03	6.221e-01	5.498e-01	4.055e-03	5.731e-01	3.503e-03
	3.187e-01	3.532e-03	1.336e-02	**3.515e-03**	3.519e-03	3.524e-03	6.489e-01	5.787e-01	7.089e-03	6.019e-01	4.309e-03

Best performance is shown in bold

DB Deterministic Back, *RB* Random Back, *Z* Zero, *A* Adjust and *U* Unmodified

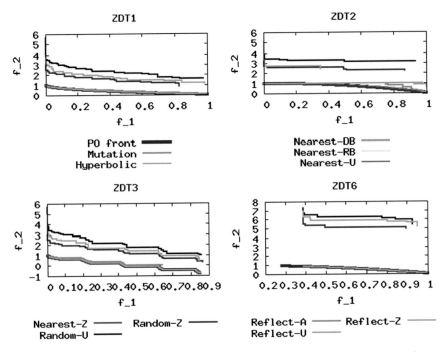

Fig. 1 0 % attainment plots for ZDT problems using MOPSO with bound handling techniques

perform much better than their counterpart Reflect U because in these two cases the velocity vector is modified (reversed or made zero) which helps the particle to remain in the feasible region and eventually reach near the true Pareto front. However the results acquired with them still have a scope of further improvement. They may be used (although some other methods are much better) to solve problems with uniform or discrete convex fronts like ZDT1 and ZDT3 and non-uniform problems like ZDT6 but fail miserably in solving non-convex problems like ZDT2 problem. (See the IGD values for ZDT2). The mutation based method also fails at a general level in the context of the shape of the Pareto front, poor convergence and getting stuck in the local optimum. One straight forward explanation is that the small perturbation from the unbounded position does not ensure feasibility. Hyperbolic, though not the best, can still be used with all problems mentioned in the paper except ZDT2. Its performance can be attributed to the fact that it never allows the particle to leave the feasible region and hence has proven to be quite competitive. As mentioned in [2] the velocities when using Hyperbolic are very small, which reduces the exploration capability of the swarm and it is highly probable that it prematurely converges at a local optimum. This explains its poor performance in solving ZDT2 problem.

All the Nearest methods have performed well. Nearest RB method has proven to be the most consistent. The method never fails as depicted by the IGD values which are the best or very close to the best compared to other methods even in the worst cases.

It is closely followed by Nearest Z and Nearest DB methods that also give good results most of the time. Nearest U method does not perform as good as the rest of them due to the unmodified velocity vector.

An important reason to be noted here, which explains the superior performance of all the Nearest methods is that with ZDT problems, part of the Pareto front lies at the bound, and since the Nearest method keeps the particle at the bound to prevent infeasibility, often the particles are able to reach a local optimum and then they diverge from there to form the Pareto front. Nearest RB performs the best, as it sets the velocity randomly compared to the other Nearest versions where the velocity is predetermined which limits the exploration capability of the swarm.

From our analysis in this section we can say that the Nearest Methods and to some extent the Hyperbolic method, are the most reliable methods for bound handling. Finally we have identified that the Nearest RB method, may not produce the best results in all cases, but it is very close to the best in all the cases. Hence we recommend that, Nearest RB method should be used for bound handling with various problem types.

5 Conclusion

In this paper we implemented several methods for bound handling of decision variables for multi-objective particle swarm optimization, and showed that MOPSO coupled with boundary handling techniques performs better than without the usage of a well-defined boundary handling technique. The Nearest methods and the Hyperbolic method perform well at handling the bounds, particularly the Nearest RB method beats them all. The reflect method also gives fair results. Mutation based method disappoints given its popularity in the literature. The algorithm works almost perfectly for most of the 2-D benchmark problems including ZDT1, ZDT2, ZDT3 and ZDT6. The future scope of research and study lies in the area of development of further efficient bound handling methods for MOPSO that can be used for multi and many objective optimization problems.

References

1. Kennedy, J., Eberhart, R.: Particle swarm optimization. In: Proceedings of IEEE International Conference on Neural Networks. vol. 4, pp 1942–1948 (1995)
2. Helwig, S., Branke, J., Mostaghim, S.: Experimental analysis of bound handling techniques in particle swarm optimization. IEEE Trans. Evol. Comput. 17(2), 259–271 (2013)
3. Padhye, N., Deb, K., Mittal, P.: Boundary handling approaches in particle swarm optimization. In Proceedings of Seventh International Conference on Bio-Inspired

Computing: Theories and Applications (BIC-TA 2012), Advances in Intelligent Systems and Computing, vol. 201, pp. 287–298 (2013)
4. Helwig, S., Wanka, R.: Particle swarm optimization in high dimensional bounded search spaces. In: Proceedings IEEE Swarm Intelligence Symposium, pp. 198–205 (2007)
5. Helwig, S., Wanka, R.: Theoretical analysis of initial particle swarm behavior. In Proceedings of 10th International Conference PPSN, pp. 889–898 (2008)
6. Coello, C.A.C., Pulido, G.T., Lechuga, M.S.: Handling multiple objectives with particle swarm optimization. IEEE Trans. Evol. Comput. **8**(3), 256–279 (2004)
7. Wang, Y., Li, B., Weise, T., Wang, J., Yuan, B., Tian, Q.: Self-adaptive learning based particle swarm optimization. Inf. Sci. **181**(20), 4515–4538 (2011)
8. Li, F., Xie, S., Ni, Q.: A novel boundary based multiobjective particle swarm optimization. Adv. Swarm Comput. Intell. **9140**, 153–163 (2015)
9. Padhye, N., Branke, J., Mostaghim, S.: Empirical comparison of MOSPO methods—guide selection and diversity preservation. In: Proceedings of Congress of Evolutionary Computation, pp. 2516–2523 (2009)
10. Zitzler, E., Deb, K., Thiele, L.: Comparison of multiobjective evolutionary algorithms: empirical results. Evol. Comput. **8**(2), 173–195 (2000)
11. Deb, K., Pratap, A., Agarwal, S., Meyarivan, T.: A fast and elitist multiobjective genetic algorithm: NSGA-II. IEEE Trans. Evol. Comput. **6**(2), 182–197 (2002)
12. Jenkins, W.K., Mather, B., Munson, D.C., Jr.: Nearest neighbor and generalized inverse distance interpolation for fourier domain image reconstruction. In: Acoustics, Speech, and Signal Processing, IEEE International Conference on ICASSP'85, vol. 10, pp. 1069–1072 (1985)

Role of Bio-Inspired Optimization in Disaster Operations Management Research

R. Dhveya and T. Amudha

Abstract Disaster Operations Management (DOM) is a Non deterministic polynomial, Multi-objective optimization problem. DOM is a cyclic process and holds various phases. It is impossible to prevent natural disasters completely but optimized solutions can be suggested to avoid/mitigate the impact of disasters. India is a developing country and natural disaster is one among the major issues in India. Due to various geo-climatic conditions, frequent natural disaster events like earthquake, flood, drought, landslides, tropical cyclone and storm strikes according to the vulnerability of the hazard prone area especially in the region of Tamil Nadu, India. DOM prevents human and economic losses by applying computer algorithms, computer based decision making tools and software solutions to handle disasters effectively. Still more improvisation is needed in an optimized pattern for acquiring best solutions. This paper presents the optimization perspective of DOM and highlights the need for bio inspired optimization techniques to reduce the impact of disasters, towards a social focus.

Keywords Disaster operations management · Bio inspired algorithms · Optimization · Computer based tools and techniques

1 Introduction

Nature in the broadest sense can be defined as the natural, physical, material world or universe. Nature has its own positive and negative forces. Disaster is one among the negative forces and can be defined as a sudden accident or a natural catastrophe

---

R. Dhveya (✉) · T. Amudha
Department of Computer Applications, Bharathiar University, Coimbatore,
Tamil Nadu, India
e-mail: dhveya7@gmail.com

T. Amudha
e-mail: amudha.swamynathan@gmail.com

© Springer International Publishing Switzerland 2016
V. Snášel et al. (eds.), *Innovations in Bio-Inspired Computing and Applications*,
Advances in Intelligent Systems and Computing 424,
DOI 10.1007/978-3-319-28031-8_50

that causes great damage or loss of life. It can be classified at a high level as either manmade disaster or natural disaster. At global level there has been considerable concern over natural disasters. Around last 30 years of time span, India has faced more than 430 disasters resulting majorly in loss of life and economy [10]. India has an area of 3,287,263 km^2 and a coastline of 7516 km, with the official census 2001 reference. Tamil Nadu covers an area of 1,30,058 km^2 and has a coastline of around 1,076 kms which is about 15 % of the coastline of India [28]. Thus the geographical setting of Tamil Nadu makes the state vulnerable to natural disasters such as cyclones, floods, earthquake induced Tsunami etc. [28]. Though quite few measures are considered in giving a better solution for reducing natural disaster related problems, there is always a requirement for improved techniques.

2 Disaster Operations Management (DOM)

Due to the complex nature of the problem, DOM is considered as a multi objective optimization problem. Researchers on disaster operations management classify DOM into different activities such as Mitigation, Preparedness, Response and Recovery. In this modern era, new technologies, use of tools, methodologies like Operations Research (OR), advance enhancement in algorithms along with hardware, software developments are making excellent outlines for the optimization system and lot of attentions have been given in recent years [18, 26]. Disaster Operations Management can be broadly defined into three categories, Pre-disaster phase, Disaster phase and Post-disaster phase. Pre-disaster phase includes activities of Mitigation and Preparedness. Disaster phase includes activities of Response. Post-disaster phase includes activities of Recovery [8]. Through disaster research review, it was identified that 22 % of the disaster research was focused on mitigation phase, 28 % on preparedness phase, 46 % on response phase and 4 % on recovery phase [17].

2.1 Mitigation

Mitigation phase in general involves activities such as public education and awareness for reducing the effect of disaster events [8]. National Institute of Disaster Management (NIDM) includes trainings like face to face, Web based, Satellite based and self study courses to get awareness from natural disasters [11]. Capacity building is the important factor in mitigation [26].

Optimization can be framed with the help of GIS, GPS and ICT technologies. Still optimal focus in terms of bio-inspired computing strategy is less when compared to other phases of DOM.

2.2 Preparedness

Preparedness phase in general involves pre-disaster activities, training and early warning systems (EWS) [8]. Quite few incidents related to natural disasters revealed that there is no specific and timely warning within agencies like India Meteorological Department (IMD), Central Water Commission (CWC), National Disaster Management Authority (NDMA), State Disaster Management Authority (SDMA) and others. Decision Support System (DSS) with optimal solutions are always useful, mainly because of flexibility in nature [23].

Optimization can be framed in the usage of technologies like Mobile Technology, GPS Maps, where people can easily get warning messages and can navigate to safe places [8]. Real time monitoring of natural disaster over large areas can be archived with the development of Geographical Information System (GIS) and GPS, which helps to take effective actions during damages [22]. Table 1 shows the improvement in disaster management with the usage of ICT [9].

2.3 Response

Response phase in general involves disaster situation activities like search, evacuation, rescue and relief [8]. Channels for communication like either Network media (TV, Telephone, Internet, Mobile, Satellite radio and others) or Agencies (Local/State/Central government, Private guarding agencies, International agencies) play key role in response phase. Emergency Evacuation is the evacuation of inhabitants from affected area (Hazard area) to safe area (Shelter in large capacity areas like Educational Institutions, Stadiums, Halls, and Places of worship). Algorithms help to evacuate maximum number of people in a minimum travel time from hazard zone to safe zone.

Optimization mainly focuses on three aspects of supply distribution namely Economics (minimizing total cost), Effectiveness (minimizing total travel time) and Equity (maximizing fairness). With the help of mentioned technologies and by the usage of bio-inspired computing techniques better solutions can be given optimally.

Table 1 Impact of ICT in disaster management [9]

	Region	Impact of ICT—EWS	Year	Deceased count
Cyclone	Andhra	Before ICT Impact–EWS	1979	10,000
Hit	Pradesh	After ICT Impact—EWS	1990	<1,000
	Bangladesh	Before ICT Impact—EWS	1970	3,00,000
		After ICT Impact—EWS	Recent	3,000

2.4   Recovery

Recovery phase in general involves post disaster activities such as medical care, temporary housing [8].

Optimization with the usage of GIS in recovery phase is becoming an up-coming trend. Still research interest in accordance with the funding mechanisms, financial cycle which involve details of Insurers, Government and Private sectors/NGOs has to be focused, so that improvisation is possible. Table 2 shows Bio inspired algorithms and other techniques used in various phases of DOM.

3   Optimization Factors in DOM

DOM has certain factors, that are considered for optimizing natural disasters and they are as follows.

3.1   Climate Prediction

Number of rain days per year decreases but the intensity of precipitation increases. Palmer Drought Severity Index (PDSI) is used to determine whether the changes in rainfall regimes may lead to more floods or droughts. Global Climate Models (GCMs) are widely considered as the most acceptable tool for studying changes in climate [28]. It is one of the important factors in optimization.

3.2   Transportation

In most of the developing countries, DOM mechanism sounds difficult mainly due to lack of collaboration and communication between management responsible entities like Police, Fire brigade, Ambulance etc. [8]. NDMA has noted some gaps in the communication between agencies [9]. Table 3 lists the optimization problems and algorithms in transportation.

3.3   Population Distribution

In India, environment is mainly affected due to the increase in population and migration into urban areas. Key factors for optimizations are the number of people in affected areas, capacities in safe areas, spatial location of areas and the distance between hazard and safe areas [7].

Table 2 Bio inspired algorithms and other techniques used in various DOM phases

DOM Phase	Bio inspired algorithms/other techniques used	Year and reference
Mitigation	Information and communication technology (ICT)	2015 [9]
	Global positioning system (GPS)	2014 [14]
	Geographical information system (GIS)	2014 [26]
	Disaster behavior analysis and probabilistic Reasoning system (DBAPRS)	2014 [14]
Preparedness	Information and communication technology (ICT)	2012 [23], 2014 [8]
	Particle swarm optimization (PSO)	2015 [3]
	Improved PSO (IPSO)	2015 [3]
	Artificial neural network (ANN)	2012 [23], 2015 [9] 2015 [27]
	Back propagation neural network (BPNN)	2015 [9]
	Geographical information system (GIS)	2010 [22], 2015 [27]
	Adaptive neuro fuzzy inference system (ANFIS)	2015 [27]
	Wireless sensor network (WSN)	2009 [16]
	Geographical positioning system (GPS)	2010 [22], 2014 [8]
	Stochastic optimization model	2010 [32]
	Computer simulation model	2011 [2]
Response	Wireless sensor network (WSN)	2015 [5]
	Particle swarm optimization (PSO)	2015 [5], 2015 [33]
	Geographical information system (GIS)	2010 [30], 2014 [14]
	Evolution strategies and programming (ES and EP)	2015 [33]
	Ant colony optimization (ACO)	2015 [33]
	Biogeography based optimization (BBO)	2015 [33]
	Artificial immune system (AIS)	2015 [33]
	Hybrid algorithms (HA)	2015 [33]
	Mobile technologies	2014 [14]
	National emergency management network, Sahana	2010 [30]
Recovery	Geographical information system (GIS)	2009 [20]
	Non dominated sorting genetic algorithm (NSGA-II)	2015 [6]
	Seed block algorithm (SBA)	2015 [6]
	Computer aided disaster recovery planning (CADRP)	2015 [19]
	RecHADS	2014 [15]

Table 3 Optimization problems and algorithms in transportation [18]

Focused classes	Optimization problems and algorithms
General transportation planning	Integer programming (IP): Non-linear IP, Mixed integer programming (MIP): Non-linear MIP, Particle swarm optimization (PSO), Genetic algorithm (GA), Artificial immune system (AIS), Multi-objective evolutionary algorithm (MOEA)
Location and routing	Heuristic genetic algorithm (GA), Vehicle routing problem (VRP), Linear integer programming (LIP), Hybrid ant colony optimization (ACO), Immune ant colony optimization (IACO) = ais + aco, Location routing problem (LRP), Artificial neural network (ANN), Hill climbing
Roadway repair	Hybrid genetic algorithm (GA), Ant colony optimization (ACO), Time-space network model, Multi-objective scheduling model, Multi-objective model of emergency roadway repair
Integrated problems	Particle swarm optimization (PSO), Ant colony optimization (ACO), Modified genetic algorithm (GA), Network flow model = Roadway repair + Relief distribution

3.4   Hospital Location

Locations of hospitals in rural and urban area in an optimal pattern is very important and GIS helps in finding optimal locations of hospitals as well as separation of emergency hospital with the help of population data, in reducing risks among factors like minimizing cost, time, distance and with the high concentration in preventing human loss [12, 15]. Table 4 shows the applied computational techniques of optimization factors.

3.5   Community and Humanitarian Networks (CH-N/W)

A web-based prototype has been implemented for building Business Continuity Information Network (BCIN) which helps in providing a collaboration and communication in networks [30]. In the distribution of emergency goods to a population, Humanitarian Logistics considers factors like time, cost reliability, security and demand satisfaction for optimal solutions [15].

3.6   Power Supply

Limited power transmission in case of quick rescue operations creates emergency situation [5]. A strategic optimization is studied in restoration of electrical parts in a power system and Robust or Stochastic optimization is used in determining how best to repair the power system with the supply [21].

Table 4 Optimization factors and applied computational techniques in DOM

Optimization factors	Computational techniques applied	Year and reference
Climate prediction	Artificial neural network (ANN)	2012 [23], 2015 [9] 2011 [13]
	Wireless sensor network (WSN)	2009 [16]
	Back propagation neural network (BPNN)	2009 [16]
	Automatic weather station (AWS)	2012 [29]
	Apriori association rule mining algorithm	2011 [1]
	WEKA, a data mining tool	2011 [1]
	Palmer drought severity index (PDSI)	1997 [34]
	Global climate models (GCMs)	1997 [34]
	ARIMA	2010 [22]
Transportation	Artificial bee colony (ABC)	2014 [18]
	Wireless sensor network (WSN)	2015 [4]
	Vehicular adhoc network (VANET)	2014 [14]
	Intelligent transport system (ITS)	2014 [14]
	Genetic algorithm (GA)	2014 [18]
	Particle swarm optimization (PSO)	2014 [18]
	Ant colony optimization (ACO)	2014 [18]
	Biogeography based optimization (BBO)	2014 [18]
Population distribution	Genetic algorithm (GA)	2010 [32]
	Simulated annealing (SA)	2010 [32]
	Back propagation neural network (BPNN)	2011 [13]
	Mat Lab's GA and simulated annealing	2010 [32]
Hospital location	Geographical information system (GIS)	2013 [12], 2014 [15]
CH N/W	Sahana	2010 [30]
Power supply	Wireless sensor network (WSN)	2015 [5]
	NS2 simulator	2010 [32]
	Artificial bee colony (ABC)	2015 [5]

4 Major Technologies Used in DOM

Now in the age of technology, we can manage disasters by using various features of Information Technology (IT) in the form of Internet, GIS, Remote Sensing, etc. Table 5 shows the purpose of GIS and RS.

We can optimize the issues caused in disaster environment with the help of GIS [21, 27, 22, 24, 31] and RS [31, 25]. GPS and Mobile Technology can optimize the issues caused in disaster environment with the help of mobile device like smart phones using Global Positioning System (GPS) coordinates, Latitude and Longitude [26]. Internet can optimize the issues caused in disaster environment especially in collecting real time data and in communication [29].

Table 5 Purpose of GIS and RS in disaster management

Nature of disaster	Purpose of GIS and RS
Drought	To develop early warnings of drought conditions and to detectground water sites for well digging program
Earthquake	To prepare seismic hazard maps for assessing exact risk
Floods	To map and monitor flood areas, damage assessments, hazardzoning and post flood survey of rivers, protection works
Landslides	To prepare landslide hazard zonation map
Search/rescue	To identify disaster prone zone according to risk magnitudes

5   Disasters in India

India is a large country and prone to a number of natural hazards. In India, out of 35 states and union territories in the country, 27 of them are disaster prone. Almost 58.6 per cent of the landmass is prone to earthquakes of moderate to very high intensity; over 40 million hectares (12 per cent of land) are prone to floods and river erosion; Among 7,516 km long coastline, close to 5,700 km is prone to cyclones and tsunamis. Cyclonic activities are more severe on east coast than west coast and mostly occurs between the months April–May and October–November [28]; 68 per cent of the cultivable area is vulnerable to drought and hilly areas are at risk from landslides and avalanches [9, 29]. The last occurrence of massive Tsunami on 26/12/2004 and the occurrence of Uttarakhand calamities on 15–17 June, 2013 have worsened the situation [34]. In India, database for the occurrence of various disaster types is yet to be developed. At the global level, the available three main sources in collecting and updating disaster data are EM-DAT, Natcat and Sigma. Table 6 shows the major disasters occurred in Tamil Nadu from 2000 to 2014 [35].

Table 6 Disasters in Tamil Nadu (2000–2014) [35]

Year	Disaster type: sub type	Total deaths	Total affected
2004	Earthquake: tsunami	16,389	6,54,512
2005	Flood: river line	30	2,00,000
2007	Flood: river line	29	50,000
2008	Flood	37	10,278
2008	Flood: slide (land/mud)	1063	79,00,000
2008	Flood: river line	–	50,000
2008	Flood: flash	54	8,03,740
2009	Flood: slide (land/mud)	70	8
2012	Storm: tropical cyclone	40	70,000

The main objective of this study is to identify natural disasters related problems in India, specifically in Tamilnadu context and to suggest best possible solutions to address the issues by using bio-inspired optimization methods aiming to assist the society in pre-disaster as well as post disaster phases, since the significance of bio-inspired algorithms are going to be a new revolution in computer science.

6 Conclusion

DOM is a NP-hard problem, which aims to reduce human and economic losses. In this study, it is clearly identified that most of the research works dealt with Disaster Response and Preparedness phases and very few research works dealt with Mitigation and Recovery phases. In India, trends like GIS, Remote Sensing in DOM are not vastly developed when compared to other countries. Compared with worldwide disaster management, India still lacks in preparedness towards managing disasters. Bio inspired techniques combined with GIS and RS can reliably be applied in DOM for effective optimization of the needy factors addressed in this paper. This study has analyzed the role of computer based methods, algorithms, tools etc. in disaster management, the significance of bio inspired algorithms in DOM, and highlighted the requirement of improved optimization techniques that could facilitate effective handling of disasters without much loss to the humanity.

References

1. Rajput, A., Soni, R., Aharwal, R.P., Sharma, R.: Impact of data mining in drought monitoring. IJCSI Int. J. Comput. Sci. Issues 8(6), 2 (2011). ISSN (Online):1694-0814
2. Landström, C., Whatmore, S.J., Lane, S.N.: Virtual engineering: computer simulation modelling for flood risk management in England. J. Sci. Stud. 24(2), 3–22 (2011)
3. Chavan, S.D., Kulkarni, A.V., Adgokar, N.P.: Efficient bio-inspired method for disaster monitoring in wireless sensor networks using improved PSO. Int. Res. J. Eng. Technol. (IRJET) 2(4) (2015) e-ISSN:2395-0056, p-ISSN:2395-0072
4. Chavan, S.D., Kulkarni, A.V., Khot, T.S.: An efficient routing algorithm for lifetime enhancement in wireless sensor network using artificial bee colony algorithm. Int. Res. J. Eng. Technol. (IRJET) 2(4) (2015)
5. Chavan, S.D., Khot, T.S.: Efficient and reliable routing algorithm to enhance connectivity in disaster scenario: ABC algorithm. Int. J. Sci. Res. (IJSR) 4(5) (2013) ISSN (Online):2319-7064
6. Deepa, S., Ramachandran, G.: Disaster recovery system using seed block algorithm in cloud computing environment. Int. J. Adv. Res. Comput. Sci. Softw. Eng. 5(2) (2015) ISSN:2277 128X
7. Kaisa, E.I., Hess, L., E.I., Alicia Benazir Portal Palomo, E.I.: An Emergency Evacuation planning model for special needs populations. J. Public Transport. 15(2) (2012)
8. Shafiq, F., Ahsan, K.: An ICT based early warning system for flood disasters in Pakistan. Res. J. Recent Sci. 3(9), 108–118 (2014) ISSN:2277-2502

9. GopalDatt.,Ashutosh Kumar Bhatt., Sunil Kumar: Disaster management information system framework using feed forward back propagation neural network. Int. J. Adv. Res. Comput. Commun. Eng. **4**(3), (2015) ISSN(Online):2278-1021, ISSN(Print):2319-5940

10. Government of India National Institute of Disaster Management New Delhi. Report of Human Resource and Capacity Development Plan for Disaster Management and Risk Reduction in India (2013)

11. Government of India Ministry of Home Affairs. Report of Disaster Management in India (2011)

12. Han, M., Hakansson, J., Rebreyend, P.: How do different densities in a network affect the optimal location of service centres. Eur. J. Oper. Res. 15, (2013) ISSN:1650-5581

13. Wang, H., Li, X., Long, H., Qiao, Y., Li, Y.: Development and application of a simulation model for changes in land-use patterns under drought scenarios. Elsevier Comput. Geosci. **37**, 831–843 (2011)

14. Jeeva, V.R., Jibi J Puthiyidam.: A survey on disaster management system. Int. J. Adv. Res. Comput. Sci. Manage. Stud. **2**(11), (2014) ISSN:2321-7782

15. Liberatore, F., Ortun, M.T., Tirado, G., Vitoriano, B., Scaparra, M.P.: A hierarchical compromise model for the joint optimization of recovery operations and distribution of emergency goods in humanitarian logistics. Elsevier Comput. Oper. Res. **42**, 3–31 (2014)

16. Ramesh, M.V.: Real-time wireless sensor network for landslide detection. In: Third International Conference on Sensor Technologies and Applications (2009)

17. CamilaHoyos, M., Morales, R.S., Akhavan-Tabatabaei, R.: OR models with stochastic components in disaster operations management: A literature survey. Elsevier Comput. Ind. Eng. **82**, 183–197 (2015)

18. Zhang, M.-X., Zhang, B., Zheng, Y.-J.,: Bio-Inspired metaheuristics for emergency transportation problems. Algorithms-Open Access J. **7**, 15–31 (2014) ISSN:1999-4893

19. Alhazmi, O.H.: Computer-aided disaster recovery planning tools (CADRP). Int. J. Comput. Sci. Secur. (IJCSS) **9**(3) (2015)

20. El-Anwar, O., El-Rayes, K., Elnashai, A.: An automated system for optimizing post-disaster temporary housing allocation. Elsevier Autom. Constr. **18**, 983–993 (2009)

21. van Hentenryck, P.: Computational disaster management. In: Proceedings of the Twenty-Third International Joint Conference on Artificial Intelligence (2011)

22. Han, P., Wang, P.X., Zhang, S.Y., Zhu, D.: Drought forecasting based on the remote sensing data using ARIMA models. Elsevier Math. Comput. Model. **51**, 1398–1403 (2010)

23. Subramanian, S.: Implementation of neural networks in flood forecasting. Int. J. Sci. Res. Publ. **2**(10), (2012). ISSN:2250-3153

24. Renuka Devi, M., SanthoshBaboo, S.: Land use and land cover change detection for three decades: case study for Coimbatore area. Int. J. Eng. Res. Appl. (IJERA) **2**(1), 1013-1028 (2012). ISSN:2248-9622

25. Renuka Devi, M., SanthoshBaboo, S.: Land use and land cover for one decade in Coimbatore dist using historical and recent high resolution satellite data. Int. J. Sci. Eng. Res. **3**(2) (2012). ISSN:2229-5518

26. Saravana Kumar, J., Veeramani, R.: GPS Location alert system. IOSR J. Comput. Eng. (IOSR-JCE) **16**(2), 36–42 (2014). Ver. IV, e-ISSN:2278-0661, p-ISSN:2278-8727

27. SonamLhamuBhutia, RatikaPradhan, Ghose, M.K.: A Survey on landslide susceptibility mapping using soft computing techniques. IOSR J. Appl. Geol. Geophys. (IOSR-JAGG) **3**(1), 16–20 (2015). Ver. I, e-ISSN:2321-0990, p-ISSN:2321-0982

28. Stephen, A: Natural disasters in India with special reference to Tamil Nadu. J Acad. Ind. Res **1** (2), (2012) ISSN:2278-521

29. Vyas, T., Desai, A.: Information technology for disaster management. In: Proceedings of National Conference on Computing for Nation Development, pp. 23–24 (2007)

30. Hristidis, V., Chen, S.-C., Li, T., Luis, S., Deng, Y.: Survey of data management and analysis in disaster situations. Elsevier J. Syst. Softw. **83**, 1701–1714 (2010)

31. Varadharajan, A., Iyappan, L., Kasinathapandian, P.: Assessment on landuse changes in Coimbatore North taluk using image processing and geospatial techniques. Int. J. Eng. Res. Appl. (IJERA). **2**(4), 233–237 (2012) ISSN:2248-9622
32. Boñgolan, V.P., Ballesteros., F.C., Jr., Banting, J.A.M., Olaes, A.M.Q., Aquino, C.R.: Metaheuristics in flood disaster management and risk assessment. In: Proceedings of 8th National conference on Information Technology Edu-cation (NCITE), pp. 20–23 (2010)
33. Zhenga, Y.-J., Chena, S.-Y., Ling, H.-F..: Evolutionary optimization for disaster relief operations: a survey. Elsevier Appl. Soft Comput. **27**, 553–566 (2015)
34. Kothavala, Z.: Extreme precipitation events and the applicability of global climate models to the study of floods and droughts. Elsevier Math. Comput. Simul. **43**, 261–268 (1997)
35. http://emdat.be/database

Optimal Reservoir Release for Hydropower Generation Maximization Using Particle Swarm Optimization

D. Kiruthiga and T. Amudha

Abstract Hydropower generation is one of the significant reservoir operations, especially in a large multi-purpose reservoir and optimizing the hydropower generation is complex but necessary task. Deriving optimal operation rules for reservoir release to maximize the hydropower generation is quite indispensable among the diverse roles of the reservoir purposes like irrigation, hydropower and flood control. The primary objective of this work is to optimize the reservoir release and in turn maximize the hydropower generation in a consistent manner. In this research, reservoir release is optimized by using Non-Linear Programming (NLP) technique and Particle Swarm Optimization (PSO), and it was observed that PSO gave the optimal best release. Hence, the optimal release given by PSO is used to maximize the hydropower generation at Aliyar reservoir in Coimbatore district of TamilNadu. Various release patterns are also framed and tested in this research, and the results have indicated that application of suitable release patterns optimized by PSO offer good scope for hydropower generation maximization in the future.

Keywords Maximization of hydropower generation · Reservoir release optimization · Particle swarm optimization · Non linear programming technique · Optimal release patterns

D. Kiruthiga (✉)
Department of Computer Applications, Bharathiar University,
Coimbatore, Tamilnadu, India
e-mail: kiruthigadevaraj04@gmail.com

T. Amudha (✉)
Department of Computer Applications, Bharathiar University,
Coimbatore, Tamilnadu, India
e-mail: amudha.swamynathan@gmail.com

© Springer International Publishing Switzerland 2016
V. Snášel et al. (eds.), *Innovations in Bio-Inspired Computing and Applications*,
Advances in Intelligent Systems and Computing 424,
DOI 10.1007/978-3-319-28031-8_51

1   Introduction

Hydropower is one of the clean and green schemes among many types of power generation schemes. The advantages of hydropower generation are quick response to meet demands, absence of pollution during operation and free of fuel cost. To satisfy the power demand of a country, hydropower generation maximization is a vital task [1]. In India, most of the reservoirs are having hydropower plants. However, as per Asian Development Bank (ADB) report, nearly 78 % of the hydropower potential remains without any plan for exploitation and only less than 23 % of the hydropower potential had been utilized for development [2] due to several reasons. Some of them are (i) In some reservoirs in India, hydropower is produced only when irrigation release is made, and when the irrigation release is not properly planned, hydropower production will not be done properly (ii) In some other reservoirs, release for hydropower and release for irrigation are performed separately, and hence the total available water is not utilized for hydropower, which reduces the prospective generation (iii) the disputes in sharing the available water among different stakeholders (different state governments). Thus, there is a need to exploit the full potential of the existing hydropower plants by optimizing the reservoir operations [3].

Hydropower generation optimization is quite complex involving various non-linear functional relationships, large number of decision variables, satisfying equality constraints and several other physical, technical and operational constraints [4]. Bio-inspired techniques are used to solve several optimization problems, and they outperformed other conventional optimization techniques such as, linear programming (LP), non-linear programming (NLP), and dynamic programming (DP) [5]. In this work, reservoir release optimization is formulated as NLP model and is solved by using LINGO 14.0/Global solver [12] and also by using the swarm intelligence technique, PSO. The optimal release obtained from both the techniques for different years are compared. The better optimal release obtained by PSO technique is applied to optimize the hydropower generation optimization in Aliyar reservoir.

The Aliyar reservoir in TamilNadu, India is considered as study area. The Aliyar Reservoir is essentially a part of the comprehensive Parambikulam—Aliyar Project (PAP), which is an interstate project of TamilNadu and Kerala. As per the interstate government agreement [6], Government of TamilNadu should release water to Tamilnadu and Kerala for irrigation. The Aliyar Reservoir would act as a balancing reservoir to receive surplus waters through Aliyar Feeder Canal diverted from Nirar, Sholayar and Parambikulam Basin (from Power House). The power house was designed to harness the capacity of 2×1250 kW. For this research, historical data of Aliyar reservoir was collected in person from the offices of Public Works Department (PWD) in Coimbatore.

2 Reservoir Release Optimization for Hydropower Generation

Optimum utilization of scarce water resources plays an important role in the sustainable development of any region. In water resources system, reservoir operation is one of the challenging problems that involves lot of complexities and create difficulties for water managers to release water optimally based upon demands. Reservoir operation model is formulated with an objective of maximizing the annual hydropower production subject to various physical and operational constraints. Among various purpose of reservoir, hydropower is very important for our country due to increasing power demand from various sectors like industry, domestic and agriculture. Hydropower production depends upon hydraulic head and flow through turbine. The hydropower production during any time period t for any reservoir relies on the installed capacity of plant, flow through turbines, average storage head and the plant operating duration [1]. Thus power produced in any time period t is expressed as,

$$PH_t = 2725 * R_t * NH_t * \eta \tag{1}$$

where, R_t is the release to the powerhouse during the time period t, NH_t is the net head of water available during the time period t, η is the overall plant efficiency and 2725 is the constant for converting the product of flow, head and plant efficiency to electrical energy in terms of kilowatt hours (kWh) [7]. In Eq. (1) net head (NH_t) during time period t is estimated by deducting the tail water level from average head (H_t). The average head H_t available during a particular time period t, expressed as the second order function of the storage is given in Eq. (2). C_1, C_2, C_3 are regression coefficients and S_t is the initial storage during time period t.

$$H_t = C_1 S_t + C_2 S_t^2 + C_3 \tag{2}$$

The objective function of the model is to maximize the hydropower production in a year. Mathematically the objective function is expressed as Eq. (3):

$$Max\ Z = \sum_{t=1}^{n} PH_t \tag{3}$$

where, n is expressed as number of years, PH_t is the power produced from the powerhouse during time period t. The above objective function is subjected to various constraints, as given in Eqs. (4) and (5).

The net head (NH_t) (m) of the reservoir should be greater than or equal to the minimum drawdown level $(MDDL)$ of the powerhouse, as given in Eq. (4).

$$NH_t \geq MDDL \tag{4}$$

Depending upon the power plant capacity and reliability, the maximum and minimum production has to be fixed for any power plants. This is given in Eq. (5), where, P_{min} and P_{max} are the minimum and maximum power production limits in kWh and PH_t is the hydropower produced in the time period t.

$$P_{min} \leq PH_t \leq P_{max} \tag{5}$$

The water release at any time period t (Mm^3) denoted as R_t should lie between the minimum (R_{min}) and maximum (R_{max}) limits of water release from the reservoir in a year, as given in Eq. (6) below.

$$R_{min} \leq R_t \leq R_{max} \tag{6}$$

The reservoir storage S_t during any time period t is given in Eq. (7) where S_{dead} is the minimum storage of the reservoir and S_{max} is the maximum storage.

$$S_{dead} \leq S_t \leq S_{max} \tag{7}$$

The evaporation loss (EV_t) during the period t is also an important component, which is expressed as a function of initial and final storage. The constants a_t and b_t are estimated by regression analysis from average water spread area and initial and final storage of reservoir in a particular time period t. The regression equation is expressed as:

$$EV_t = a_t + b_t \left(\frac{S_t + S_{t+1}}{2} \right) \tag{8}$$

Reservoir system should follow the mass balance principle which is given in Eq. (9), where, S_{t+1}, S_t, I_t, R_t, EV_t and O_t are the final storage, initial storage, inflow, release, evaporation and overflow in the reservoir during the time period t (Mm^3).

$$S_{t+1} = S_t + I_t + R_t - O_t - EV_t \tag{9}$$

The reservoir system may have overflow during monsoon period, when the final storage exceeds the reservoir capacity. This overflow constraint is given by:

$$R(t) = D(t) \left[\frac{\sqrt{S_t^2 + I_t^2} - \sqrt{S_{dead}^2 + I_{min}^2}}{\sqrt{S_{max}^2 + I_{max}^2} - \sqrt{S_{dead}^2 + I_{min}^2}} \right]^m \tag{10}$$

where, S_t is initial storage, $D(t)$ is demand of water during time period t. I_{min} and I_{max} are minimum and maximum inflow values in the year respectively. It may be noted that the above equation is defined such that minimum storage combined with minimum inflow provide no release ($R(t) = 0$); and maximum storage combined with maximum inflow provide maximum release ($R(t) = D(t)$) [5].

Many reservoirs in India have less inflow in dry seasons compared to wet seasons. A year was considered dry when at least 70 % of its inflow values are zero. For such years, the releases $R(t)$ were conditioned only on the storage $S(t)$. For dry years, the regression equation becomes:

$$R(t) = D(t) \left[\frac{S_t - S_{dead}}{S_{max} - S_{dead}} \right]^m \tag{11}$$

The task is thus to find, for each year, the value of the parameter m that fits the above equations to the data from the optimization model.

3 Particle Swarm Optimization

The PSO algorithm is population-based random search technique for optimization which was inspired by social behavior of bird flocking or fish schooling.PSO was originally proposed by Eberhart and Kennedy [8]. PSO algorithm involves random initialization of population and generation of new solutions by using swarm updating rules. It searches for best solution by improving its objective function value continuously over several iterations. In PSO, to extract a new generation of candidate solution, the operators are not inspired by nature evolution as in other evolutionary algorithms, rather, the velocities are dynamically adjusted according to their historical behavior. The best thing about PSO is that it relies on exchange of information among particles in the population so that each particle adjusts its path towards its own previous best and towards the current best position attained by other particles in its neighborhood [9]. Figure 1 gives the pseudo code of the PSO technique.

```
1    Begin (Initialization, i=1,...,N)
2    Generate Initial position X_i (0)
3    Generate Initial velocity V_i (0)
4    End
5    Set n=0 (n is the iteration number)
6    While (termination criteria not met) do
7    For (i=1,...,N)
8        Compute the fitness function value
9        Compute (pbest)
10   End For
11       Compute (gbest)
12   For (i=1,...,N)
13       Calculate V_i^{n+1} using equation (12)
14       Calculate X_i^{n+1} using equation (13)
15   End For
16       Set next iteration (n=n+1)
17   End While
```

Fig. 1 Pseudo—code of PSO algorithm

Depending upon the number of decision variables involved in the problem, each variable is represented by a single dimension. At the beginning, a set of solution is randomly generated for each particle in the swarm. In D-dimensional search space, the ith particle position can be represented as $X_i = (x_{i1}, x_{i2}, ..., x_{iD})$. The position of the particle changes based upon the velocity, represented as $V_i = (v_{i1}, v_{i2}, ..., v_{iD})$. During the execution of the model, the best position visited by the particle is stored. The best position of the ith particle is represented as $P_i = (p_{i1}, p_{i2}, ..., p_{iD})$ [6]. The following Eqs. (12) and (13) give the updating rules for velocity and position of the particle.

$$V_{id}^{n+1} = V_{id}^n + C1 * Rnd(0,1) * \left(pb_{id}^n - X_{id}^n \right) + C2 * Rnd(0,1) * gb_{gd}^n + X_{id}^n \quad (12)$$

$$X_{id}^{n+1} = X_{id}^n + V_{id}^n \quad (13)$$

where, C1 and C2 are the cognitive and social parameters of PSO; $d = 1, 2, ..., D$, the index for decision variables; $i = 1, 2, ..., N$ the index for swarm population; N is the size of the swarm; gd the index of best particle among all particles in the population; n is the generation number; $Rnd(0, 1)$ is a random function which uniformly generated random numbers between 0 and 1.

4   Implementation Results and Discussions

The objective function of this research work is to maximize the hydropower production. Aliyar reservoir is a multi-purpose reservoir, and there is a need to satisfy the demand of the irrigation and other purposes. In this research, reservoir optimization is modeled as (i) Non Linear Programming (NLP) and is solved by using LINGO 14.0/Global Solver and (ii) regression based ISO and is solved by using PSO algorithm coded in Java language. NLP is the process of solving an optimization problem defined by a system of equalities and inequalities, collectively termed constraints, over a set of unknown real variables, along with an objective function to be maximized or minimized, where some of the constraints or the objective functions are nonlinear. It is the sub-field of mathematical optimization that deals with problems that are not linear [10]. Aliyar Reservoir data was collected in person from the office of the PWD for 5 years, 2009 to 2013, which are used to test the models.

To solve the reservoir operation problem by using PSO algorithm, proper selection of algorithm parameters is essential, which helps in getting quicker optimal solutions. The parameters values used in this work are given in Table 1. The algorithm gave its best performance with 5 particles and 100 runs, beyond which it converged. Aliyar reservoir has a standard annual irrigation demand, domestic supply demand and interstate water demand, which are listed in Table 2. The optimally released water should be supplied based on the respective demands.

Table 1 PSO parameter setting

Variables	Values/ranges
Number of runs	100
Number of particles	3–5
Acceleration constants	[1.0, 0.5]

Table 2 Annual demands of Aliyar reservoir

Nature of demand	Demand in (Mm^3)
Kerala demand (As per interstate agreement [7])	205
Irrigation—old ayacut	70
Irrigation—new ayacut	35
Drinking water and industrial purpose	45

Table 3 gives the optimal release of water from Aliyar reservoir for the years 2009 to 2013, calculated through NLP and PSO. The actual and optimal release obtained through NLP and PSO are compared with actual release of reservoir and represented in Fig. 1.

It is observed from the results that PSO yields better optimal release for the given demand than NLP and hence, the optimal release given by PSO is used to estimate the improvement in power production. Presently, in Aliyar reservoir, the Kerala demand water alone is released through the powerhouse. In order to improve the hydropower production, two release patterns were framed in this research based on the optimal release by PSO, which is given below (Fig. 2).

Release Pattern I: Diverting only the Kerala demand water through power house. In this case, 58 % of the optimal release (PSO) is diverted to powerhouse for power generation.

Release Pattern II: Diverting both the Kerala demand and the irrigation demand of New Ayacut water through power house. In this case, 68 % of the optimal release (PSO) is diverted to powerhouse for power generation.

In case of release pattern I, it was observed from the results that power production could be considerably improved with the help of optimally released water. Release pattern II is framed in this work for further improvement in maximizing the power generation. The maximum power production estimated in this work is presented in Table 4 and Fig. 3.

Table 3 Optimal release of water (Mm^3) obtained through NLP and PSO

Year	Demand	Actual release	Optimal release NLP	Optimal release PSO
2009	354.5	338.28	328.94	353.57
2010	354.5	290.11	311.30	338.53
2011	354.5	301.31	310.52	316.51
2012	354.5	247.90	293.17	341.42
2013	354.5	282.51	294.21	344.8

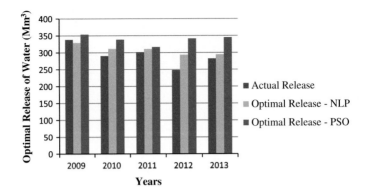

Fig. 2 Comparison optimal release of water (Mm^3) obtained through NLP and PSO

Table 4 Comparison of actual and estimated power production using PSO

Year	Actual power production (GWh)	Estimated power production for release pattern I (GWh)	Estimated power production for release pattern II (GWh)
2009	7.43	13.00	15.08
2010	7.77	12.83	15.00
2011	8.91	11.74	13.73
2012	6.46	12.38	14.48
2013	4.57	11.03	12.90

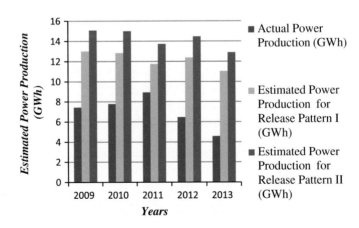

Fig. 3 Comparison of actual and estimated power production using PSO

The actual capacity of the Aliyar reservoir power plant is 14 Giga Watt hour (GWh). This research work has estimated a maximal power production within the confined plant capacity in Release Pattern I, whereas the estimated maxima in

Release Pattern II have exceeded the plant capacity. It could be suggested that there is a possibility of maximizing the power production not only through the optimal release of available water, but also through increasing the plant capacity.

5 Conclusion

This research work was focused on optimizing the reservoir release in order to maximize the hydropower generation at Aliyar reservoir in Coimbatore District of Tamilnadu is taken as the study area. Non-Linear Programming (NLP) technique and Particle Swarm Optimization (PSO) were used to optimize the reservoir release, and it was observed from the results that PSO gave the optimal best reservoir release. Two different reservoir release patterns were framed to maximize the power production and tested in this research, and the results have indicated that the application of suitable release patterns optimized by PSO leads to a considerable maximization of hydropower generation. In general, the large diversion of water gives more power production for all release patterns, constrained to the plant capacity. In future, this research will focus on forecasting the inflow to the reservoir using bio-inspired algorithms and to predict the corresponding optimal release that could further improve the power generation.

References

1. Arunkumar, R., Jothiprakash, V.: Optimal reservior operation for hydropower generation using non-linear programming model. J. Inst. Eng. India Ser. A **93**(2), 111–120 (2012)
2. ADB.: Hydropower Development in India: A Sector Assessment. Asian Development Bank, Publication Stock No. 031607, Philippines (2007)
3. Arunkumar, R., Jothiprakash, V.: Multi—reservoir optimization for hydropower production using NLP technique. KSCE J. Civil Eng. **18**(1) (2014)
4. Nagesh, Kumar, Janga, D., Reddy, M.: Ant colony optimization for multi-purpose reservoir operation. Water Resour. Manage **20**, 879–898 (2006)
5. Yeh, W.W.G.: Reservoir management and operation models: a state-of-the-art review. Water Resour. Res. **21**(12), 1797–1818 (1985)
6. Public Works Department: Aliyar Reservoir. Water Resource Organization, Pollachi Region, Coimbatore (2013)
7. Loucks, D.P., Stedinger, J.R., Haith, D.A.: Water Resources Systems Planning and Analysis. Prentice Hall Inc, Englewood Cliffs, New Jersey (1981)
8. http://www.swarmintelligence.org/
9. Ghimire, B.N.S., Reddy, M.J.: Optimalreservior operation for hydropower production using particle swarm optimization and sustainability analysis of hydropower. ISH. J. Hydraul. Eng. **19**(3), 196–210 (2013)
10. Avriel, M.: Nonlinear Programming: Analysis and Methods. Dover Publishing. (2003). ISBN 0-486-43227-0

Author Index

A

Abdelwahab, Sara, 151
Abdullah, Rezhna Mirza, 323
Abdullah, Saman M., 323
Abraham, Ajith, 93, 151, 347, 427
Agarwal, Devang, 555
Aggarwal, Charu, 393
Amudha, T., 481, 565, 577
Anitha, S. Pillai, 523
Anjana, K.P., 163
Arreola, Julio, 285

B

Bedi, Punam, 383, 393
Bhasin, Veenu, 393
Bhuvan, Nikhila T., 221
Buhari, Seyed M., 1

C

Calderón, Juan M., 263, 499, 511
Chandra, P. Helen, 43, 55, 187, 337
Chinnaswamy, Arunkumar, 229
Choo, Yun-Huoy, 347

D

Deekshatulu, B.L., 187, 337
Desai, Kamalakar D., 447
Dhivya, M., 359
Dhveya, R., 565

E

Eapen, Madhuri Elsa, 437
Ezugwu, Absalom E., 1

F

Frincu, Marc E., 1

G

Gajdoš, Petr, 209
Gautam, K.S., 275
George, Renu, 117
Gnanamalar David, N., 251
González, Saúl, 285
Gupta, Amit Kumar, 297
Gupta, Naveen Kumar, 297

H

Haritha, H., 275
Hosseinabadi, Ali Asghar Rahmani, 93

I

Ibitoye, Ayodeji O.J., 491
Issac, Biju, 105

J

Jabbar, M.A., 187, 337
Jadhav, Priya S., 427
Jayakrishna, V, 417
Jayakumar, T.V., 313
Jayaseeli, A. Mary Imelda, 43
Ježowicz, Tomáš, 209
Jindal, Vinita, 383
Jisha, G, 241
Joseph, Sumy, 403
Junaidu, Sahalu B., 1

K

Kalavathy, S.M. Saroja T., 43, 55
Kalita, Hemanta Kr, 105
Kalyani, M. Nithya, 55
Kardgar, Maryam, 93
Karoline, J. Philomenal, 43
Kiruthiga, D., 577

Kulkarni, Anand J., 427
Kumar, Rakesh, 297
Kumar, T. Senthil, 275
Kumari, S. Neetu, 523

L
López, Jorge, 263
Lozano, Fernando, 499

M
Manoj Ray, D., 175
Margain, Lourdes, 285
Mary, Robson, 313
Mathew, Prabha Susy, 543
Mohanan, Murali, 197
MohanaPriya, P., 463
Moreno, Wilfrido, 511
Muda, Azah Kamilah, 347, 473
Mulay, Preeti, 29

N
Nagar, Atulya K., 251
Nair, Jisha M., 437
Nebu, Cynthia Marea, 403

O
Ochang, Paschal A., 1
Ochoa, Alberto, 285
Ojha, Varun Kumar, 151
Onifade, Olufade F.W., 491

P
Padilla, Teresa, 285
Pandey, Tulika, 463
Panicker, Athira V, 241
Pérez, Katherín, 263
Pillai, Anitha S., 543
Pradeep, C., 437
Pratama, Satrya Fajri, 347
Preetha, K.G., 139, 163
Puri, Karuna, 29

Q
Quintero, Carlos, 263, 499

R
Ramesh, Dharavath, 371
Rashid, Tarik A., 323
Rawat, Sanyog, 535
Ray, Kanad, 535

Remesh Babu, K.R., 15, 67
Rhine, Reema, 221
Rodas, Jorge, 285
Rodriguez, Saith, 263
Rojas, Eyberth, 263
Roy, Hima Anns, 417

S
Samdanielthompson, G., 251
Samuel, Philip, 15, 67, 117, 129, 175, 197
Sandhya, N., 129
Sangeetha, A., 481
Sankhe, Manoj S., 447
Saranya, S., 15
Saw, Yee Ching, 473
Sethi, Krishan Kumar, 371
Shalinie, S. Mercy, 463
Shamshirband, Shahab, 93
Sharma, Deepak, 555
Sharma, Rupam Kumar, 105
Sharma, Somesh, 535
Shastri, Apoorva S., 427
Shojafar, Mohammad, 93
Singh, Neha, 393
Singh, Pushpendra, 535
Snášel, Václav, 209
Sreelakshmi, S., 139
Srinivasan, Ramakrishnan, 229
Srivastava, Prakash, 297
Subramanian, K.G., 251

T
Tokudded, Odalid, 285
Toshniwal, Sandeep, 535

U
Uher, Vojtěch, 209
Umamaheswari, K., 359
Uribe, Reinaldo, 499

V
Varghese, Anila Ann, 437
Varghese, Paul, 79
Vinu, R., 79

W
Weitzenfeld, Alfredo, 511

Y
Yakmut, Daniel I., 1

Printed in the United States
By Bookmasters